Inverse Problem Theory
and Methods for Model Parameter Estimation

Inverse Problem Theory
and Methods for Model Parameter Estimation

Albert Tarantola

Institut de Physique du Globe de Paris
Université de Paris 6
Paris, France

Society for Industrial and Applied Mathematics
Philadelphia

10 9 8 7 6 5 4 3 2 1

Library of Congress Cataloging-in-Publication Data

Tarantola, Albert.
    Inverse problem theory and methods for model parameter estimation / Albert Tarantola.
        p. cm.
    Includes bibliographical references and index.
    ISBN 0-89871-572-5 (pbk.)
    1. Inverse problems (Differential equations)   I. Title.

    QA371.T357 2005
    515'.357—dc22

                                                                2004059038

siam is a registered trademark.

To my parents,
Joan and Fina

Contents

Preface

Physical theories allow us to make predictions: given a complete description of a physical system, we can predict the outcome of some measurements. This problem of predicting the result of measurements is called the *modelization problem*, the *simulation problem*, or the *forward problem*. The *inverse problem* consists of using the actual result of some measurements to infer the values of the parameters that characterize the system.

While the forward problem has (in deterministic physics) a unique solution, the inverse problem does not. As an example, consider measurements of the gravity field around a planet: given the distribution of mass inside the planet, we can uniquely predict the values of the gravity field around the planet (forward problem), but there are different distributions of mass that give *exactly* the same gravity field in the space outside the planet. Therefore, the inverse problem — of inferring the mass distribution from observations of the gravity field — has multiple solutions (in fact, an infinite number).

Because of this, in the inverse problem, one needs to make explicit any available a priori information on the model parameters. One also needs to be careful in the representation of the data uncertainties.

The most general (and simple) theory is obtained when using a probabilistic point of view, where the a priori information on the model parameters is represented by a probability distribution over the 'model space.' The theory developed here explains how this a priori probability distribution is transformed into the a posteriori probability distribution, by incorporating a physical theory (relating the model parameters to some observable parameters) and the actual result of the observations (with their uncertainties).

To develop the theory, we shall need to examine the different types of parameters that appear in physics and to be able to understand what a total absence of a priori information on a given parameter may mean.

Although the notion of the inverse problem could be based on conditional probabilities and Bayes's theorem, I choose to introduce a more general notion, that of the 'combination of states of information,' that is, in principle, free from the special difficulties appearing in the use of conditional probability densities (like the well-known Borel paradox).

The general theory has a simple (probabilistic) formulation and applies to any kind of inverse problem, including linear as well as strongly nonlinear problems. Except for very simple examples, the probabilistic formulation of the inverse problem requires a resolution in terms of 'samples' of the a posteriori probability distribution in the model space. This, in particular, means that the solution of an inverse problem is not a model but a collection of models (that are consistent with both the data and the a priori information). This is why Monte Carlo (i.e., random) techniques are examined in this text. With the increasing availability of computer power, Monte Carlo techniques are being increasingly used.

Some special problems, where nonlinearities are weak, can be solved using special, very efficient techniques that do not differ essentially from those used, for instance, by Laplace in 1799, who introduced the 'least-absolute-values' and the 'minimax' criteria for obtaining the best solution, or by Legendre in 1801 and Gauss in 1809, who introduced the 'least-squares' criterion.

The first part of this book deals exclusively with discrete inverse problems with a finite number of parameters. Some real problems are naturally discrete, while others contain functions of a continuous variable and can be discretized if the functions under consideration are smooth enough compared to the sampling length, or if the functions can conveniently be described by their development on a truncated basis. The advantage of a discretized point of view for problems involving functions is that the mathematics is easier. The disadvantage is that some simplifications arising in a general approach can be hidden when using a discrete formulation. (Discretizing the forward problem and setting a discrete inverse problem is not always equivalent to setting a general inverse problem and discretizing for the practical computations.)

The second part of the book deals with general inverse problems, which may contain such functions as data or unknowns. As this general approach contains the discrete case in particular, the separation into two parts corresponds only to a didactical purpose.

Although this book contains a lot of mathematics, it is not a mathematical book. It tries to explain how a method of acquisition of information can be applied to the actual world, and many of the arguments are heuristic.

This book is an entirely rewritten version of a book I published long ago (Tarantola, 1987). Developments in inverse theory in recent years suggest that a new text be proposed, but that it should be organized in essentially the same way as my previous book. In this new version, I have clarified some notions, have underplayed the role of optimization techniques, and have taken Monte Carlo methods much more seriously.

I am very indebted to my colleagues (Bartolomé Coll, Georges Jobert, Klaus Mosegaard, Miguel Bosch, Guillaume Évrard, John Scales, Christophe Barnes, Frédéric Parrenin, and Bernard Valette) for illuminating discussions. I am also grateful to my collaborators at what was the *Tomography Group* at the Institut de Physique du Globe de Paris.

Albert Tarantola
Paris, June 2004

Chapter 1
The General Discrete Inverse Problem

Far better an approximate answer to the right question,
which is often vague,
than an exact answer to the wrong question,
which can always be made precise.

John W. Tukey, 1962

Central to this chapter is the concept of the 'state of information' over a parameter set. It is postulated that the most general way to describe such a state of information is to define a probability density over the parameter space. It follows that the results of the measurements of the observable parameters (data), the a priori information on model parameters, and the information on the physical correlations between observable parameters and model parameters can all be described using probability densities. The general inverse problem can then be set as a problem of 'combining' all of this information. Using the point of view developed here, the solution of inverse problems, and the analysis of uncertainty (sometimes called 'error and resolution analysis'), can be performed in a fully nonlinear way (but perhaps with a large amount of computing time). In all usual cases, the results obtained with this method reduce to those obtained from more conventional approaches.

1.1 Model Space and Data Space

Let $\mathfrak{S}$ be the *physical system* under study. For instance, $\mathfrak{S}$ can be a galaxy for an astrophysicist, Earth for a geophysicist, or a quantum particle for a quantum physicist.

The scientific procedure for the study of a physical system can be (rather arbitrarily) divided into the following three steps.

i) *Parameterization of the system*: discovery of a minimal set of *model parameters* whose values completely characterize the system (from a given point of view).

1

ii) *Forward modeling*: discovery of the *physical laws* allowing us, for given values of the model parameters, to make predictions on the results of measurements on some *observable parameters*.

iii) *Inverse modeling*: use of the actual results of some measurements of the observable parameters to infer the actual values of the model parameters.

Strong feedback exists between these steps, and a dramatic advance in one of them is usually followed by advances in the other two. While the first two steps are mainly inductive, the third step is deductive. This means that the rules of thinking that we follow in the first two steps are difficult to make explicit. On the contrary, the mathematical theory of logic (completed with probability theory) seems to apply quite well to the third step, to which this book is devoted.

1.1.1 Model Space

The choice of the model parameters to be used to describe a system is generally not unique.

Example 1.1. *An anisotropic elastic sample $\mathfrak{S}$ is analyzed in the laboratory. To describe its elastic properties, it is possible to use the tensor $c^{ij}{}_{k\ell}(\mathbf{x})$ of elastic* stiffnesses *relating stress, $\sigma^{ij}(\mathbf{x})$, to strain, $\varepsilon^{ij}(\mathbf{x})$, at each point $\mathbf{x}$ of the solid:*

$$\sigma^{ij}(\mathbf{x}) \;=\; c^{ij}{}_{k\ell}(\mathbf{x})\,\varepsilon^{k\ell}(\mathbf{x}) \quad . \tag{1.1}$$

Alternatively, it is possible to use the tensor $s^{ij}{}_{k\ell}(\mathbf{x})$ of elastic compliances *relating strain to stress,*

$$\varepsilon^{ij}(\mathbf{x}) \;=\; s^{ij}{}_{k\ell}(\mathbf{x})\,\sigma^{k\ell}(\mathbf{x}) \quad , \tag{1.2}$$

where the tensor $\mathbf{s}$ is the inverse of $\mathbf{c}$, $c^{ij}{}_{k\ell}\,s^{k\ell}{}_{mn} = \delta^i_m\,\delta^j_n$. The use of stiffnesses or of compliances is completely equivalent, and there is no 'natural' choice.

A particular choice of model parameters is a *parameterization* of the system. Two different parameterizations are *equivalent* if they are related by a bijection (one-to-one mapping).

Independently of any particular parameterization, it is possible to introduce an abstract space of points, a *manifold*,[1] each point of which represents a conceivable model of the system. This manifold is named the *model space* and is denoted $\mathfrak{M}$. Individual models are points of the model space manifold and could be denoted $\mathcal{M}_1$, $\mathcal{M}_2$, ... (but we shall use another, more common, notation).

For quantitative discussions on the system, a particular parameterization has to be chosen. To define a parameterization means to define a set of experimental procedures allowing, at least in principle, us to measure a set of physical quantities that characterize the system. Once a particular parameterization has been chosen, with each point $\mathcal{M}$ of the

---

[1]The reader interested in the theory of differentiable manifolds may refer, for instance, to Lang (1962), Narasimhan (1968), or Boothby (1975).

model space $\mathfrak{M}$ a set of numerical values $\{m^1, \dots, m^n\}$ is associated. This corresponds to the definition of a system of *coordinates* over the model manifold $\mathfrak{M}$.

Example 1.2. *If the elastic sample mentioned in Example* 1.1 *is, in fact, isotropic and homogeneous, the model manifold* $\mathfrak{M}$ *is two-dimensional (as such a medium is characterized by two elastic constants). As parameters to characterize the sample, one may choose, for instance,* $\{m^1, m^2\} = \{$ Young modulus , Poisson ratio $\}$ *or* $\{m^1, m^2\} = \{$ bulk modulus , shear modulus $\}$. *These two possible choices define two different coordinate systems over the model manifold* $\mathfrak{M}$.

Each point $\mathcal{M}$ of $\mathfrak{M}$ is named a *model*, and, to conform to usual notation, we may represent it using the symbol $\mathbf{m}$. By no means is $\mathbf{m}$ to be understood as a vector, i.e., as an element of a linear space. For the manifold $\mathfrak{M}$ may be linear or not, and even when the model space $\mathfrak{M}$ is linear, the coordinates being used may not be a set of Cartesian coordinates.

Example 1.3. *Let us choose to characterize the elastic samples mentioned in Example* 1.2 *using the bulk modulus and the shear modulus,* $\{m^1, m^2\} = \{\kappa, \mu\}$. *A convenient[2] definition of the distance between two elastic media is*

$$d = \sqrt{\left(\log \frac{\kappa_2}{\kappa_1}\right)^2 + \left(\log \frac{\mu_2}{\mu_1}\right)^2} \quad . \tag{1.3}$$

This clearly shows that the two coordinates $\{m^1, m^2\} = \{\kappa, \mu\}$ *are* not *Cartesian. Introducing the logarithmic bulk modulus* $\kappa^* = \log(\kappa/\kappa_0)$ *and the logarithmic shear modulus* $\mu^* = \log(\mu/\mu_0)$ *(where* κ_0 *and* μ_0 *are arbitrary constants) gives*

$$d = \sqrt{(\kappa_2^* - \kappa_1^*)^2 + (\mu_2^* - \mu_1^*)^2} \quad . \tag{1.4}$$

The logarithmic bulk modulus and the logarithmic shear modulus are *Cartesian coordinates over the model manifold* $\mathfrak{M}$.

The number of model parameters needed to completely describe a system may be either finite or infinite. This number is infinite, for instance, when we are interested in a property $\{m(\mathbf{x}) ; \mathbf{x} \in V\}$ that depends on the position $\mathbf{x}$ inside some volume V.

The theory of infinite-dimensional manifolds needs a greater technical vocabulary than the theory of finite-dimensional manifolds. In what follows, and in all of the first part of this book, I assume that the model space is *finite dimensional*. This limitation to systems with a finite number of parameters may be severe from a mathematical point of view. For instance, passing from a continuous field $m(\mathbf{x})$ to a discrete set of quantities $m^\alpha = m(\mathbf{x}^\alpha)$ by discretizing the space will only make sense if the considered fields are smooth. If this is indeed the case, then there will be no practical difference between the numerical results given by functional approaches and those given by discrete approaches to

---

[2] This definition of distance is invariant of form when changing these positive elastic parameters by their inverses, or when multiplying the values of the elastic parameters by a constant. See Appendix 6.3 for details.

inverse problem theory (although the numerical algorithms may differ considerably, as can be seen by comparing the continuous formulation in sections 5.6 and 5.7 and the discrete formulation in Problem 7.3).

Once we agree, in the first part of this book, to deal only with a finite number of parameters, it remains to decide if the parameters may take continuous or discrete values (i.e., in fact, if the quantities are real numbers or integer numbers). For instance, if a parameter m^α represents the mass of the Sun, we can assume that it can take any value from zero to infinity; if m^α represents the spin of a quantum particle, we can assume a priori that it can only take discrete values. As the use of 'delta functions' allows us to consider parameters taking discrete values as a special case of parameters taking continuous values, we shall, to simplify the discussion, use the terminology corresponding to the assumption that all the parameters under consideration take their values in a continuous set. If this is not the case in a particular problem, the reader will easily make the corresponding modifications.

When a particular parameterization of the system has been chosen, each point of $\mathfrak{M}$ (i.e., each model) can be represented by a particular set of values for the model parameters $\mathbf{m} = \{m^\alpha\}$, where the index α belongs to some discrete finite index set. As we have interpreted any particular parameterization of the physical system $\mathfrak{S}$ as a choice of coordinates over the manifold $\mathfrak{M}$, the variables m^α can be named the *coordinates* of $\mathbf{m}$, but not the 'components' of $\mathbf{m}$, unless a linear space can be introduced. But, more often than not, the model space is not linear. For instance, when trying to estimate the geographical coordinates $\{\theta, \varphi\}$ of the (center of the) meteoritic impact that killed the dinosaurs, the model space $\mathfrak{M}$ is the surface of Earth, which is intrinsically curved.

When it can be demonstrated that the model manifold $\mathfrak{M}$ has no curvature, to introduce a linear (vector) space still requires a proper definition of the 'components' of vectors. When such a structure of linear space has been introduced, then we can talk about the *linear model space*, denoted $\mathbb{M}$, and, by definition, the *sum of two models*, $\mathbf{m}_1$ and $\mathbf{m}_2$, corresponds to the sum of their *components*, and the *multiplication of a model by a real number* corresponds to the multiplication of all its components:[3]

$$(\mathbf{m}_1 + \mathbf{m}_2)^\alpha = m_1{}^\alpha + m_2{}^\alpha \quad , \quad (\lambda\,\mathbf{m}\,)^\alpha = \lambda\,m^\alpha \quad . \tag{1.5}$$

Example 1.4. *For instance, in the elastic solid considered in Example 1.3, to have a structure of linear (vector) space, one must select an arbitrary point of the manifold $\{\kappa_0, \mu_0\}$ and define the vector $\mathbf{m} = \{m^1, m^2\}$ whose components are*

$$m^1 = \log(\kappa/\kappa_0) \quad , \quad m^2 = \log(\mu/\mu_0) \quad . \tag{1.6}$$

Then, the distance between two models, as defined in Example 1.3, equals $\| \mathbf{m}_2 - \mathbf{m}_1 \|$, the norm here being understood in its ordinary sense (for vectors in a Euclidean space).

One must keep in mind, however, that the basic definitions of the theory developed here will not depend in any way on the assumption of the linearity of the model space. We are about to see that the only mathematical objects to be defined in order to deal with the most general formulation of inverse problems are probability distributions over the model space

---

[3]The index α in equation (1.5) may just be a shorthand notation for a multidimensional index (see an example in Problem 7.3). For details of array algebra see Snay (1978) or Rauhala (2002).

manifold. A probability over $\mathfrak{M}$ is a mapping that, with any subset $\mathcal{A}$ of $\mathfrak{M}$, associates a nonnegative real number, $P(\mathcal{A})$, named the probability of $\mathcal{A}$, with $P(\mathfrak{M}) = 1$. Such probability distributions can be defined over any finite-dimensional manifold $\mathfrak{M}$ (curved or linear) and irrespective of any particular parameterization of $\mathfrak{M}$, i.e., independently of any particular choice of coordinates. But if a particular coordinate system $\{m^\alpha\}$ has been chosen, it is then possible to describe a probability distribution using a probability density (and we will make extensive use of this possibility).

1.1.2 Data Space

To obtain information on model parameters, we have to perform some observations during a physical experiment, i.e., we have to perform a measurement of some observable parameters.[4]

Example 1.5. *For a nuclear physicist interested in the structure of an atomic particle, observations may consist in a measurement of the flux of particles diffused at different angles for a given incident particle flux, while for a geophysicist interested in understanding Earth's deep structure, observations may consist in recording a set of seismograms at Earth's surface.*

We can thus arrive at the abstract idea of a *data space*, which can be defined as the space of all conceivable instrumental responses. This corresponds to another manifold, the *data manifold* (or data space), which we may represent by the symbol $\mathfrak{D}$. Any conceivable (exact) result of the measurements then corresponds to a particular point $\mathcal{D}$ on the manifold $\mathfrak{D}$.

As was the case with the model manifold, it shall sometimes be possible to endow the data space with the structure of a linear manifold. When this is the case, then we can talk about the *linear data space*, denoted by $\mathbb{D}$; the coordinates $\mathbf{d} = \{d^i\}$ (where i belongs to some discrete and finite index set) are then *components*,[5] and, as usual,

$$(\mathbf{d}_1 + \mathbf{d}_2)^i = d_1{}^i + d_2{}^i \quad , \quad (r\,\mathbf{d})^i = r\,d^i \quad . \tag{1.7}$$

Each possible realization of $\mathbf{d}$ is then named a *data vector*.

1.1.3 Joint Manifold

The separation suggested above between the model parameters $\{m^\alpha\}$ and the data parameters $\{d^i\}$ is sometimes clear-cut. In other circumstances, this may require some argumentation, or may not even be desirable. It is then possible to introduce one single manifold $\mathfrak{X}$ that represents all the parameters of the problem. A point of the manifold $\mathfrak{X}$ can be represented by the symbol $\mathcal{X}$ and a system of coordinates by $\{x^A\}$.

---

[4]The task of experimenters is difficult not only because they have to perform measurements as accurately as possible, but, more essentially, because they have to *imagine* new experimental procedures allowing them to measure observable parameters that carry a maximum of information on the model parameters.

[5]As mentioned above for the model space, the index i here may just be a shorthand notation for a multidimensional index (see an example in Problem 7.3).

As the quantities $\{d^i\}$ were termed observable parameters and the quantities $\{m^\alpha\}$ were termed model parameters, we can call $\{x^A\}$ the *physical parameters* or simply the *parameters*. The manifold $\mathfrak{X}$ is then named the *parameter manifold*.

1.2 States of Information

The probability theory developed here is self-sufficient. For good textbooks with some points in common with the present text, see Jeffreys (1939) and Jaynes (2003).

1.2.1 Definition of Probability

We are going to work with a finite-dimensional manifold $\mathfrak{X}$ (for instance, the model or the data space) and the field of all its subsets $\mathcal{A}$, $\mathcal{B}$, ... These subsets can be individual points, disjoint collections of points, or contiguous collections of points (whole regions of the manifold $\mathfrak{X}$). As is traditional in probability theory, a subset $\mathcal{A} \subseteq \mathfrak{X}$ is called an *event*. The *union* and the *intersection* of two events $\mathcal{A}$ and $\mathcal{B}$ are respectively denoted $\mathcal{A} \cup \mathcal{B}$ and $\mathcal{A} \cap \mathcal{B}$.

The field of events is called, in technical terms, a σ-field, meaning that the complement of an event is also an event. The notion of a σ-field could allow us to introduce probability theory with great generality, but we limit ourselves here to probabilities defined over a finite-dimensional manifold.

By definition, a *measure* over the manifold $\mathfrak{X}$ is an application $P(\cdot)$ that with any event $\mathcal{A}$ of $\mathfrak{X}$ associates a real positive number $P(\mathcal{A})$, named the *measure of* $\mathcal{A}$, that satisfies the following two properties (Kolmogorov axioms):

- If $\mathcal{A}$ and $\mathcal{B}$ are two disjoint events, then

$$P(\mathcal{A} \cup \mathcal{B}) = P(\mathcal{A}) + P(\mathcal{B}) \quad . \tag{1.8}$$

- There is *continuity at zero*, i.e., if a sequence $\mathcal{A}_1 \supseteq \mathcal{A}_2 \supseteq \cdots$ tends to the empty set, then $P(\mathcal{A}_i) \to 0$.

This last condition implies that the probability of the empty event is zero,

$$P(\emptyset) = 0 \quad , \tag{1.9}$$

and it immediately follows from condition (1.8) that if the two events $\mathcal{A}$ and $\mathcal{B}$ are not necessarily disjoint, then

$$P(\mathcal{A} \cup \mathcal{B}) = P(\mathcal{A}) + P(\mathcal{B}) - P(\mathcal{A} \cap \mathcal{B}) \quad . \tag{1.10}$$

The probability of the whole manifold, $P(\mathfrak{X})$, is not necessarily finite. If it is, then P is termed a *probability* over $\mathfrak{X}$. In that case, P is usually normalized to unity: $P(\mathfrak{X}) = 1$. In what follows, the term 'probability' will be reserved for a value, like $P(\mathcal{A})$ for the probability of $\mathcal{A}$. The function $P(\cdot)$ itself will rather be called a *probability distribution*.

An important notion is that of a sample of a distribution, so let us give its formal definition. A randomly generated point $\mathcal{P} \in \mathfrak{X}$ is a *sample* of a probability distribution

$P(\cdot)$ if the probability that the point $\mathcal{P}$ is generated inside any $A \subset \mathcal{X}$ equals $P(A)$, the probability of A. Two points $\mathcal{P}$ and $\mathcal{Q}$ are *independent samples* if (i) both are samples and (ii) the generation of the samples is independent (i.e., if the actual place where each point has materialized is, by construction, independent of the actual place where the other point has materialized).[6]

Let P be a probability distribution over a manifold $\mathcal{X}$ and assume that a particular coordinate system $\mathbf{x} = \{x^1, x^2, \dots\}$ has been chosen over $\mathcal{X}$. For any probability distribution P, there exists (Radon–Nikodym theorem) a positive function $f(\mathbf{x})$ such that, for any $A \subseteq \mathcal{X}$, $P(A)$ can be obtained as the integral

$$P(A) = \int_A d\mathbf{x}\, f(\mathbf{x}) \quad , \tag{1.11}$$

where

$$\int_A d\mathbf{x} \equiv \underbrace{\int dx^1 \int dx^2 \cdots}_{\text{over } A} \quad . \tag{1.12}$$

Then, $f(\mathbf{x})$ is termed the *probability density* representing P (with respect to the given coordinate system). The functions representing probability densities may, in fact, be distributions, i.e., generalized functions containing in particular Dirac's delta function.

Example 1.6. *Let $\mathcal{X}$ be the 2D surface of the sphere endowed with a system of spherical coordinates $\{\theta, \varphi\}$. The probability density*

$$f(\theta, \varphi) = \frac{\sin \theta}{4\pi} \tag{1.13}$$

associates with every region A of $\mathcal{X}$ a probability that is proportional to the surface of A. Therefore, the probability density $f(\theta, \varphi)$ is 'homogeneous' (although the function does not take constant values).

Example 1.7. *Let $\mathcal{X} = \mathbb{R}^+$ be the positive part of the real line, and let $f(x)$ be the function $1/x$. The integral $P(x_1 < x < x_2) = \int_{x_1}^{x_2} dx\, f(x)$ then defines a measure over $\mathcal{X}$, but not a probability (because $P(0 < x < \infty) = \infty$). The function $f(x)$ is then a measure density but not a probability density.*

To develop our theory, we will effectively need to consider nonnormalizable measures (i.e., measures that are not a probability). These measures cannot describe the probability of a given event A: they can only describe the *relative probability* of two events A_1 and A_2. We will see that this is sufficient for our needs. To simplify the discussion, we will sometimes use the linguistic abuse of calling probability a nonnormalizable measure.

It should be noticed that, as a probability is a real number, and as the parameters $x^1, x^2, \dots$ in general have physical dimensions, the physical dimension of a probability

---

[6]Many of the algorithms used to generate samples in large-dimensional spaces (like the Gibbs sampler of the Metropolis algorithm) do *not* provide independent samples.

density is a *density* of the considered space, i.e., it has as physical dimensions the inverse of the physical dimensions of the volume element of the considered space.

Example 1.8. *Let v be a velocity and m be a mass. The respective physical dimensions are LT^{-1} and M. Let $f(v, m)$ be a probability density on (v, m). For the probability*

$$P(v_1 \le v \le v_2 \text{ and } m_1 \le m \le m_2) = \int_{v_1}^{v_2} dv \int_{m_1}^{m_2} dm \ f(v, m) \qquad (1.14)$$

to be a real number, the physical dimensions of f have to be $M^{-1}L^{-1}T$.

Let P be a probability distribution over a manifold $\mathfrak{X}$ and $f(\mathbf{x})$ be the probability density representing P in a given coordinate system. Let

$$\mathbf{x}^* = \mathbf{x}^*(\mathbf{x}) \qquad (1.15)$$

represent a *change of coordinates* over $\mathfrak{X}$, and let $f^*(\mathbf{x}^*)$ be the probability density representing P in the new coordinates:

$$P(\mathcal{A}) = \int_{\mathcal{A}} d\mathbf{x}^* f^*(\mathbf{x}^*) \ . \qquad (1.16)$$

By definition of $f(\mathbf{x})$ and $f^*(\mathbf{x}^*)$, for any $\mathcal{A} \subseteq \mathfrak{X}$,

$$\int_{\mathcal{A}} d\mathbf{x} \ f(\mathbf{x}) = \int_{\mathcal{A}} d\mathbf{x}^* f^*(\mathbf{x}^*) \ . \qquad (1.17)$$

Using the elementary properties of the integral, the following important property (called the *Jacobian rule*) can be deduced:

$$\boxed{f^*(\mathbf{x}^*) = f(\mathbf{x}) \left| \frac{\partial \mathbf{x}}{\partial \mathbf{x}^*} \right| \ ,} \qquad (1.18)$$

where $|\partial \mathbf{x}/\partial \mathbf{x}^*|$ represents the absolute value of the Jacobian of the transformation. See Appendix 6.3 for an example of the use of the Jacobian rule.

Instead of introducing a probability *density*, we could have introduced a *volumetric probability* that would be an invariant (not subjected to the Jacobian rule). See Appendix 6.1 for some details.

1.2.2 Interpretation of a Probability

It is possible to associate more than one intuitive meaning with any mathematical theory. For instance, the axioms and theorems of a three-dimensional vector space can be interpreted as describing the physical properties of the sum of forces acting on a material particle as well as the physiological sensations produced in our brain when our retina is excited by a light composed of a mixing of the three fundamental colors. Hofstadter (1979) gives some examples of different valid intuitive meanings that can be associated with a given formal system.

There are two different usual intuitive interpretations of the axioms and definitions of probability as introduced above.

The first interpretation is purely statistical: when some physical random process takes place, it leads to a given realization. If a great number of realizations have been observed, these can be described in terms of probabilities, which follow the axioms above. The physical parameter allowing us to describe the different realizations is termed a *random variable*. The mathematical theory of statistics is the natural tool for analyzing the outputs of a random process.

The second interpretation is in terms of a *subjective degree of knowledge* of the 'true' value of a given physical parameter. By subjective we mean that it represents the knowledge of a given individual, obtained using rigorous reasoning, but that this knowledge may vary from individual to individual because each may possess different information.

Example 1.9. What is the mass of Earth's metallic core? *Nobody knows exactly. But with the increasing accuracy of geophysical measurements and theories, the* information *we have on this parameter improves continuously. The opinion maintained in this book is that the most general (and scientifically rigorous) answer it is possible to give at any moment to that question consists of defining the probability of the actual value of the mass* m *of Earth's core being within* m_1 *and* m_2 *for any couple of values* m_1 *and* m_2. *That is to say, the most general answer consists of the definition of a* probability density *over the physical parameter* m *representing the mass of the core.*

This subjective interpretation of the postulates of probability theory is usually named *Bayesian*, in honor of Bayes (1763). It is not in contradiction with the statistical interpretation. It simply applies to different situations.

One of the difficulties of the approach is that, given a state of information on a set of physical parameters, it is not always easy to decide which probability models it best. I hope that the examples in this book will help to show that it is possible to use some commonsense rules to give an adequate solution to this problem.

I set forth explicitly the following principle:

Let $\mathcal{X}$ *be a finite-dimensional manifold representing some physical parameters. The most general way we have to describe any state of information on* $\mathcal{X}$ *is by defining a probability distribution (or, more generally, a measure distribution) over* $\mathcal{X}$.

Let $P(\cdot)$ denote the probability distribution corresponding to a given state of information over a manifold $\mathcal{X}$ and $\mathbf{x} \mapsto f(\mathbf{x})$ denote the associated probability density:

$$P(A) = \int_A d\mathbf{x}\, f(\mathbf{x}) \qquad (\text{for any } A \subseteq \mathcal{X}) \quad . \tag{1.19}$$

The probability distribution $P(\cdot)$ or the probability density $f(\cdot)$ is said to *represent* the corresponding state of information.

1.2.3 Delta Probability Distribution

Consider a manifold $\mathcal{X}$ and denote as $\mathbf{x} = \{x^1, x^2, \dots\}$ any of its points. If we definitely know that only $\mathbf{x} = \mathbf{x}_0$ is possible, we can represent this state of information by a (Dirac) delta function centered at point $\mathbf{x}_0$:

$$f(\mathbf{x}) = \delta(\mathbf{x}; \mathbf{x}_0) \qquad (1.20)$$

(in the case where the manifold $\mathfrak{X}$ is a linear space $\mathbb{X}$, we can more simply write $f(\mathbf{x}) = \delta(\mathbf{x} - \mathbf{x}_0)$).

This probability density gives null probability to $\mathbf{x} \neq \mathbf{x}_0$ and probability 1 to $\mathbf{x} = \mathbf{x}_0$. In typical inference problems, the use of such a state of information does not usually make sense in itself, because all our knowledge of the real world is subject to uncertainties, but it is often justified when a certain type of uncertainty is negligible when compared to another type of uncertainty (see, for instance, Examples 1.34 and 1.35, page 34).

1.2.4 Homogeneous Probability Distribution

Let us now assume that the considered manifold $\mathfrak{X}$ *has a notion of volume*, i.e., that independently of any probability defined over $\mathfrak{X}$, we are able to associate with every domain $\mathcal{A} \subseteq \mathfrak{X}$ its *volume* $V(\mathcal{A})$. Denoting by

$$dV(\mathbf{x}) = v(\mathbf{x})\, d\mathbf{x} \qquad (1.21)$$

the volume element of the manifold in the coordinates $\mathbf{x} = \{x^2, x^2, \dots\}$, we can write the volume of a region $\mathcal{A} \subseteq \mathfrak{X}$ as

$$V(\mathcal{A}) = \int_{\mathcal{A}} d\mathbf{x}\, v(\mathbf{x}) \quad . \qquad (1.22)$$

The function $v(\mathbf{x})$ can be called the *volume density* of the manifold in the coordinates $\mathbf{x} = \{x^1, x^2, \dots\}$.

Assume first that the total volume of the manifold, say V, is finite, $V = \int_{\mathfrak{X}} d\mathbf{x}\, v(\mathbf{x})$. Then, the probability density

$$\mu(\mathbf{x}) = v(\mathbf{x})\, /\, V \qquad (1.23)$$

is normalized and it associates with any region $\mathcal{A} \subseteq \mathfrak{X}$ a probability

$$M(\mathcal{A}) = \int_{\mathcal{A}} d\mathbf{x}\, \mu(\mathbf{x}) \qquad (1.24)$$

that is proportional to the volume $V(\mathcal{A})$. We shall reserve the letter M for this probability distribution. The probability M, and the associated probability density $\mu(\mathbf{x})$, shall be called *homogeneous*. The reader should always remember that the homogeneous probability density does not need to be constant (see Example 1.6 on page 7).

Once a notion of volume has been introduced over a manifold $\mathfrak{X}$, one usually requires that any probability distribution $P(\cdot)$ to be considered over $\mathfrak{X}$ satisfy one consistency requirement: that the probability $P(\mathcal{A})$ of any event $\mathcal{A} \subseteq \mathfrak{X}$ that has zero volume, $V(\mathcal{A}) = 0$, must have zero probability, $P(\mathcal{A}) = 0$. On the probability densities, this imposes at any point $\mathbf{x}$ the condition

$$\mu(\mathbf{x}) = 0 \quad \Rightarrow \quad f(\mathbf{x}) = 0 \quad . \qquad (1.25)$$

Using mathematical jargon, all the probability densities $f(\mathbf{x})$ to be considered must be *absolutely continuous* with respect to the homogeneous probability density $\mu(\mathbf{x})$.

If the manifold under consideration has an infinite volume, then equation (1.23) cannot be used to define a probability density. In this case, we shall simply take $\mu(\mathbf{x})$ proportional to $v(\mathbf{x})$, and the homogeneous probability distribution is not normalizable. As we shall see, this generally causes no problem.

To define a notion of volume over an abstract manifold, one may use some invariance considerations, as the following example illustrates.

Example 1.10. *The elastic properties of an isotropic homogeneous medium were mentioned in Example* 1.3 *using the bulk modulus (or incompressibility modulus) and the shear modulus,* $\{\kappa, \mu\}$, *and a distance over the 2D manifold was proposed that is invariant of form when changing these positive elastic parameters by their inverses, or when multiplying the values of the elastic parameters by a constant. Associated with this definition of distance is the (2D) volume element[7]* $dV(\kappa, \mu) = (d\kappa/\kappa)(d\mu/\mu)$, *i.e., the (2D) volume density* $v(\kappa, \mu) = 1/(\kappa\,\mu)$. *Therefore, the (nonnormalizable) homogeneous probability density is*

$$\mu(\kappa, \mu) = 1/(\kappa\,\mu) \quad . \tag{1.26}$$

Changing the two parameters by their inverses, $\gamma = 1/\kappa$ *and* $\varphi = 1/\mu$, *and using the Jacobian rule (equation* (1.18)*), we obtain the new expression for the homogeneous probability density:*

$$\mu^*(\gamma, \varphi) = 1/(\gamma\,\varphi) \quad . \tag{1.27}$$

Of course, the invariance of form of the distance translates into this invariance of form for the homogeneous probability density.

In Bayesian inference for continuous parameters, the notion of 'noninformative probability density' is commonly introduced (see Jaynes, 1939; Box and Tiao, 1973; Rietsch, 1977; Savage, 1954, 1962), which usually results in some controversy. Here I claim that, more often than not, a homogeneous probability density can be introduced. Selecting this one as an a priori probability distribution to be used in a Bayesian argument is a choice that must be debated. I acknowledge that this choice is as informative as any other, and the 'noninformative' terminology is, therefore, not used here.[8]

Example 1.10 suggests that the probability density

$$f(x) = 1/x \tag{1.28}$$

plays an important role. Note that taking the logarithm of the parameter,

$$x^* = \alpha \log(x/x_0) \quad , \tag{1.29}$$

transforms the probability density into a constant one,

$$f^*(x^*) = \text{const} \quad . \tag{1.30}$$

---

[7] See Appendix 6.3 for details.
[8] It was used in the first edition of this book, which was a mistake.

It is shown in Appendix 6.7 that the probability density $1/x$ is a particular case of the log-normal probability density. Parameters accepting probability densities like the log-normal or its limit, the $1/x$ density, were discussed by Jeffreys (1957, 1968). These parameters have the characteristic of being positive and of being as popular as their inverses. We call them *Jeffreys parameters*. For more details, see Appendix 6.2.

No coherent inverse theory can be set without the introduction of the homogeneous probability distribution. From a practical point of view, it is only in highly degenerated inverse problems that the particular form of $\mu(\mathbf{x})$ plays a role.

If $f(\mathbf{x})$ is a probability density and $\mu(\mathbf{x})$ is the homogeneous probability density, the ratio

$$\varphi(\mathbf{x}) = f(\mathbf{x}) / \mu(\mathbf{x}) \tag{1.31}$$

plays an important role.[9] The function $\varphi(\mathbf{x})$, which is not a probability density,[10] shall be called a *likelihood* or a *volumetric probability*.

1.2.5 Shannon's Measure of Information Content

Given two normalized probability densities $f_1(\mathbf{x})$ and $f_2(\mathbf{x})$, *the relative information content* of $\mathbf{f}_1$ with respect to $\mathbf{f}_2$ is defined by

$$I(\mathbf{f}_1; \mathbf{f}_2) = \int_x d\mathbf{x}\, f_1(\mathbf{x}) \log \frac{f_1(\mathbf{x})}{f_2(\mathbf{x})} \quad . \tag{1.32}$$

When the logarithm base is 2, the unit of information is termed a *bit*; if the base is $e = 2.71828\ldots$, the unit is the *nep*; if the base is 10, the unit is the *digit*.

When the homogeneous probability density $\mu(\mathbf{x})$ is normalized, one can define the relative information content of a probability density $f(\mathbf{x})$ with respect to $\mu(\mathbf{x})$:

$$I(\mathbf{f}; \boldsymbol{\mu}) = \int d\mathbf{x}\, f(\mathbf{x}) \log \frac{f(\mathbf{x})}{\mu(\mathbf{x})} \quad . \tag{1.33}$$

We shall simply call this the *information content* of $f(\mathbf{x})$. This expression generalizes Shannon's (1948) original definition for discrete probabilities, $\sum_i P_i \log P_i$, to probability densities.[11] It can be shown that the information content is always positive,

$$I(\mathbf{f}; \boldsymbol{\mu}) \geq 0 \quad , \tag{1.34}$$

and that it is null only if $f(\mathbf{x}) \equiv \mu(\mathbf{x})$, i.e., if $f(\mathbf{x})$ is the homogeneous state of information.

---

[9]For instance, the maximum likelihood point is the point where $\varphi(\mathbf{x})$ is maximum (see section 1.6.4), and the Metropolis sampling method, when used to sample $f(\mathbf{x})$ (see section 2.3.5), depends on the values of $\varphi(\mathbf{x})$.

[10]When changing variables, the ratio of two probability densities is an invariant not subject to the Jacobian rule.

[11]Note that the expression $\int_x d\mathbf{x}\, f(\mathbf{x}) \log f(\mathbf{x})$ could not be used as a definition. Besides the fact that the logarithm of a dimensional quantity is not defined, a bijective change of variables $\mathbf{x}^* = \mathbf{x}^*(\mathbf{x})$ would alter the information content, which is not the case with the right definition (1.33). For let $f(\mathbf{x})$ be a probability density representing a given state of information on the parameters $\mathbf{x}$. The information content of $f(\mathbf{x})$ has been defined by equation (1.33), where $\mu(\mathbf{x})$ represents the homogeneous state of information. If instead of the parameters $\mathbf{x}$ we decide to use the parameters $\mathbf{x}^* = \mathbf{x}^*(\mathbf{x})$, the *same* state of information is described in the new variables by (expression (1.18)), $f^*(\mathbf{x}^*) = f(\mathbf{x}) |\partial\mathbf{x}/\partial\mathbf{x}^*|$, while the reference state of information is described by $\mu^*(\mathbf{x}^*) = \mu(\mathbf{x}) |\partial\mathbf{x}/\partial\mathbf{x}^*|$, where $|\partial\mathbf{x}/\partial\mathbf{x}^*|$ denotes the absolute value of the Jacobian of the transformation. A computation of the information content in the new variables gives $I(\mathbf{f}^*; \boldsymbol{\mu}^*) = \int d\mathbf{x}^*\, f^*(\mathbf{x}^*) \log(f^*(\mathbf{x}^*)/\mu^*(\mathbf{x}^*)) = \int d\mathbf{x}^* |\partial\mathbf{x}/\partial\mathbf{x}^*|\, f(\mathbf{x}) \log(f(\mathbf{x})/\mu(\mathbf{x}))$, and, using $d\mathbf{x}^* |\partial\mathbf{x}/\partial\mathbf{x}^*| = d\mathbf{x}$, we directly obtain $I(\mathbf{f}^*; \boldsymbol{\mu}^*) = I(\mathbf{f}; \boldsymbol{\mu})$.

1.2.6 Two Basic Operations on Probability Distributions

Inference theory is usually developed by first introducing the notion of conditional probability, then demonstrating a trivial theorem (the Bayes theorem), and then charging this
theorem with a (nontrivial) semantic content involving 'prior' and 'posterior' probabilities.
Although there is nothing wrong with that approach, I prefer here to use the alternative
route of introducing some basic structure to the space of all probability distributions (the
space characterized by the Kolmogorov axioms introduced in section 1.2.1). This structure
consists of defining two basic operations among probability distributions that are a generalization of the logical 'or' and 'and' operations among propositions. Although this approach
is normal in the theory of fuzzy sets, it is not usual in probability theory.[12]

Then, letting $\mathcal{X}$ be a finite-dimensional manifold, and given two probability distributions P_1 and P_2 over $\mathcal{X}$, we shall define the *disjunction* $P_1 \vee P_2$ (to be read P_1 *or*
P_2) and the *conjunction* $P_1 \wedge P_2$ (to be read P_1 *and* P_2). Taking inspiration from the
operations between logical propositions, we shall take as the first set of defining properties
the condition that, for any event $A \subseteq \mathcal{X}$,

$$\begin{aligned}
P_1(A) \neq 0 \quad \text{or} \quad P_2(A) \neq 0 \quad &\Rightarrow \quad (P_1 \vee P_2)(A) \neq 0 \quad, \\
P_1(A) = 0 \quad \text{or} \quad P_2(A) = 0 \quad &\Rightarrow \quad (P_1 \wedge P_2)(A) = 0 \quad.
\end{aligned} \tag{1.35}$$

In other words, for the disjunction $(P_1 \vee P_2)(A)$ to be different from zero, it is enough
that any of the two (or both) $P_1(A)$ or $P_2(A)$ be different from zero. For the conjunction
$(P_1 \wedge P_2)(A)$ to be zero, it is enough that any of the two (or both) $P_1(A)$ or $P_2(A)$ be
zero.

The two operations must be *commutative*,[13]

$$P_1 \vee P_2 = P_2 \vee P_1 \quad, \qquad P_1 \wedge P_2 = P_2 \wedge P_1 \quad, \tag{1.36}$$

and the homogeneous measure distribution M must be neutral for the conjunction operation,
i.e., for any P,

$$P \wedge M = P \quad. \tag{1.37}$$

As suggested in Appendix 6.17, if $f_1(\mathbf{x})$, $f_2(\mathbf{x})$, and $\mu(\mathbf{x})$ are the probability densities representing P_1, P_2, and M, respectively,

$$P_1(A) = \int_A d\mathbf{x}\, f_1(\mathbf{x}) \quad, \quad P_2(A) = \int_A d\mathbf{x}\, f_2(\mathbf{x}) \quad, \quad M(A) = \int_A d\mathbf{x}\, \mu(\mathbf{x}) \quad, \tag{1.38}$$

and $(f_1 \vee f_2)(\mathbf{x})$ and $(f_1 \wedge f_2)(\mathbf{x})$ are the probability densities representing respectively
$P_1 \vee P_2$ and $P_1 \wedge P_2$,

$$(P_1 \vee P_2)(A) = \int_A d\mathbf{x}\, (f_1 \vee f_2)(\mathbf{x}) \quad, \quad (P_1 \wedge P_2)(A) = \int_A d\mathbf{x}\, (f_1 \wedge f_2)(\mathbf{x}) \quad, \tag{1.39}$$

---

[12]A fuzzy set (Zadeh, 1965) is characterized by a *membership function* $f(\mathbf{x})$ that is similar to a probability
density, except that it takes values in the interval $[0, 1]$ (and its interpretation is different).

[13]The compact writing of these equations of course means that the properties are assumed to be valid for any
$A \subseteq \mathcal{X}$. For instance, the expression $P_1 \vee P_2 = P_2 \vee P_1$ means that, for any A, $(P_1 \vee P_2)(A) = (P_2 \vee P_1)(A)$.

then the simplest solution to the axioms above is[14]

$$(f_1 \vee f_2)(\mathbf{x}) = \tfrac{1}{2}(f_1(\mathbf{x}) + f_2(\mathbf{x})) \quad ; \quad (f_1 \wedge f_2)(\mathbf{x}) = \frac{1}{\nu} \frac{f_1(\mathbf{x})\, f_2(\mathbf{x})}{\mu(\mathbf{x})} \quad ,$$

(1.40)

where ν is the normalization constant[15] $\nu = \int_{\mathfrak{X}} d\mathbf{x}\, \frac{f_1(\mathbf{x})\, f_2(\mathbf{x})}{\mu(\mathbf{x})}$.

These two operations bear some resemblance to the union and intersection of fuzzy sets[16] and to the 'or' and 'and' operations introduced in multivalued logic,[17] but are not identical (in particular, there is nothing like $\mu(\mathbf{x})$ in fuzzy sets or in multivalued logic). The notion of the conjunction of states of information is related to the problem of aggregating expert opinions (Bordley, 1982; Genest and Zidek, 1986; Journel, 2002).

The conjunction operation is naturally associative, and one has

$$\frac{(f_1 \wedge f_2 \wedge \cdots \wedge f_n)(\mathbf{x})}{\mu(\mathbf{x})} = \frac{1}{\nu} \frac{f_1(\mathbf{x})}{\mu(\mathbf{x})} \frac{f_2(\mathbf{x})}{\mu(\mathbf{x})} \cdots \frac{f_2(\mathbf{x})}{\mu(\mathbf{x})} \quad ,$$

(1.41)

where $\nu = \int_{\mathfrak{X}} d\mathbf{x}\, \frac{f_1(\mathbf{x})}{\mu(\mathbf{x})} \frac{f_2(\mathbf{x})}{\mu(\mathbf{x})} \cdots \frac{f_2(\mathbf{x})}{\mu(\mathbf{x})}$. But, under the normalized form proposed in equation (1.40), the disjunction is not associative. So, for more generality, we can take the definition

$$(f_1 \vee f_2 \vee \cdots \vee f_n)(\mathbf{x}) = \tfrac{1}{n}(f_1(\mathbf{x}) + f_2(\mathbf{x}) + \cdots + f_n(\mathbf{x})) \quad .$$

(1.42)

Example 1.11. Disjunction of probabilities (making histograms). *Let r and φ be polar coordinates on a cathodic screen. Imagine that a device projects an electron onto the screen and we do our best to measure the (polar) coordinates of the impact point. Because of the finite accuracy of our measuring instrument, we cannot exactly know the coordinates of the actual impact point, but we can propose a probability density $f(r, \varphi)$ for the coordinates of this impact point. Now, instead of a single electron, the device sequentially (and randomly) projects a large number of electrons, according to a probability density $g(r, \varphi)$ that is unknown to us. For each impact point, our measuring instrument provides a probability density, so we have the (large) collection $f_1(r, \varphi)$, $f_2(r, \varphi)$, ..., $f_n(r, \varphi)$ of probability densities. The normalized disjunction $h(r, \varphi) = \tfrac{1}{n}(f_1(r, \varphi) \vee f_2(r, \varphi) \vee \cdots \vee f_n(r, \varphi))$, i.e., the sum $h(r, \varphi) = \tfrac{1}{n}(f_1(r, \varphi) + f_2(r, \varphi) + \cdots + f_n(r, \varphi))$, is a rough estimation of the unknown probability density $g(r, \varphi)$ in much the same way as an ordinary histogram (where the impact points are counted inside some ad hoc "boxes") would be. In the limit when $n \to \infty$ and the functions $f_i(r, \varphi)$ tend to be infinitely sharp, $h(r, \varphi) \to g(r, \varphi)$.*

Example 1.12. Conjunction of probabilities (I). *An estimation of the position of a floating object at the surface of the sea by an airplane navigator gives a probability distribution for*

---

[14]The conjunction of states of information was first introduced by Tarantola and Valette (1982a).

[15]This is only defined if the product $f_1(\mathbf{x})\, f_2(\mathbf{x})$ is not zero everywhere.

[16]If $f_1(\mathbf{x})$ and $f_2(\mathbf{x})$ are the membership functions characterizing two fuzzy sets, their *union* and *intersection* are respectively defined (Zadeh, 1965) by the membership functions max($f_1(\mathbf{x})$, $f_2(\mathbf{x})$) and min($f_1(\mathbf{x})$, $f_2(\mathbf{x})$).

[17]Multivalued logic typically uses the notion of *triangular conorm* (associated with the "or" operation) and the *triangular norm* (associated with the "and" operation). They were introduced by Schweizer and Sklar (1963).

the position of the object corresponding to the probability density $f(\varphi, \lambda)$, where $\{\varphi, \lambda\}$ are the usual geographical coordinates (longitude and latitude). An independent, simultaneous estimation of the position by another airplane navigator gives a probability distribution corresponding to the probability density $g(\varphi, \lambda)$. How should the two probability densities $f(\varphi, \lambda)$ and $g(\varphi, \lambda)$ be combined to obtain a resulting probability density? The answer is given by the conjunction of the two probability densities,

$$(f \wedge g)(\varphi, \lambda) = \frac{1}{\nu} \frac{f(\varphi, \lambda) g(\varphi, \lambda)}{\mu(\varphi, \lambda)} \quad , \tag{1.43}$$

where $\mu(\varphi, \lambda) = (\cos \lambda)/(4\pi R^2)$ is the homogeneous probability density on the surface of the sphere and ν is the normalization constant $\nu = \int_{-\pi}^{+\pi} d\varphi \int_{-\pi/2}^{+\pi/2} d\lambda \, \frac{f(\varphi, \lambda) g(\varphi, \lambda)}{\mu(\varphi, \lambda)}$.

Example 1.13. Conjunction of probabilities (II). *Consider a situation similar to that in Example 1.11, but where the actual probability density for the impact points $g(r, \varphi)$ is exactly known (or has been estimated as suggested in the example). We are interested in knowing, as precisely as possible, the coordinates of the next impact point. Again, when the point materializes, the finite accuracy of our measuring instrument only provides the probability density $f(r, \varphi)$. How can we combine the two probability densities $f(r, \varphi)$ and $g(r, \varphi)$ in order to have better information on the impact point? For the same reasons that the notion of conditional probabilities is used to update probability distributions (see below), here the updated version of the probability density $f(r, \varphi)$ is the normalized conjunction (expression at right in equation (1.40))*

$$f'(r, \varphi) = k \frac{f(r, \varphi) g(r, \varphi)}{\mu(r, \varphi)} \quad , \tag{1.44}$$

where k is a normalization constant and $\mu(r, \varphi)$ is the homogeneous probability density in polar coordinates[18] *($\mu(r, \varphi) = \text{const.} \, r$). A numerical illustration of this example is developed in Problem 7.9.*

While the Kolmogorov axioms define the space $\mathcal{E}$ of all possible probability distributions (over a given manifold), these two basic operations, conjunction and disjunction, furnish $\mathcal{E}$ with the structure to be used as the basis of all inference theory.

1.2.7 Conditional Probability

Rather than introduce the notion of conditional probability as a primitive notion of the theory, I choose to obtain it here as a special case of the conjunction of probability distributions. To make this link, we need to introduce a quite special probability distribution, the 'probability-event.'

An event $\mathcal{A}$ corresponds to a region of the manifold $\mathfrak{X}$. If P, Q, ... are probability distributions over $\mathfrak{X}$, characterized by the probability densities $f(\mathbf{x})$, $g(\mathbf{x})$, ..., the probabilities $P(\mathcal{A}) = \int_{\mathcal{A}} d\mathbf{x} \, f(\mathbf{x})$, $Q(\mathcal{A}) = \int_{\mathcal{A}} d\mathbf{x} \, g(\mathbf{x})$, ... are defined. To any event $\mathcal{A} \subseteq \mathfrak{X}$ we can attach a particular probability distribution that we shall denote $M_{\mathcal{A}}$. It can

---

[18]The surface element of the Euclidean 2D space in polar coordinates is $dS(r, \varphi) = r \, dr \, d\varphi$, from which $\mu(r, \varphi) = k \, r$ follows using expression (1.23) (see also the comments following that equation).

be characterized by a probability density $\mu_A(\mathbf{x})$ defined as follows (k being a possible normalization constant):

$$\mu_A(\mathbf{x}) = \begin{cases} k\,\mu(\mathbf{x}) & \text{if } \mathbf{x} \in A \ , \\ 0 & \text{otherwise} \ . \end{cases} \qquad (1.45)$$

In other words, $\mu_A(\mathbf{x})$ equals zero everywhere except inside A, where it is proportional to the homogeneous probability density $\mu(\mathbf{x})$. The probability distribution M_A so defined associates with any event $B \subseteq \mathcal{X}$ the probability $M_A(B) = \int_B d\mathbf{x}\,\mu_A(\mathbf{x})$ (because $\mu(\mathbf{x})$ is related to the volume element of $\mathcal{X}$, the probability $M_A(B)$ is proportional to the volume of $A \cap B$). While $A \subseteq \mathcal{X}$ is called an event, the probability distribution M_A shall be called a probability-event, or, for short, a *p-event*. See a one-dimensional illustration in Figure 1.1.

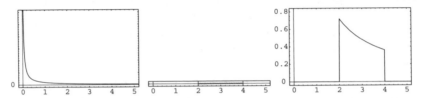

Figure 1.1. *The homogeneous probability density for a Jeffreys parameter is* $f(x) = 1/x$ *(left). In the middle is the event* $2 \le x \le 4$. *At the right is the probability-event (p-event) associated with this event. While the homogeneous probability density (at left) cannot be normalized, the p-event (at right) has been normalized to one.*

We can now set the following definition.

Definition. Let B be an event of the manifold $\mathcal{X}$ and M_B be the associated p-event. Let P be a probability distribution over $\mathcal{X}$. The conjunction of P and M_A, i.e., the probability distribution $(P \wedge M_B)$, shall be called the *conditional probability distribution* of P given B.

Using the characterization (at right in equation (1.40)) for the conjunction of probability distributions, it is quite easy to find an expression for $(P \wedge M_B)$. For the given $B \subseteq \mathcal{X}$, and for any $A \subseteq \mathcal{X}$, one finds

$$\boxed{(P \wedge M_B)(A) = \frac{P(A \cap B)}{P(B)}} \ , \qquad (1.46)$$

an expression that is valid provided $P(B) \neq 0$. The demonstration is provided as a footnote.[19]

---

[19]Let us introduce the probability density $f(\mathbf{x})$ representing P, the probability density $\mu_B(\mathbf{x})$ representing M_B, and the probability density $g(\mathbf{x})$ representing $P \wedge M_B$. It then follows, from the expression at right in equation (1.40), that $g(\mathbf{x}) = k\,f(\mathbf{x})\,\mu_B(\mathbf{x})\,/\,\mu(\mathbf{x})$, i.e., because of the definition of a p-event (equation (1.45)), $g(\mathbf{x}) = k\,f(\mathbf{x})$ if $\mathbf{x} \in B$ and $g(\mathbf{x}) = 0$ if $\mathbf{x} \notin B$. The normalizing constant k is (provided the expression is finite) $k = 1/\int_B d\mathbf{x}\,f(\mathbf{x}) = 1/P(B)$. We then have, for any $A \subseteq \mathcal{X}$ (and for the given $B \subseteq \mathcal{X}$), $(P \wedge M_B)(A) = \int_A d\mathbf{x}\,g(\mathbf{x}) = k\int_{A \cap B} d\mathbf{x}\,f(\mathbf{x}) = k\,P(A \cap B)$, from which expression (1.46) follows (using the value of k just obtained).

The expression on the right-hand side is what is usually taken as the definition of conditional probability density and is usually denoted $P(\mathcal{A}|\mathcal{B})$:

$$P(\mathcal{A}|\mathcal{B}) \equiv \frac{P(\mathcal{A}\cap\mathcal{B})}{P(\mathcal{B})} \quad . \tag{1.47}$$

Therefore, we have arrived at the property that, for any $\mathcal{A}$ and $\mathcal{B}$,

$$(P \wedge M_\mathcal{B})(\mathcal{A}) = P(\mathcal{A}|\mathcal{B}) \quad . \tag{1.48}$$

This shows that the notion of the conditional probability distribution is a special case of the notion of the conjunction of probability distributions. The reverse is not true (one cannot obtain the conjunction of probability distributions as a special case of conditional probabilities). As suggested in Figure 1.2, while conditioning a probability distribution to a

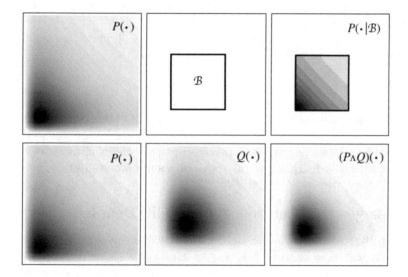

Figure 1.2. *In this figure, probability distributions are assumed to be defined over a two-dimensional manifold and are represented by the level lines of their probability densities. In the top row, left, is a probability distribution $P(\cdot)$ that with any event $\mathcal{A}$ associates the probability $P(\mathcal{A})$. In the middle is a particular event $\mathcal{B}$, and at the right is the conditional probability distribution $P(\cdot|\mathcal{B})$ (that with any event $\mathcal{A}$ associates the probability $P(\mathcal{A}|\mathcal{B})$). The probability density representing $P(\cdot|\mathcal{B})$ is the same as that representing $P(\cdot)$, except that values outside $\mathcal{B}$ are set to zero (and it has been renormalized). In the bottom row are a probability distribution $P(\cdot)$, a second probability distribution $Q(\cdot)$, and their conjunction $(P \wedge Q)(\cdot)$. Should we have chosen for the probability distribution $Q(\cdot)$ the p-event associated with $\mathcal{B}$, the two right panels would have been identical. The notion of the conjunction of probability distribution generalizes that of conditional probability in that the conditioning can be made using soft bounds. To generate this figure, the two probability densities $P(\cdot)$ and $Q(\cdot)$ have been chosen log-normal (see Appendix 6.7). Then, $(P \wedge Q)(\cdot)$ happens also to be log-normal.*

given event corresponds to adding some 'hard bounds,' the conjunction of two probability distributions allows the possible use of 'soft' bounds. This better corresponds to typical inference problems, where uncertainties may be present everywhere. In section 1.5.1, the conjunction of states of information is used to combine information obtained from measurements with information provided by a physical theory and is shown to be the basis of the inverse problem theory.

Equation (1.47) can, equivalently, be written $P(A \cap B) = P(A|B) P(B)$. Introducing the conditional probability distribution $P(A|B)$ would lead to $P(A \cap B) = P(B|A) P(A)$, and equating the two last expressions leads to the *Bayes theorem*

$$P(B|A) = \frac{P(A|B) P(B)}{P(A)} \quad . \tag{1.49}$$

It is sometimes said that this equation expresses the *probability of the causes*.[20] Again, although inverse theory could be based on Bayes theorem, we will use here the notion of the conjunction of probabilities.

To be complete, let me mention here that two events A and B are said to be *independent* with respect to a probability distribution $P(\cdot)$ if

$$P(A \cap B) = P(A) P(B) \quad . \tag{1.50}$$

It then immediately follows that $P(A|B) = P(A)$ and $P(B|A) = P(B)$: the conditional probabilities equal the unconditional ones (hence the term "independent" for A and B). Of course, if A and B are independent with respect to a probability distribution $P(\cdot)$, they will not, in general, be independent with respect to another probability distribution $Q(\cdot)$.

1.2.8 Marginal and Conditional Probability Density

Let $\mathfrak{U}$ and $\mathfrak{V}$ be two finite-dimensional manifolds with points respectively denoted $\mathbf{u} = \{u^1, u^2, \ldots\}$ and $\mathbf{v} = \{v^1, v^2, \ldots\}$, and let $\mathfrak{W} = \mathfrak{U} \times \mathfrak{V}$ be the Cartesian product of the two manifolds, i.e., the manifold whose points are of the form $\mathbf{w} = \{\mathbf{u}, \mathbf{v}\} = \{u^1, u^2, \ldots, v^2, v^2, \ldots\}$. A probability over $\mathfrak{W}$ is represented by a probability density $f(\mathbf{w}) = f(\mathbf{u}, \mathbf{v})$ that is here assumed to be normalized, $\int_{\mathfrak{U}} d\mathbf{u} \int_{\mathfrak{V}} d\mathbf{v}\, f(\mathbf{u}, \mathbf{v}) = \int_{\mathfrak{V}} d\mathbf{v} \int_{\mathfrak{U}} d\mathbf{u}\, f(\mathbf{u}, \mathbf{v}) = 1$.

In this situation, one can introduce the two *marginal probability densities*

$$f_{\mathfrak{V}}(\mathbf{v}) = \int_{\mathfrak{U}} d\mathbf{u}\, f(\mathbf{u}, \mathbf{v}) \quad , \quad f_{\mathfrak{U}}(\mathbf{u}) = \int_{\mathfrak{V}} d\mathbf{v}\, f(\mathbf{u}, \mathbf{v}) \quad , \tag{1.51}$$

which are normalized via

$$\int_{\mathfrak{V}} d\mathbf{v}\, f_{\mathfrak{V}}(\mathbf{v}) = 1 \quad , \quad \int_{\mathfrak{U}} d\mathbf{u}\, f_{\mathfrak{U}}(\mathbf{u}) = 1 \quad . \tag{1.52}$$

---

[20] Assume we know the (unconditional) probabilities $P(A)$ and $P(B)$ and the conditional probability $P(A|B)$ for the effect A given the cause B (these are the three terms at the right in expression (1.49)). If the effect A is observed, the Bayes formula gives the probability $P(B|A)$ for B being the cause of the effect A.

The variables $\mathbf{u}$ and $\mathbf{v}$ are said to be *independent* if the joint probability density equals the product of the two marginal probability densities:

$$f(\mathbf{u}, \mathbf{v}) = f_{\mathfrak{U}}(\mathbf{u})\, f_{\mathfrak{V}}(\mathbf{v}) \quad . \tag{1.53}$$

The interpretation of the marginal probability densities is as follows. Assume there is a probability density $f(\mathbf{w}) = f(\mathbf{u}, \mathbf{v})$ from which a (potentially infinite) sequence of random points (samples) $\mathbf{w}_1 = \{\mathbf{u}_1, \mathbf{v}_1\}$, $\mathbf{w}_2 = \{\mathbf{u}_2, \mathbf{v}_2\}, \ldots$ is generated. By definition, this sequence defines the two sequences $\mathbf{u}_1, \mathbf{u}_2, \ldots$ and $\mathbf{v}_1, \mathbf{v}_2, \ldots$. Then, the first sequence constitutes a set of samples of the marginal probability density $f_{\mathfrak{U}}(\mathbf{u})$, while the second sequence constitutes a set of samples of the marginal probability density $f_{\mathfrak{V}}(\mathbf{v})$. Note that generating a set $\mathbf{u}_1, \mathbf{u}_2, \ldots$ of samples of $f_{\mathfrak{U}}(\mathbf{u})$ and (independently) a set $\mathbf{v}_1, \mathbf{v}_2, \ldots$ of samples of $f_{\mathfrak{V}}(\mathbf{v})$, and then building the sequence $\mathbf{w}_1 = \{\mathbf{u}_1, \mathbf{v}_1\}$, $\mathbf{w}_2 = \{\mathbf{u}_2, \mathbf{v}_2\}, \ldots$, does not provide a set of samples of the joint probability density $f(\mathbf{w}) = f(\mathbf{u}, \mathbf{v})$ unless the two variables are independent, i.e., unless the property (1.53) holds.

To distinguish the original probability density $f(\mathbf{u}, \mathbf{v})$ from its two marginals $f_{\mathfrak{U}}(\mathbf{u})$ and $f_{\mathfrak{V}}(\mathbf{v})$, one usually calls $f(\mathbf{u}, \mathbf{v})$ the *joint probability density*.

The introduction of the notion of 'conditional probability density' is more subtle and shall be done here only in a very special situation. Consider again the joint probability density $f(\mathbf{u}, \mathbf{v})$ introduced above (in the same context), and let $\mathbf{u} \mapsto \mathbf{v}(\mathbf{u})$ represent an application from $\mathfrak{U}$ into $\mathfrak{V}$ (see Figure 1.3). The general idea is to 'condition' the joint probability density $f(\mathbf{u}, \mathbf{v})$, i.e., to forget all values of $f(\mathbf{u}, \mathbf{v})$ for which $\mathbf{v} \neq \mathbf{v}(\mathbf{u})$, and to retain only the information on the values of $f(\mathbf{u}, \mathbf{v})$ for which $\mathbf{v} = \mathbf{v}(\mathbf{u})$.

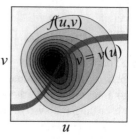

Figure 1.3. *A conditional probability density can be defined as a limit of a conditional probability (when the region suggested around the function $v = v(u)$ collapses into a line). The particular type of limit matters, as the conditional probability density essentially depends on it. In this elementary theory, we are only interested in the simple case where, with an acceptable approximation, one may take $f_{u|v=v(u)}(u) = \frac{f(u,v(u))}{\int du\, f(u,v(u))}$.*

To do this, one starts with the general definition of conditional probability (equation (1.47)), and then one takes the limit toward the hypersurface $\mathbf{v} = \mathbf{v}(\mathbf{u})$. The problem is that there are many possible ways of taking this limit, each producing a different result. Examining the detail of this problem is outside the scope of this book (the reader is referred, for instance, to the text by Mosegaard and Tarantola, 2002). Let us simply admit here that the situations we shall consider are such that (i) the application $\mathbf{v} = \mathbf{v}(\mathbf{u})$ is only mildly nonlinear (or it is linear), and (ii) the coordinates $\{u^1, u^2, \ldots\}$ and $\{v^1, v^2, \ldots\}$ are not too far from being 'Cartesian coordinates' over approximately linear manifolds. Under these restrictive conditions, one may define the *conditional probability density*[21]

$$f_{\mathfrak{U}|\mathfrak{V}}(\mathbf{u}\,|\,\mathbf{v}(\mathbf{u})) = \frac{f(\mathbf{u}, \mathbf{v}(\mathbf{u}))}{\int_{\mathfrak{U}} d\mathbf{u}'\, f(\mathbf{u}', \mathbf{v}(\mathbf{u}'))} \quad , \tag{1.54}$$

which obviously satisfies the normalization condition $\int_{\mathfrak{U}} d\mathbf{u}\, f_{\mathfrak{U}|\mathfrak{V}}(\mathbf{u}\,|\,\mathbf{v}(\mathbf{u})) = 1$.

---

[21] We could use the simpler notation $f_{\mathfrak{U}|\mathfrak{V}}(\mathbf{u})$ for the conditional probability density, but some subsequent notation then becomes more complicated.

A special case of this definition corresponds to the case where the conditioning is not made on a general relation $\mathbf{v} = \mathbf{v}(\mathbf{u})$, but on a constant value $\mathbf{v} = \mathbf{v}_0$. Then, equation (1.54) becomes $f_{\mathfrak{U}|\mathfrak{V}}(\mathbf{u} \,|\, \mathbf{v}_0) = f(\mathbf{u}, \mathbf{v}_0) / \int_{\mathfrak{U}} d\mathbf{u}' \, f(\mathbf{u}', \mathbf{v}_0)$, or, dropping the index 0 in $\mathbf{v}_0$, $f_{\mathfrak{U}|\mathfrak{V}}(\mathbf{u} \,|\, \mathbf{v}) = f(\mathbf{u}, \mathbf{v}) / \int_{\mathfrak{U}} d\mathbf{u}' \, f(\mathbf{u}', \mathbf{v})$. We recognize in the denominator the marginal probability density introduced in equation (1.51), so one can finally write

$$f_{\mathfrak{U}|\mathfrak{V}}(\mathbf{u} \,|\, \mathbf{v}) = \frac{f(\mathbf{u}, \mathbf{v})}{f_{\mathfrak{V}}(\mathbf{v})} \quad . \tag{1.55}$$

This shows, in particular, that under the present hypotheses, one can write the joint probability density as the product of the conditional and the marginal:

$$f(\mathbf{u}, \mathbf{v}) = f_{\mathfrak{U}|\mathfrak{V}}(\mathbf{u} \,|\, \mathbf{v}) \, f_{\mathfrak{V}}(\mathbf{v}) \quad . \tag{1.56}$$

Using instead the conditional of $\mathbf{v}$ with respect to $\mathbf{u}$, we can write $f(\mathbf{u}, \mathbf{v}) = f_{\mathfrak{V}|\mathfrak{U}}(\mathbf{v} \,|\, \mathbf{u}) \, f_{\mathfrak{U}}(\mathbf{u})$, and, combining the last two equations, we arrive at the *Bayes theorem*,

$$f_{\mathfrak{U}|\mathfrak{V}}(\mathbf{u} \,|\, \mathbf{v}) = \frac{f_{\mathfrak{V}|\mathfrak{U}}(\mathbf{v} \,|\, \mathbf{u}) \, f_{\mathfrak{U}}(\mathbf{u})}{f_{\mathfrak{V}}(\mathbf{v})} \quad , \tag{1.57}$$

that allows us to write the conditional for $\mathbf{u}$ given $\mathbf{v}$ in terms of the conditional for $\mathbf{v}$ given $\mathbf{u}$ (and the two marginals). This version of the Bayes theorem is less general (and more problematic) than the one involving events (equation (1.49)). We shall not make any use of this expression in this book.

1.3　Forward Problem

Experiments suggest physical theories, and physical theories predict the outcome of experiments. The comparison of the predicted outcome and the observed outcome allows us to ameliorate[22] the theory. If in the 'physical theory' we include the physical parameters describing the system under study, then inverse problem theory is about the quantitative rules to be used for this comparison between predictions and observations.

Taking first a naïve point of view, to solve a 'forward problem' means to predict the error-free values of the observable parameters $\mathbf{d}$ that would correspond to a given model $\mathbf{m}$. This theoretical prediction can be denoted

$$\boxed{\mathbf{m} \;\mapsto\; \mathbf{d} = \mathbf{g}(\mathbf{m})} \quad , \tag{1.58}$$

where $\mathbf{d} = \mathbf{g}(\mathbf{m})$ is a short notation for the set of equations $d^i = g^i(m^1, m^2, \dots)$ ($i = 1, 2, \dots$). The (usually nonlinear) operator $\mathbf{g}(\cdot)$ is called the *forward operator*. It expresses our mathematical model of the physical system under study.

Example 1.14. *A geophysicist may be interested in the coordinates $\{r, \theta, \varphi\}$ of the point where an earthquake starts, as well as in its origin time T. Then, the model parameters are $\mathbf{m} = \{r, \theta, \varphi, T\}$. The data parameters may be the arrival times $\mathbf{d} = \{t^1, t^2, \dots, t^n\}$ of*

---

[22]Or, taking an extreme Popperian point of view, to refute the theory if the disagreement is unacceptably large (Popper, 1959).

the elastic waves (generated by the earthquake) at some seismological observatories. If the velocities of propagation of the elastic waves inside Earth are known, it is possible, given the model parameters $\mathbf{m} = \{r, \theta, \varphi, T\}$, *to predict the arrival times* $\mathbf{d} = \{t^1, t^2, \ldots, t^n\}$, *which involves the use of some algorithm (a ray-based algorithm using Fermat's principle or a finite-differencing algorithm modeling the propagation of waves). Then, the functions* $d^i = g^i(\mathbf{m})$ *are defined. In some simple cases, these functions have a simple analytical expression. Sometimes they are implicitly defined through a complex algorithm. If the velocity model is itself not perfectly known, the parameters describing it also enter the model parameters* $\mathbf{m}$. *To compute the arrival times at the seismological observatories, the coordinates of the observatories must be used. If they are perfectly known, they can just be considered as fixed constants. If they are uncertain, they must also enter the parameter set* $\mathbf{m}$. *In fact, the set of model parameters is practically defined as all the parameters we must put on the right-hand side of the equation* $\mathbf{d} = \mathbf{g}(\mathbf{m})$ *in order for this to correspond to the usual prediction of the result of some observations — given a complete description of a physical system.*

The predicted values cannot, in general, be identical to the observed values for two reasons: measurement uncertainties and modelization imperfections. These two very different sources of error generally produce uncertainties with the *same order of magnitude*, because, due to the continuous progress of scientific research, as soon as new experimental methods are capable of decreasing the experimental uncertainty, new theories and new models arise that allow us to account for the observations more accurately. For this reason, it is generally not possible to set inverse problems properly without a careful analysis of modelization uncertainties.

The way to describe experimental uncertainties will be studied in section 1.4; this is mostly a well-understood matter. But the proper way of putting together measurements and physical predictions — each with its own uncertainties — is still a matter in progress. In this book, I propose to treat both sources of information symmetrically and to obey the postulate mentioned above, stating that the more general way of describing any state of information is to define a probability density. Therefore, the error-free equation (1.58) is replaced with a probabilistic correlation between model parameters $\mathbf{m}$ and observable parameters $\mathbf{d}$. Let us see how this is done.

Let $\mathfrak{M}$ be the model space manifold, with some coordinates (model parameters) $\mathbf{m} = \{m^\alpha\} = \{m^1, m^2, \ldots\}$ and with homogeneous probability density $\mu_M(\mathbf{m})$, and let $\mathfrak{D}$ be the data space manifold, with some coordinates (observable parameters) $\mathbf{d} = \{d^1\} = \{d^1, d^2, \ldots\}$ and with homogeneous probability density $\mu_D(\mathbf{d})$. Let $\mathfrak{X}$ be the joint manifold built as the Cartesian product of the two manifolds, $\mathfrak{D} \times \mathfrak{M}$, with coordinates $\mathbf{x} = \{\mathbf{d}, \mathbf{m}\} = \{d^1, d^2, \ldots, m^1, m^2, \ldots\}$ and with homogeneous probability density that, by definition, is $\mu(\mathbf{x}) = \mu(\mathbf{d}, \mathbf{m}) = \mu_D(\mathbf{d}) \mu_M(\mathbf{m})$.

From now on, the notation $\Theta(\mathbf{d}, \mathbf{m})$ is reserved for the joint probability density describing the correlations that correspond to our physical theory, together with the inherent uncertainties of the theory (due to an imperfect parameterization or to some more fundamental lack of knowledge).

Example 1.15. *If the data manifold* $\mathfrak{D}$ *and the model manifold* $\mathfrak{M}$ *are two linear spaces (respectively denoted* $\mathbb{D}$ *and* $\mathbb{M}$ *), then both homogeneous probability distributions are* *(unnormalizable) constants:* $\mu_{\mathrm{D}}(\mathbf{d}) = \text{const.}$, $\mu_{\mathrm{M}}(\mathbf{m}) = \text{const.}$ *The (singular) probability density*

$$\Theta(\mathbf{d}, \mathbf{m}) = \text{const.}\,\delta(\mathbf{d} - \mathbf{G}\,\mathbf{m}) \quad , \tag{1.59}$$

where $\mathbf{G}$ *is a linear operator (in fact, a matrix), clearly imposes the linear constraint* $\mathbf{g} = \mathbf{G}\,\mathbf{m}$ *between model parameters and observable parameters. The 'theory' is here assumed to be exact (or, more precisely, its uncertainties are assumed negligible compared to other sources of uncertainty). The marginal probability densities* $\Theta_{\mathrm{D}}(\mathbf{d}) = \int_{\mathrm{M}} \Theta(\mathbf{d}, \mathbf{m})$ *and* $\Theta_{\mathrm{M}}(\mathbf{m}) = \int_{\mathbb{D}} \Theta(\mathbf{d}, \mathbf{m})$ *are constant, meaning that the theory, although it carries information on the correlations between* $\mathbf{d}$ *and* $\mathbf{m}$*, does not carry any information on the* $\mathbf{d}$ *or the* $\mathbf{m}$ *themselves.*

Example 1.16. *In the same context of the previous example, replacing the probability density in equation (1.59) with (Gaussian probability densities are examined in Appendix 6.5)*

$$\Theta(\mathbf{d}, \mathbf{m}) = \text{const.}\,\exp\!\left(-\tfrac{1}{2}(\mathbf{d} - \mathbf{G}\,\mathbf{m})^{t}\,\mathbf{C}_{\mathrm{T}}^{-1}\,(\mathbf{d} - \mathbf{G}\,\mathbf{m})\right) \tag{1.60}$$

corresponds to assuming that the theoretical relation is still linear, $\mathbf{d} \approx \mathbf{G}\,\mathbf{m}$*, but has 'theoretical uncertainties' that are described by a Gaussian probability density with a covariance matrix* $\mathbf{C}_{\mathrm{T}}$*. Of course, the uncertainties can be described using other probabilistic models than the Gaussian one.*

If the two examples above are easy to understand (and, I hope, to accept), nontrivial complications arise when the relation between $\mathbf{d}$ and $\mathbf{m}$ is not linear. These complications are those appearing when trying to properly define the notion of conditional probability density (an explicit definition of a limit is required). I do not make any effort here to enter that domain: the reader is referred to Mosegaard and Tarantola (2002) for a quite technical introduction to the topic.

In many situations, one may, for every model $\mathbf{m}$, do slightly better than to exactly predict an associated value $\mathbf{d}$: one may, for every model $\mathbf{m}$, exhibit a probability density for $\mathbf{d}$ that we may denote $\theta(\mathbf{d}\,|\,\mathbf{m})$ (see Figure 1.4). A joint probability density can be written as the product of a conditional and a marginal (equation (1.56)). Taking for the marginal for the model parameters the homogeneous probability density then gives

$$\boxed{\Theta(\mathbf{d}, \mathbf{m}) = \theta(\mathbf{d}\,|\,\mathbf{m})\,\mu_{\mathrm{M}}(\mathbf{m}) \quad .} \tag{1.61}$$

But there is a major difference between this case and the two Examples 1.15 and 1.16: while in the two examples above both marginals of $\Theta(\mathbf{d}, \mathbf{m})$ correspond to the homogeneous probability distributions for $\mathbf{d}$ and $\mathbf{m}$, respectively, expression (1.61) only ensures that the marginal for $\mathbf{m}$ is homogeneous, not necessarily that the marginal for $\mathbf{d}$ is. This problem is implicit in all Bayesian formulations of the inverse problem, even if it is not mentioned explicitly. In this elementary text, I just suggest that equation (1.61) can be used in all situations where the dependence of $\mathbf{d}$ on $\mathbf{m}$ is only mildly nonlinear.

Example 1.17. *Assume that the data manifold* $\mathfrak{D}$ *is, in fact, a linear space denoted* $\mathbb{D}$ *with vectors denoted* $\mathbf{d} = \{d^{1}, d^{2}, \dots\}$*. Because this is a linear space, the homogeneous*

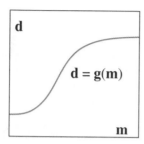 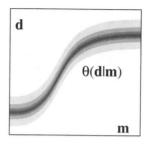

Figure 1.4. a) *If uncertainties in the forward modelization can be neglected, a functional relationship* $\mathbf{d} = \mathbf{g}(\mathbf{m})$ *gives, for each model* $\mathbf{m}$, *the predicted (or calculated) data values* $\mathbf{d}$. b) *If forward-modeling uncertainties cannot be neglected, they can be described, giving, for each value of* $\mathbf{m}$, *a probability density for* $\mathbf{d}$ *that we may denote* $\theta(\mathbf{d}|\mathbf{m})$. *Roughly speaking, this corresponds to putting vertical uncertainty bars on the theoretical relation* $\mathbf{d} = \mathbf{g}(\mathbf{m})$.

probability density $\mu_D(\mathbf{d})$ *is constant. Let* $\mathfrak{M}$ *be an arbitrary model manifold with coordinates (model parameters) denoted* $\mathbf{m} = \{m^1, m^2, \dots\}$ *and with a homogeneous probability density* $\mu_M(\mathbf{m})$. *The choice*

$$\theta(\mathbf{d}|\mathbf{m}) = \text{const. } \exp\left(-\tfrac{1}{2}(\mathbf{d} - \mathbf{g}(\mathbf{m}))^t \, \mathbf{C}_T^{-1} \, (\mathbf{d} - \mathbf{g}(\mathbf{m})) \right) \quad, \tag{1.62}$$

where $\mathbf{g}(\mathbf{m})$ *is a (mildly) nonlinear function of the model parameters* $\mathbf{m}$, *imposes on* $\mathbf{d} \approx \mathbf{g}(\mathbf{m})$ *some uncertainties assumed to be Gaussian with covariance operator* $\mathbf{C}_T$. *Equation (1.61) then leads to the joint probability density*

$$\Theta(\mathbf{d}, \mathbf{m}) = \text{const. } \exp\left(-\tfrac{1}{2}(\mathbf{d} - \mathbf{g}(\mathbf{m}))^t \, \mathbf{C}_T^{-1} \, (\mathbf{d} - \mathbf{g}(\mathbf{m})) \right) \mu_M(\mathbf{m}) \quad. \tag{1.63}$$

If the 'theoretical uncertainties' (as described by the covariance matrix $\mathbf{C}_T$ *) are to be neglected, then the limit of this probability density is*

$$\Theta(\mathbf{d}, \mathbf{m}) = \delta(\mathbf{d} - \mathbf{g}(\mathbf{m})) \, \mu_M(\mathbf{m}) \quad. \tag{1.64}$$

The $\Theta(\mathbf{d}, \mathbf{m})$ in equation (1.64) is the (singular) probability density we shall take in this book to exactly impose the (mildly) nonlinear constraint $\mathbf{d} = \mathbf{g}(\mathbf{m})$. When theoretical uncertainties cannot be neglected, the Gaussian model in equation (1.63) can be used, or any other simple probabilistic model, or still better, a realistic account of the modelization uncertainties.

There is a class of problems where the correlations between $\mathbf{d}$ and $\mathbf{m}$ are not predicted by a formal theory, but result from an accumulation of observations. In this case, the joint probability density $\Theta(\mathbf{d}, \mathbf{m})$ appears naturally in the description of the available information.

Example 1.18. *The data parameters* $\mathbf{d} = \{d^i\}$ *may represent the current state of a volcano (intensity of seismicity, rate of accumulation of strain, . . .). The model parameters* $\mathbf{m} = \{m^\alpha\}$ *may represent, for instance, the time interval to the next volcanic eruption, the* *magnitude of this eruption, etc. Our present knowledge of volcanoes does not allow us to relate these parameters realistically using physical laws, so that, at present, the only scientific description is statistical. Provided that in the past we were able to observe a significant number of eruptions of this volcano, we can construct a histogram in the space $\mathfrak{D} \times \mathfrak{M}$ that describes all our information correlating the parameters (see Tarantola, Trygvasson, and Nercession, 1985, for an example). This histogram can directly be identified with $\Theta(\mathbf{d}, \mathbf{m})$.*

Of course, the simple scheme developed here may become considerably more complicated when the details concerning particular problems are introduced, as the following example suggests.

Example 1.19. *A rock may primarily be described by some petrophysical parameters $\mathbf{m}$, like mineral content, porosity, permeability, etc. Information on these parameters can be obtained by propagating elastic waves in the rock to generate some waveform data $\mathbf{d}$, but the theory of elastic wave propagation may only use a set $\boldsymbol{\mu}$ of higher order parameters, like bulk modulus, shear modulus, attenuation, etc. Using the definition of conditional and marginal probability density, any joint probability density $f(\mathbf{d}, \boldsymbol{\mu}, \mathbf{m})$ can be decomposed as $f(\mathbf{d}, \boldsymbol{\mu}, \mathbf{m}) = f(\mathbf{d}|\boldsymbol{\mu}, \mathbf{m})\, f(\boldsymbol{\mu}|\mathbf{m})\, f(\mathbf{m})$. In the present problem, this suggests replacing equation (1.61) with*

$$\Theta(\mathbf{d}, \boldsymbol{\mu}, \mathbf{m}) = \theta(\mathbf{d}|\boldsymbol{\mu}, \mathbf{m})\, \theta(\boldsymbol{\mu}|\mathbf{m})\, \mu_{\mathrm{M}}(\mathbf{m}) \quad . \tag{1.65}$$

Furthermore, if the waveform data $\mathbf{d}$ are assumed to depend on the petrophysical parameters $\mathbf{m}$ mainly through the higher order parameter $\boldsymbol{\mu}$, one can use the approximation $\theta(\mathbf{d}|\boldsymbol{\mu}, \mathbf{m}) = \theta(\mathbf{d}|\boldsymbol{\mu})$, in which case we can write

$$\Theta(\mathbf{d}, \boldsymbol{\mu}, \mathbf{m}) = \theta(\mathbf{d}|\boldsymbol{\mu})\, \theta(\boldsymbol{\mu}|\mathbf{m})\, \mu_{\mathrm{M}}(\mathbf{m}) \quad . \tag{1.66}$$

Here, the conditional probability density $\theta(\boldsymbol{\mu}|\mathbf{m})$ may represent the correlations observed in the laboratory (using a large number of different rocks) between the petrophysical parameters $\mathbf{m}$ and the higher order parameters $\boldsymbol{\mu}$, while $\theta(\mathbf{d}|\boldsymbol{\mu})$ is the prediction of waveform data $\mathbf{d}$ given the higher order parameters $\boldsymbol{\mu}$. For this, one may take something as simple as (see equation (1.64)) $\theta(\mathbf{d}|\boldsymbol{\mu}) = \delta(\mathbf{d} - \mathbf{g}(\mathbf{m}))$.

1.4   Measurements and A Priori Information

1.4.1   Results of the Measurements

All physical measurements are subjected to uncertainties. Therefore, the result of a measurement act is not simply an 'observed value' (or a set of 'observed values') but a 'state of information' acquired on some observable parameter. If $\mathbf{d} = \{d^1, d^2, \ldots, d^n\}$ represents the set of observable parameters, the result of the measurement act can be represented by a probability density $\rho_{\mathrm{D}}(\mathbf{d})$ defined over the data space $\mathfrak{D}$.

Example 1.20. *In the simplest situation, when measuring an n-dimensional data vector $\mathbf{d}$ (considered an element of a linear space $\mathbb{D}$), we may obtain the observed values $\mathbf{d}_{\mathrm{obs}}$, with* *uncertainties assumed to be of the Gaussian type, described by a covariance matrix $\mathbf{C}_D$. Then, $\rho_D(\mathbf{d})$ is a Gaussian probability density centered at $\mathbf{d}_{obs}$:*

$$\rho_D(\mathbf{d}) = ((2\pi)^n \det \mathbf{C}_D)^{-1/2} \exp\left(-\tfrac{1}{2} (\mathbf{d} - \mathbf{d}_{obs})^t \, \mathbf{C}_D^{-1} \, (\mathbf{d} - \mathbf{d}_{obs}) \right) \quad . \tag{1.67}$$

See also Example 1.25.

Example 1.21. *Consider a measurement made to obtain the arrival time of a given seismic wave recorded by a seismograph (see Figure 1.5). Sometimes, the seismogram is simple enough to give a simple result, but sometimes, due to strong noise (with unknown statistics), the measurement is not trivial. The figure suggests a situation where it is difficult to obtain a numerical value, say t_{obs}, for the arrival time. The use of a probability density (bottom of the figure) allows us to describe the information we actually have on the arrival time with a sufficient degree of generality (using here a bimodal probability density). With these kinds of data, it is clear that the subjectivity of the scientist plays a major role. It is indeed the case,* whichever inverse method is used, *that results obtained by different scientists from similar data sets are different. Objectivity can only be attained if the data redundancy is great enough that differences in data interpretation among different observers do not significantly alter the models obtained.*

Figure 1.5. *At the top, a seismogram showing the arrival of a wave. Due to the presence of noise, it is difficult to pick the first arrival time of the wave. Here, in particular, one may hesitate between the "big arrival" and the "small arrival" before, which may or may not just be noise. In this situation, one may give to the variable arrival time a bimodal probability density (bottom). The width of each peak represents the uncertainty of the reading of each of the possible arrivals, the area of each peak represents the probability for the arrival time to be there, and the separation of the peaks represents the overall uncertainty.*

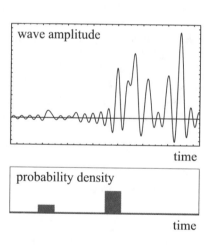

Example 1.22. Observations are the output of an instrument with known statistics. *Let us place ourselves under the hypothesis that the data space is a linear space (so the use of conditional probability densities is safe). To simplify the discussion, I will refer to "the instrument" as if all the measurements could result from a single reading on a large apparatus, although, more realistically, we generally have some readings from several apparatuses. Assume that when making a measurement the instrument delivers a given value of $\mathbf{d}$, denoted $\mathbf{d}_{out}$. Ideally, the supplier of the apparatus should provide a statistical analysis of the uncertainties of the instrument. The most useful and general way of giving the results of the statistical analysis is to define the probability density for the value of the output, $\mathbf{d}_{out}$, when the actual input is $\mathbf{d}$. Let $\nu(\mathbf{d}_{out}|\mathbf{d})$ be this conditional probability*

density. If $f(\mathbf{d}_{out}, \mathbf{d})$ denotes the joint probability density for $\mathbf{d}_{out}$ and $\mathbf{d}$, and if we don't use any information on the input, we have (equation (1.56)) $f(\mathbf{d}_{out}, \mathbf{d}) = \nu(\mathbf{d}_{out}|\mathbf{d})\,\mu_D(\mathbf{d})$, where $\mu_D(\mathbf{d})$ is the homogeneous probability density. As the data space is linear, the homogeneous probability density is constant, and we simply have

$$f(\mathbf{d}_{out}, \mathbf{d}) = \text{const. } \nu(\mathbf{d}_{out}|\mathbf{d}) \quad . \tag{1.68}$$

If the actual result of a measurement is $\mathbf{d}_{out} = \mathbf{d}_{obs}$, then we can identify $\rho_D(\mathbf{d})$ with the conditional probability density for $\mathbf{d}$ given $\mathbf{d}_{out} = \mathbf{d}_{obs}$: $\rho_D(\mathbf{d}) = f(\mathbf{d}\,|\,\mathbf{d}_{out} = \mathbf{d}_{obs})$. This gives $\rho_D(\mathbf{d}) = f(\mathbf{d}_{obs}, \mathbf{d})\,/\,\int_D d\mathbf{d}\,f(\mathbf{d}_{obs}, \mathbf{d})$, i.e.,

$$\rho_D(\mathbf{d}) = \text{const. } \nu(\mathbf{d}_{obs}|\mathbf{d}) \quad . \tag{1.69}$$

This relates the instrument characteristic $\nu(\mathbf{d}_{out}|\mathbf{d})$, the observed value $\mathbf{d}_{obs}$, and the probability density $\rho_D(\mathbf{d})$ in the data space that represents the information found by the measurement.

Example 1.23. Perfect instrument. *In the context of the previous example, a perfect instrument corresponds to $\nu(\mathbf{d}_{out}|\mathbf{d}) = \delta(\mathbf{d}_{out} - \mathbf{d})$. Then, if we observe the value $\mathbf{d}_{obs}$, $\rho_D(\mathbf{d}) = \delta(\mathbf{d}_{obs} - \mathbf{d})$, i.e.,*

$$\rho_D(\mathbf{d}) = \delta(\mathbf{d} - \mathbf{d}_{obs}) \quad . \tag{1.70}$$

The assumption of a perfect instrument may be made when measuring uncertainties are negligible compared to modelization uncertainties.

Example 1.24. *In the context of Example 1.22, assume that the uncertainties due to the measuring instrument are independent of the input. Assume that the output $\mathbf{d}_{out}$ is related to the input $\mathbf{d}$ through the simple relation*

$$\mathbf{d}_{out} = \mathbf{d} + \boldsymbol{\epsilon} \quad , \tag{1.71}$$

where $\boldsymbol{\epsilon}$ is an unknown error with known statistics described by the probability density $f(\boldsymbol{\epsilon})$. In that case, if we observe the value $\mathbf{d}_{obs}$,

$$\rho_D(\mathbf{d}) = \nu(\mathbf{d}_{obs}|\mathbf{d}) = f(\boldsymbol{\epsilon}) = f(\mathbf{d}_{obs} - \mathbf{d}) \quad . \tag{1.72}$$

This result is illustrated in Figure 1.6.

Example 1.25. Gaussian uncertainties. *In the context of the previous example, assume that the probability density for the error $\boldsymbol{\epsilon}$ is Gaussian with zero mean and covariance $\mathbf{C}_D$: $f(\boldsymbol{\epsilon}) = \text{Gaussian}(\boldsymbol{\epsilon}, \mathbf{0}, \mathbf{C}_D)$. Equation (1.72) then gives $\rho_D(\mathbf{d}) = \text{Gaussian}(\mathbf{d}_{obs} - \mathbf{d}, \mathbf{0}, \mathbf{C}_D)$, i.e., $\rho_D(\mathbf{d}) = \text{Gaussian}(\mathbf{d} - \mathbf{d}_{obs}, \mathbf{0}, \mathbf{C}_D)$. Equivalently,*

$$\rho_D(\mathbf{d}) = \text{Gaussian}(\mathbf{d}, \mathbf{d}_{obs}, \mathbf{C}_D) \quad , \tag{1.73}$$

corresponding to the result already suggested in Example 1.20.

Example 1.26. Outliers in a data set. *Some data sets contain outliers that are difficult to eliminate, in particular when the data space is highly dimensioned, because it is difficult to visualize such data sets. Problem 7.7 shows that a single outlier in a data set can lead*

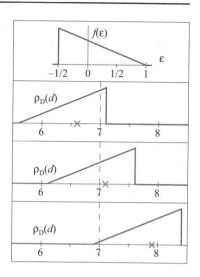

Figure 1.6. *As an illustration of Example 1.24, consider that the statistics of the (scalar) error ε is represented by the probability density $f(\varepsilon)$ at the top of the figure. Imagine that the true (unknown) value of the measured quantity is $d = 7$. This figure shows the three probability densities we should infer for d (according to equation (1.72)) in the three different situations where the output of the instrument is, respectively, $d_{\mathrm{obs}} = 6.6$, $d_{\mathrm{obs}} = 7.1$, and $d_{\mathrm{obs}} = 7.9$. Observe that the shape of the probability density $\rho_{\mathrm{D}}(d)$ is the mirror image of the shape of $f(\varepsilon)$.*

to unacceptable inverse results if the Gaussian assumption is used. That problem suggests using long-tailed *probability densities to represent uncertainties on this kind of data sets. A simple example of a long-tailed density function is the symmetric exponential (Laplace) function $f(x) = \frac{1}{2\sigma} \exp(-|x - x_0|/\sigma)$, where σ is the mean deviation of the distribution. Using this probability model gives (uncertainties are assumed independent)*

$$\rho_{\mathrm{D}}(\mathbf{d}) = \left(\prod_i \frac{1}{2\,\sigma^i} \right) \exp\left(-\sum_i \frac{|d^i - d^i_{\mathrm{obs}}|}{\sigma^i} \right) \ . \tag{1.74}$$

Example 1.27. *Assume that the only instrument we have for measuring a given observable d is a buzzer that responds when the true value d is in the range $d_{\mathrm{min}} \leq d \leq d_{\mathrm{max}}$. We make the measurement, and the buzzer does* not *respond. The corresponding probability density is then*

$$\rho_{\mathrm{D}}(d) = \begin{cases} 0 & \text{for } d_{\mathrm{min}} \leq d \leq d_{\mathrm{max}} \ , \\ \text{const. } \mu_{\mathrm{D}}(d) & \text{otherwise} \ , \end{cases} \tag{1.75}$$

where $\mu_{\mathrm{D}}(d)$ is the homogeneous probability density for the observable parameter d.

1.4.2 A Priori Information on Model Parameters

By *a priori information* (or prior information) we shall mean information that is obtained independently of the results of measurements. The probability density representing this a priori information will be denoted by $\rho_{\mathrm{M}}(\mathbf{m})$.

Example 1.28. *We have no a priori information other than the basic definition of the model parameters. In that case,*

$$\rho_{\mathrm{M}}(\mathbf{m}) = \mu_{\mathrm{M}}(\mathbf{m}) \ , \tag{1.76}$$

where $\mu_{\mathrm{M}}(\mathbf{m})$ is the homogeneous probability density for model parameters, as introduced in section 1.2.4.

Example 1.29. *For a given parameter m^α we have only the information that it is strictly bounded by the two values $m^\alpha_{\min}$ and $m^\alpha_{\max}$. We can take*

$$\rho_M(\mathbf{m}) = \prod_{\alpha \in I_M} \rho_\alpha(m^\alpha) \quad , \tag{1.77}$$

where

$$\rho_\alpha(m^\alpha) = \begin{cases} \text{const. } \mu_\alpha(m^\alpha) & \text{for } m^\alpha_{\min} \leq m^\alpha \leq m^\alpha_{\max} \quad , \\ 0 & \text{otherwise} \quad , \end{cases} \tag{1.78}$$

and where $\mu_\alpha(m^\alpha)$ represents the homogeneous probability density for m^α.

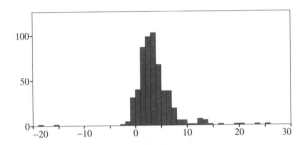

Figure 1.7. *Histogram of disintegration periods (half-lives) of the first 580 atomic nuclei in the CRC Handbook of Chemistry and Physics (1984). The horizontal axis represents the logarithm of the half-life: $T^* = \log_{10}(T/T_0)$, with $T_0 = 1$ s. It is very difficult to show the histogram using a nonlogarithmic time axis, because observed disintegration periods span 45 orders of magnitude. Note that the use of a logarithmic period axis allows the histogram to be roughly approximated by a Gaussian probability density. This implies a log-normal probability density for the half-lives.*

Example 1.30. *Figure 1.7 shows a histogram of the half-lives of the first 580 atomic nuclei quoted in the CRC Handbook of Chemistry and Physics (1984). Denoting by T the half-life, the abscissa of the figure is*

$$T^* = \log_{10} \frac{T}{T_0} \qquad (T_0 = 1\,\text{s}) \quad . \tag{1.79}$$

The logarithmic scale has been chosen for the time axis because, as the half-lives span many orders of magnitude, it is difficult to show the histogram using a linear time axis. With this logarithmic scale, the histogram may conveniently be approximated by a Gaussian function,

$$f^*(T^*) = \frac{1}{(2\pi)^{1/2} \sigma^*} \exp\left(-\frac{1}{2} \frac{(T^* - T_0^*)^2}{(\sigma^*)^2} \right) \quad , \tag{1.80}$$

with $T_0^* \simeq 3$ and $\sigma^* \simeq 3$. This implies a log-normal probability density (see Appendix 6.7) for the half-life T (which is adequate for a Jeffreys quantity). If we are going to experiment with one such atomic nucleus, but we ignore which one, we can use this probability distribution to represent our a priori information on its half-life.

Example 1.31. *The left of Figure* 1.8 *shows the histogram of densities of different known minerals in Earth's crust (independently of their relative abundance). At the right of the figure, the same histogram is shown on a logarithmic scale. If one has some mineral on which to perform some measurements, one may well take these (equivalent) probability densities to represent the a priori state of information on the mass density. If you do not have this figure in mind, the homogeneous probability density will represent your ignorance well. In mass density, this is* $\mu(\rho) = 1/\rho$ *, and, in logarithmic mass density,* $\mu^*(\rho^*) = \text{const.}$

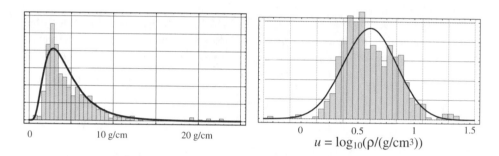

Figure 1.8. *Histogram of mass densities of the* 571 *different rock types quoted by Johnson and Olhoeft* (1984). *The histogram is very asymmetric due to the existence of very heavy minerals* $(\rho \simeq 20\,\text{g cm}^{-3})$. *Also shown is a best-fitting log-normal probability density. The same histogram of the previous figure in a logarithmic horizontal scale has* $\rho^* = \log_{10}(\rho/\text{g cm}^{-3})$. *In a logarithmic scale, it is much more symmetric. Also shown is the best-fitting normal (i.e., Gaussian) probability density.*

Example 1.32. *Assume that we are going to solve an inverse problem involving a one-dimensional (layered) Earth model, with each layer describing a thickness and a value of the mass density. We are given the a priori information that the layer thickness has the probability density displayed at the left in Figure* 1.9 *and the mass density has the probability density displayed at the right in Figure* 1.9, *all the variables being independent. It is then possible (and quite easy, as all the variables are independent) to randomly generate models that represent this a priori information. Figure* 1.10 *shows three such models. This random generation of models is at the basis of the strategy proposed in chapter 2 for solving inverse problems.*

Example 1.33. *Assume that the parameters* m^α *represent a discretization of a continuous function. We have reason to believe that the particular function under study is a random realization of a Gaussian random field, with given mean* $\mathbf{m}_{\text{prior}}$ *and given covariance* $\mathbf{C}_M$ *(examples of realizations of such random fields are displayed in chapter 2 (Figure 2.4) and chapter 5 (Figures 5.7 and 5.8)). This a priori information is then represented by the Gaussian probability density*

$$\rho_M(\mathbf{m}) = \left((2\pi)^n \det \mathbf{C}_M\right)^{-1/2} \exp\left(-\tfrac{1}{2}(\mathbf{m} - \mathbf{m}_{\text{prior}})^t \, \mathbf{C}_M^{-1} \, (\mathbf{m} - \mathbf{m}_{\text{prior}})\right) \quad , \quad (1.81)$$

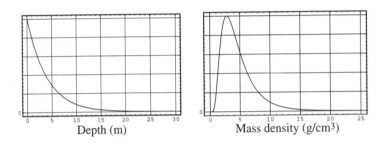

Depth (m)   Mass density (g/cm³)

Figure 1.9. *When using the probability density at the left for the thickness of the layers of a geological model and the probability density at the right for the mass density inside each layer, one may generate random models, like the three displayed in Figure 1.10.*

Mass density (g/cm³)

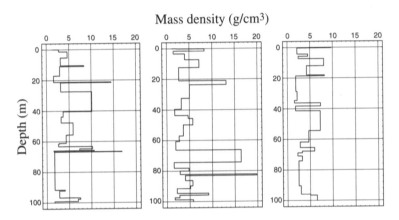

Figure 1.10. *Three randomly generated layered models of Earth. The layer thicknesses are generated according to an exponential probability density and the mass densities according to a log-normal probability density (that of Figure 1.8, page 29).*

where n is the number of discretization points. This probability density gives a high probability density to models **m** *that are close to* **m**$_{prior}$ *in the sense of the covariance operator* **C**$_M$*, i.e., models in which the difference* **m** − **m**$_{prior}$ *is small at each point (with respect to standard deviations in* **C**$_M$ *) and smooth (with respect to correlations in* **C**$_M$ *). Covariance operators are defined in Appendix 6.5.*

The Gaussian assumption of the previous example is quite elementary. More realistic assumptions are these days being used by geostatisticians[23] (see, for instance, Caers, Srinivasan, and Journel, 2000). A well-designed geophysical inverse problem should use the concepts of geostatistics to introduce the a priori information and the equations of this book to formulate the inverse problem.

---

[23]For some general information on geostatistics, see Journel and Huijbregts, 1978, Goovaerts, 1997, and Deutsch and Journel, 1998.

The examples in this section show how it is possible to use probability densities to describe prior information. I have never found a state of information (in the intuitive sense) that cannot be very precisely stated using a probability density. On the other hand, it may seem that probability densities have too many degrees of freedom to allow a definite choice that represents a given state of information. In fact, only a few characteristics of a probability density are usually relevant, such as, for instance, the position of the center, the degree of asymmetry, the size of the uncertainty bounds, the correlations between different parameters, and the behavior of the probability density far from the center.

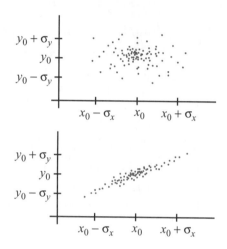

Figure 1.11. *The random points (x_i, y_i) of these diagrams have been generated using two-dimensional probability densities $f_1(x, y)$ (top) and $f_2(x, y)$ (bottom). The two probability densities have identical mean values and standard deviations. Only the covariance C_{xy} is different. On the top, the covariance is small; on the bottom, it is large. The probability density $f_2(x, y)$ is more informative than $f_1(x, y)$ because it demarcates a smaller region in the space. This example suggests that if off-diagonal elements of a covariance operator are difficult to estimate, setting them to zero corresponds to neglecting information.*

If hesitation exists in choosing the a priori uncertainty bars, it is of course best to be overconservative and to choose them very large. A conservative choice for correlations is to neglect them (see Figure 1.11 for an example). The behavior of the probability densities far from the center is only crucial if outliers may exist: the choice of functions tending too rapidly to zero (boxcar functions or even Gaussian functions) may lead to inconsistencies; the solution to the problem (as defined in the next section) may not exist, or may be senseless.

Usually the a priori states of information have the form of soft bounds; the normal or log-normal probability densities generally apply well to that case. If the normal function is thought to vanish too rapidly when the parameter's value tends to infinity, longer tailed functions may be used such as, for instance, the symmetric-exponential function (see Appendix 6.6).

1.4.3 Joint Prior Information

By definition, the a priori information on model parameters is independent of observations. The information we have in both model parameters and observable parameters can then be described in the manifold $\mathcal{D} \times \mathcal{M}$ by the joint probability density (this is the independence mentioned in equation (1.53))

$$\rho(\mathbf{d}, \mathbf{m}) = \rho_D(\mathbf{d})\,\rho_M(\mathbf{m}) \quad . \tag{1.82}$$

It may happen that part of the "a priori information" has been obtained from a first, rough, analysis of the data set. Rigorously, then, there exist correlations between $\mathbf{d}$ and $\mathbf{m}$ in $\rho(\mathbf{d}, \mathbf{m})$, and equation (1.82) no longer holds. For a maximum of generality, we thus have to assume the existence of a general probability density $\rho(\mathbf{d}, \mathbf{m})$, not necessarily satisfying (1.82), and representing all the information we have in data and model parameters *independently* of the use of any theoretical information (which is described by the probability density $\Theta(\mathbf{d}, \mathbf{m})$ introduced in section 1.3).

1.5   Defining the Solution of the Inverse Problem

1.5.1   Combination of Experimental, A Priori, and Theoretical Information

We have seen in the previous section that the *prior probability density* $\rho(\mathbf{d}, \mathbf{m})$, defined in the space $\mathfrak{D} \times \mathfrak{M}$, represents both information obtained on the observable parameters (data) $\mathbf{d}$ and a priori information on the model parameters $\mathbf{m}$. We have also seen that the *theoretical probability density* $\Theta(\mathbf{d}, \mathbf{m})$ represents the information on the physical correlations between $\mathbf{d}$ and $\mathbf{m}$, as obtained from a physical law, for instance.

These two states of information combine to produce the *a posteriori state of information*. I propose that the method of the previous sections to introduce the a priori and the theoretical states of information is such that the a posteriori state of information is given by the *conjunction* of these two states of information. Using (1.40), the probability density $\sigma(\mathbf{d}, \mathbf{m})$ representing the a posteriori information is then

$$\sigma(\mathbf{d}, \mathbf{m}) = k \, \frac{\rho(\mathbf{d}, \mathbf{m}) \, \Theta(\mathbf{d}, \mathbf{m})}{\mu(\mathbf{d}, \mathbf{m})} \quad , \tag{1.83}$$

where $\mu(\mathbf{d}, \mathbf{m})$ represents the homogeneous state of information and where k is a normalization constant.

This is justified by the correctness of the consequences we shall obtain, as the rest of this book is based on (1.83). It will be seen that the conclusions obtained from this equation, although more general than those obtained from more traditional approaches, reduce to them in all particular cases. Equation (1.83) first appeared in Tarantola and Valette (1982a).

Once the a posteriori information in the $\mathfrak{D} \times \mathfrak{M}$ space has been defined, the a posteriori information in the model space is given by the marginal probability density

$$\sigma_M(\mathbf{m}) = \int_{\mathfrak{D}} d\mathbf{d} \, \sigma(\mathbf{d}, \mathbf{m}) \quad , \tag{1.84}$$

while the a posteriori information in the data space is given by

$$\sigma_D(\mathbf{d}) = \int_{\mathfrak{M}} d\mathbf{m} \, \sigma(\mathbf{d}, \mathbf{m}) \quad . \tag{1.85}$$

Figure 1.12 illustrates the determination of $\sigma_M(\mathbf{m})$ and $\sigma_D(\mathbf{d})$ from $\rho(\mathbf{d}, \mathbf{m})$ and $\Theta(\mathbf{d}, \mathbf{m})$ geometrically.

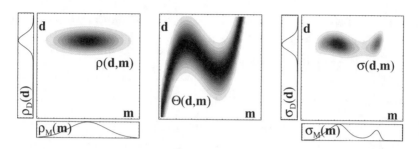

Figure 1.12. *Left: The probability densities* $\rho_D(\mathbf{d})$ *and* $\rho_M(\mathbf{m})$ *respectively represent the information on observable parameters (data) and the prior information on model parameters. As the prior information on model parameters is, by definition, independent of the information on observable parameters (measurements), the joint probability density in the space* $\mathfrak{D} \times \mathfrak{M}$ *representing both states of information is* $\rho(\mathbf{d}, \mathbf{m}) = \rho_D(\mathbf{d})\,\rho_M(\mathbf{m})$. *Center:* $\Theta(\mathbf{d}, \mathbf{m})$ *represents the information on the physical correlations between* $\mathbf{d}$ *and* $\mathbf{m}$, *as predicted by a (nonexact) physical theory. Right: Given the two states of information represented by* $\rho(\mathbf{d}, \mathbf{m})$ *and* $\Theta(\mathbf{d}, \mathbf{m})$, *their conjunction is* $\sigma(\mathbf{d}, \mathbf{m}) = k\,\rho(\mathbf{d}, \mathbf{m})\,\Theta(\mathbf{d}, \mathbf{m})\,/\,\mu(\mathbf{d}, \mathbf{m})$ *and represents the combination of the two states of information. From* $\sigma(\mathbf{d}, \mathbf{m})$ *it is possible to obtain the marginal probability densities* $\sigma_M(\mathbf{m}) = \int d\mathbf{d}\,\sigma(\mathbf{d}, \mathbf{m})$ *and* $\sigma_D(\mathbf{d}) = \int d\mathbf{m}\,\sigma(\mathbf{d}, \mathbf{m})$. *By comparison of the posterior probability density* $\sigma_M(\mathbf{m})$ *with the prior one,* $\rho_M(\mathbf{m})$, *we see that some information has been gained on the model parameters thanks to the data* $\rho_D(\mathbf{d})$ *and the theoretical information* $\Theta(\mathbf{d}, \mathbf{m})$.

1.5.2 Resolution of Inverse Problems

Equation (1.83) solves a very general problem. Inverse problems correspond to the particular case where the spaces $\mathfrak{D}$ and $\mathfrak{M}$ have fundamentally different physical meaning and where we are interested in translating information from the data space $\mathfrak{D}$ into the model space $\mathfrak{M}$. Let us make the usual assumptions in this sort of problem.

First, as discussed in section 1.3, it is assumed that a reasonable approximation for representing the physical theory relating the model parameters $\mathbf{m}$ to the observable parameters $\mathbf{d}$ can be written under the form of a probability density for $\mathbf{d}$ given any possible $\mathbf{m}$ (equation (1.61)):

$$\Theta(\mathbf{d}, \mathbf{m}) = \theta(\mathbf{d} \mid \mathbf{m})\,\mu_M(\mathbf{m}) \quad , \tag{1.86}$$

where, as usual, $\mu_M(\mathbf{m})$ is the homogeneous probability density over the model space manifold $\mathfrak{M}$. Second, the prior information in the joint manifold $\mathfrak{D} \times \mathfrak{M}$ takes the special form

$$\rho(\mathbf{d}, \mathbf{m}) = \rho_D(\mathbf{d})\,\rho_M(\mathbf{m}) \quad , \tag{1.87}$$

which means that information in the space of observable parameters (data) has been obtained (from measurements) independently of the prior information in the model space (see section 1.4). In particular, the homogeneous limit of this last equation is

$$\mu(\mathbf{d}, \mathbf{m}) = \mu_D(\mathbf{d})\,\mu_M(\mathbf{m}) \quad . \tag{1.88}$$

Using equations (1.83)–(1.84), this gives, for the posterior information in the model space,

$$\sigma_M(\mathbf{m}) = k\,\rho_M(\mathbf{m}) \int_{\mathcal{D}} d\mathbf{d}\,\frac{\rho_D(\mathbf{d})\,\theta(\mathbf{d} \mid \mathbf{m})}{\mu_D(\mathbf{d})} \ . \tag{1.89}$$

Sometimes, we shall write equation (1.89) as

$$\sigma_M(\mathbf{m}) = k\,\rho_M(\mathbf{m})\,L(\mathbf{m}) \ , \tag{1.90}$$

where k is a constant and $L(\mathbf{m})$ is the *likelihood function*

$$L(\mathbf{m}) = \int_{\mathcal{D}} d\mathbf{d}\,\frac{\rho_D(\mathbf{d})\,\theta(\mathbf{d} \mid \mathbf{m})}{\mu_D(\mathbf{d})} \ , \tag{1.91}$$

which gives a measure of how good a model $\mathbf{m}$ is in explaining the data.

Equation (1.89) gives *the* solution of the general inverse problem. From $\sigma_M(\mathbf{m})$ it is possible to obtain any sort of information we wish on model parameters: mean values, median values, maximum likelihood values, uncertainty bars, etc. More importantly, from $\sigma_M(\mathbf{m})$ one may compute the probability for a model $\mathbf{m}$ to satisfy some characteristic (by integrating the probability density over the region of the model manifold made by all the models satisfying the given characteristic). Finally, from $\sigma_M(\mathbf{m})$ one may obtain a sequence $\mathbf{m}_1$, $\mathbf{m}_2, \dots$ of *samples* that may provide a good intuitive understanding of the information one actually has on the model parameters. See section 1.6 for a discussion.

The existence of the solution simply means that $\sigma_M(\mathbf{m})$, as defined by (1.89), is not identically null. If this were the case, it would indicate the incompatibility of the experimental results, the a priori hypothesis on model parameters, and the theoretical information, thus showing that some uncertainty bars have been underestimated.

The uniqueness of the solution is evident when by solution we mean the probability density $\sigma_M(\mathbf{m})$ itself and is simply a consequence of the uniqueness of the conjunction of states of information. Of course, $\sigma_M(\mathbf{m})$ may be very pathological (nonnormalizable, multimodal, etc.), but that would simply mean that such is the information we possess on the model parameters. The information itself is uniquely defined.

Example 1.34. Negligible modelization uncertainties. *If* (i) *the data space* $\mathbb{D}$ *is a linear space (so, in particular, the homogeneous probability density over* $\mathbb{D}$ *is constant,* $\mu_D(\mathbf{d}) = \text{const.}$ *) and* (ii) *modelization uncertainties are negligible compared to observational uncertainties, we can take*[24]

$$\theta(\mathbf{d} \mid \mathbf{m}) = \delta(\mathbf{d} - \mathbf{g}(\mathbf{m})) \ , \tag{1.92}$$

where $\mathbf{d} = \mathbf{g}(\mathbf{m})$ *denotes the (exact) resolution of the forward problem. Equation* (1.89) *then gives*

$$\sigma_M(\mathbf{m}) = \frac{1}{\nu}\,\rho_M(\mathbf{m})\,\rho_D(\mathbf{g}(\mathbf{m})) \ , \tag{1.93}$$

---

[24]The hypothesis $\theta(\mathbf{d} \mid \mathbf{m}) = \delta(\mathbf{d} - \mathbf{g}(\mathbf{m}))$ may be plainly wrong if the data space is not a linear space, as a properly defined delta function may not simply be a function of the difference $\mathbf{d} - \mathbf{g}(\mathbf{m})$.

where v is the normalization constant $v = \int_{\mathfrak{M}} dm\, \rho_M(\mathbf{m})\, \rho_D(\mathbf{g}(\mathbf{m}))$. Although we have obtained the result (1.93) via the notion of conjunction of states of information, we could have arrived here by introducing the conditional probability density of $\rho(\mathbf{d}, \mathbf{m}) = \rho_D(\mathbf{d})\, \rho_M(\mathbf{m})$ given relation $\mathbf{d} = \mathbf{g}(\mathbf{m})$ (using equation (1.54)). Sometimes, we shall write equation (1.93) as

$$\sigma_M(\mathbf{m}) = k\, \rho_M(\mathbf{m})\, L(\mathbf{m}) \quad , \tag{1.94}$$

where k is a constant and $L(\mathbf{m})$ is the likelihood function

$$L(\mathbf{m}) = \rho_D(\mathbf{g}(\mathbf{m})) \quad , \tag{1.95}$$

which, as in equation (1.91), gives a measure of how good a model $\mathbf{m}$ is in explaining the data.

Equation (1.93) is of considerable practical importance: a majority of inverse problems can be solved using it directly (under the assumption that the data space is a linear space[25]). There are three elements in the equation: (i) the a priori probability density in the model space, $\rho_M(\mathbf{m})$ (which can be replaced by its homogeneous limit $\mu_M(\mathbf{m})$ if no other a priori information is available), (ii) the probability density $\rho_D(\mathbf{d})$ describing the result of the measurements (with the attached uncertainties), and (iii) the nonlinear function $\mathbf{m} \mapsto \mathbf{g}(\mathbf{m})$ solving the forward problem.

Example 1.35. Negligible observational uncertainties. *Assume that the data space is a linear space. Letting $\mathbf{d}_{obs}$ denote the observed data values, the hypothesis of negligible observational uncertainties (with respect to modelization uncertainties) is written*

$$\rho_D(\mathbf{d}) = \delta(\mathbf{d} - \mathbf{d}_{obs}) \quad . \tag{1.96}$$

Equation (1.89) then gives $\sigma_M(\mathbf{m}) = k\, \rho_M(\mathbf{m})\, \theta(\mathbf{d}_{obs} \mid \mathbf{m})$, or, under normalized form,

$$\sigma_M(\mathbf{m}) = \frac{\rho_M(\mathbf{m})\, \theta(\mathbf{d}_{obs} \mid \mathbf{m})}{\int_{\mathfrak{M}} dm\, \rho_M(\mathbf{m})\, \theta(\mathbf{d}_{obs} \mid \mathbf{m})} \quad . \tag{1.97}$$

Example 1.36. Gaussian modelization and observational uncertainties. *This corresponds respectively to (equation (1.62) of Example 1.17)*

$$\theta(\mathbf{d} \mid \mathbf{m}) = \text{const. } \exp\left(-\tfrac{1}{2}(\mathbf{d} - \mathbf{g}(\mathbf{m}))^t\, \mathbf{C}_T^{-1}\,(\mathbf{d} - \mathbf{g}(\mathbf{m})) \right) \quad , \tag{1.98}$$

and, provisionally using the notation $\mathbf{C}_d$ for the covariance matrix representing the measurement uncertainties (equation (1.73) of Example 1.25),

$$\rho_D(\mathbf{d}) = \text{const. } \exp\left(-\tfrac{1}{2}(\mathbf{d} - \mathbf{d}_{obs})^t\, \mathbf{C}_d^{-1}\,(\mathbf{d} - \mathbf{d}_{obs}) \right) \quad . \tag{1.99}$$

As demonstrated in Appendix 6.21, equation (1.89) then gives

$$\sigma_M(\mathbf{m}) = \text{const. } \rho_M(\mathbf{m})\, \exp\left(-\tfrac{1}{2}(\mathbf{g}(\mathbf{m}) - \mathbf{d}_{obs})^t\, \mathbf{C}_D^{-1}\,(\mathbf{g}(\mathbf{m}) - \mathbf{d}_{obs}) \right) \quad , \tag{1.100}$$

---

[25]For a slightly more general setting, see Mosegaard and Tarantola, 2002.

where

$$\mathbf{C}_D = \mathbf{C}_d + \mathbf{C}_T \quad . \tag{1.101}$$

This result is important because it shows that, in the Gaussian assumption, observational uncertainties and modelization uncertainties simply combine by addition of the respective covariance operators, even when the forward problem is nonlinear.

Example 1.37. Gaussian model. *In the context of Example* 1.36, *further assume that the model space is linear and that the a priori information on the model parameters is also Gaussian:*

$$\rho_M(\mathbf{m}) = \text{const. } \exp\left(-\tfrac{1}{2}(\mathbf{m} - \mathbf{m}_{\text{prior}})^t \, \mathbf{C}_M^{-1} \, (\mathbf{m} - \mathbf{m}_{\text{prior}}) \right) \quad . \tag{1.102}$$

Then,

$$\sigma_M(\mathbf{m}) = k \exp(-S(\mathbf{m})) \quad , \tag{1.103}$$

where k is a constant and where the misfit function $S(\mathbf{m})$ *is the sum of squares:*

$$S(\mathbf{m}) = \tfrac{1}{2}\left((\mathbf{g}(\mathbf{m}) - \mathbf{d}_{\text{obs}})^t \, \mathbf{C}_D^{-1} \, (\mathbf{g}(\mathbf{m}) - \mathbf{d}_{\text{obs}}) + (\mathbf{m} - \mathbf{m}_{\text{prior}})^t \, \mathbf{C}_M^{-1} \, (\mathbf{m} - \mathbf{m}_{\text{prior}}) \right). \tag{1.104}$$

Example 1.38. Gaussian linear model. *If the relation between model parameters and data parameters is linear,*

$$\mathbf{d} = \mathbf{g}(\mathbf{m}) = \mathbf{G}\,\mathbf{m} \quad , \tag{1.105}$$

then the posterior probability density $\sigma_M(\mathbf{m})$ (equation (1.103)) is also Gaussian (see chapter 3 for details) with mean

$$\begin{aligned}
\tilde{\mathbf{m}} &= (\mathbf{G}^t \, \mathbf{C}_D^{-1} \, \mathbf{G} + \mathbf{C}_M^{-1})^{-1} \, (\mathbf{G}^t \, \mathbf{C}_D^{-1} \, \mathbf{d}_{\text{obs}} + \mathbf{C}_M^{-1} \, \mathbf{m}_{\text{prior}}) \\
&= \mathbf{m}_{\text{prior}} + (\mathbf{G}^t \, \mathbf{C}_D^{-1} \, \mathbf{G} + \mathbf{C}_M^{-1})^{-1} \, \mathbf{G}^t \, \mathbf{C}_D^{-1} \, (\mathbf{d}_{\text{obs}} - \mathbf{G}\,\mathbf{m}_{\text{prior}}) \\
&= \mathbf{m}_{\text{prior}} + \mathbf{C}_M \, \mathbf{G}^t \, (\mathbf{G}\,\mathbf{C}_M\,\mathbf{G}^t + \mathbf{C}_D)^{-1} \, (\mathbf{d}_{\text{obs}} - \mathbf{G}\,\mathbf{m}_{\text{prior}})
\end{aligned} \tag{1.106}$$

and covariance

$$\tilde{\mathbf{C}}_M = (\mathbf{G}^t \, \mathbf{C}_D^{-1} \, \mathbf{G} + \mathbf{C}_M^{-1})^{-1} = \mathbf{C}_M - \mathbf{C}_M \, \mathbf{G}^t \, (\mathbf{G}\,\mathbf{C}_M\,\mathbf{G}^t + \mathbf{C}_D)^{-1} \, \mathbf{G}\,\mathbf{C}_M \quad . \tag{1.107}$$

Example 1.39. Generalized Gaussian model. *In the case of independent uncertainties, the Gaussian model of Example* 1.37 *can be generalized (replacing the Gaussian model of uncertainties with the generalized Gaussian [see Appendix 6.6]). With obvious definitions, one arrives at*

$$\sigma_M(\mathbf{m}) = \text{const. } \exp\left(-\frac{1}{p}\left(\sum_i \frac{|g^i(\mathbf{m}) - d^i_{\text{obs}}|^p}{(\sigma^i_D)^p} + \sum_\alpha \frac{|m^\alpha - m^\alpha_{\text{prior}}|^p}{(\sigma^\alpha_M)^p} \right) \right) \quad . \tag{1.108}$$

In particular, for $p = 1$, one has $\sigma_M(\mathbf{m}) = k \exp(-S(\mathbf{m}))$, where the misfit function $S(\mathbf{m})$ is now a sum of absolute values:

$$S(\mathbf{m}) = \sum_i \frac{|g^i(\mathbf{m}) - d^i_{obs}|}{\sigma^i_D} + \sum_\alpha \frac{|m^\alpha - m^\alpha_{prior}|}{\sigma^\alpha_M} \quad . \tag{1.109}$$

1.6 Using the Solution of the Inverse Problem

1.6.1 Computing Probabilities

We have set the inverse problem as a problem of the conjunction of states of information. Then, the solution of the inverse problem is the posterior probability distribution over the model space manifold, represented by the probability density $\sigma_M(\mathbf{m})$. The more general questions we may ask in an inverse problem are more basic in probability theory: which is the probability of a certain event, i.e., the probability that a model $\mathbf{m}$ belongs to a given region $A \subseteq \mathfrak{M}$ of the model space manifold?

Example 1.40. *If the system under study is a galaxy and among the parameters used to describe it are the respective masses m_G and m_{BH} of the entire galaxy and of its central black hole, an event $A \subseteq \mathfrak{M}$ may be $m_{BH} < 10^{-5} m_G$. The probability of the event $A : m_{BH} < 10^{-5} m_G$ is*

$$P(A) = \int_A d\mathbf{m} \, \sigma_M(\mathbf{m}) \quad . \tag{1.110}$$

We can then say that the probability that the mass of the central black hole is less than 10^{-5} times the mass of the total galaxy equals $P(A)$.

In chapter 2 it is explained how these probabilities can be computed in high-dimensional spaces, where Monte Carlo sampling methods are often necessary.

Sometimes, the inverse problem is solved as an intermediate one in a more general decision problem in which one has to combine information obtained from the inverse problem with economic considerations. As an example, the 30 years of production data of an oil field were used to obtain information on the 3D distribution of porosity and permeability, solving in fact a huge inverse problem involving 3D simulations of fluid flow. The problem was solved in a Monte Carlo way, so the solution to the problem was a large set of possible solutions, samples of the posterior probability density in the model space, $\sigma_M(\mathbf{m})$. For each of those samples, it was possible to run the fluid flow simulation forward in time to evaluate the total future production of the oil field. As each sampled model predicts a different total production, one obtains a histogram that, in fact, represents the probability density of what the total future production can be (See Figure 1.13).

In a related domain, readers interested in Bayesian decision theory can refer to Box and Tiao (1973), Morgan (1968), Schmitt (1969), or Winkler (1972).

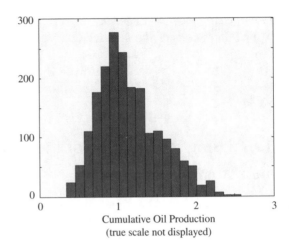

Figure 1.13. *The histogram of total future oil production of an oil field, as obtained by solving an inverse problem (for porosity and permeability) using as data 30 years of production on the field, from Landa and Guyaguler (2003), with permission (the actual values of oil production are confidential).*

1.6.2 Analysis of Uncertainties

Should the model space manifold be a linear space, and the probability density $\sigma_M(\mathbf{m})$ be reasonably close to a Gaussian distribution, the mean model $\langle \mathbf{m} \rangle$ may be of interest,

$$\langle \mathbf{m} \rangle = \int_{\mathfrak{M}} d\mathbf{m} \; \mathbf{m} \; \sigma_M(\mathbf{m}) \quad , \tag{1.111}$$

together with the covariance matrix

$$\tilde{\mathbf{C}}_M = \int_{\mathfrak{M}} d\mathbf{m} \; (\mathbf{m} - \langle \mathbf{m} \rangle)(\mathbf{m} - \langle \mathbf{m} \rangle)^t \; \sigma_M(\mathbf{m}) \quad . \tag{1.112}$$

The comparison of this posterior covariance matrix with the prior one,

$$\mathbf{C}_M = \int_{\mathfrak{M}} d\mathbf{m} \; (\mathbf{m} - \langle \mathbf{m} \rangle)(\mathbf{m} - \langle \mathbf{m} \rangle)^t \; \rho_M(\mathbf{m}) \quad , \tag{1.113}$$

corresponds to what is usually called the analysis of uncertainties: for each individual parameter one may see how the initial uncertainty has been reduced.[26]

But it must be clear that this simple method of analyzing uncertainties will not make sense for complex problems, where the posterior probability density may itself be complex. Then, the direct examination of the prior and posterior probabilities of some events (as suggested above) is the only means we may have.

1.6.3 Random Exploration of the Model Space

If the number of model parameters is very small (say, less than 10) and if the computation of the numerical value of $\sigma_M(\mathbf{m})$ for an arbitrary $\mathbf{m}$ is inexpensive (i.e., not consuming too much computer time), we can define a grid over the model space, compute $\sigma_M(\mathbf{m})$

---

[26]Remember that the square root of the diagonal elements of a covariance matrix is the standard deviation.

everywhere in the grid, and directly use these results to discuss the information obtained on the model parameters. This is certainly the most general way of solving the inverse problem. Problem 7.1 gives an illustration of the method.

If the number of parameters is not small, and if the computation of $\sigma_M(\mathbf{m})$ at any point $\mathbf{m}$ is not expensive, the systematic exploration of the model space can advantageously be replaced with a random (Monte Carlo) exploration. Monte Carlo methods are discussed in chapter 2. Also, the computation of the mathematical expectation and of the posterior covariance operator can be made by evaluating the sums (1.111) and (1.112) by a Monte Carlo method of numerical integration.

1.6.4 Maximum Likelihood Point

Let $\mathfrak{X}$ be a manifold with some coordinates $\mathbf{x} = \{x^1, x^2, \dots\}$ and with volume element

$$dV(\mathbf{x}) = v(\mathbf{x}) \, dx^1 \, dx^2 \dots \quad . \tag{1.114}$$

Let $f(\mathbf{x})$ be a probability density over $\mathfrak{X}$. How is the *maximum likelihood point* to be defined? A common mistake is to 'define' it as the point where $f(\mathbf{x})$ is maximum. This cannot be: under a change of variables, a probability density has its values multiplied by the Jacobian of the transformation (see equation (1.18)), so the point at which a probability density achieves its maximum depends as much on the probability distribution per se as on the variables being used.

The maximum likelihood point must be defined as follows: considering around every point $\mathbf{x}$ a small region with fixed volume dV_0, which is the point $\mathbf{x}$ at which the probability $dP(\mathbf{x})$ is maximum? By definition of probability density, $dP(\mathbf{x}) = f(\mathbf{x}) \, dx^1 \, dx^2 \dots$. As we wish fixed $dV_0 = v(\mathbf{x}) \, dx^1 \, dx^2 \dots$, we can write $dP(\mathbf{x}) = (f(\mathbf{x})/v(\mathbf{x})) \, dV_0$, and, therefore, the maximum of $dP(\mathbf{x})$ is obtained when

$$\frac{f(\mathbf{x})}{v(\mathbf{x})} \quad \text{is maximum} \quad . \tag{1.115}$$

The homogeneous probability density was introduced in section 1.2.4 where we arrived (see equation (1.23) there) at the relation $\mu(\mathbf{x}) = \text{const.} \, v(\mathbf{x})$. Therefore, the condition characterizing the maximum likelihood point can also be written

$$\boxed{\text{maximum likelihood point} \quad : \quad \frac{f(\mathbf{x})}{\mu(\mathbf{x})} \quad \text{maximum} \quad .} \tag{1.116}$$

The ratio $f(\mathbf{x})/\mu(\mathbf{x})$ was called the likelihood function above (see equation (1.31)).

Example 1.41. *The Fisher probability density over the surface of the sphere is (when centered at the "North pole"), using spherical coordinates* $\{\theta, \varphi\}$,

$$f(\theta, \varphi) = \frac{\kappa \, \sin\theta}{4\pi \, \sinh\kappa} \exp(\kappa \, \cos\theta) \quad , \tag{1.117}$$

where the parameter $1/\kappa$ *measures the dispersion of the distribution. It is normalized, as one has* $\int_0^\pi d\theta \int_{-\pi}^\pi d\varphi \, f(\theta, \varphi) = 1$ *(remember that a probability density is integrated using* *the capacity element $d\theta \, d\varphi$, not using the surface element $dS(\theta, \varphi)$). The points where the probability density $f(\theta, \varphi)$ reaches its maximum have no special meaning. As the surface element on the surface of the sphere is $dS(\theta, \varphi) = \sin\theta \, d\theta \, d\varphi$, the homogeneous probability density is $\mu(\theta, \varphi) = \frac{1}{4\pi} \sin\theta$ (it could also be obtained as the limit of $f(\theta, \varphi)$ when $\kappa \to 0$). The maximum likelihood point is defined by the condition*

$$\frac{f(\theta, \varphi)}{\mu(\theta, \varphi)} \quad \text{maximum} \quad , \tag{1.118}$$

i.e., the condition that $\frac{\kappa}{\sinh\kappa} \exp(\kappa \cos\theta)$ is maximum. This happens for $\theta = 0$ as it should.

Returning now to our inverse problems, given a model space manifold $\mathfrak{M}$ with homogeneous probability density $\mu_M(\mathbf{m})$, and given the posterior probability density $\sigma_M(\mathbf{m})$, we see that the point of maximum (posterior) likelihood is that satisfying

$$\frac{\sigma_M(\mathbf{m})}{\mu_M(\mathbf{m})} \quad \text{maximum} \quad . \tag{1.119}$$

To find the maximum likelihood point, one typically uses gradient methods.

In that respect, two major comments need to be made.

- Probability distributions tend to be bell shaped, and gradient methods tend to work very inefficiently far from the maximum likelihood point. It is generally much better to define

$$\Psi(\mathbf{m}) \;=\; \log \frac{\sigma_M(\mathbf{m})}{\mu_M(\mathbf{m})} \tag{1.120}$$

and to use standard gradient techniques to find the point fulfilling the equivalent condition

$$\Psi(\mathbf{m}) \quad \text{maximum} \quad . \tag{1.121}$$

The function $S(\mathbf{m}) = -\Psi(\mathbf{m})$ corresponds to the usual misfit function (to be minimized).

- "Gradient" is not synonymous with "direction of steepest ascent," and an explicit definition of distance over the model space manifold has to be introduced in order to effectively use gradient methods. Some examples are given in chapters 3 and 4.

As $\sigma_M(\mathbf{m})$ is, in general, an arbitrarily complicated function of $\mathbf{m}$, there is no guarantee that the maximum likelihood point is unique, or that a given point that is locally maximum is the absolute maximum. Only a full exploration of the space would give the proof, but this is generally too expensive to make (when the number of dimensions is large). Unless solid qualitative arguments can be made, one is only left with the possibility of more or less extensive explorations of the model space, random or not.

Chapter 2

Monte Carlo Methods

Now and then, luck brings confusion
in the biological order established by selection.
It periodically shifts its too restrictive barriers,
and allows natural evolution to change its course.
Luck is anti-conservative.

Jacques Ruffié, 1982

We have seen in chapter 1 that the most general solution of an inverse problem provides a probability distribution over the model space. It is only when the probability distribution in the model space is very simple (for instance, when it has only one maximum) that analytic techniques can be used to characterize it.

For more general probability distributions, one needs to perform an extensive exploration of the model space. Except for problems with a very small number of dimensions, this exploration cannot be systematic (as the number of required points grows too rapidly with the dimension of the space). Well-designed random (or pseudorandom) explorations can solve many complex problems. These random methods were jokingly called *Monte Carlo methods* by the team at Los Alamos that was at the origin, among others, of the Metropolis sampling algorithm, and the name "Monte Carlo" has now become established.

2.1 Introduction

That Monte Carlo (i.e., random) methods can be used for computation has been known for centuries. For instance, one can use a Monte Carlo method to evaluate the number π: on a regular floor, made of strips of equal width w, one throws needles of length $w/2$. The probability that a needle will intersect a groove in the floor equals $1/\pi$ (Georges Louis Leclerc, Comte de Buffon [1707–1788]). A series of 50 observations, each with 100 trials, made by Wolff in Zurich in 1850 led to a value for π of 3.1596 ± 0.0524. In numerical methods, the throwing of needles is replaced with a pseudorandom generation of numbers by a computer code.

One domain where Monte Carlo computations are usual is for the numerical evaluation of integrals in large-dimensional spaces: the systematic evaluation of the function at a regular grid is impossible (too many points would be required), and a Monte Carlo sampling of the function can provide an estimation of the result, together with an estimation of the error (see Appendix 6.9 or, for more details, Kalos and Whitlock, 1986).

The use of Monte Carlo methods for the solution of inverse problems was initiated by Keilis-Borok and Yanovskaya (1967) and Press (1968, 1971). More recent interesting works are Anderssen and Seneta (1971, 1972), Rothman (1985a, 1985b, 1986), and Dahl-Jensen et al. (1998).

From the perspective of this book, where probability distributions over parameter spaces are central, we face the problem of how to use them. The definition of 'central estimators' (like the mean or the median) and of 'estimators of dispersion' (like the covariance matrix) lacks generality, as it is quite easy to find examples (like multimodal distributions in high-dimensional spaces) where these estimators fail to have any interesting meaning.

When a probability distribution has been defined over a space of low dimension (say, from one to four dimensions), we can directly represent the associated probability density. This is trivial in one or two dimensions. It is easy in three dimensions, and some tricks may allow us to represent a four-dimensional probability distribution. Moreover, the probability of an event A can directly be evaluated by an integral, using standard (nonrandom) numerical techniques.

Figure 2.1. *The sampling of a probability density allows us to perform any of the computations introduced in probability theory (computation of the probability of an event, estimation of some moments, etc.) using simple statistics.*

When the model space has a large number of dimensions, representing a probability density is impossible, but we can, at least in principle, do something that is largely equivalent: we can sample[27] the probability density, as suggested in Figure 2.1. The advantage of considering a set of samples of a probability distribution is that the individual 'points' can be represented, typically as images, as those in Figure 2.2 (see also the images at the right in Figure 2.5).

There is one problem with large-dimensional spaces that it is easy to underestimate: *they tend to be terribly empty.* Figure 2.3 shows that the probability of hitting by chance the (maximum-size) hypersphere inscribed in a hypercube rapidly decreases to zero when the dimension of the space grows. When the goal is not to hit a large sphere, but a small region

---

[27]To sample a probability density means to generate (independent) points that are samples of it, i.e., such that the probability of any of the points being inside any domain A equals the probability of the domain A.

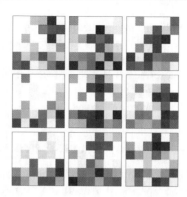

Figure 2.2. *Nine random realizations of a probability distribution over a 36-dimensional space. The 36 values are in fact values on a 6 × 6 array, and when the values are represented using grades of gray, each realization is an image. Note that most of the images present a "Greek cross" (highlighted). Also, the pixels in the two bottom rows tend to correspond to higher values of the random variable (darker grades of gray). Given enough of these samples, the probability of different events can be evaluated, for instance,* (i) *the probability that a Greek cross may be present,* (ii) *the probability of a Greek cross at* the left side of the image *(they tend to be at the right side),* (iii) *the probability of having a Greek cross and, at the same time, low values of the variable at the bottom of the figure, etc.*

Figure 2.3. *Large-dimensional spaces tend to be terribly empty. Hitting by chance the circle inscribed in a square is easy. Hitting by chance the sphere inscribed in a cube is a little bit more difficult. When the dimension of the space grows, the probability of hitting the hypersphere inscribed in a hypercube rapidly tends to zero. At the top, the volume of the hypersphere and of the hypercube is given as a function of the dimension* n *, and at the bottom the ratio of the volumes is displayed. This figure explains why the random exploration of large-dimensional spaces is always difficult and why Brownian motion types of random exploration are used.*

Volume hypersphere: $\dfrac{2\,\pi^{n/2}\,r^n}{n\,\Gamma(n/2)}$

Volume hypercube: $(2r)^n$

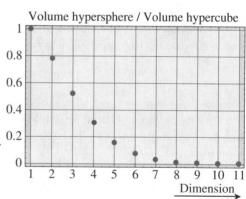

of significant probability, with not much idea of where it may be, one easily understands that Monte Carlo methods that may work on large-dimensional spaces are far from trivial.

In fact, there are two problems in a Monte Carlo sampling of a probability distribution in a large-dimensional space: (i) locating the region(s) of significant probability, and (ii) sampling the whole of the region(s) densely enough. Discovering the location of the regions is the most difficult problem, and mathematics alone cannot solve it (because of the great emptiness of large-dimensional spaces): it is the particular physics (or geometry) of the problem at hand that may help on this. Once one has been able to come close to one of these regions, the techniques described below (Gibbs sampler or Metropolis algorithm) are able to perform a random walk, a sort of Brownian motion that is efficient in exploring the region, and avoid leaving it (thus entering the vast empty regions of the space).

2.2 The Movie Strategy for Inverse Problems

We have seen in chapter 1 that two typical inputs to the inverse problem are a probability density $\rho_M(\mathbf{m})$, describing the a priori information on the model parameters, and a probability density $\rho_D(\mathbf{d})$, describing the information we have on the data parameters, gained through some measurements. The solution to the inverse problem is given by a (posterior) probability density $\sigma_M(\mathbf{m})$ that equals the (normalized) product of the prior probability density $\rho_M(\mathbf{m})$ times a likelihood function $L(\mathbf{m})$:

$$\sigma_M(\mathbf{m}) \;=\; k \, \rho_M(\mathbf{m}) \, L(\mathbf{m}) \quad . \tag{2.1}$$

The likelihood function is a measure of how good the model $\mathbf{m}$ is in fitting the data. The normalizing constant can be written $k = 1 / \int_{\mathfrak{M}} d\mathbf{m} \, \rho_M(\mathbf{m}) \, L(\mathbf{m})$.

In the most general setting, the relation between data and model parameters is probabilistic and is represented by a conditional probability density $\theta(\mathbf{d}|\mathbf{m})$. Then, the likelihood function is (see equations (1.89)–(1.91))

$$L(\mathbf{m}) \;=\; \int_{\mathfrak{D}} d\mathbf{d} \, \frac{\rho_D(\mathbf{d}) \, \theta(\mathbf{d} \mid \mathbf{m})}{\mu_D(\mathbf{d})} \quad , \tag{2.2}$$

where $\mu_D(\mathbf{d})$ is the homogeneous probability density in the data manifold.

Sometimes, the relation between data and model parameters is functional, $\mathbf{d} = \mathbf{g}(\mathbf{m})$, and in this case the likelihood function is (see equations (1.93)–(1.95))

$$L(\mathbf{m}) \;=\; \rho_D(\,\mathbf{g}(\mathbf{m})\,) \quad . \tag{2.3}$$

A couple of examples of such a likelihood function are given as a footnote.[28]

As the methods about to be developed require the computation of the value of the likelihood at many points, that the expression of the likelihood takes the form (2.2) or (2.3) is far from irrelevant.

In this chapter, we are going to describe methods that allow us, first, to obtain samples $\{\mathbf{m}_1, \mathbf{m}_2, \dots\}$ of the prior probability density $\rho_M(\mathbf{m})$, then samples $\{\mathbf{m}_1', \mathbf{m}_2', \dots\}$ of the posterior probability density $\sigma_M(\mathbf{m})$. As, typically, each sample (i.e., each model) can be represented as an image, the display of many samples corresponds to the display of a 'movie.' Let us start discussing the generation of the 'prior movie.'

---

[28]For instance, if d_{obs}^i represents the observed data values and σ^i the estimated mean deviations, assuming double exponentially distributed observational errors gives $L(\mathbf{m}) = \exp(-\sum_i |g^i(\mathbf{m}) - d_{\text{obs}}^i| / \sigma^i)$. If $\mathbf{C}_D$ represents the covariance operator describing estimated data uncertainties and uncertainty correlations, assuming a Gaussian distribution gives (equation (1.101)) $L(\mathbf{m}) = \exp(-\frac{1}{2}(\mathbf{g}(\mathbf{m}) - \mathbf{d}_{\text{obs}})^t \mathbf{C}_D^{-1}(\mathbf{g}(\mathbf{m}) - \mathbf{d}_{\text{obs}}))$. Some other examples are given in chapter 1.

The display (and study) of the samples from $\rho_M(\mathbf{m})$ allows us to verify that we are using adequate a priori information. It may help convey to others which kind of a priori information we have in mind (and may perhaps allow these persons to criticize the a priori information we are trying to use). We already saw one example of three samples from an a priori probability distribution in Example 1.32, where three layered Earth models were displayed (Figure 1.10, page 30). Let us see another simple example.

Example 2.1. Gaussian random field. *A Gaussian random field is characterized by its mean field $m_{\mathrm{mean}}(\mathbf{x})$ and its covariance $C_M(\mathbf{x}, \mathbf{x}')$. Given these, it is possible to generate as many random realizations of the random field as one may wish (using, for instance, the method suggested in Example 2.2 below). Figure 2.4 displays three random realizations of a Gaussian random field of zero mean and spherical covariance[29] (with values coded in a color scale). Given a large enough number of these realizations, the mean field and the covariance are easy to estimate (using simple statistics) so a large enough number of realizations completely characterizes the random field. Because the covariance of this random field happens to be stationary, three images already convey a good idea of the random field itself. More images would be needed for a more general covariance. The display of these samples of the a priori probability distribution allows other scientists to evaluate the correctness of the a priori information being input into the inverse problem (that the actual (unknown) field is a random realization of such a random field). Displaying the mean of the Gaussian random field and plotting the covariance is not an alternative to displaying a certain number of realizations, because mean and covariance do not relate in an intuitive way to the realizations.*

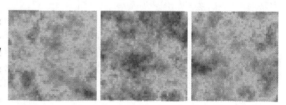

Figure 2.4. *Three random realizations of a 2D Gaussian random field. The mean field is zero, and the (stationary) covariance is the sum of a white noise plus a spherical covariance. (A. Boucher, pers. comm.)*

When the problem of displaying samples of the prior probability density $\rho_M(\mathbf{m})$ has been solved (and agreement has been reached about the existence and suitability of such a priori information), one may produce samples of the posterior probability density $\sigma_M(\mathbf{m})$. The general methods of producing these samples of the posterior distribution are only to be introduced in the sections below, but we can examine here a very simple example of an inverse problem, where the generation of these samples can be done using a simple theory.

Example 2.2. Sampling a conditional Gaussian random field. *Let us briefly examine here a simple inverse problem that has a high pedagogical value. We face an unknown 2D field that is assumed to be a random realization of the Gaussian random field introduced in Example 2.1 (three random realizations of the random field were displayed in Figure 2.4).*

---

[29]The spherical covariance, a commonly used model of covariance in geostatistics, is described in Example 5.7, page 113.

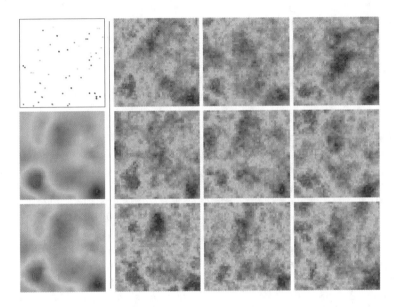

Figure 2.5. *From a random realization of the Gaussian random field described in Figure 2.4, the 50 values shown at the top left were extracted to be used as data. The particular random realization that was used to obtain these data is unknown to us, but the mean and the covariance of the Gaussian random field are given. The prior Gaussian field (as defined by its mean and covariance) is then conditioned with the given data values (corresponding to the solution of a simple kind of inverse problem). In this way, a posterior Gaussian field is defined whose mean $\widetilde{m}_{\mathrm{mean}}(\mathbf{x})$ and covariance $\widetilde{C}_M(\mathbf{x}, \mathbf{x})$ can be expressed (see text for details). Rather than becoming directly interested in this posterior mean and covariance, it is much better to generate some random samples of the posterior Gaussian random field of which we know the mean and covariance. Nine such random realizations are displayed in the right part of the figure. All these realizations satisfy the 50 initial data values. The actual realization from which the 50 data values were extracted could have resembled any of these realizations (in fact, it was the one at the left in Figure 2.4). These nine panels convey a clear idea of the variability in the solution to the problem. Note that the variability is greater in the regions where there is not much data. The statistical mean of 50 such random realizations is displayed at the bottom left, together with the (theoretical) posterior mean $\widetilde{m}_{\mathrm{mean}}(\mathbf{x})$ (just above the statistical mean), and they are virtually indistinguishable. The computations here have been performed by A. Boucher (pers. comm.).*

We are given, as data, the k values $\{m(\mathbf{x}^1), m(\mathbf{x}^2), \ldots, m(\mathbf{x}^k)\}$ of the unknown realization $m(\mathbf{x})$ at k points $\{\mathbf{x}^1, \mathbf{x}^2, \ldots, \mathbf{x}^k\}$. If these values are assumed to be exactly known, we can pass from the prior random field to the posterior random field just using the notion of conditional probability. If the values are only known with some uncertainties, the posterior random field can be obtained setting an inverse problem, essentially using the third of equations (1.106) and the second of equations (1.107) (see Example 5.25 on page 135 for an explicit formulation of this problem). In either case we end up with a posterior Gaussian *random field, of which we can express the mean $\widetilde{m}_{\text{mean}}(\mathbf{x})$ and the covariance $\widetilde{C}_{\text{M}}(\mathbf{x}, \mathbf{x}')$. But, again, instead of representing the mean and analyzing the covariance, it is better to generate realizations of the random field, which can be done as follows. Take randomly a point $\mathbf{x}_a$ in the space. At this point we have a Gaussian random variable with mean $\widetilde{m}_{\text{mean}}(\mathbf{x}_a)$ and variance $\widetilde{C}_{\text{M}}(\mathbf{x}_a, \mathbf{x}_a)$. It is then possible to generate a random realization of this one-dimensional random variable.[30] This will give some value $m(\mathbf{x}_a)$. We started the problem with the values of the field given at k points. We now have the values of the field given at $k + 1$ points. The problem can then be reformulated using the $k + 1$ points, and the same method can be used to randomly generate a new value at some point $\mathbf{x}_b$, and so on until we have realized the value of the random field in as many points as we may wish. Figure 2.5 presents an actual implementation[31] of this algorithm in two space dimensions, starting with the values of the field given at 50 points.*

From the example above, the reader should keep in mind that the solution of an inverse problem, being a probability distribution, is never one image, but a set of images, samples of the posterior probability density $\sigma_{\text{M}}(\mathbf{m})$. The common practice of plotting the 'best image' or the 'mean image' should be abandoned, even if it is accompanied by some analysis of error and resolution. For instance, when using least-squares methods to formulate the problem described in the example, what is called the solution is the mean of the posterior Gaussian distribution, i.e., the smooth image in the middle of the left column of Figure 2.5. This is not the solution; it is, rather, the mean of all possible solutions (hence its smoothness). Looking at this mean provides much less information than looking at a movie of realizations. Note that, by construction, each of the realizations captures the essential random fluctuations of the actual field from which the data were extracted (at the left in Figure 2.4).

For another example of the generation of such a movie (in a geophysical context), see Koren et al. (1991).[32]

When such a (hopefully large) collection of random models is available, we can also answer quite interesting questions. For instance, in a tectonic model, one may ask, *At which depth is the subsurface structure?* To answer this, we can make a histogram of the depth of the given geological structure over the collection of random models, and the histogram *is* the answer to the question. *What is the probability of having a low velocity zone around a given depth?* The ratio of the number of models presenting such a low velocity zone over the total number of models in the collection gives the answer (if the collection of models is large enough). This is essentially what must be proposed: to look at a large number of randomly generated models (first, of the prior distributon, and then, of the posterior distribution) in order to intuitively apprehend the basic properties of the probability distribution, followed by calculation of the probabilities of all interesting events.

Sometimes, it may nevertheless be necessary to estimate some moments (mean, covariance, etc.) of the distribution. Of course, they can also be evaluated using the samples. The relevant formulas are given as a footnote.[33] Appendix 6.9 gives some details on Monte Carlo methods of numerical integration.

---

[30]See, for instance, the inversion method explained in section 2.3.1.

[31]This implementation is based on the Gaussian sequential simulation algorithm (Goovaerts, 1997).

[32]The movie itself may be watched at `http://www.ccr.jussieu.fr/tarantola/`.

[33]Assume that the model space manifold is a linear space. If the posterior probability density $\sigma_{\text{M}}(\mathbf{m})$ is not too far from Gaussian, one may wish to compute the mean model and the covariance of the distribution (one can

The sampling method suggested in the example above (sequential random realization) is not easy to use in general inverse problems, since it involves the consideration of one-dimensional conditional and marginal probability distributions that are usually not available, except for very simple problems, as when we have linear relations between data and model parameters. We must then resort to much less efficient, but much more general, methods, like those based on the Metropolis algorithm, described later in this chapter.

2.3 Sampling Methods

2.3.1 The Inversion Method

Consider a probability density $f(x)$ depending on only one (scalar) variable x. This may occur when we really have one single random variable or, more often, when on a multidimensional manifold we consider a conditional distribution along a line (along which x is a parameter). The *inversion method* consists of introducing the cumulative probability

$$y = F(x) = \int_{x_{\min}}^{x} dx'\, f(x') \quad , \tag{2.4}$$

which takes values in the interval $[0, 1]$, and the inverse function $x = F^{-1}(y)$. It is easy to see[34] that if one randomly generates values of y with constant probability density in the interval $[0, 1]$, then the values $x = F^{-1}(y)$ are random samples of the probability density $f(x)$. Provided the function F^{-1} is available, the method is simple and efficient.

Example 2.3. *Let y_1, y_2, ... be samples of a random variable with constant probability density in the interval $[0, 1]$, and let erf^{-1} be the inverse error function.[35] The numbers $\mathrm{erf}^{-1}(y_1)$, $\mathrm{erf}^{-1}(y_2)$, ... are then normally distributed, with zero mean and unit variance (see Figure 2.6).*

Figure 2.6. *Use of the inversion method to produce samples of a two-dimensional Gaussian probability density.*

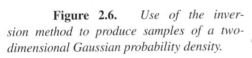
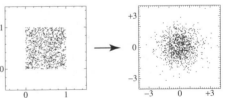

---

compute this event for distributions that are far from Gaussian, but these estimators may not be very meaningful). If one has K samples $\mathbf{m}_1, \mathbf{m}_2, \ldots, \mathbf{m}_K$ of $\sigma_M(\mathbf{m})$, the mean model (of $\sigma_M(\mathbf{m})$) is approximately given by $\langle \mathbf{m} \rangle = \frac{1}{K} \sum_{n=1}^{K} \mathbf{m}_n$ and the covariance matrix by $\mathbf{C} = \frac{1}{K} \sum_{n=1}^{K} (\mathbf{m}_n - \langle \mathbf{m} \rangle)(\mathbf{m}_n - \langle \mathbf{m} \rangle)^t$, or, equivalently, $\mathbf{C} = \frac{1}{K} \sum_{n=1}^{K} \mathbf{m}_n \mathbf{m}_n^t - \langle \mathbf{m} \rangle \langle \mathbf{m} \rangle^t$.

[34]This immediately results from the Jacobian rule, as $dy/dx = f(x)$.

[35]The error function $\mathrm{erf}(x)$ is the integral between $-\infty$ and x of a normalized Gaussian with zero mean and unit variance (be careful, there are different definitions). One may find in the literature different series expressions for erf^{-1}.

2.3.2 The Rejection Method

The *rejection method* starts by generating samples x_1, x_2, ... of the homogeneous probability density $\mu(x)$, which is usually a simple problem. Then, each sample is submitted to the possibility of a rejection, the probability that the sample x_k is accepted being taken equal to

$$P = \frac{f(x_k)/\mu(x_k)}{(f/\mu)_{\max}} \quad , \tag{2.5}$$

where $(f/\mu)_{\max}$ stands for the maximum of all the values $f(x)/\mu(x)$, or any larger number (the larger the number, the less efficient the method). It is then easy to prove that any accepted point is a sample of the probability density $f(x)$.

This method works reasonably well in one dimension or two dimensions (Figure 2.1 was generated using the rejection method) and could, in principle, be applicable in any number of dimensions. But, as was already mentioned, large-dimensional spaces tend to be very empty, and the chances that this method accepts a point may be dramatically low when working with multidimensional spaces.

2.3.3 Sequential Realization

In this method, one uses the property that a general n-dimensional probability density $f_n(m^1, m^2, \ldots, m^n)$ can always be decomposed as the product of a one-dimensional marginal and a series of one-dimensional conditionals (see Appendix 6.10):

$$\begin{aligned} &f_n(m^1, m^2, \ldots, m^n) \\ &= f_1(m^1) \, f_{1|1}(m^2|m^1) \, f_{1|2}(m^3|m^1, m^2) \, \ldots \, f_{1|n-1}(m^n|m^1, \ldots, m^{n-1}) \quad . \end{aligned} \tag{2.6}$$

All these marginal and conditional probability densities are contained in the original n-dimensional joint probability density $f_n(m^1, m^2, \ldots, m^n)$ and can, at least in principle, be evaluated from it using integrals. Assume that they are all known, and let us see how an n-dimensional sample could be generated.

One starts generating a (one-dimensional) sample for the variable m^1, using the one-dimensional marginal $f_1(m^1)$, giving a value m_0^1. With this value at hand, one generates a (one-dimensional) sample for the variable m^2, using the conditional $f_{1|1}(m^2|m_0^1)$, giving a value m_0^2. Then, one generates a (one-dimensional) sample for the variable m^3, using the conditional $f_{1|2}(m^3|m_0^1, m_0^2)$, giving a value m_0^3, and so on, until one generates a sample for the variable m^n, using the conditional $f_{1|n-1}(m^n|m_0^1, \ldots, m_0^{n-1})$, giving a value m_0^n. In this manner, a point $\{m_0^1, m_0^2, \ldots, m_0^n\}$ is generated that is a sample of the original $f_n(m^1, m^2, \ldots, m^n)$.

2.3.4 The Gibbs Sampler

The so-called *Gibbs sampler* (Geman and Geman, 1984) corresponds to performing a random walk in an n-dimensional parameter manifold that is very similar to a Metropolis random walk, except that no rejection is used (as discussed below, this advantage is more virtual than real). First, we must assume that the parameter space is a linear space, so that the notion of direction at a given point makes sense.

Let $f(x_k^1, x_k^2, \ldots, x_k^n)$ be the probability density we wish to sample, and let $\mathbf{x}_k = \{x_k^1, x_k^2, \ldots, x_k^n\}$ be the last point visited. Define randomly[36] a line in the parameter space that passes through the current point $\mathbf{x}_k$. Along that line one has a one-dimensional (conditional) probability density. One sample is then generated along this one-dimensional probability distribution, giving a new point $\mathbf{x}_{k+1}$. It can be demonstrated that iterating this procedure actually produces a series of samples of the joint probability distribution $f(x_k^1, x_k^2, \ldots, x_k^n)$.

When an analytical, explicit expression is available for the probability density $f(x_k^1, x_k^2, \ldots, x_k^n)$, this method works well and is efficient. When solving inverse problems, we wish to sample the posterior probability density in the model parameter space, $\sigma_M(\mathbf{m})$, and except in very simple problems, the evaluation of the value of σ_M at a given point $\mathbf{m}$ requires extensive computations. Then, the requirement of the Gibbs sampler method, that we know the conditional probability density along given directions, is not immediately satisfied. I have never been convinced that, in complex problems, the numerical estimation of the conditional probability density along a given direction, plus an exact (nonrejection) generation of a sample along that direction, gives superior results to the Metropolis (rejection) method presented in the next section.

2.3.5 The Metropolis Algorithm

The Metropolis (or Metropolis–Hastings) algorithm was developed by Metropolis and Ulam (1949), Metropolis et al. (1953), and Hastings (1970). It is a Markov chain Monte Carlo (MCMC) method, i.e., it is random (Monte Carlo) and has no memory, in the sense that each step depends only on the previous step (Markov chain).

The basic idea is to perform a random walk, a sort of Brownian motion, that, if unmodified, would sample some initial probability distribution, then, using a probabilistic rule, to modify the walk (some proposed moves are accepted, some are rejected) in such a way that the modified random walk samples the target distribution. While it is quite easy to invent probabilistic rules that would satisfy the goal, the Metropolis rule is the most efficient (it accepts the maximum of the proposed moves, reducing the computational requirements). I follow here the presentation of the Metropolis algorithm made by Mosegaard and Tarantola (2002).

Consider the following problem. We have two probability densities $f(\mathbf{x})$ and $g(\mathbf{x})$ together with their homogeneous limit (see chapter 1) $\mu(\mathbf{x})$. We have an algorithm that is able to generate samples of $f(\mathbf{x})$. How should we modify the algorithm in order to obtain samples of the conjunction of the two probability densities

$$h(\mathbf{x}) = k \frac{f(\mathbf{x}) g(\mathbf{x})}{\mu(\mathbf{x})} \quad ? \tag{2.7}$$

The criteria to be used will not depend on the values of the probability density $g(\mathbf{x})$, but on the values of the associated likelihood function (see equation (1.31)), so let us introduce it explicitly:

$$\gamma(\mathbf{x}) = g(\mathbf{x}) / \mu(\mathbf{x}) \quad . \tag{2.8}$$

---

[36]Often, instead of choosing directions at random, the n directions defined by the n coordinates x^i are used sequentially.

By hypothesis, some random rules define a random walk that samples the probability density $f(\mathbf{x})$. At a given step, the random walker is at point $\mathbf{x}_i$, and the application of the rules would lead to a transition to point $\mathbf{x}_j$. When all such proposed transitions $\mathbf{x}_i \to \mathbf{x}_j$ are accepted, the random walker will sample the probability density $f(\mathbf{x})$. Instead of always accepting the proposed transition $\mathbf{x}_i \to \mathbf{x}_j$, we reject it sometimes by using the following rule (to decide if the random walker is allowed to move to $\mathbf{x}_j$ or if it must stay at $\mathbf{x}_i$):

- If $\gamma(\mathbf{x}_j) \geq \gamma(\mathbf{x}_i)$, then accept the proposed transition to $\mathbf{x}_j$.

- If $\gamma(\mathbf{x}_j) < \gamma(\mathbf{x}_i)$, then decide randomly to move to $\mathbf{x}_j$, or to stay at $\mathbf{x}_i$, with the following probability of accepting the move to $\mathbf{x}_j$:

$$P_{i \to j} = \gamma(\mathbf{x}_j) / \gamma(\mathbf{x}_i) \quad . \tag{2.9}$$

Then one has the following theorem:[37] *The random walker samples the conjunction $h(\mathbf{x})$ (equation (2.7)) of the probability densities $f(\mathbf{x})$ and $g(\mathbf{x})$.* We shall call the above acceptance rule the *Metropolis rule*. Note that to run the algorithm, there is no need to know the normalizing constant k in equation (2.7).

As a special case of the proposed algorithm, we can take $f(\mathbf{x}) = \mu(\mathbf{x})$, in which case $h(\mathbf{x}) = g(\mathbf{x})$. Then, starting with a random walk that, if left unperturbed, would sample the homogeneous probability density $\mu(\mathbf{x})$, we end up with a random walk that samples any desired probability density $g(\mathbf{x})$.

In Appendix 6.11, the cascaded use of the algorithm is described, which allows us to sample the conjunction

$$h(\mathbf{x}) = k \, \mu(\mathbf{x}) \, \frac{f_1(\mathbf{x})}{\mu(\mathbf{x})} \, \frac{f_2(\mathbf{x})}{\mu(\mathbf{x})} \, \cdots \, \frac{f_p(\mathbf{x})}{\mu(\mathbf{x})} \tag{2.10}$$

of a sequence of probability densities.

2.3.6 Genetic Algorithms

One usually finds in the literature another type of Monte Carlo techniques, based on a biological analogy, called the *genetic algorithms* (Goldberg, 1989). Unfortunately, genetic algorithms lack the basic theorem of the Metropolis algorithm ("if you do this and this, then you generate samples of the distribution, in the precise, technical sense of *sample*") and are not described here.

2.4 Monte Carlo Solution to Inverse Problems

As mentioned at the beginning of section 2.2, the a posteriori probability density in the model manifold is expressed as

$$\sigma_M(\mathbf{m}) = k \, \rho_M(\mathbf{m}) \, L(\mathbf{m}) \quad , \tag{2.11}$$

where the probability density $\rho_M(\mathbf{m})$ represents the a priori information on the model parameters and the likelihood function $L(\mathbf{m})$ is a measure of the goodness of the model $\mathbf{m}$ in fitting the data. Two possible expressions for $L(\mathbf{m})$ are given in equations (2.2) and (2.3).

---

[37] See Mosegaard and Tarantola (2002) for the demonstration.

2.4.1 Sampling the Prior Probability Distribution

The movie strategy proposed above requires that we start by generating samples of the prior probability density $\rho_M(\mathbf{m})$. In typical inverse problems, the probability density $\rho_M(\mathbf{m})$ is quite simple (the contrary happens with the posterior probability density $\sigma_M(\mathbf{m})$). Therefore, the sampling of $\rho_M(\mathbf{m})$ can often be done using simple methods. We have seen two examples of this (Example 1.32 on page 29 and Example 2.1 on page 45). The sampling of the prior probability density usually involves a sequential use of one-dimensional sampling methods, like those described above. Sometimes, a Gibbs sampler or even a Metropolis algorithm may be needed, but these are usually simple to develop.

So, let us assume that we are able to obtain samples of the prior probability density $\rho_M(\mathbf{m})$, and let us move to the difficult problem of obtaining samples of the posterior probability density $\sigma_M(\mathbf{m})$.

2.4.2 Sampling the Posterior Probability Distribution

The adaptation of the Metropolis algorithm (presented in section 2.3.5) to the problem of sampling the posterior probability density (equation (2.11)),

$$\sigma_M(\mathbf{m}) \; = \; k \, \rho_M(\mathbf{m}) \, L(\mathbf{m}) \quad , \tag{2.12}$$

is immediate. As just discussed, assume that we are able to obtain as many samples of the prior probability density $\rho_M(\mathbf{m})$ as we wish. At a given step, the random walker is at point $\mathbf{m}_i$, and the application of the rules would lead to a transition to point $\mathbf{m}_j$. Sometimes, we reject this proposed transition by using the following rule:

- If $L(\mathbf{m}_j) \geq L(\mathbf{m}_i)$, then accept the proposed transition to $\mathbf{m}_j$.

- If $L(\mathbf{m}_j) < L(\mathbf{m}_i)$, then decide randomly to move to $\mathbf{m}_j$, or to stay at $\mathbf{m}_i$, with the following probability of accepting the move to $\mathbf{m}_j$:

$$P_{i \rightarrow j} \; = \; \frac{L(\mathbf{m}_j)}{L(\mathbf{m}_i)} \quad . \tag{2.13}$$

Then, *the random walker samples the a posteriori probability density* $\sigma_M(\mathbf{m})$.

2.4.3 Designing the Random Walk

The goal is to obtain samples of the posterior probability density $\sigma_M(\mathbf{m})$ that are *independent*. One easy way to obtain independency of the posterior samples is to present to the Metropolis algorithm independent samples of the prior probability density $\rho_M(\mathbf{m})$. Except for problems where the model manifold has a very small number of dimensions, this will not work, because of the emptiness of large-dimensional spaces (mentioned in section 2.1). Therefore, the sampling of the prior probability distribution has to be done jumping from point to point making small jumps. This kind of sampling, called a random walk, is a sort of Brownian motion that is far from producing independent samples. Then, if the samples of the prior distribution presented to the Metropolis algorithm are not independent, the samples of the posterior distribution produced by the Metropolis algorithm will not be independent.

There is only one solution to this problem: instead of taking all the samples produced by the Metropolis algorithm, after taking one sample, wait until a sufficient number of moves have been made, so that the algorithm has "forgotten" that sample. How many moves have we to wait, after one sample, in order to have some confidence that the next sample we consider is independent of the previous one? No general rule can be given, as this will strongly depend on the particular problem at hand.

The second important point is the following: only a small fraction of the infinitely many random walks that would sample the prior distribution will allow the Metropolis algorithm to have a reasonable efficiency.

The basic rule is the following: among the many possible random walks that can sample the prior probability density $\rho_M(\mathbf{m})$, select one that, when jumping from one sample of the prior probability density to the next, the perturbation of the likelihood function $L(\mathbf{m})$ is as small as possible (in order to increase the acceptance rate of the Metropolis rule). To be more precise, the *type* of perturbations in the model space has to be such that large perturbations (in the model space) only produce small perturbations of the predicted data. When the type of perturbations in the model space satisfies this requirement, it remains to decide the *size* of the perturbations to be made. There is a compromise between our wish to move rapidly in the model space and the need for the Metropolis algorithm to find some of the proposed moves acceptable. So, the size of the perturbations in the model space has to be such that the acceptance rate of the Metropolis criterion is, say, 30–50%. If the acceptance rate is larger, we are not moving fast enough in the model space; if it is much smaller, we are wasting computer resources to test models that are not accepted. In this way, we strike a balance between extensive exploration of the model space (large steps but many rejects) and careful sampling of located probability maxima (small steps, few rejects, but slow walk).

These remarks show that considerable ingenuity is required in designing the random walk that is to sample $\rho_M(\mathbf{m})$. For instance, in a problem involving a model of mass density distribution, the data consisting of values of the gravity field, Mosegaard and Tarantola (1995) chose to make large perturbations of the mass density distribution but kept the total mass approximately constant.

The last point to be examined concerns the decision to stop the random walk when the posterior probability density $\sigma_M(\mathbf{m})$ has been sufficiently sampled. There are two subproblems here, an easy one and a difficult one. The easy problem is to decide, when exploring a given maximum of the probability density, that this maximum has conveniently been sampled. The literature contains some good rules of thumb for this.[38] The difficult problem, of course, is about the possibility that we may be completely missing some region of significant probability of $\sigma_M(\mathbf{m})$, an isolated maximum, for instance. This problem is inherent in all Monte Carlo methods and is very acute in highly nonlinear inverse problems.[39] Unfortunately, nothing can be said here that would be applicable to any large class of inverse problems: each problem has its own physics, and the experience of the implementer is crucial here. This issue must be discussed every time an inverse problem is solved using Monte Carlo methods.

---

[38] For instance, see Geweke (1992) or Raftery and Lewis (1992).

[39] In the case where one has a relation $\mathbf{d} = \mathbf{g}(\mathbf{m})$, this means that the function $\mathbf{g}(\cdot)$ is highly nonlinear.

A final comment about the cost of using the Metropolis algorithm for solving inverse problems: Each step of the algorithm requires the evaluation of $L(\mathbf{m})$. This requires the resolution of a forward problem (in the case where the likelihood function is given by expression (2.3)) or the evaluation of an integral (in the case where the likelihood function is given by expression (2.2)). This may be very demanding on computational resources.

2.5   Simulated Annealing

The simulated annealing technique is designed to obtain the maximum likelihood point of any probability density, in particular for the posterior probability density $\sigma_M(\mathbf{m})$. But at the core of the simulated annealing there is a Metropolis algorithm that is able to sample $\sigma_M(\mathbf{m})$. My point of view is that if we are able to sample the probability density $\sigma_M(\mathbf{m})$, we should not be interested in the maximum likelihood point. As any central estimator (like the mean or the median), the maximum likelihood point is of very little interest when dealing with complex probability distributions.

The simulated annealing technique is described here for completeness of the theory, not because it is an important element of it.

Annealing consists of heating a solid until thermal stresses are released, then in cooling it very slowly to the ambient temperature. Ideally, the substance is heated until it melts, and then cooled very slowly until a perfect crystal is formed. The substance then reaches the state of lowest energy. If the cooling is not slow enough, a metastable glass can be formed.

Simulated annealing (Kirkpatrick et al., 1983, Geman and Geman, 1984) is a numerical method, using an analogy between the process of physical annealing and the mathematical problem of obtaining the global minimum of a function (assimilated to an energy) that may have local minima (metastable states). It has been introduced in the framework of inverse theory by Rothman (1985a, 1985b, 1986).

We are about to see that simulated annealing has, at its core, a Metropolis algorithm. And we have seen above that in using a Metropolis algorithm one may sample the probability distribution in the model space. Distorting it until it peaks at the maximum likelihood point (this is what simulated annealing does) is not necessarily an elegant strategy. Let us, nevertheless, mention the basic tools here.

Let $\sigma_M(\mathbf{m})$ be the (not necessarily normalized) posterior probability density in the model space manifold, and let $\mu_M(\mathbf{m})$ be its homogeneous limit. We wish to obtain the maximum likelihood point $\mathbf{m}_{ML}$:

$$\frac{\sigma_M(\mathbf{m})}{\mu_M(\mathbf{m})} \qquad \text{maximum for} \qquad \mathbf{m} = \mathbf{m}_{ML} \ . \tag{2.14}$$

As explained in section 1.6.4, the maximum of a probability density does not (invariantly) define a point; the maximum of the ratio of the probability density to the homogeneous probability density does.

To locate this maximum, we may define the *energy function*

$$S(\mathbf{m}) = -T_0 \log \frac{\sigma_M(\mathbf{m})}{\mu_M(\mathbf{m})} \ , \tag{2.15}$$

where T_0 is an arbitrary, but fixed, real (adimensional) positive number, termed the *ambient temperature* (for instance, $T_0 = 1$). This gives

$$\sigma_M(\mathbf{m}) \;=\; \mu_M(\mathbf{m}) \; e^{-S(\mathbf{m})/T_0} \quad . \tag{2.16}$$

Define now, for arbitrary temperature T , the new function

$$\sigma_M(\mathbf{m}, T) \;=\; \mu_M(\mathbf{m}) \; e^{-S(\mathbf{m})/T} \quad , \tag{2.17}$$

and, for any fixed value of T , start a Metropolis algorithm that produces samples of $\sigma_M(\mathbf{m}, T)$. If, while the Metropolis algorithm is at work, one very slowly changes the value of T toward zero, it is clear that the sampling algorithm will end up sampling only models that are in the immediate vicinity of the maximum value of the energy function $S(\mathbf{m})$, i.e., in the immediate vicinity of the maximum likelihood point $\mathbf{m}_{ML}$.

   If the temperature T is brought to zero too rapidly, then the system will converge to a 'metastable' solution, i.e., to a local maximum of $\sigma_M(\mathbf{m})/\mu_M(\mathbf{m})$ instead of to the global maximum.

   For $\mu_M(\mathbf{m}) = \text{const.}$, equation (2.17) clearly resembles the Gibbs distribution (also called the canonical distribution), giving the probability of a state $\mathbf{m}$ with energy $S(\mathbf{m})$ of a statistical system at temperature T (the Boltzmann constant k is taken here equal to 1). This justifies the name "energy function" for $S(\mathbf{m})$ and "temperature" for T . The factor $\mu_M(\mathbf{m})$ slightly generalizes the Gibbs distribution: the probability density at infinite temperature is the homogeneous probability density $\mu_M(\mathbf{m})$.

Example 2.4. *In the Gaussian case examined in Example 1.37, we have arrived at the posterior probability density*

$$\sigma_M(\mathbf{m}) \;=\; k \, \exp(-S(\mathbf{m})) \quad , \tag{2.18}$$

where k is a normalization constant and $S(\mathbf{m})$ is the least-squares misfit function

$$S(\mathbf{m}) = \tfrac{1}{2} \Big((\mathbf{g}(\mathbf{m}) - \mathbf{d}_{obs})^t \, \mathbf{C}_D^{-1} \, (\mathbf{g}(\mathbf{m}) - \mathbf{d}_{obs}) + (\mathbf{m} - \mathbf{m}_{prior})^t \, \mathbf{C}_M^{-1} \, (\mathbf{m} - \mathbf{m}_{prior}) \Big). \tag{2.19}$$

We see that the energy function introduced in equation (2.15) is identical here to the misfit function. Changing the temperature simply means replacing equation (2.18) with (taking $T_0 = 1$)

$$\sigma_M(\mathbf{m}) \;=\; k \, \exp\left(-\frac{S(\mathbf{m})}{T} \right) \quad , \tag{2.20}$$

i.e., multiplying the misfit function by a constant.

Chapter 3

The Least-Squares Criterion

> If we know that our individual errors and fluctuations
> follow the magic bell-shaped curve *exactly*,
> then the resulting estimates are known to have
> almost all the nice properties that people have been able to think of.
>
> *John W. Tukey*, 1965

Least squares are popular for solving inverse problems because they lead to the easiest computations. Their only drawback is their lack of robustness, i.e., their strong sensitivity to a small number of large errors (outliers) in a data set.

In this book, the least-squares criterion is justified by the hypothesis that all initial uncertainties in the problem can be modeled using Gaussian distributions. Covariance operators play a central role in the method; the underlying mathematics is simple and beautiful.

When the equation solving the forward problem is linear, posterior uncertainties are also Gaussian, and an explicit expression is obtained for the posterior probability distribution. When the forward equation is nonlinear, the posterior probability isn't Gaussian, but, if nonlinearities are not too severe, finding the maximum likelihood point of the distribution and estimating the shape of the distribution around this point (i.e., estimating the covariance matrix of the distribution) may satisfactorily solve the problem.

3.1 Preamble: The Mathematics of Linear Spaces

We are about to see that least-squares methods are intimately linked to Gaussian probability distributions. These are only defined over linear spaces (which is why on the surface of the sphere one uses the Fisher distribution, as there is no Gaussian defined).

Later in this section, therefore, the model space manifold $\mathfrak{M}$ is assumed to be a linear space (denoted $\mathbb{M}$), and the data manifold $\mathfrak{D}$ is also assumed to be linear (and is denoted $\mathbb{D}$). Sums and difference of vectors and multiplication of a vector by a real number are, therefore, defined. For some problems, the model parameter space is not linear, but one may work only in a small region around some initial point. Then, one uses the linear tangent space. Let us start by recalling some basic terminology of linear spaces.

3.1.1 Dual of a Linear Space

Let $\mathbb{V}$ be an n-dimensional linear space[40] with vectors denoted $\mathbf{v}, \mathbf{w}, \dots$. When a basis $\{\mathbf{e}_1, \dots, \mathbf{e}_n\}$ is chosen, any vector $\mathbf{v}$ can be decomposed as $\mathbf{v} = v^i\,\mathbf{e}_i$, defining the components $\{v^i\}$ of the vector. In a linear space, the sum of two elements $\mathbf{v} + \mathbf{w}$ is defined (and equals the sum of the components), as is defined the product of a vector by a real number:

$$(\mathbf{u} + \mathbf{v})^i \;=\; u^i + v^i \qquad , \qquad (\lambda\,\mathbf{v})^i \;=\; \lambda\,v^i \;\;. \tag{3.1}$$

The *dual* of $\mathbb{V}$, denoted $\mathbb{V}^*$, is the space of all the *linear forms* over $\mathbb{V}$, i.e., the space of all the linear applications mapping $\mathbb{V}$ into $\mathfrak{R}$. If ω is an element of $\mathbb{V}^*$, the real number it associates with an element $\mathbf{v}$ of $\mathbb{V}$ is denoted $\langle\,\omega\,,\,\mathbf{v}\,\rangle$, so one may write

$$\mathbf{v} \quad \mapsto \quad \lambda = \langle\,\omega\,,\,\mathbf{v}\,\rangle \;\;. \tag{3.2}$$

One says that $\langle\,\omega\,,\,\mathbf{v}\,\rangle$ is the *duality product* of ω by $\mathbf{v}$. As Figure 3.1 suggests, a linear form can be represented by a "mille-feuilles."

Figure 3.1. *While a vector can be represented as an "arrow," a linear form can be represented by a mille-feuille. The number associated by a linear form with a vector equals the number of "feuilles" (layers) traversed by the vector. Although only four layers are represented, there is an infinity of them (and they are infinitely large).*

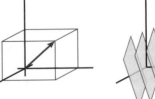
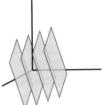

The sum of two linear forms and the product of a form by a real number are defined through

$$\langle\,(\omega + \nu)\,,\,\mathbf{v}\,\rangle \;=\; \langle\,\omega\,,\,\mathbf{v}\,\rangle + \langle\,\nu\,,\,\mathbf{v}\,\rangle \qquad , \qquad \langle\,\alpha\,\omega\,,\,\mathbf{v}\,\rangle = \alpha\,\langle\,\omega\,,\,\mathbf{v}\,\rangle \quad , \tag{3.3}$$

and it is easy to see that, with these definitions, the dual $\mathbb{V}^*$ of a linear space $\mathbb{V}$ is also a linear space. Taking a basis $\{\epsilon^1, \dots, \epsilon^n\}$ in $\mathbb{V}^*$ allows us to write any element $\omega \in \mathbb{V}^*$ as $\omega = \omega_i\,\epsilon^i$, defining the components $\{\omega_i\}$ of ω. The place of the indices (in the upper or lower position) is traditional in tensor notation and allows us to use the *implicit sum*

---

[40] In a linear space, the sum $\mathbf{v}_1 + \mathbf{v}_2$ of two elements of $\mathbb{V}$ and the multiplication $\lambda\,\mathbf{v}$ of a real number by an element of $\mathbb{V}$ are defined and satisfy the following conditions. For any $\mathbf{v}_1$ and $\mathbf{v}_2$, $\mathbf{v}_1 + \mathbf{v}_2 = \mathbf{v}_2 + \mathbf{v}_1$ (commutativity). For any $\mathbf{v}_1$, $\mathbf{v}_2$, and $\mathbf{v}_3$, $(\mathbf{v}_1 + \mathbf{v}_2) + \mathbf{v}_3 = \mathbf{v}_1 + (\mathbf{v}_2 + \mathbf{v}_3)$ (associativity of the sum). There is an element $\mathbf{0}$ such that, for any $\mathbf{v}$, $\mathbf{v} + \mathbf{0} = \mathbf{v}$ (existence of zero). For any $\mathbf{v}$, there exists $(-\mathbf{v})$ such that $\mathbf{v} + (-\mathbf{v}) = \mathbf{0}$ (existence of opposite). For any $\mathbf{v}_1$, $\mathbf{v}_2$ and any real λ, $\lambda\,(\mathbf{v}_1 + \mathbf{v}_2) = \lambda\,\mathbf{v}_1 + \lambda\,\mathbf{v}_2$ (first distributivity). For any $\mathbf{v}$ and any reals λ and μ, $(\lambda + \mu)\,\mathbf{v} = \lambda\,\mathbf{v} + \mu\,\mathbf{v}$ (second distributivity). For any $\mathbf{v}$ and any reals λ and μ, $(\lambda\,\mu)\,\mathbf{v} = \lambda\,(\mu\,\mathbf{v})$ (associativity of the product). Finally, for any $\mathbf{v}$, $1\,\mathbf{v} = \mathbf{v}$.

convention, defined as follows. In normal expressions, the sums over indices involve an index in the lower position and an index in the upper position, as in $r = \sum_i \omega_i v^i$ or $d^i = \sum_\alpha G^i_{\ \alpha} m^\alpha$. By convention, the sum sign is not written, and one simplifies the expressions into

$$\omega_i v^i \equiv \sum_i \omega_i v^i \quad, \quad G^i_{\ \alpha} m^\alpha \equiv \sum_\alpha G^i_{\ \alpha} m^\alpha \qquad \text{(implicit sum convention)} . \quad (3.4)$$

The bases in $\mathbb{V}$ and $\mathbb{V}^*$ can always be chosen so as to have

$$\langle\, \epsilon^i \,,\, \mathbf{e}_j \,\rangle = \delta^i_j \quad. \qquad (3.5)$$

When this is the case, one says that the bases are *mutually dual*. Then, the duality product of $\omega \in \mathbb{V}^*$ and $\mathbf{v} \in \mathbb{V}$ can be successively written $\langle\, \omega \,,\, \mathbf{v} \,\rangle = \langle\, \omega_i \epsilon^i \,,\, v^j \mathbf{e}_j \,\rangle = \omega_i v^j \langle\, \epsilon^i \,,\, \mathbf{e}_j \,\rangle = \omega_i v^j \delta^i_j = \omega_i v^i$. Therefore, the duality product can simply be expressed as (note that the implicit sum convention is being used)

$$\langle\, \omega \,,\, \mathbf{v} \,\rangle = \omega_i v^i \quad. \qquad (3.6)$$

For the same reason that when a basis $\{\mathbf{e}_i\}$ is chosen over the linear space $\mathbb{V}$ an element $\mathbf{v} \in \mathbb{V}$ can just be seen as a sequence of n quantities $\{v^1, \dots, v^n\}$, when a basis $\{\epsilon^i\}$ is chosen over $\mathbb{V}^*$, an element $\omega \in \mathbb{V}^*$ can just be seen as a sequence of n quantities $\{\omega_1, \dots, \omega_n\}$. From this perspective, the dual of a linear space is just an ad hoc space, closely resembling the original space, except that if the quantities $\{v^1, \dots, v^n\}$ have physical dimensions, the quantities $\{\omega_1, \dots, \omega_n\}$ must have the reciprocal physical dimensions in order for the expression $\lambda = \omega_i v^i$ to make sense.

It is useful to also introduce the notation $\langle\, \omega \,,\, \mathbf{v} \,\rangle = \omega^t \mathbf{v}$: if the elements $\mathbf{v}$ and ω are seen as column matrices, the expression $\omega^t \mathbf{v}$ can be interpreted as a matrix product. We then have the three equivalent expressions

$$\boxed{\langle\, \omega \,,\, \mathbf{v} \,\rangle = \omega^t \mathbf{v} = \omega_i v^i} \quad. \qquad (3.7)$$

3.1.2 Transpose of a Linear Operator

Let $\mathbb{M}$ and $\mathbb{D}$ be two linear spaces and $\mathbb{M}^*$ and $\mathbb{D}^*$ be the respective dual spaces. The duality product of $\delta \in \mathbb{D}^*$ and $\mathbf{d} \in \mathbb{D}$ (resp., of $\mu \in \mathbb{M}^*$ and $\mathbf{m} \in \mathbb{M}$) is denoted $\langle\, \delta \,,\, \mathbf{d} \,\rangle_{\mathrm{D}}$ (resp., $\langle\, \mu \,,\, \mathbf{m} \,\rangle_{\mathrm{M}}$). Let $\mathbf{G}$ be a linear operator mapping $\mathbb{M}$ into $\mathbb{D}$. We can write

$$\mathbf{d} = \mathbf{G}\,\mathbf{m} \qquad (3.8)$$

or, in terms of the components (implicit sum convention used throughout this chapter),

$$d^i = G^i_{\ \alpha} m^\alpha \quad. \qquad (3.9)$$

The *transpose* of $\mathbf{G}$, denoted $\mathbf{G}^t$, is a linear operator, mapping $\mathbb{D}^*$ into $\mathbb{M}^*$, defined by the condition that, for any $\delta \in \mathbb{D}^*$ and any $\mathbf{m} \in \mathbb{M}$,

$$\boxed{\langle\, \mathbf{G}^t \delta \,,\, \mathbf{m} \,\rangle_{\mathrm{M}} = \langle\, \delta \,,\, \mathbf{G}\,\mathbf{m} \,\rangle_{\mathrm{D}}} \quad. \qquad (3.10)$$

Using the pseudomatricial notation introduced above for the duality product, the definition (3.10) can be written

$$(\mathbf{G}^t \, \delta)^t \, \mathbf{m} \; = \; \delta^t \, (\mathbf{G} \, \mathbf{m}) \tag{3.11}$$

or, in terms of the components, $(\mathbf{G}^t \, \delta)_\alpha \, m^\alpha = \delta_i \, (\mathbf{G} \, \mathbf{m})^i = \delta_i \, G^i{}_\alpha \, m^\alpha$. Writing the components of $\mathbf{G}^t$ as $(\mathbf{G}^t)_\alpha{}^i$ gives

$$(\mathbf{G}^t)_\alpha{}^i \, \delta_i \, m^\alpha \; = \; \delta_i \, G^i{}_\alpha \, m^\alpha \tag{3.12}$$

and, as this must hold for any δ and any $\mathbf{m}$, we arrive at

$$(\mathbf{G}^t)_\alpha{}^i \; = \; G^i{}_\alpha \quad . \tag{3.13}$$

This expresses that the two operators $\mathbf{G}$ and $\mathbf{G}^t$ have the same components, except in that they are 'transposed': the matrix representing the operator $\mathbf{G}^t$ is the transpose of the matrix representing $\mathbf{G}$ (although we could have taken this as the definition of transpose of a matrix, the definition given above generalizes, as we will see in chapter 5, to linear operators that are not matrices).

The result just demonstrated suggests a simplification in the notation:

$$G_\alpha{}^i \; \equiv \; (\mathbf{G}^t)_\alpha{}^i \quad . \tag{3.14}$$

Then, while $G^i{}_\alpha$ maps an element of $\mathbb{M}$ into an element of $\mathbb{D}$, as in $d^i = G^i{}_\alpha \, m^\alpha$, $G_\alpha{}^i$ maps an element of $\mathbb{D}^*$ into an element of $\mathbb{M}^*$, as in $\mu_\alpha = G_\alpha{}^i \, \delta_i$.

In the special circumstance where a linear operator $\mathbf{L}$ is considered that maps a linear space $\mathbb{V}$ into its dual $\mathbb{V}^*$, then, by definition, the transpose $\mathbf{L}^t$ also maps $\mathbb{V}$ into $\mathbb{V}^*$. If, in that case, $\mathbf{L} = \mathbf{L}^t$, the operator is called *symmetric*.

3.1.3 Scalar Product

Let $\mathbb{V}$ be a linear space and $\mathbf{W}$ be a *weighting operator* over $\mathbb{V}$, i.e., a *linear, symmetric, and positive definite*[41] *operator* mapping $\mathbb{V}$ into its dual $\mathbb{V}^*$. Given such a weighting operator, the *scalar product* of two elements of $\mathbb{V}$ is defined — via the duality product — as

$$\boxed{(\, \mathbf{u} \, , \, \mathbf{v} \,) \; = \; \langle \, \mathbf{W} \mathbf{u} \, , \, \mathbf{v} \, \rangle} \quad . \tag{3.15}$$

Using matricial notations for the duality product gives $(\, \mathbf{u} \, , \, \mathbf{v} \,) = (\mathbf{W} \mathbf{u})^t \, \mathbf{v} = \mathbf{u}^t \, \mathbf{W}^t \, \mathbf{v} = \mathbf{u}^t \, \mathbf{W} \mathbf{v}$, i.e.,

$$(\, \mathbf{u} \, , \, \mathbf{v} \,) \; = \; \mathbf{u}^t \, \mathbf{W} \mathbf{v} \quad . \tag{3.16}$$

The inverse of a weighting operator, $\mathbf{C} = \mathbf{W}^{-1}$, is called a *covariance operator*, which is also symmetric and positive definite. (For a demonstration that the usual probabilistic definition of a covariance operator defines a symmetric and positive definite operator,

---

[41] $\mathbf{W}$ is positive definite if, for any $\mathbf{v} \neq \mathbf{0}$, $\langle \, \mathbf{W} \mathbf{v} \, , \, \mathbf{v} \, \rangle > 0$.

see, for instance, Pugachev, 1965.) In terms of the covariance operator, the scalar product is written

$$(\mathbf{u} , \mathbf{v}) = \mathbf{u}^t \, \mathbf{C}^{-1} \, \mathbf{v} \quad . \tag{3.17}$$

When a weighting operator $\mathbf{W}$ has been defined, there is a bijection between a linear space and its dual. Denoting by $\hat{\mathbf{v}}$ the element of $\mathbb{V}^*$ associated with $\mathbf{v} \in \mathbb{V}$ by $\mathbf{W}$, we may write

$$\boxed{\hat{\mathbf{v}} = \mathbf{W} \mathbf{v}} \tag{3.18}$$

or, using components, $\hat{v}_i = W_{ij} \, v^j$. When there is no ambiguity about the operator $\mathbf{W}$ defining this bijection, the "hat" in $\hat{v}_i$ is dropped, and one simply writes

$$v_i = W_{ij} \, v^j \quad . \tag{3.19}$$

The equation $\mathbf{W} \mathbf{C} = \mathbf{I}$ defining the covariance operator becomes, using components,

$$W_{ij} \, C^{jk} = \delta_i^k \quad , \tag{3.20}$$

and the reciprocal of equation (3.19) is

$$v^i = C^{ij} \, v_j \quad . \tag{3.21}$$

As the position of the indices clearly designates W_{ij} and C^{ij}, one could use a common letter for both (as is done in differential geometry, where the same symbol is used for the 'covariant metric' g_{ij} and the 'contravariant metric' g^{ij}), but we shall not do this here.

Given a scalar product, the *norm* of a vector $\mathbf{v}$ is defined[42] as $\| \mathbf{v} \| = (\mathbf{v} , \mathbf{v})^{1/2}$. Therefore,

$$\| \mathbf{v} \|^2 = (\mathbf{v} , \mathbf{v}) = \mathbf{v}^t \, \mathbf{C}^{-1} \, \mathbf{v} = \mathbf{v}^t \, \mathbf{W} \mathbf{v} = v^i \, W_{ij} \, v^j \quad . \tag{3.22}$$

One should keep in mind that a covariance operator maps a space into its dual and that in usual problems one must simultaneously deal with two quite different spaces.

Example 3.1. *Sometimes it happens that the components of* $\mathbf{v}$ *represent digitized values of some continuous field (see, for instance, Example 1.33 of chapter 1). The covariance operator* $\mathbf{C}$ *is then generally a smoothing operator (see, for instance, Pugachev, 1965), and the elements of* $\mathbb{V}$ *are smooth (i.e., are the discretized versions of smooth functions). The inverse* $\mathbf{C}^{-1}$ *is then a "roughing" operator. The elements of* $\mathbb{V}^*$ *, obtained as images of those of* $\mathbb{V}$ *through* $\mathbf{C}^{-1}$ *, are "rough" functions.*

3.1.4 Adjoint of a Linear Operator

When defining the transpose of a linear operator, it is not assumed that the vector spaces in consideration have a scalar product. If they do, then it is possible to define the 'adjoint' of a linear operator.

---

[42] See Appendix 6.12 for the demonstration that this actually defines a norm.

As above, let $\mathbf{G}$ be a linear operator mapping $\mathbb{M}$ into $\mathbb{D}$, and let $(\ , \ )_D$ and $(\ , \ )_M$ represent the scalar products in $\mathbb{D}$ and $\mathbb{M}$, respectively. The *adjoint* of $\mathbf{G}$ is denoted by $\mathbf{G}^*$ and is the linear operator mapping $\mathbb{D}$ into $\mathbb{M}$, defined by the condition that, for any $\mathbf{d} \in \mathbb{D}$ and any $\mathbf{m} \in \mathbb{M}$, the following property holds:

$$(\mathbf{G}^* \mathbf{d} , \ \mathbf{m})_M = (\mathbf{d} , \ \mathbf{G} \mathbf{m})_D \ . \tag{3.23}$$

Let $\mathbf{C_M}$ and $\mathbf{C_D}$ be the covariance operators defining the respective scalar products in $\mathbb{M}$ and $\mathbb{D}$. Using successively the definitions of scalar product and of the transpose of an operator, we can write $(\mathbf{G}^* \mathbf{d} , \ \mathbf{m})_M = (\mathbf{d} , \ \mathbf{G} \mathbf{m})_D = \langle \mathbf{C_D}^{-1} \mathbf{d} , \ \mathbf{G} \mathbf{m} \rangle_D = \langle \mathbf{G}^t \mathbf{C_D}^{-1} \mathbf{d} , \ \mathbf{m} \rangle_D = (\mathbf{C_M} \mathbf{G}^t \mathbf{C_D}^{-1} \mathbf{d} , \ \mathbf{m})_M$, so, as this must hold for any $\mathbf{m}$ and any $\mathbf{d}$, we arrive at the relation between adjoint and transpose:

$$\mathbf{G}^* = \mathbf{C_M} \mathbf{G}^t \mathbf{C_D}^{-1} \ . \tag{3.24}$$

Sometimes the terms *adjoint* and *transpose* are incorrectly used as synonyms. The last equation shows that they are not.

In the special circumstance where a linear operator $\mathbf{L}$ is considered that maps a linear space into itself, then, by definition, the adjoint $\mathbf{L}^*$ also maps the space into itself. If, in that case, $\mathbf{L} = \mathbf{L}^*$, the operator is called *self-adjoint*.

Another special circumstance is when an operator maps a space into its dual, as is the case with (the inverse of) a covariance operator. In Problem 7.10, it is demonstrated that, while a covariance operator is symmetric, $\mathbf{C}^t = \mathbf{C}$, it is not self-adjoint. Rather, the adjoint of a covariance operator equals its inverse, $\mathbf{C}^* = \mathbf{C}^{-1}$ (it is anti-self-adjoint).

3.2   The Least-Squares Problem

3.2.1   Formulation of the Problem

The model space manifold and the data manifold were introduced in chapter 1. They are here assumed to be linear spaces and are respectively denoted $\mathbb{M}$ and $\mathbb{D}$. Expressions like

$$\mathbf{m}_2 + \mathbf{m}_1 \quad , \quad \mathbf{m}_2 - \mathbf{m}_1 \quad , \quad \lambda \mathbf{m} \quad , \quad \mathbf{d}_2 + \mathbf{d}_1 \quad , \quad \mathbf{d}_2 - \mathbf{d}_1 \quad , \quad \lambda \mathbf{d} \tag{3.25}$$

are assumed to make (invariant)[43] sense.

Example 3.2. *If one parameter is, say, an electric resistance R, one should not try to define a vector having the quantity R as one of its components. For the sum of two electric resistances cannot be made compatible with the sum of two electric conductances (a conductance is the inverse of a resistance). The logarithmic parameter $R^* = \log(R/R_0)$, where R_0 is an arbitrary constant, can be used as one of the components of a vector (the sum of logarithmic resistances is equivalent to the sum of logarithmic conductances). Note that*

---

[43] Defining these sums as sums of 'coordinates' of a nonlinear manifold would not make invariant sense, as the sums would change meaning under a change of coordinates.

to take a Gaussian distribution to model the a priori information on a positive parameter is not coherent because a Gaussian function gives a nonvanishing probability to negative values of the parameter.[44]

We are going to manipulate probability densities over $\mathbb{M}$ and $\mathbb{D}$. As mentioned in chapter 1, the homogeneous probability density over a linear space is always constant. Therefore, everywhere in this chapter, the two homogeneous probability densities over the model space and the data space are constant:

$$\mu_M(\mathbf{m}) = \text{const.} \quad , \quad \mu_D(\mathbf{d}) = \text{const.} \tag{3.26}$$

It is clear that these constant probability densities can be interpreted as the limit of the Gaussians to be considered below when their uncertainties are taken to be infinite.

Least-squares techniques arise when all the 'input' probability densities are assumed to be Gaussian. Let us see this with some detail.

The elements of our problem are as follows:

- The a priori information that the (unknown) model $\mathbf{m}$ is a sample of a known Gaussian probability density whose mean is $\mathbf{m}_{\text{prior}}$ and whose covariance matrix is $\mathbf{C}_M$. The a priori probability density over the model space $\mathbb{M}$ is, therefore,

$$\rho_M(\mathbf{m}) = \text{const. } \exp\left(-\tfrac{1}{2} (\mathbf{m} - \mathbf{m}_{\text{prior}})^t \, \mathbf{C}_M^{-1} \, (\mathbf{m} - \mathbf{m}_{\text{prior}}) \right) \quad . \tag{3.27}$$

  This probability density is assumed to be a priori in the sense that it is independent of the result of the measurements on the observable parameters $\mathbf{d}$ (considered below).

- A relation

$$\mathbf{d} = \mathbf{g}(\mathbf{m}) \tag{3.28}$$

  that solves the 'forward problem,' i.e., that predicts the values of the observable parameters $\mathbf{d}$ that should correspond to the model $\mathbf{m}$. This theoretical prediction is assumed to be error free (see below for the introduction of a simple form of 'theoretical uncertainties').

- Some measurements on the observable parameters $\mathbf{d}$ whose results can be represented by a Gaussian probability density centered at $\mathbf{d}_{\text{obs}}$ and with covariance matrix $\mathbf{C}_D$:

$$\rho_D(\mathbf{d}) = \text{const. } \exp\left(-\tfrac{1}{2} (\mathbf{d} - \mathbf{d}_{\text{obs}})^t \, \mathbf{C}_D^{-1} \, (\mathbf{d} - \mathbf{d}_{\text{obs}}) \right) \quad . \tag{3.29}$$

The combination of these three types of information was considered in Example 1.34, where it was shown to lead to the a posteriori probability density in the model space (equation (1.93))

$$\sigma_M(\mathbf{m}) = \frac{\rho_M(\mathbf{m}) \, \rho_D(\mathbf{g}(\mathbf{m}))}{\int_{\mathfrak{M}} d\mathbf{m}' \, \rho_M(\mathbf{m}') \, \rho_D(\mathbf{g}(\mathbf{m}'))} \quad . \tag{3.30}$$

---

[44]Sometimes, a least-squares criterion is used for such parameters, completed with a positivity constraint. This is not the most rigorous nor the easiest way to attack this sort of problem. As suggested in section 1.2.4, these parameters usually accept a log-normal function as a prior probability density. Taking the *logarithm* of the positive parameter defines a new (unbounded) parameter whose a priori probability density is Gaussian and for which standard least-squares techniques apply.

This gives

$$\sigma_M(\mathbf{m}) = \text{const. } \exp(-S(\mathbf{m})) \quad , \tag{3.31}$$

where (twice) the *misfit function*[45] $S(\mathbf{m})$ is defined as

$$2\,S(\mathbf{m}) = \parallel \mathbf{g}(\mathbf{m}) - \mathbf{d}_{\text{obs}} \parallel_D^2 + \parallel \mathbf{m} - \mathbf{m}_{\text{prior}} \parallel_M^2$$

$$= (\mathbf{g}(\mathbf{m}) - \mathbf{d}_{\text{obs}})^t \, \mathbf{C}_D^{-1} \, (\mathbf{g}(\mathbf{m}) - \mathbf{d}_{\text{obs}}) + (\mathbf{m} - \mathbf{m}_{\text{prior}})^t \, \mathbf{C}_M^{-1} \, (\mathbf{m} - \mathbf{m}_{\text{prior}}) \; .$$

$$\tag{3.32}$$

If instead of an exact theoretical prediction $\mathbf{d} = \mathbf{g}(\mathbf{m})$ one assumes that modelization uncertainties can be described using Gaussian statistics, with a covariance matrix $\mathbf{C}_T$ (see Example 1.17), then one should replace $\mathbf{C}_D$ with $\mathbf{C}_D + \mathbf{C}_T$ in equation (3.32) (see Example 1.36). Thanks to the simplicity of this result, when using the Gaussian models for uncertainties, we can forget that there are two different sources of uncertainties in the data space. All happens as if the forward modelization were exact and the observational uncertainties were those represented by the covariance matrix $\mathbf{C}_D + \mathbf{C}_T$.

If the relation $\mathbf{d} = \mathbf{g}(\mathbf{m})$ is linear, the misfit function (3.32) is quadratic and the posterior probability density $\sigma_M(\mathbf{m})$ is Gaussian (this case is analyzed in section 3.2.2). The further the relation $\mathbf{d} = \mathbf{g}(\mathbf{m})$ is from being linear, the further the posterior probability density $\sigma_M(\mathbf{m})$ is from being a Gaussian. Figure 3.2 suggests the different 'regimes' of nonlinearity usually encountered in inverse problems.

We shall see that the analysis of nonlinear problems usually involves finding the maximum (posterior) likelihood point, i.e., the point that maximizes the posterior probability density (3.31). Maximizing $\sigma_M(\mathbf{m})$ is equivalent to minimizing the misfit function (3.32). As the misfit function is here a sum of squares,[46] this justifies using the terminology 'least-squares' for the kinds of problems here examined (i.e., problems based on Gaussian input uncertainties). The first references to the method of least squares are due to Laplace (1812) and Gauss (ca. 1820).

Linear and nonlinear problems are separately analyzed in sections 3.2.2 and 3.2.3 below.

3.2.2 Linear Problems

If the equation $\mathbf{d} = \mathbf{g}(\mathbf{m})$ solving the forward problem is linear, one writes, instead,

$$\mathbf{d} = \mathbf{G}\,\mathbf{m} \quad . \tag{3.33}$$

The two equations (3.31)–(3.32) giving the posterior probability density in the model space then become

$$\sigma_M(\mathbf{m}) = \text{const. } \exp(-S(\mathbf{m})) \quad , \tag{3.34}$$

---

[45] Also called the *cost* function, *objective* function, *least-squares* function, or *chi-squared* function.

[46] For uncorrelated uncertainties, $(\mathbf{C}_D)^{ij} = (\sigma_D^i)^2 \delta^{ij}$, $(\mathbf{C}_M)^{\alpha\beta} = (\sigma_M^\alpha)^2 \delta^{\alpha\beta}$, (twice) the misfit function $S(\mathbf{m})$ becomes $2\,S(\mathbf{m}) = \sum_i \frac{(g^i(\mathbf{m}) - d_{\text{obs}}^i)^2}{(\sigma_D^i)^2} + \sum_\alpha \frac{(m^\alpha - m_{\text{prior}}^\alpha)^2}{(\sigma_M^\alpha)^2}$, which is a sum of squares.

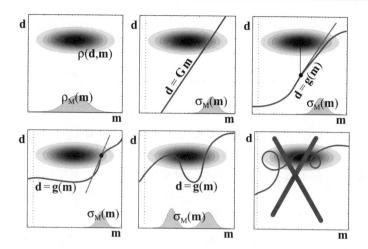

Figure 3.2. *The first sketch is a representation of the probability density $\rho(\mathbf{d}, \mathbf{m})$ representing both the information we have on the observable parameters $\mathbf{d}$ (data) and the a priori information on the model parameters $\mathbf{m}$ (the marginal $\rho_M(\mathbf{m})$). In the second sketch, the forward equation $\mathbf{d} = \mathbf{G}\,\mathbf{m}$ is linear. The posterior probability density $\sigma_M(\mathbf{m})$ is then also Gaussian. In the third sketch, the forward equation $\mathbf{d} = \mathbf{g}(\mathbf{m})$ can be linearized around $\mathbf{m}_{\mathrm{prior}}$, giving $\mathbf{g}(\mathbf{m}) \simeq \mathbf{g}(\mathbf{m}_{\mathrm{prior}}) + \mathbf{G}(\mathbf{m} - \mathbf{m}_{\mathrm{prior}})$, where $\mathbf{G}$ represents the derivative matrix with elements $G^i{}_\alpha = (\partial g^i / \partial m^\alpha)_{\mathbf{m}_{\mathrm{prior}}}$. The posterior probability density $\sigma_M(\mathbf{m})$ is approximately Gaussian. In the fourth sketch, the forward equation $\mathbf{d} = \mathbf{g}(\mathbf{m})$ can be linearized around the maximum likelihood point, $\mathbf{m}_{\mathrm{ML}}$: $\mathbf{g}(\mathbf{m}) \simeq \mathbf{g}(\mathbf{m}_{\mathrm{ML}}) + \mathbf{G}(\mathbf{m} - \mathbf{m}_{\mathrm{ML}})$, where now $\mathbf{G}$ represents the derivative operator with elements $G^i{}_\alpha = (\partial g^i / \partial m^\alpha)_{\mathbf{m}_{\mathrm{ML}}}$. The point $\mathbf{m}_{\mathrm{ML}}$ has to be obtained by the nonquadratic minimization of $S(\mathbf{m}) = \frac{1}{2}((\mathbf{g}(\mathbf{m}) - \mathbf{d}_{\mathrm{obs}})^t \, \mathbf{C}_\mathrm{D}^{-1}\,(\mathbf{g}(\mathbf{m}) - \mathbf{d}_{\mathrm{obs}}) + (\mathbf{m} - \mathbf{m}_{\mathrm{prior}})^t\,\mathbf{C}_\mathrm{D}^{-1}\,(\mathbf{m} - \mathbf{m}_{\mathrm{prior}}))$. In the fifth sketch, the forward equation $\mathbf{d} = \mathbf{g}(\mathbf{m})$ cannot be linearized, so the a posteriori probability density may be far from a Gaussian and special methods must be used (see text). In the last sketch, the nonlinearities between the parameters are so strong that the methods proposed in this elementary text cannot be used.*

with

$$2\,S(\mathbf{m}) \;=\; \|\,\mathbf{G}\,\mathbf{m} - \mathbf{d}_{\mathrm{obs}}\,\|_\mathrm{D}^2 + \|\,\mathbf{m} - \mathbf{m}_{\mathrm{prior}}\,\|_\mathrm{M}^2$$

$$= (\mathbf{G}\,\mathbf{m} - \mathbf{d}_{\mathrm{obs}})^t\,\mathbf{C}_\mathrm{D}^{-1}\,(\mathbf{G}\,\mathbf{m} - \mathbf{d}_{\mathrm{obs}}) + (\mathbf{m} - \mathbf{m}_{\mathrm{prior}})^t\,\mathbf{C}_\mathrm{M}^{-1}\,(\mathbf{m} - \mathbf{m}_{\mathrm{prior}}) \quad .$$

$$(3.35)$$

As the misfit function is quadratic in $\mathbf{m}$, the posterior probability density $\sigma_M(\mathbf{m})$ is, in fact, a Gaussian probability density, so there must be a point $\widetilde{\mathbf{m}}$ and a covariance matrix $\widetilde{\mathbf{C}}_M$ such that the posterior probability density can be written

$$\sigma_M(\mathbf{m}) \;=\; \text{const. } \exp\!\left(-\tfrac{1}{2}\,(\mathbf{m} - \widetilde{\mathbf{m}})^t\,\widetilde{\mathbf{C}}_\mathrm{M}^{-1}\,(\mathbf{m} - \widetilde{\mathbf{m}})\right) \quad . \qquad (3.36)$$

The basic problem, in this linear case, is then the evaluation of the center $\widetilde{\mathbf{m}}$ and the covariance $\widetilde{\mathbf{C}}_M$ of the (Gaussian) posterior covariance probability density $\sigma_M(\mathbf{m})$.

The demonstration shall be carried out in a moment. Let us first give the solution: the center of the posterior Gaussian is given by any of the three equivalent expressions

$$
\begin{aligned}
\tilde{\mathbf{m}} &= (\mathbf{G}^t\,\mathbf{C}_D^{-1}\,\mathbf{G} + \mathbf{C}_M^{-1})^{-1}\,(\mathbf{G}^t\,\mathbf{C}_D^{-1}\,\mathbf{d}_{\text{obs}} + \mathbf{C}_M^{-1}\,\mathbf{m}_{\text{prior}}) \\
&= \mathbf{m}_{\text{prior}} + (\mathbf{G}^t\,\mathbf{C}_D^{-1}\,\mathbf{G} + \mathbf{C}_M^{-1})^{-1}\,\mathbf{G}^t\,\mathbf{C}_D^{-1}\,(\mathbf{d}_{\text{obs}} - \mathbf{G}\,\mathbf{m}_{\text{prior}}) \\
&= \mathbf{m}_{\text{prior}} + \mathbf{C}_M\,\mathbf{G}^t\,(\mathbf{G}\,\mathbf{C}_M\,\mathbf{G}^t + \mathbf{C}_D)^{-1}\,(\mathbf{d}_{\text{obs}} - \mathbf{G}\,\mathbf{m}_{\text{prior}})
\end{aligned} \tag{3.37}
$$

and its covariance by either of the two equivalent expressions

$$
\begin{aligned}
\tilde{\mathbf{C}}_M &= (\mathbf{G}^t\,\mathbf{C}_D^{-1}\,\mathbf{G} + \mathbf{C}_M^{-1})^{-1} \\
&= \mathbf{C}_M - \mathbf{C}_M\,\mathbf{G}^t\,(\mathbf{G}\,\mathbf{C}_M\,\mathbf{G}^t + \mathbf{C}_D)^{-1}\,\mathbf{G}\,\mathbf{C}_M \quad.
\end{aligned} \tag{3.38}
$$

The point $\tilde{\mathbf{m}}$ has been defined here as the center of the posterior Gaussian. It could have been defined as the point realizing the minimum of the least-squares misfit function in equation (3.35), hence the usual name "least-squares estimator." From this perspective, $\tilde{\mathbf{m}}$ is the 'best point' in the sense that, at the same time, it is close to the 'prior point' $\mathbf{m}_{\text{prior}}$, and the predicted data $\mathbf{G}\,\tilde{\mathbf{m}}$ are close to the observed data $\mathbf{d}_{\text{obs}}$.

Demonstrating that equations (3.37)–(3.38) actually give the mean and the covariance of the posterior probability density $\sigma_M(\mathbf{m})$ amounts to demonstrating that the misfit function (3.35) can be written

$$
2\,S(\mathbf{m}) = (\mathbf{m} - \tilde{\mathbf{m}})^t\,\tilde{\mathbf{C}}_M^{-1}\,(\mathbf{m} - \tilde{\mathbf{m}}) + K \quad, \tag{3.39}
$$

where K is a constant term (independent of $\mathbf{m}$) that is absorbed in the constant into equation (3.36). This is easily verified by introducing in (3.39) the first of expressions (3.37) and using the first of expressions (3.38). We are left with the problem of demonstrating the equivalence between the three expressions in (3.37) and the two expressions in (3.38).

Using the first of expressions (3.37) and the first of expressions (3.38), we can write $\tilde{\mathbf{m}} = \tilde{\mathbf{C}}_M\,(\mathbf{G}^t\,\mathbf{C}_D^{-1}\,\mathbf{d}_{\text{obs}} + \mathbf{C}_M^{-1}\,\mathbf{m}_{\text{prior}}) = \tilde{\mathbf{C}}_M\,\mathbf{G}^t\,\mathbf{C}_D^{-1}\,\mathbf{d}_{\text{obs}} + \tilde{\mathbf{C}}_M\,\mathbf{C}_M^{-1}\,\mathbf{m}_{\text{prior}} = \tilde{\mathbf{C}}_M\,\mathbf{G}^t\,\mathbf{C}_D^{-1}\,\mathbf{d}_{\text{obs}} + \tilde{\mathbf{C}}_M\,(\mathbf{G}^t\,\mathbf{C}_D^{-1}\,\mathbf{G} + \mathbf{C}_M^{-1} - \mathbf{G}^t\,\mathbf{C}_D^{-1}\,\mathbf{G})\,\mathbf{m}_{\text{prior}} = \tilde{\mathbf{C}}_M\,\mathbf{G}^t\,\mathbf{C}_D^{-1}\,\mathbf{d}_{\text{obs}} + \tilde{\mathbf{C}}_M\,(\tilde{\mathbf{C}}_M^{-1} - \mathbf{G}^t\,\mathbf{C}_D^{-1}\,\mathbf{G})\,\mathbf{m}_{\text{prior}} = \tilde{\mathbf{C}}_M\,\mathbf{G}^t\,\mathbf{C}_D^{-1}\,\mathbf{d}_{\text{obs}} + (\mathbf{I} - \tilde{\mathbf{C}}_M\,\mathbf{G}^t\,\mathbf{C}_D^{-1}\,\mathbf{G})\,\mathbf{m}_{\text{prior}}$, i.e., $\tilde{\mathbf{m}} = \mathbf{m}_{\text{prior}} + \tilde{\mathbf{C}}_M\,\mathbf{G}^t\,\mathbf{C}_D^{-1}\,(\mathbf{d}_{\text{obs}} - \mathbf{G}\,\mathbf{m}_{\text{prior}})$, that is, the second of expressions (3.37). The remaining two expressions are then easily obtained using the matricial identity demonstrated in Problem 6.30.

So, in the linear case $\mathbf{d} = \mathbf{G}\,\mathbf{m}$, we have started with the a priori information in the model space represented by a Gaussian centered at $\mathbf{m}_{\text{prior}}$ with covariance $\mathbf{C}_M$ and have ended with the a posteriori information in the model space represented by a Gaussian centered at $\tilde{\mathbf{m}}$ with covariance $\tilde{\mathbf{C}}_M$. The comparison of the posterior uncertainties (as represented by $\tilde{\mathbf{C}}_M$) with the prior uncertainties (as represented by $\mathbf{C}_M$) shows which parameters have been 'resolved' and by how much. Of course, if the posterior variance of a parameter is identical to the prior variance, no information has been brought by the data on this parameter per se (but its correlations with the other parameters may have changed).

More details on the interpretation and use of the posterior covariance matrix are in section 3.3.

Example 3.3. *In sections 5.6 and 5.7, the tomographic inverse problem is set, where the unknowns are functions (representing a three-dimensional medium). In all rigor, the covariance functions used there should only represent actual a priori information. One may also decide to use smooth "a priori" covariance functions with the sole objective of forcing the final solution to be smooth. Note that when using smooth "a priori" covariance functions, if the smoothness length is larger than the typical separation between the rays of a tomographic inversion, no low-resolution holes are left between the rays.*

Example 3.4. *Some observable parameters $\mathbf{d} = \{d^i\}$ are related to some model parameters $\mathbf{m} = \{m^\alpha\}$ by the linear relation $\mathbf{d} = \mathbf{G}\,\mathbf{m}$. The measurement of the observable parameters has produced the (vector) value $\mathbf{d}_{\mathrm{obs}}$, with uncertainties that can be assumed to be Gaussian, with covariance matrix $\mathbf{C}_{\mathrm{D}}$. Estimate the model parameters $\mathbf{m}$ assuming that there is no a priori information available. The solution can be obtained[47] using the first of equations (3.37), taking the limit $\mathbf{C}_{\mathrm{M}}^{-1} \to 0$ (as there is no a priori information). This gives*

$$\widetilde{\mathbf{m}} \;=\; (\mathbf{G}^t\,\mathbf{C}_{\mathrm{D}}^{-1}\,\mathbf{G})^{-1}\,(\mathbf{G}^t\,\mathbf{C}_{\mathrm{D}}^{-1}\,\mathbf{d}_{\mathrm{obs}}) \quad , \tag{3.40}$$

the uncertainties being given by (second of equations (3.38))

$$\widetilde{\mathbf{C}}_{\mathrm{M}} \;=\; (\mathbf{G}^t\,\mathbf{C}_{\mathrm{D}}^{-1}\,\mathbf{G})^{-1} \quad . \tag{3.41}$$

These equations only make sense when the matrix $\mathbf{G}^t\,\mathbf{C}_{\mathrm{D}}^{-1}\,\mathbf{G}$ is invertible, which happens when the linear system $\mathbf{d} = \mathbf{G}\,\mathbf{m}$ is overdetermined. If, furthermore, the number of data and of unknowns is identical (i.e., if the matrix $\mathbf{G}$ is squared), the solution (3.40) simplifies to

$$\widetilde{\mathbf{m}} \;=\; \mathbf{G}^{-1}\,\mathbf{d}_{\mathrm{obs}} \quad , \tag{3.42}$$

that is, the Cramer solution of the linear system $\mathbf{d}_{\mathrm{obs}} = \mathbf{G}\,\mathbf{m}$.

In Appendix 6.18, equations are given that allow us to invert data sequentially, the prior model vector and prior covariance matrix to be used in one inversion step being the posterior model vector and posterior covariance matrix of the previous step. This bears some resemblance to the Kalman filter, used in similar kinds of problems.

So far, we have only been interested in the posterior probability density for model parameters. It is easy to see that the posterior probability density in the data space, as defined in equation (1.85), is a Gaussian here,

$$\sigma_{\mathrm{D}}(\mathbf{d}) \;=\; \mathrm{const.}\,\exp\!\left(-\tfrac{1}{2}\,(\mathbf{d} - \widetilde{\mathbf{d}})^t\,\widetilde{\mathbf{C}}_{\mathrm{D}}^{-1}\,(\mathbf{d} - \widetilde{\mathbf{d}}) \right) \quad , \tag{3.43}$$

with

$$\widetilde{\mathbf{d}} \;=\; \mathbf{G}\,\widetilde{\mathbf{m}} \quad \text{and} \quad \widetilde{\mathbf{C}}_{\mathrm{D}} \;=\; \mathbf{G}\,\widetilde{\mathbf{C}}_{\mathrm{M}}\,\mathbf{G}^t \quad . \tag{3.44}$$

---

[47] That the data space is a linear space is implicit in the use of a Gaussian distribution for data uncertainties. The typical context in this type of problems makes the parameter space a linear space, so we are covered by the hypotheses made for the use of the least-squares formulation.

Quite often, the least-squares solution is justified using a statistical point of view. In this case, $\mathbf{d}$ and $\mathbf{m}$ are viewed as random variables with known covariance operators $\mathbf{C_D}$ and $\mathbf{C_M}$ and unknown means $\mathbf{d}_{true}$ and $\mathbf{m}_{true}$. Then, $\mathbf{d}_{obs}$ and $\mathbf{m}_{prior}$ are interpreted as two particular *realizations* of the random variables $\mathbf{d}$ and $\mathbf{m}$, and the problem is to obtain an estimator of $\mathbf{m}_{true}$, which is, in some sense, optimum. The Gauss–Markoff theorem (see, for instance, Plackett, 1972, or Rao, 1973) shows that, for *linear problems*, the least-squares estimator has *minimum variance* among all the estimators that are linear functions of $\mathbf{d}_{obs}$ and $\mathbf{m}_{prior}$, *irrespective of the particular form of the probability density functions of the random variables* $\mathbf{d}$ *and* $\mathbf{m}$. This is not as good as it may seem: minimum variance may be a bad criterion when the probability densities are far from Gaussian, as, for instance, when a small number of large, uncontrolled errors are present in a data set (see Problem 7.7). As the general approach developed in chapter 1 justifies the least-squares criterion only when all uncertainties (modelization uncertainties, observational uncertainties, uncertainties in the a priori model) are Gaussian, I urge the reader to limit the use of the techniques described in this chapter to the cases where this assumption is not too strongly violated.

3.2.3 Nonlinear Problems

If the equation $\mathbf{d} = \mathbf{g}(\mathbf{m})$ solving the forward problem is actually nonlinear, there is no simplification in equations (3.31)–(3.32) giving the posterior probability density in the model space,

$$\sigma_M(\mathbf{m}) = \text{const. } \exp(-S(\mathbf{m})) \quad , \tag{3.45}$$

with

$$2\,S(\mathbf{m}) = \| \mathbf{g}(\mathbf{m}) - \mathbf{d}_{obs} \|_D^2 + \| \mathbf{m} - \mathbf{m}_{prior} \|_M^2$$

$$= (\mathbf{g}(\mathbf{m}) - \mathbf{d}_{obs})^t \, \mathbf{C_D}^{-1} \, (\mathbf{g}(\mathbf{m}) - \mathbf{d}_{obs}) + (\mathbf{m} - \mathbf{m}_{prior})^t \, \mathbf{C_M}^{-1} \, (\mathbf{m} - \mathbf{m}_{prior}) \quad . \tag{3.46}$$

If $\mathbf{g}(\mathbf{m})$ is not a linear function of $\mathbf{m}$, $\sigma_M(\mathbf{m})$ is not Gaussian. The more nonlinear $\mathbf{g}(\mathbf{m})$ is, the more remote is $\sigma_M(\mathbf{m})$ from a Gaussian function.

The weakest case of nonlinearity arises when the function $\mathbf{g}(\mathbf{m})$ can be linearized around $\mathbf{m}_{prior}$ (third of the sketches in Figure 3.2),

$$\mathbf{g}(\mathbf{m}) \simeq \mathbf{g}(\mathbf{m}_{prior}) + \mathbf{G}\,(\mathbf{m} - \mathbf{m}_{prior}) \quad , \tag{3.47}$$

where

$$G^i{}_\alpha = \left(\frac{\partial g^i}{\partial m^\alpha} \right)_{\mathbf{m}_{prior}} \quad . \tag{3.48}$$

The symbol $\simeq$ in equation (3.47) means precisely that second-order terms can be neglected compared to observational and modelization uncertainties (i.e., compared with standard deviations and correlations in $\mathbf{C_D}$). Replacing (3.47) in equations (3.45)–(3.46), one sees that the a posteriori probability density is then approximately Gaussian, the center being given by

$$\tilde{\mathbf{m}} \simeq \mathbf{m}_{\text{prior}} + \left(\mathbf{G}^t\,\mathbf{C}_D^{-1}\,\mathbf{G} + \mathbf{C}_M^{-1}\right)^{-1}\mathbf{G}^t\,\mathbf{C}_D^{-1}\left(\mathbf{d}_{\text{obs}} - \mathbf{g}(\mathbf{m}_{\text{prior}})\right)$$
$$= \mathbf{m}_{\text{prior}} + \mathbf{C}_M\,\mathbf{G}^t\left(\mathbf{G}\,\mathbf{C}_M\,\mathbf{G}^t + \mathbf{C}_D\right)^{-1}\left(\mathbf{d}_{\text{obs}} - \mathbf{g}(\mathbf{m}_{\text{prior}})\right) \tag{3.49}$$

and the a posteriori covariance operator by

$$\tilde{\mathbf{C}}_M \simeq \left(\mathbf{G}^t\,\mathbf{C}_D^{-1}\,\mathbf{G} + \mathbf{C}_M^{-1}\right)^{-1} = \mathbf{C}_M - \mathbf{C}_M\,\mathbf{G}^t\left(\mathbf{G}\,\mathbf{C}_M\,\mathbf{G}^t + \mathbf{C}_D\right)^{-1}\mathbf{G}\,\mathbf{C}_M \quad. \tag{3.50}$$

These are basically equations (3.37)–(3.38), so we see that solving a linearizable problem is in fact equivalent to solving a linear problem.

In the fourth of the sketches of Figure 3.2, the case is suggested where the linearization (3.47) is no longer acceptable, but the function $\mathbf{g}(\mathbf{m})$ is still quasilinear inside the region of the $\mathbb{M} \times \mathbb{D}$ space of significant posterior probability density. The right strategy for these problems is to use some iterative algorithm to obtain the maximum likelihood point of $\sigma_M(\mathbf{m})$, say $\mathbf{m}_{\text{ML}}$, and then to use a linearization of $\mathbf{g}(\mathbf{m})$ around $\mathbf{m}_{\text{ML}}$ to estimate the a posteriori covariance operator. As the homogeneous probability density is here constant, the maximum likelihood point $\mathbf{m}_{\text{ML}}$ is just the point that maximizes $\sigma_M(\mathbf{m})$ (see discussion in section 1.6.4). As the point maximizing $\sigma_M(\mathbf{m})$ is the point minimizing the sum of squares in equation (3.46), we face here the typical problem of 'nonlinear least-squares' minimization.

Using, for instance, a quasi-Newton method (see section 3.4 and Appendix 6.22 for details on optimization techniques), the iterative algorithm

$$\boxed{\mathbf{m}_{n+1} = \mathbf{m}_n - \mu_n\left(\mathbf{G}_n^t\,\mathbf{C}_D^{-1}\,\mathbf{G}_n + \mathbf{C}_M^{-1}\right)^{-1}\left(\mathbf{G}_n^t\,\mathbf{C}_D^{-1}\left(\mathbf{d}_n - \mathbf{d}_{\text{obs}}\right) + \mathbf{C}_M^{-1}\left(\mathbf{m}_n - \mathbf{m}_{\text{prior}}\right)\right),}$$
$$\tag{3.51}$$

where $\mathbf{d}_n = \mathbf{g}(\mathbf{m}_n)$, $(\mathbf{G}_n)^i{}_\alpha = (\partial g^i/\partial m^\alpha)_{\mathbf{m}_n}$, and $\mu_n \lesssim 1$,[48] when initialized at an arbitrary point $\mathbf{m}_0$, converges to a local optimal point. If there is one global optimum, then the algorithm converges to it. If not, the algorithm must be initiated at a point $\mathbf{m}_0$ close enough to the global optimum. In many practical applications, the simple choice

$$\mathbf{m}_0 = \mathbf{m}_{\text{prior}} \tag{3.52}$$

is convenient. The number of iterations required for a quasi-Newton algorithm to provide a sufficiently good approximation of the maximum likelihood point is typically between one and one dozen. Once the maximum likelihood point $\mathbf{m}_{\text{ML}}$ has been conveniently approached, the a posteriori covariance operator can be estimated as

$$\boxed{\tilde{\mathbf{C}}_M \simeq \left(\mathbf{G}^t\,\mathbf{C}_D^{-1}\,\mathbf{G} + \mathbf{C}_M^{-1}\right)^{-1} = \mathbf{C}_M - \mathbf{C}_M\,\mathbf{G}^t\left(\mathbf{G}\,\mathbf{C}_M\,\mathbf{G}^t + \mathbf{C}_D\right)^{-1}\mathbf{G}\,\mathbf{C}_M \quad,}$$
$$\tag{3.53}$$

---

[48] As explained in sections 3.4.1 and 3.4.2, gradient-based methods need a parameter defining the length of the 'jump' to be performed at each step. If it is taken too small, the algorithm converges too slowly; if it is taken too large, the algorithm may diverge. In most situations, for the quasi-Newton algorithm, one may just take $\mu_n = 1$. In Appendix 6.22, some suggestions are made for choosing adequate values for this parameter.

where, this time, $\mathbf{G}$ are the partial derivatives taken at the convergence point, $(\mathbf{G}_n)^i{}_\alpha = (\partial g^i / \partial m^\alpha)_{\mathbf{m}_{\mathrm{ML}}}$. The main computational difference between this 'nonlinear' solution and the linearized solution mentioned above is that here, $\mathbf{g}(\mathbf{m})$, the predicted data for the current model, has to be computed at each iteration without using any linear approximation. In usual problems, it is more difficult to compute $\mathbf{g}(\mathbf{m})$ than $\mathbf{g}(\mathbf{m}_0) + \mathbf{G}(\mathbf{m} - \mathbf{m}_0)$: nonlinear problems are in general more expensive to solve than linearizable problems.

Nonlinearities may be stronger and stronger. Many inverse problems correspond to the case illustrated in the fifth sketch of Figure 3.2: there may be some local maxima of the posterior probability density $\sigma_{\mathrm{M}}(\mathbf{m})$. If the number of local optima is small, all of them can be visited, using the iterative algorithm just mentioned, and around each local optimum, the (local) covariance matrix is to be estimated as above.

If the number of local optimal points is very large, then it is better to directly make use of the Monte Carlo methods developed in chapter 2 (as, in that case, no advantage is taken of the Gaussian hypothesis, we do better to drop it and use a more realistic uncertainty modelization).

Finally, there are problems (suggested in the last sketch of Figure 3.2) where nonlinearities are so strong that some of the assumptions made here break (see the comments made in section 1.2.8 about the definition of conditional probability density and the notion of the 'vertical' uncertainty bars in the theoretical relation $\mathbf{d} = \mathbf{g}(\mathbf{m})$, represented in Figure 1.4). In these circumstances, more general methods, directly based on the notion of the 'conjunction of states of information' (see section 1.5.1) are necessary.

The quasi-Newton iterative algorithm of equation (3.51) is not the only one possible. For instance, in section 3.4 we arrive at the steepest descent algorithm (equation (3.89))

$$\mathbf{m}_{n+1} = \mathbf{m}_n - \mu_n \left(\mathbf{C}_{\mathrm{M}} \, \mathbf{G}_n^t \, \mathbf{C}_{\mathrm{D}}^{-1} \, (\mathbf{d}_n - \mathbf{d}_{\mathrm{obs}}) + (\mathbf{m}_n - \mathbf{m}_{\mathrm{prior}}) \right) \quad , \qquad (3.54)$$

where, again, μ_n is an ad hoc parameter defining the size of the jump to be performed. Contrary to quasi Newton, this algorithm does not require the resolution of a linear system at each iteration,[49] but, of course, it requires more iterations to converge. The philosophy behind the steepest descent and the Newton algorithms is explained in section 3.4 (where the 'variable metric methods' are mentioned).

Concerning equation (3.54), one may note that the operator $\mathbf{C}_{\mathrm{M}} \, \mathbf{G}_n^t \, \mathbf{C}_{\mathrm{D}}^{-1}$ is, in fact, the adjoint of $\mathbf{G}_n$ as defined in section 3.1.4,

$$\mathbf{G}_n^* = \mathbf{C}_{\mathrm{M}} \, \mathbf{G}_n^t \, \mathbf{C}_{\mathrm{D}}^{-1} \quad , \qquad (3.55)$$

so the algorithm can be written $\mathbf{m}_{n+1} = \mathbf{m}_n - \mu_n \left(\mathbf{G}_n^* (\mathbf{d}_n - \mathbf{d}_{\mathrm{obs}}) + (\mathbf{m}_n - \mathbf{m}_{\mathrm{prior}}) \right)$.

3.3 Estimating Posterior Uncertainties

3.3.1 Posterior Covariance Operator

We have seen that if the relation $\mathbf{m} \mapsto \mathbf{g}(\mathbf{m})$ is nonlinear enough, the probability density $\sigma_{\mathrm{M}}(\mathbf{m})$, as given by equations (3.31)–(3.32), may be far from a Gaussian. It has already

---

[49] It is well known in numerical analysis that, given the vector $\mathbf{y}$ and the matrix $\mathbf{A}$, the computation of $\mathbf{x} = \mathbf{A}^{-1} \mathbf{y}$ is not to be done by actually computing the inverse of the matrix $\mathbf{A}$, but by rewriting the equation as $\mathbf{A} \mathbf{x} = \mathbf{y}$ and using any of the many efficient methods existing to solve a linear system.

been mentioned that if the probability is multimodal, but has a small number of maxima, the maxima can all be searched, and a covariance matrix adjusted around each optimum point. If the number of maxima is large, the more general Monte Carlo techniques of chapter 2 must be used.

Let us assume here that $\sigma_M(\mathbf{m})$ is reasonably close to a Gaussian. In that case, we have seen that this posterior covariance matrix can be approximated by either of the two expressions (equation (3.53))

$$\widetilde{\mathbf{C}}_M \simeq \left(\mathbf{G}^t\, \mathbf{C}_D^{-1}\, \mathbf{G} + \mathbf{C}_M^{-1}\right)^{-1} = \mathbf{C}_M - \mathbf{C}_M\, \mathbf{G}^t \left(\mathbf{G}\, \mathbf{C}_M\, \mathbf{G}^t + \mathbf{C}_D\right)^{-1} \mathbf{G}\, \mathbf{C}_M \quad , \qquad (3.56)$$

where $\mathbf{G}$ is the matrix of partial derivatives taken at the convergence point, $(\mathbf{G}_n)^i{}_\alpha = (\partial g^i / \partial m^\alpha)_{\mathbf{m}_{ML}}$.

The most trivial use of the posterior covariance operator $\widetilde{\mathbf{C}}_M$ is to interpret the square roots of the diagonal elements (variances) as 'uncertainty bars' on the posterior values of the model parameters.

A direct examination of the off-diagonal elements (covariances) of a covariance operator is not easy, and it is much better to introduce the *correlations*

$$\rho^{\alpha\beta} = \frac{C^{\alpha\beta}}{\sqrt{C^{\alpha\alpha}}\sqrt{C^{\beta\beta}}} \qquad \text{(no sums involved)} \quad , \qquad (3.57)$$

which have the well-known property

$$-1 \leq \rho^{\alpha\beta} \leq +1 \quad . \qquad (3.58)$$

If the posterior correlation between parameters m^α and m^β is close to zero, the posterior uncertainties are uncorrelated (in the intuitive sense). If the correlation is close to $+1$ (resp., close to -1), the uncertainties are highly correlated (resp., anticorrelated). A strong correlation on uncertainties means that the two parameters have not been independently resolved by the data set and that only some linear combination of the parameters is resolved.

Sometimes, the parameters $m^1, m^2, \dots$ represent the discretized values of some spatial (or temporal) function. Each row (or column) of the posterior covariance operator can then directly be interpreted in terms of spatial (or temporal) correlations. See Figure 5.17 for an example.

The human brain is not very good at interpreting covariances in high-dimensional problems. But it is very good at comparing random samples of a probability distribution. Knowing this, the usual presentation of 'the solution' of a least-squares problem (in fact, the mean of the posterior Gaussian), together with the covariances (as an expression of 'uncertainties' in the solution), should systematically be replaced with a better presentation. Given the mean $\widetilde{\mathbf{m}}$ and the covariance $\widetilde{\mathbf{C}}_M$ of the posterior Gaussian, one should generate pseudorandom samples $\mathbf{m}_1, \mathbf{m}_2, \dots, \mathbf{m}_K$ of the probability density $\sigma_M(\mathbf{m}) = \text{Gaussian}(\mathbf{m}, \widetilde{\mathbf{m}}, \widetilde{\mathbf{C}}_M)$ and present the samples $\{\mathbf{m}_1, \mathbf{m}_2, \dots, \mathbf{m}_K\}$ instead. How many? A quantity of models large enough to convey all the subtleties present in the probability distribution is needed. In any case, given the samples $\{\mathbf{m}_1, \mathbf{m}_2, \dots, \mathbf{m}_K\}$, information about $\widetilde{\mathbf{m}}$ could be obtained as $\widetilde{\mathbf{m}} \approx \frac{1}{K} \sum_{n=1}^{K} \mathbf{m}_n$, and information about $\widetilde{\mathbf{C}}_M$ could be obtained as $\widetilde{\mathbf{C}}_M \approx \frac{1}{K} \sum_{n=1}^{K} (\mathbf{m}_n - \widetilde{\mathbf{m}})(\mathbf{m}_n - \widetilde{\mathbf{m}})^t$.

This presentation of the results of a least-squares inversion as a sequence of samples would not only help the human observer to grasp the actual information obtained in the inversion process but also help in the confrontation of the results obtained by different teams working on the same inverse problem using different data sets. It is not because one is working inside a least-squares context that the 'movie strategy' presented in chapter 2 ceases to be valid.

3.3.2 Resolution Operator

In the approaches to inversion not directly based on probabilistic concepts, it is usual to introduce the 'resolution operator.' In order to make the link with these methods, let me briefly introduce this concept.

Assume we face a linear problem, $\mathbf{d} = \mathbf{G}\,\mathbf{m}$, and that we have the a priori information $\{\mathbf{m}_{\text{prior}}, \mathbf{C}_M\}$ and the observations $\{\mathbf{d}_{\text{obs}}, \mathbf{C}_D\}$. The least-squares solution is (third of equations (3.37))

$$\widetilde{\mathbf{m}} \;=\; \mathbf{m}_{\text{prior}} + \mathbf{C}_M\,\mathbf{G}^t\left(\mathbf{G}\,\mathbf{C}_M\,\mathbf{G}^t + \mathbf{C}_D\right)^{-1}\left(\mathbf{d}_{\text{obs}} - \mathbf{G}\,\mathbf{m}_{\text{prior}}\right) \quad, \tag{3.59}$$

where (second of equations (3.38))

$$\widetilde{\mathbf{C}}_M \;=\; \mathbf{C}_M - \mathbf{C}_M\,\mathbf{G}^t\left(\mathbf{G}\,\mathbf{C}_M\,\mathbf{G}^t + \mathbf{C}_D\right)^{-1}\mathbf{G}\,\mathbf{C}_M \quad. \tag{3.60}$$

Assume that instead of using as input the the observed values $\mathbf{d}_{\text{obs}}$, which are uncertain, we create some exact data $\mathbf{d}_{\text{exact}}$ from an artificial "exact" model $\mathbf{m}_{\text{exact}}$:

$$\mathbf{d}_{\text{exact}} \;=\; \mathbf{G}\,\mathbf{m}_{\text{exact}} \quad. \tag{3.61}$$

The solution $\widetilde{\mathbf{m}}$ produced by the algorithm would then clearly satisfy

$$\widetilde{\mathbf{m}} - \mathbf{m}_{\text{prior}} \;=\; \mathbf{R}\,(\mathbf{m}_{\text{exact}} - \mathbf{m}_{\text{prior}}) \quad, \tag{3.62}$$

where $\mathbf{R} = \mathbf{C}_M\,\mathbf{G}^t\left(\mathbf{G}\,\mathbf{C}_M\,\mathbf{G}^t + \mathbf{C}_D\right)^{-1}\mathbf{G}$, or, using the expression (3.60) for the posterior covariance operator,

$$\mathbf{R} \;=\; \mathbf{I} - \widetilde{\mathbf{C}}_M\,\mathbf{C}_M^{-1} \quad. \tag{3.63}$$

Equation (3.62) suggests calling $\mathbf{R}$ (following Backus and Gilbert, 1968) the *resolution operator*, as it can be interpreted as a filter: the corrections to the prior model that we obtain, $\widetilde{\mathbf{m}} - \mathbf{m}_{\text{prior}}$, are both identical to the exact deviations $\mathbf{m}_{\text{exact}} - \mathbf{m}_{\text{prior}}$, but are 'filtered' by $\mathbf{R}$ (equation (3.62)). (I don't like this way of thinking, but let us continue with the traditional reasoning.) If the resolution operator were the identity operator, we would have perfectly resolved the 'exact model.' The farther the resolution operator is from the identity, the worse the resolution is: we cannot see the real world; we can only see a filtered version. For more details (in linearized problems), the reader is referred to Backus and Gilbert (1968). Examples are also given by Aki and Lee (1976) and by Aki, Christofferson, and Husebye (1977).

I prefer to rewrite equation (3.63) as

$$\tilde{\mathbf{C}}_\mathrm{M} = (\mathbf{I} - \mathbf{R}) \, \mathbf{C}_\mathrm{M} \quad . \tag{3.64}$$

If the resolution operator $\mathbf{R}$ is close to the identity, the posterior covariance is close to zero, and we have resolved our parameters well.

Although these developments have been made for a linear problem, it is clear that they remain approximately valid for mildly nonlinear problems.

As a final comment, note that taking the trace of equation (3.63) gives

$$\mathrm{tr}\,\mathbf{I} = \mathrm{tr}\,\mathbf{R} + \mathrm{tr}\,(\tilde{\mathbf{C}}_\mathrm{M}\,\mathbf{C}_\mathrm{M}^{-1}) \quad , \tag{3.65}$$

an equation that can be broadly interpreted as follows:

$$\begin{bmatrix} \text{total number} \\ \text{of model parameters} \end{bmatrix} = \begin{bmatrix} \text{number of parameters} \\ \text{resolved by} \\ \text{the data set} \end{bmatrix} + \begin{bmatrix} \text{number of parameters} \\ \text{resolved by} \\ \text{the a priori information} \end{bmatrix} . \tag{3.66}$$

3.3.3 Eigenvector Analysis

Strictly speaking, a covariance operator has no eigenvalues nor eigenvectors: the eigenvector-eigenvalue decomposition is only defined for a linear 'automorphism,' i.e., for a linear operator mapping one space into itself, and we have seen that a covariance operator maps a vector space into its dual. In practical computations where the eigenvector-eigenvalue decomposition is carelessly applied to a covariance operator, inconsistencies with the physical units being manipulated may appear.

So, if we have the prior covariance operator $\mathbf{C}_\mathrm{M}$ and the posterior covariance operator $\tilde{\mathbf{C}}_\mathrm{M}$, a well-formed eigenvector-eigenvalue equation is[50]

$$\tilde{\mathbf{C}}_\mathrm{M}\,\delta\mathbf{m} = \lambda\,\mathbf{C}_\mathrm{M}\,\delta\mathbf{m} \quad . \tag{3.67}$$

To interpret this equation, let us ask the following question: Among those vector perturbations $\delta\mathbf{m}$ that have unit length when measured with respect to the prior covariance,

$$\| \, \delta\mathbf{m} \, \|_0^2 = \delta\mathbf{m}^t\,\mathbf{C}_\mathrm{M}^{-1}\,\delta\mathbf{m} = 1 \quad , \tag{3.68a}$$

which ones have extremal length when measured with respect to the posterior covariance:

$$\| \, \delta\mathbf{m} \, \|^2 = \delta\mathbf{m}^t\,\tilde{\mathbf{C}}_\mathrm{M}^{-1}\,\delta\mathbf{m} \qquad \text{extremal ?} \tag{3.68b}$$

Using the method of Lagrange's multipliers (see Appendix 6.29), the problem is solved by optimizing $S(\delta\mathbf{m}, \lambda) = \delta\mathbf{m}^t\,\tilde{\mathbf{C}}_\mathrm{M}^{-1}\,\delta\mathbf{m} - \lambda\,(\delta\mathbf{m}^t\,\mathbf{C}_\mathrm{M}^{-1}\,\delta\mathbf{m} - 1)$. The condition $\partial S/\partial\delta\mathbf{m} = 0$ directly leads to equation (3.67). Therefore, *the eigenvector solution of equation (3.67) gives the directions in the model space that have extremal ratio between the prior and the posterior length.*

---

[50]To use standard computer software, one may obviously rewrite this equation as $(\mathbf{C}_\mathrm{M}^{-1}\,\tilde{\mathbf{C}}_\mathrm{M})\,\delta\mathbf{m} = \lambda\,\delta\mathbf{m}$.

The eigenvector associated with the maximum eigenvalue is, among all vectors with a priori length given, the one with maximum a posteriori length. This means that it is directed along the shortest axis of the ellipsoid of uncertainties representing $\widetilde{\mathbf{C}}_M$ (with respect to the metric defined by $\mathbf{C}_M$). It therefore corresponds to a linear combination of parameters that is well resolved by the data. On the contrary, the eigenvector associated with the smallest eigenvalue is directed along the largest axis of the ellipsoid. It therefore corresponds to a linear combination of parameters that is poorly resolved by the data.

Wiggins (1972) emphasized the importance of the eigenvector-eigenvalue analysis for the identification of well-resolved parameters. Unfortunately, such an analysis is linear (or linearized), and the most interesting problems concerning the choice of parameters are nonlinear: which *nonlinear* change of parameters defines good parameters? The linearized eigenvector-eigenvalue analysis does not address this problem (which is a difficult problem without a known general solution).

3.3.4   Are the Residuals Too Large?

It may happen that some of the assumptions are violated. For instance, observational uncertainties or modeling uncertainties may be underestimated, or too much confidence may be given to the a priori model. Often, blunders exist in the data set: the Gaussian assumption is then not adequate, and other long-tailed distributions should have been chosen (see chapter 4).

It is generally not very easy to check the correctness of the assumptions. The examination of the residuals $\mathbf{d}_{obs} - \mathbf{g}(\mathbf{m}_{ML})$ and $\mathbf{m}_{prior} - \mathbf{m}_{ML}$ may be of some help. The most important is, of course, a qualitative examination, after a convenient display, but some easy numerical tests can be performed (see, for instance, Draper and Smith, 1998). The easiest concerns the value of the misfit function at the minimum, $S(\mathbf{m}_{ML})$.

In linear problems, it easily follows from the results exposed in Appendix 6.8 that the the minimum of (twice) the misfit function

$$\chi^2 = 2\,S = (\mathbf{G}\,\mathbf{m} - \mathbf{d}_{obs})^t\,\mathbf{C}_D^{-1}\,(\mathbf{G}\,\mathbf{m} - \mathbf{d}_{obs}) + (\mathbf{m} - \mathbf{m}_{prior})^t\,\mathbf{C}_M^{-1}\,(\mathbf{m} - \mathbf{m}_{prior})$$

$$(3.69)$$

arising in linear inverse problems is distributed following a χ^2 distribution with

$$\nu = \text{dimension of the data space } \mathbb{D} \qquad (3.70)$$

degrees of freedom. As mentioned in Appendix 6.8, the value of (twice) the misfit function at the minimum can be obtained by[51]

$$\chi^2 = 2\,S_{min} = (\mathbf{G}\,\mathbf{m}_{prior} - \mathbf{d}_{obs})^t\,(\mathbf{G}\,\mathbf{C}_M\,\mathbf{G}^t + \mathbf{C}_D)^{-1}(\mathbf{G}\,\mathbf{m}_{prior} - \mathbf{d}_{obs}) \quad . \qquad (3.71)$$

If, in the effective resolution of an inverse problem, a too large value of χ^2 is obtained ("too large" being understood with respect to the chi-squared probability density (see Appendix 6.8)), then some violation of the hypothesis has to be feared. Also, an improbably small value of χ^2 would suggest uncertainty overestimation.

---

[51]This expression can easily be transformed using the matrix identity $(\mathbf{G}\,\mathbf{C}_M\,\mathbf{G}^t + \mathbf{C}_D)^{-1} = \mathbf{C}_D^{-1} - \mathbf{C}_D^{-1}\,\mathbf{G}\,(\mathbf{G}^t\,\mathbf{C}_D^{-1}\,\mathbf{G} + \mathbf{C}_M^{-1})^{-1}\,\mathbf{G}^t\,\mathbf{C}_D^{-1}$.

The previous result applies only to linear problems. In the quasi-linear problems where least squares applies, the result remains approximately true but has to be used with caution.

3.4 Least-Squares Gradient and Hessian

3.4.1 Gradient and Direction of Steepest Ascent

Let $\mathfrak{M}$ be a manifold, $\mathbf{m}$ be one of its points, $\{m^\alpha\} = \{m^1, m^2, \dots\}$ be the coordinates of the point, and $\mu(\mathbf{m})$ be the homogeneous probability density over $\mathfrak{M}$. If $f(\mathbf{m})$ is now an arbitrary probability density over $\mathfrak{M}$, the maximum likelihood point is the point satisfying the condition (see equation (1.116))

$$\frac{f(\mathbf{m})}{\mu(\mathbf{m})} \quad \text{maximum} \quad . \tag{3.72}$$

Equivalently, defining the 'misfit function'

$$S(\mathbf{m}) = -\log \frac{f(\mathbf{m})}{\mu(\mathbf{m})} \quad , \tag{3.73}$$

the condition is

$$S(\mathbf{m}) \quad \text{minimum} \quad . \tag{3.74}$$

As explained above, gradient-like methods generally work much better for this second optimization problem.

The gradient of the misfit function — at a given point of the manifold — is the (local) linear form $\hat{\gamma}_\alpha$ whose components are

$$\hat{\gamma}_\alpha = \frac{\partial S}{\partial m^\alpha} \quad . \tag{3.75}$$

By itself, the gradient does not define any direction (at the considered point) in the manifold (the difference between a vector and a form was suggested in Figure 3.1).

Should one define a metric over the manifold, i.e., should one introduce a metric matrix $g_{\alpha\beta}$ such that the squared distance between the point $\mathbf{m} = \{m^1, m^2, \dots\}$ and the point $\mathbf{m} + \delta\mathbf{m} = \{m^1 + \delta m^1, m^2 + \delta m^2, \dots\}$ is

$$ds^2 = g_{\alpha\beta} \, dm^\alpha \, dm^\beta \quad , \tag{3.76}$$

then the inverse metric matrix $g^{\alpha\beta}$ (defined by the condition $g_{\alpha\sigma} g^{\sigma\beta} = \delta_\alpha^\beta$) allows us to define the vector

$$\gamma^\alpha = g^{\alpha\beta} \gamma_\beta \quad , \tag{3.77}$$

which can be shown (see Appendix 6.22) to correspond to the *steepest ascent vector* (characterizing the maximum increase of S for a small circle [in the sense defined by the metric $g_{\alpha\beta}$] around the considered point). The steepest ascent vector is then

$$\gamma^\alpha = g^{\alpha\beta} \frac{\partial S}{\partial m^\alpha} \quad . \tag{3.78}$$

An iterative algorithm of the form $\mathbf{m}_{n+1} = \mathbf{m}_N + \delta\mathbf{m}_n$ able to find the minimum of the function $S(\mathbf{m})$ is a *steepest descent algorithm* if it is has the form

$$\mathbf{m}_{n+1} = \mathbf{m}_n - \mu_n \,\boldsymbol{\Delta}(\mathbf{m}_n) \quad , \tag{3.79}$$

i.e.,

$$\boxed{m^\alpha_{n+1} = m^\alpha_n - \mu_n \, g^{\alpha\beta}(\mathbf{m}_n) \left(\frac{\partial S}{\partial m^\beta}\right)_{\mathbf{m}_n} \quad ,} \tag{3.80}$$

where the μ_n are ad hoc real constants (small enough to avoid divergence of the algorithm and large enough to allow the algorithm to actually advance).

One sometimes finds in the literature the pseudoalgorithm

$$m^\alpha_{n+1} = m^\alpha_n - \mu_n \left(\frac{\partial S}{\partial m^\alpha}\right)_{\mathbf{m}_n} \qquad \text{(this is wrong)} \quad , \tag{3.81}$$

where the inverse metric matrix $g^{\alpha\beta}$ is absent. This does not have the necessary mathematical invariances: it is dimensionally wrong (if the parameters have different physical dimensions), and the algorithm is coordinate dependent (if it converges at all, it may typically display a complex behavior (see Figure 3.3)).

Figure 3.3. *A function $S(\theta, \varphi)$ defined over the sphere has to be minimized. A naive use of the gradient method (equation (3.81)) would define a direction of steepest descent based on an ad hoc Euclidean distance over the map, instead of using the actual notion of distance over the sphere, where the metric is $ds^2 = g_{ij}\,dx^i\,dx^j = d\theta^2 + \sin^2\theta\,d\varphi^2$. The algorithm in equation (3.80) does exactly this.*

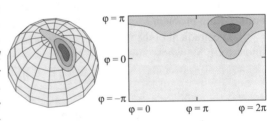

3.4.2 Newton Method of Optimization

The Newton method of optimization does not use a preexisting metric over the manifold: it uses the second derivatives of the function to be minimized to create an ad hoc metric that is optimum for the optimization problem at hand. The *Hessian matrix* is the matrix of second-order partial derivatives

$$\boxed{C_{\alpha\beta}(\mathbf{m}) = \left(\frac{\partial\hat{\gamma}_\beta}{\partial m^\alpha}\right)_{\mathbf{m}} = \left(\frac{\partial^2 S}{\partial m^\alpha \partial m^\beta}\right)_{\mathbf{m}} \quad .} \tag{3.82}$$

It is obviously symmetric. If it is positive definite, it can be used as a local metric. Introducing its inverse $\mathbf{W} = \mathbf{C}^{-1}$, one can define, from the same gradient $\hat{\gamma}_\alpha(\mathbf{m})$ as above, the new steepest ascent vector (related to the Hessian metric) $\gamma'^\alpha = W^{\alpha\beta}\,\hat{\gamma}_\beta$. The *steepest descent algorithm* then takes the form

$$m^\alpha_{n+1} = m^\alpha_n - \mu_n\,\gamma'^\alpha(\mathbf{m}_n)\quad,\tag{3.83}$$

i.e.,

$$m^\alpha_{n+1} = m^\alpha_n - \mu_n\,W^{\alpha\beta}(\mathbf{m}_n)\left(\frac{\partial S}{\partial m^\beta}\right)_{\mathbf{m}_n}.\tag{3.84}$$

This is the Newton method of optimization (although it is also a steepest descent algorithm, it is not called so). While the constants μ_n in the steepest descent algorithm (3.80) may take quite large or quite small values, the constants μ_n of the Newton algorithm are typically of the order of unity (because the Hessian metric already accounts for the local geometry of the misfit function). In a vast majority of circumstances, the Newton algorithm is just run setting $\mu_n = 1$.

As we are about to see, in the context of least squares, it is a variant of the Newton method that is often used (the so-called quasi-Newton method), where some terms arising in the computation of the Hessian (i.e., of the second-order partial derivatives of the misfit function) are neglected.

It is easy to see that the Newton method corresponds to obtaining at the current point $\mathbf{m}_n$ the 'paraboloid' that is tangent to the function $S(\mathbf{m})$ and that has the same local curvature and jumping to the point where this tangent paraboloid reaches its minimum. Figure 3.4 gives a one-dimensional illustration of this.

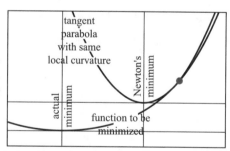

Figure 3.4. *The Newton method provides the minimum of the parabola that is tangent to the function to be minimized (and that has the same local curvature).*

For an analysis of the convergence properties of the Newton method, see, for instance, Dahlquist and Björk, 1974.

3.4.3 Variable Metric Methods and Preconditioning

We have seen that both the steepest descent method and the Newton method are steepest descent methods, the only difference being in the choice of metric to be applied to the gradient (a form) to convert the form into a direction (a vector). While in the steepest descent methods the metric is given independently of the function $S(\mathbf{m})$ to be minimized, in the Newton method the metric is defined using the second derivatives of $S(\mathbf{m})$.

This suggests that many more alternatives exist for choosing a metric. Of special practical importance are the 'variable metric methods,' where, typically, one starts iterating with the same prior metric used in the steepest descent method, and at every iteration the metric is updated in a way that it tends, as the iterations proceed, to the Newton metric. These variable metric methods are introduced in Appendix 6.22, where the algorithms obtained for the solution of least-squares problems are presented.

Gradient-based methods try to find a compromise between two somewhat contradictory pieces of information. By definition, a gradient-based method uses *local* information on the function to be optimized, i.e., information that makes full sense in a small vicinity of the current point but does not necessarily reflect the properties of the function in a large domain. But each iteration of a gradient-based method tries to make a jump as large as possible, in order to accelerate convergence. For the intended finite jumps, the local information brought by the gradient may be far from optimal. In most practical applications, the user of a gradient-based method may use physical insight to 'correct' the gradient, in order to define a direction that is much better for a finite jump. This is the idea behind the *preconditioned gradient methods* that are described in Appendix 6.22.

Choosing the right method to be used in a given least-squares inverse problem is totally problem dependent, and it is very difficult to give any suggestion at the general level. For small-sized problems, Newton methods are easy to implement and rapid to converge. For really large-sized problems, the linear system that has to be solved in the Newton method may be prohibitively expensive, and the choice of a simple steepest descent methods may sometimes work well. The experience of our research team[52] with the difficult problem of nonlinear waveform fitting, with millions of data points and model parameters, has shown that 'preconditioning operators' (i.e., metrics) can be guessed that give to simple steepest descent methods an acceptable convergence in a few iterations.

3.4.4 Steepest Descent and Quasi-Newton Method in Least Squares

If the functions $g^i(\mathbf{m})$ solving the forward problem are differentiable, i.e., if the derivatives

$$(\mathbf{G}_n)^i{}_\alpha = \left(\frac{\partial g^i}{\partial m^\alpha} \right)_{\mathbf{m}_n} \tag{3.85}$$

can be defined at any point $\mathbf{m}_n$ (or at "almost" any point), and if they can be computed,[53] then the derivatives of $S(\mathbf{m})$ can also be easily obtained, and the very powerful gradient and Newton methods can be used for minimizing $S(\mathbf{m})$.

Equation (3.46) can be written explicitly as (using the implicit sum convention)

$$2\,S(\mathbf{m}) = (\,g^i(\mathbf{m}) - d^i_{\text{obs}}\,)\,(\mathbf{C}_{\text{D}}^{-1})_{ij}\,(\,g^j(\mathbf{m}) - d^j_{\text{obs}}\,) + (\,m^\alpha - m^\alpha_{\text{prior}}\,)\,(\mathbf{C}_{\text{M}}^{-1})_{\alpha\beta}\,(\,m^\beta - m^\beta_{\text{prior}}\,)\ . \tag{3.86}$$

We obtain easily (using the implicit sum convention)

$$\left(\frac{\partial S}{\partial m^\alpha} \right)_{\mathbf{m}_n} = (\mathbf{G}_n)^i{}_\alpha\,(\mathbf{C}_{\text{D}}^{-1})_{ij}\,(\,g^j(\mathbf{m}_n) - d^j_{\text{obs}}\,) + (\mathbf{C}_{\text{M}}^{-1})_{\alpha\beta}\,(\,m^\beta - m^\beta_{\text{prior}}\,)\ , \tag{3.87}$$

---

[52]Pica, Diet, and Tarantola, 1990; Crase et al., 1990, 1992; Igel, Djikpéssé, and Tarantola, 1996; Djikpéssé et al., 1999; Charara, Barnes, and Tarantola, 2000.

[53]If they are not available analytically, one may use a finite-difference approximation.

i.e., in compact notation,

$$\left(\frac{\partial S}{\partial \mathbf{m}}\right)_n = \mathbf{G}_n^t \, \mathbf{C}_D^{-1} \, (\mathbf{g}(\mathbf{m}_n) - \mathbf{d}_{obs}) + \mathbf{C}_M^{-1} \, (\mathbf{m}_n - \mathbf{m}_{prior}) \quad . \tag{3.88}$$

The form obtained in (3.87) and (3.88) is the gradient of $S(\mathbf{m})$ at $\mathbf{m} = \mathbf{m}_n$.

We have just seen that, to use the steepest descent method, we need the metric matrix of the space (equation (3.80)). The only available metric here (that is independent of the form of the misfit function) is the covariance matrix $\mathbf{C}_M$, and, with this choice, the steepest descent algorithm (3.80) becomes, using the gradient obtained in equation (3.88) (see Appendix 6.22 for details),

$$\boxed{\mathbf{m}_{n+1} = \mathbf{m}_n - \mu_n \, (\mathbf{C}_M \, \mathbf{G}_n^t \, \mathbf{C}_D^{-1} \, (\mathbf{d}_n - \mathbf{d}_{obs}) + (\mathbf{m}_n - \mathbf{m}_{prior})) \quad ,} \tag{3.89}$$

where $\mathbf{d}_n = \mathbf{g}(\mathbf{m}_n)$.

If we wish, instead, to use the Newton method, we need to evaluate the Hessian, i.e., the matrix of second-order partial derivatives

$$\left(\frac{\partial S^2}{\partial \mathbf{m}^2}\right)_{\alpha\beta} = \frac{\partial}{\partial m^\beta} \frac{\partial S}{\partial m^\alpha} = \frac{\partial^2 S}{\partial m^\alpha \partial m^\beta} \quad . \tag{3.90}$$

From equation (3.87) we readily obtain

$$\left(\frac{\partial^2 S}{\partial \mathbf{m}^2}(\mathbf{m}_n)\right)_{\alpha\beta} = (G_n)^i{}_\alpha (C_D^{-1})_{ij} (G_n)^j{}_\beta + (C_M^{-1})_{\alpha\beta} + \left(\frac{\partial G^i{}_\alpha}{\partial m^\beta}\right)_n (C_D^{-1})_{ij} (g^j(\mathbf{m}_n) - d_{obs}^j) \quad . \tag{3.91}$$

The last term is small if (i) the residuals are small, or (ii) the nonlinearities in the forward equation $\mathbf{m} \mapsto \mathbf{g}(\mathbf{m})$ are small (as, then, the term $\partial G^i{}_\alpha / \partial m^\beta$ is small). As the last term in equation (3.91) is, in general, small, it is difficult to handle, and as descent methods work well even if the descent direction being used is not that of steepest descent, this last term is generally dropped off, thus giving the approximation $\left(\frac{\partial^2 S}{\partial \mathbf{m}^2}(\mathbf{m}_n)\right)_{\alpha\beta} \approx$ $(G_n)^i{}_\alpha (C_D^{-1})_{ij} (G_n)^j{}_\beta + (C_M^{-1})_{\alpha\beta}$, or, in compact form,

$$\left(\frac{\partial S^2}{\partial \mathbf{m}^2}\right)_{\mathbf{m}_n} \simeq \mathbf{G}_n^t \, \mathbf{C}_D^{-1} \, \mathbf{G}_n + \mathbf{C}_M^{-1} \quad . \tag{3.92}$$

With this approximation for the Hessian, the Newton algorithm (called *quasi Newton* because of this approximation) in equation (3.84) becomes the expression already suggested in equation (3.51):

$$\boxed{\mathbf{m}_{n+1} = \mathbf{m}_n - \mu_n \left(\mathbf{G}_n^t \, \mathbf{C}_D^{-1} \, \mathbf{G}_n + \mathbf{C}_M^{-1}\right)^{-1} \left(\mathbf{G}_n^t \, \mathbf{C}_D^{-1} \left(\mathbf{d}_n - \mathbf{d}_{obs}\right) + \mathbf{C}_M^{-1} \left(\mathbf{m}_n - \mathbf{m}_{prior}\right)\right) \quad .} \tag{3.93}$$

Here, $\mathbf{d}_n = \mathbf{g}(\mathbf{m}_n)$, and, as already mentioned, $\mu_n \approx 1$.

3.4.5 Comments on Numerical Implementation

As mentioned in footnote 49, an expression like $\mathbf{x} = \mathbf{A}^{-1}\mathbf{y}$ does not mean that, given $\mathbf{A}$ and $\mathbf{y}$, one needs to invert the matrix $\mathbf{A}$ in order to evaluate $\mathbf{x}$. Rather, a direct method for solving the linear system $\mathbf{A}\mathbf{x} = \mathbf{y}$ is to be used (as any book on numerical analysis will explain; see, for instance, Ciarlet, 1982).

For instance, to implement the quasi-Newton algorithm (3.93), one should evaluate

$$\begin{aligned} \mathbf{A}_n &= \mathbf{G}_n^t\,\mathbf{C}_D^{-1}\,\mathbf{G}_n + \mathbf{C}_M^{-1} \quad , \\ \mathbf{y}_n &= \mathbf{G}_n^t\,\mathbf{C}_D^{-1}\,(\mathbf{d}_n - \mathbf{d}_{\text{obs}}) + \mathbf{C}_M^{-1}\,(\mathbf{m}_n - \mathbf{m}_{\text{prior}}) \end{aligned} \tag{3.94}$$

and update the model $\mathbf{m}_n$ as

$$\mathbf{m}_{n+1} = \mathbf{m}_n - \mu_n\,\mathbf{x}_n \quad , \tag{3.95}$$

where $\mathbf{x}_n$ is to be evaluated by solving the linear system

$$\mathbf{A}_n\,\mathbf{x}_n = \mathbf{y}_n \tag{3.96}$$

directly, without inverting $\mathbf{A}_n$. The same comment applies to every occurrence of the matrices $\mathbf{C}_D^{-1}$ and $\mathbf{C}_M^{-1}$: in the expressions above, the algorithm can be rewritten in such a way that the inversion of a matrix is always replaced with the (easier) resolution of a linear system.

At the time of writing the second edition of this book, general purpose mathematical software[54] (like *Matlab* or *Mathematica*) are becoming mature enough to be useful for the resolution of medium-scale optimization problems. My own experience is that the resolution of a large-scale inverse problem still requires a lot of heavy programming to develop codes that are well adapted to the particular problem being considered.

The so-called Levenberg–Marquardt (LM) methods (due to Levenberg, 1944, and Marquardt, 1963) are not developed here. This is because I believe that the proper introduction of the a priori information, as done above, conveniently replaces the dumping philosophy behind the LM methods. For a recent textbook containing (among other things) a review of these methods, see Aster, Borchers, and Thurber (2003).

---

[54]Not to mention special purpose software, like *Math Optimizer* or *Global Optimization*.

Chapter 4

Least-Absolute-Values Criterion and Minimax Criterion

> When a traveler reaches a fork in the road,
> the ℓ_1-norm tells him to take either one way or the other,
> but the ℓ_2-norm instructs him to head off into the bushes.
>
> *John F. Claerbout and Francis Muir*, 1973

Because of its simplicity, the least-squares criterion (ℓ_2-norm criterion) is widely used for the resolution of inverse problems, even if its basic underlying hypothesis (Gaussian uncertainties) is not always satisfied. Between least-squares and general problems there is a limited class of problems that remain simple to formulate: those based on an ℓ_p-norm ($1 \leq p \leq \infty$).

As suggested in chapter 1, when outliers are suspected in a data set, long-tailed[55] probability density functions should be used to model uncertainties (see Problem 7.7). A typical long-tailed probability density is the Laplace function, i.e., the symmetric exponential function $\exp(-|x|)$. It has the advantage of leading to results intimately related to the concept of the ℓ_1-norm, so that relatively simple mathematics is available for solving the problem. The results obtained using the minimum ℓ_1-norm (least-absolute-values) criterion are known to be sufficiently insensitive to outliers (i.e., to be *robust*).

The ℓ_∞-norm criterion arises when we use boxcar functions to model the probability density for uncertainties. This assumes a strict control on errors, as for instance when they are due to *rounding* the last digit used (see Problem 7.4).

4.1 Introduction

The ℓ_1-norm criterion has been used by Laplace and Gauss. In the words of Gauss (1809), "Laplace made use of another principle for the solution of linear equations, the number

---

[55]A distribution is long tailed if it tends to zero less rapidly than the Gaussian distribution when the distance between the variable and its central value tends to infinity.

of which is greater than the number of unknown quantities, which had been previously proposed by Boscovich, namely that the differences themselves, but all of them taken positively, should make up as small a sum as possible." In modern times, Claerbout and Muir (1973) have given a detailed discussion of the robustness of the ℓ_1-norm criterion for the resolution of inverse problems and have suggested a method of resolution related to the linear programming techniques.

This chapter starts by recalling the definition of the ℓ_p-norm and by introducing a natural bijection between an ℓ_p-normed space and its dual. For $1 < p < \infty$, the methods for solving inverse problems are similar to the methods used for $p = 2$ (chapter 3). For $p = 1$ and $p = \infty$, linear programming methods can be used, but I choose to mention them only in the appendices, as my experience with large-scale inverse problems shows that gradient methods perform well (and are naturally adapted to the case where the relation between data and parameters is nonlinear). Although the minimum ℓ_1-norm and minimum ℓ_∞-norm criteria are used in almost opposite circumstances, the underlying mathematics is very similar, justifying their inclusion in the same chapter.

4.2 Preamble: ℓ_p-Norms

Contrary to what was done in the previous chapter, the implicit sum convention is not used here, as the tensor notations — characteristic of differential geometry — are not adapted to the case where the metric is not ℓ_2.

4.2.1 Definition of the Weighted ℓ_p-Norm

Let $\mathbf{x} = \{x^1, \ldots, x^n\}$ be an element of an n-dimensional linear space $\mathbb{X}$ (the x^i being the components of $\mathbf{x}$ in a given basis), and let σ^i be given positive constants such that, for any i, σ^i has the same physical dimensions as x^i (so that x^i/σ^i is an adimensional real number). For $1 \leq p \leq \infty$, the (weighted) ℓ_p-norm of $\mathbf{x}$ is denoted $\| \mathbf{x} \|_p$ and is defined by

$$\| \mathbf{x} \|_p = \left(\sum_i \frac{|x^i|^p}{(\sigma^i)^p} \right)^{1/p} . \tag{4.1}$$

For $p = 1$, one has the ℓ_1-norm

$$\| \mathbf{x} \|_1 = \sum_i \frac{|x^i|}{\sigma^i} , \tag{4.2}$$

while the ℓ_∞-norm is defined as the limit of expression (4.1) when $p \to \infty$, which gives (e.g., Watson, 1980)

$$\| \mathbf{x} \|_\infty = \max_i \frac{|x^i|}{\sigma^i} . \tag{4.3}$$

It is well known that this definition verifies the usual properties of a norm:

$$\mathbf{x} \neq 0 \implies \| \mathbf{x} \|_p > 0 \quad ,$$

$$\| \alpha \mathbf{x} \|_p = | \alpha | \| \mathbf{x} \|_p \quad , \tag{4.4}$$

$$\| \mathbf{x}_1 + \mathbf{x}_2 \|_p \leq \| \mathbf{x}_1 \|_p + \| \mathbf{x}_2 \|_p \quad .$$

A (multidimensional) *sphere* of radius R centered at $\mathbf{x}_0$ is defined as the set of points $\mathbf{x}$ such that $\| \mathbf{x} - \mathbf{x}_0 \|_p = R$. Figure 4.1 shows some two-dimensional circles corresponding to different ℓ_p-norms.

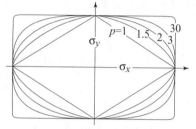

Figure 4.1. *Some unit 'circles' in the ℓ_p-norm sense. The points on the circles are such that $\frac{|x|^p}{(\sigma_x)^p} + \frac{|y|^p}{(\sigma_y)^p} = 1$ and are drawn for p respectively equal to* 1, 1.5, 2, 3, *and* 30.

4.2.2 Dual of an ℓ_p-Normed Space

In what follows, if a linear space $\mathbb{X}$ has an ℓ_p-norm defined, we denote it as $\mathbb{X}_p$.

Let χ denote a linear form over the linear space $\mathbb{X}_p$ (i.e., a linear application from $\mathbb{X}_p$ into the real line $\mathbb{R}$). The space of all linear forms over $\mathbb{X}_p$ is named the *dual* of $\mathbb{X}$ and is denoted $\mathbb{X}_q^*$ (the meaning of the index q shall become clear in a moment). The result of the action of $\chi \in \mathbb{X}_q^*$ on $\mathbf{x} \in \mathbb{X}_p$ is denoted by $\langle \chi , \mathbf{x} \rangle$ or $\chi^t \mathbf{x}$. For any form χ over a discrete space, it is possible to find constants χ_i such that

$$\langle \chi , \mathbf{x} \rangle = \chi^t \mathbf{x} = \sum_i \chi_i x^i \quad . \tag{4.5}$$

If the elements of $\mathbb{X}_p$ and $\mathbb{X}_q^*$ are represented by column matrices, the notation $\chi^t \mathbf{x}$ corresponds to the usual matrix notation; in the case where the elements of $\mathbb{X}_p$ and $\mathbb{X}_q^*$ have a more complex structure (in practice, they are represented by multidimensional arrays), the notation $\chi^t \mathbf{x}$ is still practical for analytical developments.

The space $\mathbb{X}_q^*$ can be identified with a space of vectors χ whose components are arbitrary except that the physical dimension of the ith component of χ, χ_i, is the inverse of the physical dimension of the ith component of $\mathbf{x}$, x^i (so that $\chi_i x^i$ is adimensional).

Given a particular ℓ_p-norm over a space $\mathbb{X}_p$, it is useful to define a bijection between $\mathbb{X}_p$ and its dual $\mathbb{X}_q^*$: for $1 < p < \infty$, with any given $x \in \mathbb{X}_p$, let us associate the element of $\mathbb{X}_q^*$, denoted $\hat{\mathbf{x}}$, defined by $(\hat{\mathbf{x}})_i = \frac{1}{p} \frac{\partial}{\partial x^i} (\| \mathbf{x} \|_p)^p$ or, for short,

$$\hat{\mathbf{x}} = \frac{1}{p} \frac{\partial}{\partial \mathbf{x}} (\| \mathbf{x} \|_p)^p \quad . \tag{4.6}$$

Introducing

$$\hat{\sigma}_i = \frac{1}{\sigma^i} \quad , \tag{4.7}$$

we can, for any q ($1 < q < \infty$), define an ℓ_q-norm over the dual space $\mathbb{X}_q^*$,

$$\| \hat{\mathbf{x}} \|_q = \left(\sum_i \frac{|\hat{\mathbf{x}}_i|^q}{(\hat{\sigma}_i)^q} \right)^{1/q} \quad . \tag{4.8}$$

Now, if this value q is related to p according to[56]

$$\frac{1}{p} + \frac{1}{q} = 1 \quad , \tag{4.9}$$

then the symmetric of equation (4.6) holds (see Problem 6.27.1),

$$\mathbf{x} = \frac{1}{q} \frac{\partial}{\partial \hat{\mathbf{x}}} (\| \hat{\mathbf{x}} \|_q)^q \quad , \tag{4.10}$$

and we have the following equalities (see Problem 6.27.2):

$$(\| \hat{\mathbf{x}} \|_q)^q = (\| \mathbf{x} \|_p)^p = \| \hat{\mathbf{x}} \|_q \| \mathbf{x} \|_p = \langle \hat{\mathbf{x}}, \mathbf{x} \rangle = \hat{\mathbf{x}}^t \mathbf{x} \quad . \tag{4.11}$$

Once it is understood that when the norm over the linear space $\mathbb{X}$ is an ℓ_p-norm and the norm over the dual space is the related ℓ_q-norm (with $1/p + 1/q = 1$), this set of expressions can be simplified into

$$\boxed{\| \hat{\mathbf{x}} \|^q = \| \mathbf{x} \|^p = \| \hat{\mathbf{x}} \| \| \mathbf{x} \| = \langle \hat{\mathbf{x}}, \mathbf{x} \rangle = \hat{\mathbf{x}}^t \mathbf{x} \quad .} \tag{4.12}$$

This set of (beautiful) identities justifies the bijection defined between a space and its dual and the fact that the natural norm to be considered in the dual of an ℓ_p-norm space is an ℓ_q-norm, p and q being related through expression (4.9). The cases where p or q takes the values 1 or ∞ are examined below.

   Note that the relationship (4.6) between a space and its dual is not the usual definition (e.g., Watson, 1980, exercise 1.27), $\hat{\mathbf{x}} = \frac{\partial}{\partial \mathbf{x}} \| \mathbf{x} \|_p$, in which case equation (4.12) is replaced with $\| \hat{\mathbf{x}} \|_q = 1$ and the relationship between $\mathbb{X}_p$ and $\mathbb{X}_q^*$ is no longer a bijection.

   From equations (4.6) and (4.10), we obtain the explicit representations

$$\hat{x}_i = \frac{\mathrm{sg}(x^i)}{\sigma^i} \left| \frac{x^i}{\sigma^i} \right|^{p-1} \quad \text{and} \quad x^i = \frac{\mathrm{sg}(\hat{x}_i)}{\hat{\sigma}_i} \left| \frac{\hat{x}_i}{\hat{\sigma}_i} \right|^{q-1} \quad , \tag{4.13}$$

where $\mathrm{sg}(x)$ is the sign of x (equal to 1, 0 or, -1, respectively, when $x > 0$, $x = 0$, and $x < 0$). To simplify analytical computations, we may introduce the following notation: letting u be an arbitrary scalar and r be a real positive number,

$$u^{\{r\}} \equiv \mathrm{sg}(u) \, | \, u \, |^r \quad . \tag{4.14}$$

---

[56]Equivalent expressions are $p + q = pq$ and $(p - 1)(q - 1) = 1$.

The usefulness of this notation comes from the following properties

$$u^{\{r\}} = v \qquad \Leftrightarrow \qquad u = v^{\{1/r\}} \quad ;$$

$$\frac{\partial}{\partial x} u^{\{r\}} = r \, | \, u \, |^{r-1} \frac{\partial u}{\partial x} \quad , \qquad \frac{\partial}{\partial x} | \, u \, |^r = r \, u^{\{r-1\}} \frac{\partial u}{\partial x} \quad . \tag{4.15}$$

In particular, using equation (4.15), the bijection between $\mathbb{X}_p$ and $\mathbb{X}_q^*$ can be rewritten in a more compact form:

$$\frac{\hat{x}_i}{\hat{\sigma}_i} = \left(\frac{x^i}{\sigma^i} \right)^{\{p-1\}} \quad , \qquad \frac{x^i}{\sigma^i} = \left(\frac{\hat{x}_i}{\hat{\sigma}_i} \right)^{\{q-1\}} \quad . \tag{4.16}$$

Figure 4.2 shows the function $u^{\{r\}}$ for different values of $r > 0$. For positive values of u, it simply corresponds to the function u^r. For negative values of u, it is possible to interpret $u^{\{r\}}$ as an interpolation of the functions u^r obtained when r is an odd integer.

Figure 4.2. *Some examples of the function $u^{\{r\}}$. For any r, it is a real function, symmetric (with respect to the origin), defined for any value of r (even for negative u), and, for given u, a continuous function of r.*

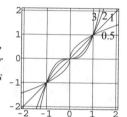

The bijection defined between $\mathbb{X}_p$ and $\mathbb{X}_q^*$ is nonlinear except for $p = 2$. In that case, $q = p = 2$ and $\hat{x}_i / \hat{\sigma}_i = x^i / \sigma^i$.

Let us now turn to the two special cases $p = 1$ and $p = \infty$, starting with $p = 1$.

If the dual space $\mathbb{X}_1^*$ is an ℓ_1-normed space (then, the primal space $\mathbb{X}_\infty$ is an ℓ_∞-normed space), the relation at the right in equation (4.13) can be used in the limit $p \to 1$ and gives the vector of the primal space

$$x^i = \frac{\text{sg}(\hat{x}_i)}{\hat{\sigma}_i} \quad , \tag{4.17}$$

where one should remember that $\text{sg}(0) \equiv 0$. This time, there is no bijection, as from the vector $\mathbf{x} = \{x^i\}$ we cannot recover the form $\hat{\mathbf{x}} = \{\hat{x}_i\}$.

If the dual space $\mathbb{X}_\infty^*$ is, instead, an ℓ_∞-normed space (then, the primal space $\mathbb{X}_1$ is an ℓ_1-normed space), the relation at the right in equation (4.13) cannot be used in the limit $p \to \infty$, as it diverges, except if the form $\hat{\mathbf{x}}$ is normed to one, $\| \hat{\mathbf{x}} \|_\infty = 1$. For in this case, some of the ratios $\hat{x}_i / \hat{\sigma}_i$ equal ± 1, the other ratios being smaller in absolute value. One then obtains, for a normed ℓ_∞-form $\hat{\mathbf{x}}$, the dual vector

$$x^i = \begin{cases} \dfrac{1}{k} \dfrac{\text{sg}(\hat{x}_i)}{\hat{\sigma}_i} & \text{if} \quad \left| \dfrac{\hat{x}_i}{\hat{\sigma}_i} \right| = 1 \quad (\,k \text{ such terms }) \quad , \\[4mm] 0 & \text{if} \quad \left| \dfrac{\hat{x}_i}{\hat{\sigma}_i} \right| < 1 \quad . \end{cases} \tag{4.18}$$

Then, $\| \mathbf{x} \|_1 = 1$. Again, the bijection between the primal and the dual case is broken here.

4.2.3 Uniqueness of the Minimization of an ℓ_p-Norm

The problem we are about to examine is, essentially, that of a minimization of an ℓ_p-norm under a constraint. This point of view helps us to gain understanding of the uniqueness of the minimum. Figure 4.3 suggests (as is indeed the case (Watson, 1980)) that *under a linear constraint* $\mathbf{F}\mathbf{x} = \mathbf{0}$ the minimization of an ℓ_p-norm $\| \mathbf{x} - \mathbf{x}_0 \|_p$ *gives a unique solution except for $p = 1$ and $p = \infty$* (the unit circle is still convex, but not strictly convex), in which cases the solution may not be unique. For a nonlinear constraint, multiple minima, secondary minima, and saddle points may exist.

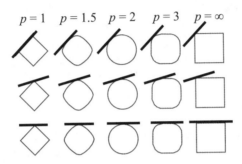

Figure 4.3. *Minimization of an ℓ_p-norm under a linear constraint. The two-dimensional problem illustrated in this figure is to obtain the point, constrained to lie on a given straight line, that is the closest to a given point. The problem can be solved geometrically by "expanding" the unit circle until it becomes tangent to the line. In the first example (top), the solution is unique except for $p = 1$; in the second example (middle), it is always unique; in the third example (bottom) it is unique except for $p = \infty$.*

4.3 The ℓ_p-Norm Problem

4.3.1 Formulation of the Problem

The formulation of the ℓ_p-norm problem is very similar to that of the least-squares problem, as done in section 3.2.1 (for easy reference, a few basic assumptions made there are repeated here).

The model space manifold and the data manifold were introduced in chapter 1. They are here assumed to be linear spaces and are respectively denoted $\mathbb{M}$ and $\mathbb{D}$. Expressions like

$$\mathbf{m}_2 + \mathbf{m}_1 \quad , \quad \mathbf{m}_2 - \mathbf{m}_1 \quad , \quad \lambda\,\mathbf{m} \quad , \quad \mathbf{d}_2 + \mathbf{d}_1 \quad , \quad \mathbf{d}_2 - \mathbf{d}_1 \quad , \quad \lambda\,\mathbf{d} \qquad (4.19)$$

are assumed to make (invariant) sense. One should remember Example 3.2, an example of coordinates where the assumption is not (directly) satisfied.

We are going to manipulate probability densities over $\mathbb{M}$ and $\mathbb{D}$. As mentioned in chapter 1, the homogeneous probability density over a linear space is always constant. Therefore, as was the case in chapter 3, the two homogeneous probability densities over the model space and the data space are constant here:

$$\mu_M(\mathbf{m}) = \text{const.} \quad , \quad \mu_D(\mathbf{d}) = \text{const.} \tag{4.20}$$

It is clear that these constant probability densities can be interpreted as the limit of the generalized Gaussians to be considered below when their 'uncertainties' are taken to be infinite.

ℓ_p-norm techniques arise when all the 'input' probability densities are assumed to be generalized Gaussians. Let us see this with some detail.

The elements of our problem are as follows:

- Some a priori information on the model parameters, represented by a generalized Gaussian probability density of order r centered at a point $\{m^\alpha_{\text{prior}}\}$ and with estimators of dispersion (of order r) $\{\sigma^\alpha_M\}$. The a priori probability density over the model space $\mathbb{M}$ is, therefore,

$$\rho_M(\mathbf{m}) = \text{const. } \exp\left(-\frac{1}{r} \sum_\alpha \frac{|m^\alpha - m^\alpha_{\text{prior}}|^r}{(\sigma^\alpha_M)^r} \right) \quad . \tag{4.21}$$

This probability density is assumed to be a priori in the sense that it is independent of the result of the measurements on the observable parameters $\mathbf{d}$ (considered below).

- A relation

$$\mathbf{d} = \mathbf{g}(\mathbf{m}) \tag{4.22}$$

that solves the forward problem, i.e., that predicts the values of the observable parameters $\mathbf{d}$ that should correspond to the model $\mathbf{m}$. This theoretical prediction is assumed to be error free (see below for the introduction of a simple form of theoretical uncertainties).

- Some measurements on the observable parameters $\mathbf{d}$ whose results can be represented by a generalized Gaussian probability density of order s centered at $\{d^i_{\text{obs}}\}$ and with estimators of dispersion (of order s) $\{\sigma^i_D\}$:

$$\rho_D(\mathbf{d}) = \text{const. } \exp\left(-\frac{1}{s} \sum_i \frac{|d^i - d^i_{\text{obs}}|^s}{(\sigma^i_D)^s} \right) \quad . \tag{4.23}$$

The combination of these three types of information was considered in Example 1.34, where it was shown to lead to the a posteriori probability density in the model space (equation (1.93))

$$\sigma_M(\mathbf{m}) = \frac{\rho_M(\mathbf{m}) \, \rho_D(\mathbf{g}(\mathbf{m}))}{\int_{\mathfrak{M}} d\mathbf{m}' \, \rho_M(\mathbf{m}') \, \rho_D(\mathbf{g}(\mathbf{m}'))} \quad . \tag{4.24}$$

This gives

$$\boxed{\sigma_M(\mathbf{m}) = \text{const. } \exp(-S(\mathbf{m})) \quad ,} \tag{4.25}$$

where the *misfit function* $S(\mathbf{m})$ is

$$S(\mathbf{m}) = \frac{1}{s} \sum_i \frac{|g^i(\mathbf{m}) - d^i_{\text{obs}}|^s}{(\sigma^i_{\text{D}})^s} + \frac{1}{r} \sum_\alpha \frac{|m^\alpha - m^\alpha_{\text{prior}}|^r}{(\sigma^\alpha_{\text{M}})^r} \quad . \tag{4.26}$$

With the posterior probability density $\sigma_{\text{M}}(\mathbf{m})$ so expressed, general methods can be used to extract information from it, like the Monte Carlo methods described in chapter 2.

A much simpler goal — although much less informative — is just to obtain the maximum likelihood model, i.e., the point at which $\sigma_{\text{M}}(\mathbf{m})$ becomes maximum.[57] Clearly, the maximization of the probability density $\sigma_{\text{M}}(\mathbf{m})$ is equivalent to the minimization of the misfit function $S(\mathbf{m})$. For this reason, the maximum likelihood model can also be called the 'best model' under the minimum ℓ_p-norm 'criterion.' Note that the ℓ_p-norm criterion is only justified if the assumption of uncertainties distributed following a generalized Gaussian of order p is acceptable (here, in fact, of order r in the model space and of order s in the data space).

The gradient of the misfit function is easy to obtain using the third of equations (4.15),

$$\hat{\gamma}_\alpha \equiv \frac{\partial S}{\partial m^\alpha} = \sum_i \frac{1}{\sigma^i_{\text{D}}} G^i{}_\alpha \left(\frac{g^i - d^i_{\text{obs}}}{\sigma^i_{\text{D}}} \right)^{\{s-1\}} + \sum_\alpha \frac{1}{\sigma^\alpha_{\text{M}}} \left(\frac{m^\alpha - m^\alpha_{\text{prior}}}{\sigma^\alpha_{\text{M}}} \right)^{\{r-1\}} \quad , \tag{4.27}$$

where, for short, the notations $\hat{\gamma}_\alpha$ and g^i are respectively used for $\hat{\gamma}_\alpha(\mathbf{m})$ and $g^i(\mathbf{m})$ and where the $G^i{}_\alpha$ are the partial derivatives

$$G^i{}_\alpha(\mathbf{m}) = \frac{\partial g^i}{\partial m^\alpha}(\mathbf{m}) \quad . \tag{4.28}$$

$\hat{\boldsymbol{\gamma}}$ is a form in the model space where an ℓ_r-norm has been considered. As demonstrated in Appendix 6.22.1, the associated steepest ascent vector is obtained using the duality introduced above (equation (4.16)), giving

$$\gamma^\alpha = (\sigma^\alpha_{\text{M}})^r (\hat{\gamma}^\alpha)^{\{r-1\}} \quad . \tag{4.29}$$

The physical dimensions of the components of the gradient being the inverse of those of the model parameters, this equation gives for the components of the steepest ascent vector the physical dimensions of the model parameters, as it should.

The simplest algorithm for seeking minima of the misfit function is, of course, the steepest descent algorithm $\mathbf{m}_{n+1} = \mathbf{m}_n - \mu_n \boldsymbol{\gamma}_n$, where the μ_n are real numbers small enough to avoid divergence (see details in Appendix 6.22). In terms of the components,

$$m^\alpha_{n+1} = m^\alpha_n - \mu_n \gamma^\alpha_n \quad . \tag{4.30}$$

The five equations (4.25), (4.26), (4.27), (4.29), and (4.30) give the four basic equations related to the formulation of ℓ_p-norm problems and their resolution using the steepest descent method.

---

[57] As explained in chapter 1, the maximum likelihood point is that maximizing the probability density, because, here, the model space is linear and, therefore, the homogeneous probability density is constant.

Of course, more sophisticated gradient methods could be used (as suggested when analyzing ℓ_2-norm problems), but their applicability would greatly depend on the particular problem at hand, so these methods are not examined here.

4.4 The ℓ_1-Norm Criterion for Inverse Problems

4.4.1 Formulation of the Problem

If uncertainties in the observed data values $\mathbf{d}_{\text{obs}}$ and uncertainties in the a priori model $\mathbf{m}_{\text{prior}}$ are assumed to be conveniently modeled using a double exponential (Laplace distribution), then the a posteriori probability density in the model space is given by (this is equation (1.109) and a special case of equations (4.25)–(4.26))

$$\sigma_{\mathrm{M}}(\mathbf{m}) = \text{const. } \exp(-S(\mathbf{m})) \quad , \tag{4.31}$$

where the ℓ_1-norm misfit function $S(\mathbf{m})$ is

$$S(\mathbf{m}) = \sum_i \frac{|g^i(\mathbf{m}) - d^i_{\text{obs}}|}{\sigma^i_{\mathrm{D}}} + \sum_\alpha \frac{|m^\alpha - m^\alpha_{\text{prior}}|}{\sigma^\alpha_{\mathrm{M}}} \quad . \tag{4.32}$$

Here, σ^i_{D} and $\sigma^\alpha_{\mathrm{M}}$ respectively represent the ℓ_1-norm estimators of dispersion (i.e., the mean deviations; see Appendix 6.5) for uncertainties in observed data and the a priori model. (In many applications, while the use of the ℓ_1-norm is adequate in the data space, to ensure robustness, one may use an ℓ_2-norm in the model space; see section 4.4.3.)

Figure 4.4 suggests the shape of the functions $\sigma_{\mathrm{M}}(\mathbf{m})$ and $S(\mathbf{m})$ (for a two-dimensional model space) when the relation $\mathbf{d} = \mathbf{g}(\mathbf{m})$ is linear. Figure 4.5 corresponds to a nonlinear relation $\mathbf{d} = \mathbf{g}(\mathbf{m})$.

As the shape of $\sigma_{\mathrm{M}}(\mathbf{m})$ is typically quite different from a Gaussian, there is no hope of obtaining an easy characterization of the dispersion (like when using the posterior covariance matrix in the ℓ_2-norm formulation).

To extract information from the probability density $\sigma_{\mathrm{M}}(\mathbf{m})$, as expressed by equations (4.31)–(4.32), one may use a random generation of models, as suggested in chapter 2. Sometimes, one may just be interested in the maximum likelihood model, that is, maximizing the probability density $\sigma_{\mathrm{M}}(\mathbf{m})$, or, equivalently, minimizing the misfit function $S(\mathbf{m})$. For obvious reasons, this model minimizing $S(\mathbf{m})$ (equation (4.32)) is called the *best* model with respect to the *least-absolute-values* (ℓ_1-norm) *criterion*.

Example 4.1. *A model* $\mathbf{m} = \{x, y\}$ *consists of two quantities* x *and* y *about which one has the a priori information*

$$x = 1 \pm 3 \quad , \quad y = 2 \pm 3 \quad . \tag{4.33}$$

The quantity $d = 2x + y$ *has been measured and one has obtained*

$$d = 5 \pm 2 \quad . \tag{4.34}$$

Give the probability density for $\mathbf{m} = \{x, y\}$*, assuming that all the mentioned uncertainties can be modeled using the double-exponential probability density. This problem is easy* to solve. A direct application of the discussion above gives the probability density

$$\sigma_M(x, y) \ = \ \text{const. } \exp(-S(x, y)) \quad , \tag{4.35}$$

with

$$S(x, y) \ = \ \frac{|5 - (2x + y)|}{2} + \frac{|x - 1|}{3} + \frac{|y - 2|}{3} \quad . \tag{4.36}$$

This misfit function is represented in Figure 4.4 together with the probability density $\sigma_M(x, y)$. If, instead, one has measured the quantity $d = 2 \sin x + y$, the misfit becomes $S(x, y) = \frac{|5 - (2 \sin x + y)|}{2} + \frac{|x-1|}{3} + \frac{|y-2|}{3}$, represented in Figure 4.5.

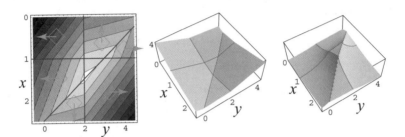

Figure 4.4. *Representation of the ℓ_1-norm misfit function defined in Example 4.1, where the relation $\mathbf{d} = \mathbf{g(m)}$ is linear. When the model space $\mathbb{M}$ is two dimensional, the misfit function $S(\mathbf{m})$ can be represented by a convex polyhedron in a 3D space (middle). The function $S(\mathbf{m})$ defines a partition of the model space $\mathbb{M}$ into convex polygons (left). The level lines of $S(\mathbf{m}) = $ const. are also convex polygons. In this left panel, the level lines indicate the gradient of $S(\mathbf{m})$ at every point. The direction of steepest ascent (with respect to the ℓ_1-norm) is indicated at some selected points (arrows). The probability density $\sigma_M(\mathbf{m}) = $ const. $\exp(-S(\mathbf{m}))$ is represented at the right.*

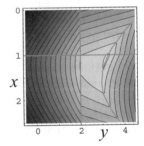

Figure 4.5. *Same as Figure 4.4, but when the relation $\mathbf{d} = \mathbf{g(m)}$ is nonlinear. To represent here the directions of steepest ascent, one should draw infinitesimally small circles.*

Linear Problems

When the relation between data and model parameters is linear, one may write

$$\mathbf{g(m)} \ = \ \mathbf{Gm} \quad , \tag{4.37}$$

where $\mathbf{G}$ represents a linear operator (in fact, a matrix).

To fix ideas, let us first consider the case where the model space has a dimension of two, i.e., we have only two parameters. The shape of the function $S(\mathbf{m})$ has been illustrated in Figure 4.4. It is easy to see that the surface representing $S(\mathbf{m})$ is then a convex polyhedron or, more precisely, the lower part of an infinite convex polyhedron. Each edge of the polyhedron clearly corresponds to a null residual in (4.32), i.e.,

$$\mathbf{m} \in \text{edge} \quad \Longleftrightarrow \quad \begin{cases} (\mathbf{G}\,\mathbf{m})^i = d^i_{\text{obs}} & \text{for a given } i \\ \text{or} \\ m^\alpha = m^\alpha_{\text{prior}} & \text{for a given } \alpha \;. \end{cases} \quad (4.38)$$

It is also clear that each vertex of the polyhedron corresponds to the points where *at least two* residuals are null. As the minimum of the convex polyhedron is necessarily attained at a vertex (or edge, face, ...), we arrive at the conclusion that at the minimum of the function $S(\mathbf{m})$ at least two of the equations

$$(\mathbf{G}\,\mathbf{m})^i = d^i_{\text{obs}} \quad , \quad m^\alpha = m^\alpha_{\text{prior}} \quad (4.39)$$

are exactly satisfied (and only two in most common situations).

It is helpful to have a visual image of what the representation of the misfit function at the top left of Figure 4.4 becomes when the dimension of the model space is greater than two. The three-dimensional case is illustrated in Figure 4.6. More generally, if the model space $\mathbb{M}$ has N dimensions, each of the equations (4.39) defines a hyperplane of dimension $N-1$ over $\mathbb{M}$ (a *line* for $N = 2$, an *ordinary plane* for $N = 3$, etc.). The whole set of equations (4.39) defines a partition of the model space into convex N-dimensional hyperpolyhedrons where the function S has constant gradient (convex polygons for $N = 2$, convex polyhedrons for $N = 3$, etc.). The minimum of S is then attained at one vertex of one of these hyperpolyhedrons (or, in the case of degeneracy, at one edge, or face, etc.). As the vertices are defined by the intersection of at least N hyperplanes, we arrive at the conclusion that, at the minimum of S, *at least N of the equations* (4.39) *are exactly satisfied* (and often only N).

Figure 4.6. *If the model space is three-dimensional, the function $S(\mathbf{m})$ defines a partition of the space into convex polyhedrons, connected by convex polygons, connected by edges, connected by vertices. In an N-dimensional model space, we have vertices, edges, convex polygons, convex polyhedrons, convex 4D hyperpolyhedrons, ... , and convex N-dimensional hyperpolyhedrons. Linear programming (see below) defines paths along edges and vertices.*

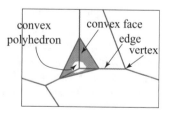

This means that the least ℓ_1-norm criterion selects among the components of d^i_{obs} and m^α_{prior} the N components whose posterior values will be identical to the prior ones. The posterior values of all other components will be computed from these N values using the linear system (4.39). The prior values of these other component values are not used (robustness results from this), except in that they have influenced the choice of the selected N components.

Example 4.2. *At different time instants* $\{t_i\} = t_1, t_2, \ldots$ *(assumed perfectly known), the quantities* $y_i = y(t_i)$ *having been measured, the results* $y_1 \pm \sigma_1$, $y_2 \pm \sigma_2 \ldots$ *have been produced, these uncertainties being understood as the mean deviations of double-exponential probability densities. Under the hypothesis that the relation between* y_i *and* t_i *is linear,* $y_i = a\,t_i + b$, *estimate the parameters* $\{a, b\}$. *This problem is solved using a special case of equations (4.31)–(4.32), without the terms defining the a priori information. The misfit function is, here,*

$$S(a, b) = \sum_i \frac{|(a\,t_i + b) - y_i|}{\sigma_i} . \tag{4.40}$$

The general problem is illustrated in Figure 4.7. Following the previous discussion, the best line in the ℓ_1-norm sense will necessarily go through at least two points. Of course it may happen that the minimum of S is attained at a horizontal edge or face (Figures 4.8 and 4.9). As in that case the minimum of S is not attained at a single point, the best model in the sense of the least-absolute-values criterion is not necessarily unique.

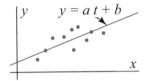

Figure 4.7. *The best straight line in the ℓ_1-norm sense minimizes* $\sum_i \frac{|(a\,t_i + b) - y_i|}{\sigma_i}$ *(the uncertainty bars are assumed to be vertical). This best line fits at least two points exactly.*

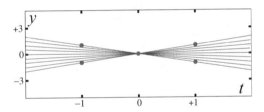

Figure 4.8. *Even for linear problems, the best solution in the ℓ_1-norm sense may not be unique. This figure illustrates a highly degenerated problem. We seek the best line $y = a\,t + b$ fitting the five points. In this example, the best line in the ℓ_1-norm sense minimizes the function* $S(a, b) = |-a+b+1| + |-a+b-1| + |b| + |a+b-1| + |a+b+1|$. $S(a, b)$ *is represented in Figure 4.9. The minimum of $S(a, b)$ is attained by a solution passing through two points (as it must always be), but also for all the straight lines represented in the figure.*

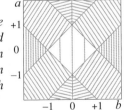

Figure 4.9. *The function $S(a, b)$ corresponding to the problem in Figure 4.8. The vertical axis is for the slope a, and the horizontal axis is for b. The minimum of S is attained at an edge (of the polyhedron representing S in a 3D space). The minimum value of S is $S = 4$, and the level lines represented are spaced with $\Delta S = 0.5$.*

Nonlinear Problems

For nonlinear problems, none of the properties discussed above necessarily remain true. In Figure 3.2 of chapter 3, the concept of the *quasi linearity* of a problem was discussed (which is more general than the concept of *linearizability*). A problem is quasi linear if the functions $\mathbf{g(m)}$ can be linearized inside the region of significant posterior probability density (but cannot necessarily be linearized with respect to the a priori model m_{prior}). All the methods to be described below will only make sense for quasi-linear problems. For strong nonlinear problems, the more general (and expensive) methods of chapters 1 and 2 should be used.

4.4.2 ℓ_1-Norm Criterion and the Method of Steepest Descent

When direct methods of minimization (like a systematic exploration of the space) are too expensive to be used, and if the derivatives

$$(\mathbf{G}_n)^i{}_\alpha = \left(\frac{\partial g^i}{\partial m^\alpha} \right)_{m_n} , \tag{4.41}$$

can be obtained, either analytically or by finite differencing, gradient methods of minimization can be used. I limit the scope of this section to the steepest descent method of minimization.

The function $S(\mathbf{m})$ does not, in general, have a continuous gradient (see, for instance, Figure 4.5), but this is not a difficulty: good experience exists in the use of gradient-based methods for large-scale inverse problems (see, for instance, Djikpéssé and Tarantola, 1999). Passing, by chance, through an edge (where the gradient is not defined) is not a practical problem: as mentioned below, the simple rule $\mathrm{sg}(0) = 0$ redefines the gradient at such a point as the average of the two facing gradients. I have never found an advantage (for large-sized nonlinear problems) in using (non)linear programming methods.

Gradient and Direction of Steepest Ascent

From now on, the terms vertex, edge, ... correspond to the representation of $S(\mathbf{m})$ in the model space (see Figures 4.5 and 4.6) and *not* to the representation of $S(\mathbf{m})$ as in the top right of Figure 4.4.

The gradient of the misfit function $S(\mathbf{m})$ can be obtained directly from equation (4.32):

$$(\hat{\boldsymbol{\gamma}}_n)_\alpha = \left(\frac{\partial S}{\partial m^\alpha} \right)_{\mathbf{m}_n} = \sum_i (\mathbf{G}_n)^i{}_\alpha \frac{1}{\sigma_D^i} \mathrm{sg}(g^i(\mathbf{m}_n) - d^i_{\mathrm{obs}}) + \frac{1}{\sigma_M^\alpha} \mathrm{sg}(m^\alpha - m^\alpha_{\mathrm{prior}}) . \tag{4.42}$$

If one or some of the signs $\mathrm{sg}(\,\cdot\,)$ in this equation happen to be zero (i.e., if the definition $\mathrm{sg}(0) = 0$ happens to be used), this would be an indication that one is at a discontinuity of the gradient. But the expression (4.42) can still be used, amounting to a redefinition of the gradient at a discontinuity point as the average of the two adjacent gradients.

The gradient of a function defined over the model space $\mathbb{M}$ is an element of the dual space $\hat{\mathbb{M}}$. As $\mathbb{M}$ is here an ℓ_1-norm space, $\hat{\mathbb{M}}$ is an ℓ_∞-norm space. As demonstrated in

Appendix 6.22.1, to pass from the gradient (element of $\hat{\mathbb{M}}$) to the steepest ascent vector (element of $\mathbb{M}$), we must evaluate the dual of the gradient.

To do this, we must use the result demonstrated at the end of section 4.2.2. The gradient is first normed to one (in the sense of the ℓ_∞-norm[58]), and then equation (4.18) is used. The result is as follows.

Among all the components $(\hat{\boldsymbol{\gamma}}_n)_\alpha$ of the gradient, we select those that attain the maximum value of the product[59] $\sigma_M^\alpha |(\hat{\boldsymbol{\gamma}}_n)_\alpha|$ and call them the *principal components of the gradient*. Usually, this maximum value is attained for a single component, but there may be more than one principal component. Letting k be the number of principal components, the components of the steepest ascent vector are then

$$(\boldsymbol{\gamma}_n)^\alpha = \begin{cases} \frac{1}{k}\,\sigma_M^\alpha\,\mathrm{sg}((\hat{\boldsymbol{\gamma}}_n)_\alpha) & \text{for the principal components} \quad, \\ 0 & \text{for the other components} \quad. \end{cases} \qquad (4.43)$$

One has $\|\boldsymbol{\gamma}_n\| = 1$ (this is here an ℓ_1-norm), as one should, because the gradient has been normalized (with respect to the ℓ_∞-norm). A look at Figure 4.10 should help us to understand this result. Some steepest ascent directions were plotted in Figure 4.4.

Figure 4.10. *A circle in an ℓ_1-normed space, and the same circle with three different linear forms over the space. The direction of steepest ascent is defined by the point on the circle that reaches the maximum value of the form. It usually points along one of the axes, although it may also point in an intermediate direction.*

Let $\mathbf{m}_n$ represent the current point we wish to update. In the previous section, we obtained the steepest ascent vector $\boldsymbol{\gamma}_n$. As usual, the steepest descent method is written

$$\mathbf{m}_{n+1} = \mathbf{m}_n - \mu_n\,\boldsymbol{\gamma}_n \quad, \qquad (4.44)$$

where μ_n is an arbitrary positive real number small enough to ensure convergence. Although one may easily imagine methods for choosing reasonably good values for this constant,[60] a direct one-dimensional search (trying different values until the minimum value of the misfit along this direction is found) is often the best method.

---

[58] Remember that the standard deviations in the dual space are the inverse of the standard deviations in the primal space (equation (4.7)).

[59] Remember that there is no implicit sum notation in this chapter.

[60] One possibility for choosing μ_n is as follows: a linearization of the misfit function $S(\mathbf{m})$ in the vicinity of $\mathbf{m}_n$ defines a partition of the model space into convex hyperpolyhedrons. The direction $-\boldsymbol{\gamma}_n$ is a direction of descent at $\mathbf{m}_n$. It is possible (and easy) to obtain the point of the intersection of $-\boldsymbol{\gamma}_n$ with the first hyperface and to take this point as the updated point $\mathbf{m}_{n+1}$ (or, in practice, a point slightly after the hyperface), thus fixing the value of μ_n. Let us consider in some detail how this works. Hyperfaces of the misfit function $S(\mathbf{m})$ correspond to changes of sign of the expressions $g^i(\mathbf{m}) - d_{\text{obs}}^i$ and $m^\alpha - m_{\text{prior}}^\alpha$, which means that the point $\mathbf{m}_{n+1}$ will be

4.4.3 ℓ_1-Norm in Data Space and ℓ_2-Norm in Model Space

As mentioned in section 4.4.1, while the use of the ℓ_1-norm may be necessary in the data space (to ensure robustness), one may use an ℓ_2-norm in the model space.

In the simplest situation (no a priori covariances), the misfit function is then

$$S(\mathbf{m}) = \sum_i \frac{|g^i(\mathbf{m}) - d^i_{\text{obs}}|}{\sigma^i_D} + \frac{1}{2} \sum_\alpha \frac{(m^\alpha - m^\alpha_{\text{prior}})^2}{(\sigma^\alpha_M)^2} , \qquad (4.45)$$

the gradient is

$$(\hat{\boldsymbol{\gamma}}_n)_\alpha = \left(\frac{\partial S}{\partial m^\alpha}\right)_{\mathbf{m}_n} = \sum_i (\mathbf{G}_n)^i{}_\alpha \frac{1}{\sigma^i_D} \text{sg}(g^i(\mathbf{m}_n) - d^i_{\text{obs}}) + \frac{1}{(\sigma^\alpha_M)^2} (m^\alpha - m^\alpha_{\text{prior}}) , \qquad (4.46)$$

and the associated steepest ascent vector is (because we now have an ℓ_2-norm in the model space)

$$(\boldsymbol{\gamma}_n)^\alpha = (\sigma^\alpha_M)^2 \sum_i (\mathbf{G}_n)^i{}_\alpha \frac{1}{\sigma^i_D} \text{sg}(g^i(\mathbf{m}_n) - d^i_{\text{obs}}) + (m^\alpha - m^\alpha_{\text{prior}}) \qquad (4.47)$$

(and the steepest descent algorithm is then, again, $\mathbf{m}_{n+1} = \mathbf{m}_n - \mu_n \boldsymbol{\gamma}_n$).

Djikpéssé and Tarantola (1999) demonstrate the applicability of this algorithm for solving a large-scale ℓ_1-norm inverse problem.

4.4.4 ℓ_1-Norm Criterion and Linear Programming

Programming techniques have been developed to obtain the solution of the problem of optimizing a real function $S(\mathbf{x})$ when the vector variable $\mathbf{x}$ is subject to a given vector constraint $\boldsymbol{\Psi}(\mathbf{x}) \leq 0$. Although this is not obvious at first sight, it can be demonstrated that an ℓ_1-norm minimization problem can be recast into a programming problem (see details in Appendix 6.23.4).

Numerous books exist on the subject of linear and nonlinear programming. The history of programming methods can be read in Dantzig (1963), who first proposed the *simplex method* in 1947. A good textbook is, for instance, Murty (1976). The reader may also refer to Luenberger (1973), Gass (1975), or Bradley, Hax, and Magnanti (1977).

When both the misfit function $S(\mathbf{x})$ and the constraints $\boldsymbol{\Psi}(\mathbf{x}) \leq 0$ are linear functions of $\mathbf{x}$, the problem is one of 'linear programming.' Linear programming techniques are often used by economists (maximizing a benefit for given limitations in available supplies) and

---

at a hyperface if $g^i(\mathbf{m}_{n+1}) - d^i_{\text{obs}} = 0$ for a given i or if $m^\alpha_{n+1} - m^\alpha_{\text{prior}} = 0$ for a given α. Linearizing $g^i(\mathbf{m})$ around $\mathbf{m}_n$ gives $g^i(\mathbf{m}_n - \mu_n \boldsymbol{\gamma}_n) \simeq g^i(\mathbf{m}_n) - \mu_n \sum_\alpha (\mathbf{G}_n)^i{}_\alpha \gamma^\alpha_n$, and, therefore, the conditions of being at a hyperface are written $g^i(\mathbf{m}_n) - d^i_{\text{obs}} = \mu_n \sum_\alpha (\mathbf{G}_n)^i{}_\alpha \gamma^\alpha_n$ for a given i or $m^\alpha_n - m^\alpha_{\text{prior}} = \mu_n \gamma^\alpha_n$ for a given α. From these equations, it is easy to compute all the values of μ_n for which we will have hyperfaces of the linearized function: $\mu^i_n = (g^i(\mathbf{m}_n) - d^i_{\text{obs}}) / (\sum_\alpha (\mathbf{G}_n)^i{}_\alpha \gamma^\alpha_n)$ (for all i) and $\mu^\alpha_n = (m^\alpha_n - m^\alpha_{\text{prior}}) / \gamma^\alpha_n$ (for all α). As $-\boldsymbol{\gamma}_n$ is a direction of descent, negative values of μ_n have to be disregarded because they would give an increase of S. Among the positive values, it is the smallest that gives the most neighboring hyperface, and the problem is solved. Taking a slightly larger value will, in general, avoid dropping *before* the true hyperface of the nonlinearized function (the chances of dropping at the hyperface are small, due to computer arithmetic).

by the military (minimizing the time needed for invading a neighboring country for given limitations on troop transportation).

The central problem in linear programming is always to obtain the minimum of a convex polyhedron in a multidimensional space, and all the methods suggested for solving the problem are very similar: first, one has to manage to obtain a vertex of the polyhedron, then one has to leave the current vertex following a descending edge until the next vertex. After a finite number of moves, the minimum is necessarily attained. The different methods only differ in the way of obtaining the first vertex, or, for a given vertex, in the choice of the edge by which the current vertex has to be left.

Although the most widely used method for solving linear programming problems is the *simplex method*, introduced in Appendix 6.23.1, the *first-in-first-out* (FIFO) method of Claerbout and Muir (1973) is sometimes simpler for solving inverse problems.

4.5 The ℓ_∞-Norm Criterion for Inverse Problems

4.5.1 Formulation of the Problem

We have seen above that if uncertainties in the observed data values, $\mathbf{d}_{\text{obs}}$, and uncertainties in the prior model, $\mathbf{m}_{\text{prior}}$, are assumed to be conveniently modeled using generalized Gaussians, then the posterior probability density in the model space is given by (equations (4.25)–(4.26))

$$\sigma_M(\mathbf{m}) \;=\; \text{const. } \exp(-S(\mathbf{m})) \quad , \tag{4.48}$$

where the misfit function is (assuming a common value, say p, for the parameters r and s)

$$S(\mathbf{m}) \;=\; \frac{1}{p}\left(\sum_i \frac{|\, g^i(\mathbf{m}) - d^i_{\text{obs}}\,|^p}{(\sigma^i_D)^p} + \sum_\alpha \frac{|\, m^\alpha - m^\alpha_{\text{prior}}\,|^p}{(\sigma^\alpha_M)^p}\right) \quad . \tag{4.49}$$

When taking the limit $p \to \infty$, the generalized Gaussian becomes a boxcar function (see Appendix 6.6), which means that this limit corresponds to the assumption of strict error bounds

$$\begin{aligned}
-\sigma^i_D \;\leq\; g^i(\mathbf{m}) - d^i_{\text{obs}} \;\leq\; +\sigma^i_D \quad , \\
-\sigma^\alpha_M \;\leq\; m^\alpha - m^\alpha_{\text{prior}} \;\leq\; +\sigma^\alpha_M \quad .
\end{aligned} \tag{4.50}$$

When $p \to \infty$, the posterior probability density gives

$$\sigma_M(\mathbf{m}) \;=\; \begin{cases} \text{const.} & \text{if the bounds (4.50) are satisfied,} \\ 0 & \text{otherwise} \quad . \end{cases} \tag{4.51}$$

It may well happen that no point of the model space satisfies the bounds. This means, in general, that the theoretical equation $\mathbf{d} = \mathbf{g}(\mathbf{m})$ does not model the real world accurately enough, or, more often, that we have been too optimistic in setting the uncertainty bars. In what follows, let us assume that $\sigma_M(\mathbf{m})$ is defined, i.e., that there exists a region $\mathbb{M}'$ of the model space satisfying the error bounds and the theoretical equation.

As $\sigma_M(\mathbf{m})$ is uniform over $\mathbb{M}'$, the maximum likelihood point is not defined, and one possibly should not bother introducing such a concept inside the ℓ_∞-norm criterion. But, in order to allow the reader to make the link with the minimax criterion of the literature, let us make some developments.

For sufficiently regular forward equations $\mathbf{m} \mapsto \mathbf{g}(\mathbf{m})$, the maximum likelihood point, say $\mathbf{m}_p$, can be defined for any value of p inside the limits $1 \leq p < \infty$. Using the results of Descloux (1963), it can be shown that the sequence $\mathbf{m}_p$ is convergent when $p \to \infty$. The convergence point $\mathbf{m}_\infty = \lim_{p\to\infty} \mathbf{m}_p$ may be termed *the strict ℓ_∞ solution.* For any finite p, the point maximizing $\sigma_M(\mathbf{m})$ also minimizes the function

$$S(\mathbf{m}) = \frac{1}{p} \left(\sum_i \frac{|g^i(\mathbf{m}) - d^i_{\text{obs}}|^p}{(\sigma^i_D)^p} + \sum_\alpha \frac{|m^\alpha - m^\alpha_{\text{prior}}|^p}{(\sigma^\alpha_M)^p} \right) , \qquad (4.52)$$

but minimizing $S(\mathbf{m})$ is equivalent to minimizing the function

$$R(\mathbf{m}) = (p\, S(\mathbf{m}))^{1/p} , \qquad (4.53)$$

i.e.,

$$R(\mathbf{m}) = \left(\sum_i \frac{|g^i(\mathbf{m}) - d^i_{\text{obs}}|^p}{(\sigma^i_D)^p} + \sum_\alpha \frac{|m^\alpha - m^\alpha_{\text{prior}}|^p}{(\sigma^\alpha_M)^p} \right)^{1/p} . \qquad (4.54)$$

The limit of $S(\mathbf{m})$ diverges when $p \to \infty$, but this is not so for the limit of $R(\mathbf{m})$: in the limit $p \to \infty$, one obtains

$$R(\mathbf{m}) = \max \left(\frac{|g^i(\mathbf{m}) - d^i_{\text{obs}}|}{\sigma^i_D} , \frac{|m^\alpha - m^\alpha_{\text{prior}}|}{\sigma^\alpha_M} \text{ (all values of } i \text{ and of } \alpha \text{)} \right) .$$

$$(4.55)$$

Figure 4.11 shows the functions $R(\mathbf{m})$ and $S(\mathbf{m})$ for a one-dimensional example.

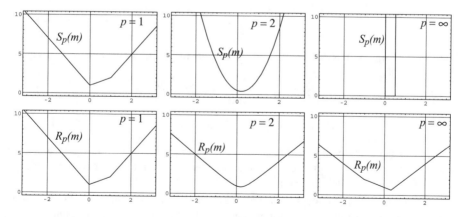

Figure 4.11. *The function* $S_p(m) = 1/p\,(|2m|^p + |m - 1|^p)$ *is shown for* p *respectively equal to* 1, 2, *and* ∞. *Also shown is the function* $R_p(m) = (p\, S(m))^{1/p}$, *which gives* $R_1(m) = S_1(m) = |2m| + |m - 1|$, $R_2(m) = (5m^2 - 2m + 1)^{1/2}$ *and* $R_\infty(m) = \max(|2m|, |m - 1|)$.

Equation 4.55 shows that the minimization of $R(\mathbf{m})$ thus corresponds to *minimizing the maximum weighted residual*. This explains the term *minimax* criterion, often used for the ℓ_∞-norm criterion. One may also remember that the terms minimax norm, uniform norm, and Chebyshev norm (in honor of the mathematician who used the ℓ_∞-norm in the 1850s) are used as synonyms of ℓ_∞-norm.

In conclusion, when boxcar probability densities can be used to model all the 'input uncertainties' in the problem (i.e., measurement uncertainties and a priori uncertainties on the model parameters), the posterior probability density in the model space, $\sigma_M(\mathbf{m})$, is also boxcar (equation (4.51)). It has no maximum likelihood point, but as far as a 'center' defined by the limit $p \rightarrow \infty$ may make sense, this center may be found by using a minimax criterion, i.e., by finding the minimum of the function $R(\mathbf{m})$ in equation (4.55). Obtaining this center is not of much help for understanding the probability density $\sigma_M(\mathbf{m})$ (see Example 4.3).

4.5.2 ℓ_∞-Norm Criterion and the Method of Steepest Descent

Let us face here solving a problem based on the minimax criterion, i.e., finding the model $\mathbf{m}$ that minimizes the function $R(\mathbf{m})$ in equation (4.55). We assume here that the derivatives

$$(\mathbf{G}_n)^i{}_\alpha = \left(\frac{\partial g^i}{\partial m^\alpha} \right)_{m_n} \tag{4.56}$$

can be obtained either analytically or by finite differencing, so gradient methods of minimization can be tried.

As at a given point $\mathbf{m}_n$, the value of $R(\mathbf{m})$ is controlled by the value of the maximum weighted residual (see equation (4.55)), the gradient of $R(\mathbf{m})$ will only depend on the corresponding term. Let us distinguish two cases:

- The maximum of $R(\mathbf{m})$ is attained for a data residual, say the term $i = i_0$. Then,

$$(\hat{\boldsymbol{\gamma}}_n)_\alpha = (\mathbf{G}_n)^{i_0}{}_\alpha \frac{1}{\sigma_D^{i_0}} \, \mathrm{sg}(g^{i_0}(\mathbf{m}_n) - d_{\mathrm{obs}}^{i_0}) \quad . \tag{4.57}$$

- The maximum of $R(\mathbf{m})$ is attained for a model residual, say the term $\alpha = \alpha_0$. Then,

$$(\hat{\boldsymbol{\gamma}}_n)_\alpha = \delta_\alpha^{\alpha_0} \frac{1}{\sigma_M^{\alpha_0}} \, \mathrm{sg}(m_n^{\alpha_0} - m_{\mathrm{prior}}^{\alpha_0}) \quad , \tag{4.58}$$

where $\delta_\alpha^{\alpha_0} = 1$ if $\alpha = \alpha_0$ and $\delta_\alpha^{\alpha_0} = 0$ if $\alpha \neq \alpha_0$.

If it happens that the maximum value is attained for more than one residual simultaneously, we are at a point where the gradient is discontinuous, and we may redefine it by taking the average of the two contiguous gradients.

To pass from the gradient to the steepest ascent vector, we must — as already done a few times above — evaluate the vector $\boldsymbol{\gamma}$ that is dual to the form $\hat{\boldsymbol{\gamma}}$. Using equation (4.17)) we obtain

$$(\boldsymbol{\gamma}_n)^\alpha = \sigma_M^\alpha \, \mathrm{sg}((\hat{\boldsymbol{\gamma}}_n)_\alpha) \quad . \tag{4.59}$$

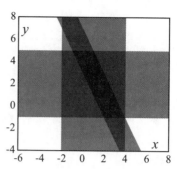

Figure 4.12. *The probability density defined in Example* 4.3. *Each of the three conditions* $| 5 - (2x + y) | < 2$, $| x - 1 | < 3$, *and* $| y - 2 | < 3$ *is represented in gray, and there is a region where the three conditions are satisfied (the darkest area). The probability density* $\sigma_M(x, y)$ *is constant inside this darkest area and vanishes outside.*

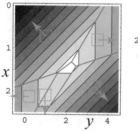

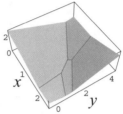

Figure 4.13. *This figure has been created with the same data as Figure* 4.4, *but using here an* ℓ_∞-*norm instead of an* ℓ_1-*norm. What is plotted here is not the misfit* $S_1(\mathbf{m})$ *(as in Figure* 4.4*), but* $R_\infty(\mathbf{m}) = \lim_{p \to \infty} (p \, S_p(\mathbf{m}))^{1/p}$. *The level lines indicate the gradient of* $R_\infty(\mathbf{m})$ *at every point. The direction of steepest ascent (with respect to the* ℓ_∞-*norm) is indicated at some selected points (arrows).*

The steepest descent algorithm then writes

$$\mathbf{m}_{n+1} = \mathbf{m}_n - \mu_n \, \boldsymbol{\gamma}_n \quad , \tag{4.60}$$

where the positive real numbers μ_n are to be chosen using a linear search.

Example 4.3. *Consider the same problem as that in Example* 4.1, *but assume now that the uncertainties can be modeled using boxcar probability densities. Here, then, one has* $\sigma_M(x, y) = $ const. *if* $| 5 - (2x + y) | < 2$ *and* $| x - 1 | < 3$ *and* $| y - 2 | < 3$, *and one has* $\sigma_M(x, y) = 0$ *if any of these conditions is not satisfied. This probability density is represented in Figure* 4.12. *The function* $R(x, y)$ *defined in the text is represented in Figure* 4.13.

4.5.3 ℓ_∞-Norm Criterion and Linear Programming

The possible use of linear programming techniques for the resolution of inverse problems based on the minimax criterion is not considered in this (main) text. Some details are given in Appendix 6.23.5.

Chapter 5

Functional Inverse Problems

When in doubt, smooth.

Sir Harold Jeffreys (Quoted by Moritz, 1980)

Many inverse problems involve functions: the data set sometimes consists of recordings as a function of time or space, and the main unknown in the parameter set sometimes consists of a function of the spatial coordinates and/or of time.

Quite often, the functions can be approximated by their discretized versions, and an infinite-dimensional problem is then reduced to a finite-dimensional one. In fact, many 'functions' are handled and displayed by digital computers, and their discretization is implicit (like, for instance, when dealing with space images of Earth).

In spite of this, there are situations where the inverse problem is better formulated using functions (i.e., using the concepts of functional analysis). One may go quite far using the functional formulation, even if, at the end, some sort of discretization is used for the actual computations.

Few developments are needed for the general inverse problem (we shall see that the very definition of random function considers successive discretizations). But in the special case where the considered random functions are Gaussian, lots of beautiful mathematics is available that provides efficient methods of resolution.

5.1 Random Functions

5.1.1 Description of a Random Function

Looking at Figures 5.1–5.2, we can understand what is generally meant by a *random function*. Each of the figures represents some *realizations* of the considered random function, each realization being an ordinary function.

101

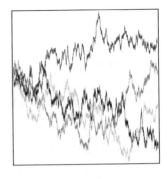

Figure 5.1. *Some realizations of a random function that is a one-dimensional random walk. At each time step (abscissa), the walker randomly chooses to walk one step up or one step down.*

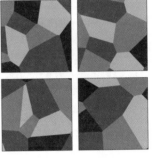

Figure 5.2. *Four random realizations of a random function defined over a 2D space. The realizations are continuous inside a finite number of blocks (these are, in fact, Voronoi cells). Contrary to the example in Figure 5.1, each realization is defined by a finite amount of information: this random function is finite dimensional.*

A particular realization can be seen as a point in an abstract space (named a *function space* or *space of functions*). To completely characterize some functions, one needs an infinite number of points, like in the example of Figure 5.1, while to characterize some other functions, a finite number of parameters is sufficient: each realization in Figure 5.2 is characterized by the coordinates of the points at the center of each domain (in fact, a Voronoi cell). Let us concentrate here on the infinite-dimensional case: if the function space is finite dimensional, the methods developed in the previous chapters directly apply.

In the first chapter, we saw the usefulness of the introduction of probability densities over finite-dimensional parameter spaces. Unfortunately, no simple generalization of the concept of probability densities exists over infinite-dimensional spaces.

To fix ideas, consider first a random function $x(t)$, where both variables t and x are scalars (one may, for instance, think of a scalar quantity x depending on time t). If $t \mapsto x(t)$ is a random function, then, by definition, for any given t, $x(t)$ is an ordinary random variable[61] whose probability density is denoted $f(x; t)$.

The knowledge of $f(x; t)$ for all t is not sufficient to characterize the random function: $f(x; t)$ only carries information on the marginal probability distribution of the variable x at a given time t, but does not carry information on the possible dependencies (or correlations) between the two random variables $x(t_1)$ and $x(t_2)$ at two instants $t = t_1$ and $t = t_2$. To characterize this, one needs the knowledge, for any two values $t = t_1$ and $t = t_2$, of the two-dimensional probability density for the two variables $x_1 = x(t_1)$ and $x_2 = x(t_2)$, a probability density that we may denote $f(x_1, x_2; t_1, t_2)$.

---

[61]Usually, a different notation is used for a random variable and for a realization of it. We do not need this sophisticated notation here.

But, again, this gives information on the two-dimensional marginals but not on the higher order joint probability densities: for a complete characterization of a random function, we need to know joint probability densities of any order. Thus,

> a random function $t \mapsto x(t)$ is characterized if the
> n-dimensional probability density $f(x_1, \ldots, x_n; t_1, \ldots, t_n)$
> is known *for any points* $t_1, \ldots, t_n$ and for *any value* of n .

Only very special random functions can be characterized by low-order marginals (for instance, a Gaussian random function is perfectly characterized by its two-dimensional marginals).

Note that when an n-dimensional probability density is given, one can deduce the lower dimensional marginals. For instance, one may pass from a two-dimensional probability density $f(x_1, x_2; t_1, t_2)$ to the one-dimensional marginal $f(x_1; t_1) = \int dx_2 \, f(x_1, x_2; t_1, t_2)$. The result $f(x_1; t_1)$ must actually be independent of the value t_2, imposing a constraint on a function $f(x_1, x_2; t_1, t_2)$ in order to accept the interpretation of being a two-dimensional probability density of a random function. The n-dimensional version of this constraint is obvious to formulate.

5.1.2 Computing Probabilities

Let $t \mapsto x(t)$ represent a random function and assume that we have a perfect description of it, i.e., that for any points $t_1, \ldots, t_n$ and for any n, the joint probability density $f(\mathbf{x}) \equiv f(x_1, \ldots, x_n; t_1, \ldots, t_n)$ of the random variables $x(t_1), \ldots, x(t_n)$ is known. We wish here to compute the probability of a realization of the random function being within certain limits. More precisely, for each value of t, let us introduce a numerical set $\mathcal{E}(t)$. We wish to compute the probability for $x(t)$ to verify (see Figure 5.3)

$$x(t) \in \mathcal{E}(t) \quad . \tag{5.1}$$

The set of all the possible realizations of the random function verifying equation (5.1) will be denoted by $\mathfrak{A}$. Following Pugachev (1965), let us first consider the subset $\mathfrak{A}_n$ of the set of all the possible realizations of the random function for which the values associated with the points $t_1, \ldots, t_n$ belong to the numerical sets $\mathcal{E}(t_1), \ldots, \mathcal{E}(t_n)$. By definition of a probability density, the probability of the set $\mathfrak{A}_n$ is clearly

$$P(\mathfrak{A}_n) = \int_{\mathcal{E}(t_1)} dx_1 \cdots \int_{\mathcal{E}(t_n)} dx_n \, f(\mathbf{x}) \quad . \tag{5.2}$$

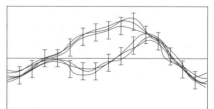

Figure 5.3. *Given a random function, it is possible to compute the probability of a realization being within given bounds (see text).*

For any positive integer value of n, we now consider a partition of the range of variation of the argument t into n subintervals such that the length of each of the n subintervals tends to zero as $n \to \infty$. Now, from each subinterval we choose a value of t such that the points $t_1, \ldots, t_n$ contain, for every n, all the points of the preceding partitions. Thus, we get a sequence $\mathfrak{A}_1, \mathfrak{A}_2, \ldots$ of sets of realizations of the random function each including all the subsequent sets. *If the realizations of the random function are sufficiently regular*, the product of all the sets $\mathfrak{A}_n$ coincides with the initial set $\mathfrak{A}$. The numbers $P(\mathfrak{A}_1), P(\mathfrak{A}_2), \ldots$ form a monotonic nonincreasing progression of nonnegative numbers. Then, the limit

$$P(\mathfrak{A}) = \lim_{n \to \infty} P(\mathfrak{A}_n) \tag{5.3}$$

exists and corresponds to the probability of a realization of the random function verifying the condition expressed by equation (5.1).

This computation shows that even if the notion of probability density does not generalize to infinite-dimensional spaces (because the equivalent of the Lebesgue integral does not exist), the actual probabilities of regions defined over infinite-dimensional manifolds can be defined (and computed).

When the considered functions belong to a linear (functional) space (we have not introduced such an assumption here), more mathematics is available. Rather than a simple discretization of the functions, one then considers arbitrary (finite-dimensional) linear subspaces. This leads to the notion of 'cylinder measures' (see, for instance, Balakrishnan, 1976), which are not considered here.

From a practical point of view, once one has been able to demonstrate that the limit defined in equation (5.3) exists, its actual evaluation, if performed using numerical methods, can sometimes be well approximated using a large, although finite, value of n (i.e., of discretization points). This means, in fact, that the random function is sufficiently characterized (from a practical point of view) by the n-dimensional probability density $f(x_1, \ldots, x_n; t_1, \ldots, t_n)$ given for, say, equally spaced points $t_1, \ldots, t_n$ and for a sufficiently high value of n (the value of n corresponds to the number of points needed for a reasonably accurate representation of any realization of the random function). This implies, of course, that the realizations are sufficiently regular. For instance, while a realization of pure 'white' noise (see section 5.3.3) cannot be described by a finite number of points, physical white noise is always somewhat 'colored' (i.e., has finite bandwidth) and can then be described by a finite (sufficiently large) number of points.

5.1.3 General Random Processes

We have seen that at each value of t, a random function $t \to x(t)$ defines a random variable. Sometimes, we have to consider two random functions $t \to x^1(t)$ and $t \to x^2(t)$ simultaneously. Each random function is individually characterized by a joint probability density $f^1(x_1^1, \ldots, x_n^1; t_1, \ldots, t_n)$ and $f^2(x_1^2, \ldots, x_m^2; t_1', \ldots, t_m')$, as indicated in section 5.1.1. In addition, we need now to characterize the dependencies (correlations) between the random variables $x^1(t)$ and $x^2(t)$.

This is made by the joint probability density

$$f^{12}(x_1^1 \ldots x_n^1, x_1^2 \ldots x_m^2; t_1 \ldots, t_n, t_1' \ldots t_m') \quad ,$$

from which the two marginal probability densities

$$f^1(x_1^1 \dots x_n^1; t_1 \dots t_n) = \int dx_1^2 \cdots \int dx_m^2 \, f^{12}(x_1^1 \dots x_n^1, x_1^2 \dots x_m^2; t_1 \dots t_n, t_1' \dots t_m') \quad,$$

$$f^2(x_1^2 \dots x_m^2; t_1' \dots t_m') = \int dx_1^1 \cdots \int dx_n^1 \, f^{12}(x_1^1 \dots x_n^1, x_1^2 \dots x_m^2; t_1 \dots t_n, t_1' \dots t_m')$$

$$(5.4)$$

can be computed.

In more general problems, instead of two random functions, we have to consider an arbitrary number. Their variables, in turn, may be multidimensional and may be different. We may have to consider not only random functions but simultaneously also some discrete random variables as well. The notation may become complicated in detail, but there is no conceptual difficulty.

5.1.4 Random Functions over Linear Spaces

The (infinite-dimensional) function spaces we consider are not necessarily linear spaces: the sum of two functions may not be defined (or may not be useful). The reasons that forced us in chapter 1 to consider that the parameter space is a general manifold, not necessarily a linear one, prevail here. A typical situation is when the electric properties of a wire may randomly depend on time: we may equivalently consider as a random function the resistance $R(t)$ of the wire or its conductance $S(t) = 1/R(t)$. The sum of two resistance functions $R_1(t) + R_2(t)$ is not equivalent to the sum of the two conductance functions $S_1(t) + S_2(t)$, and one may not wish to introduce one sum or the other. The functional space is here an infinite-dimensional manifold that is not (or not yet) a linear space.

We know that in this example it is possible to define a structure of linear space, but with a sum that, instead of being the ordinary sum, is the logarithmic sum. This amounts to introducing the logarithmic resistance $\rho(t)$ or the logarithmic resistance $\sigma(t)$ and then introducing the ordinary sum (both sums being now equivalent). It is in this sense that the hypothesis made below (that the function spaces considered are linear spaces, with the sum and the multiplication by a scalar defined) must be understood.

Consider, then, a random function $x(t)$ that is an element of a linear function space, where the sum $x_1(t) + x_2(t)$ and the multiplication by a scalar $\lambda x(t)$ are defined (and make physical sense).

If the one-dimensional (marginal) probability densities $f(x; t)$ are known for any t, central estimators or estimators of dispersion of the random function can be computed. For instance, the *mean value* of the random function is the (nonrandom) function $m(t)$ defined by

$$m(t) = \int dx \, x \, f(x; t) \quad, \tag{5.5}$$

and the standard deviation $\sigma(t)$ is defined by

$$\sigma(t)^2 = \int dx \, (x - m(t))^2 \, f(x; t) \quad. \tag{5.6}$$

If the *two-dimensional* joint probability densities $f(x_1, x_2; t_1, t_2)$ are also known for any t_1 and t_2, the correlations are known perfectly. The *covariance function* $C_2(t, t')$ of the random function is defined as the covariance (in the usual sense) between the random variables $x(t_1)$ and $x(t_2)$:

$$C(t, t') = \int dx \int dx' (x - m_2(t)) (x' - m_2(t')) f(x, x'; t, t') \quad . \tag{5.7}$$

It can be shown that a covariance function is symmetric, $C(t, t') = C(t', t)$, and definite nonnegative: for any function $\phi(t)$, one has $\int_{t_{min}}^{t_{max}} dt \int_{t_{min}}^{t_{max}} dt' \, \phi(t) \, C(t, t') \phi(t') \geq 0$. It is easy to verify[62] that the covariance between a point t and itself equals the variance: $C(t, t) = \sigma(t)^2$.

Example 5.1. Gaussian random function with given mean value and covariance function. *A random function $t \to x(t)$ is termed Gaussian with mean value $m(t)$ and covariance function $C(t, t')$ if, for any points $t_1, \ldots, t_n$ and for any n, the joint probability density of the random variables $x(t_1), \ldots, x(t_n)$ is*

$$f(\mathbf{x}) = \frac{1}{(2\pi)^{n/2} \det^{1/2} \mathbf{C}} \exp\left(-\tfrac{1}{2} (\mathbf{x} - \mathbf{m})^t \, \mathbf{C}^{-1} (\mathbf{x} - \mathbf{m})\right) \quad , \tag{5.8}$$

where

$$\mathbf{x} = \begin{pmatrix} x(t_1) \\ \cdots \\ x(t_n) \end{pmatrix}, \quad \mathbf{m} = \begin{pmatrix} m(t_1) \\ \cdots \\ m(t_n) \end{pmatrix}, \quad \mathbf{C} = \begin{pmatrix} C(t_1, t_1) & \cdots & \cdots \\ \cdots & \cdots & \cdots \\ \cdots & \cdots & C(t_n, t_n) \end{pmatrix}. \tag{5.9}$$

That the covariance function, as defined by equation (5.7), is the $C(t, t')$ appearing in the expression (5.8) of a Gaussian function is a well-known result but is not trivial to demonstrate.

A Gaussian random function is called *stationary* if the mean value $m(t)$ is, in fact, independent of t and if the covariance function $C(t, t')$ depends, in fact, only on the distance between t and t'. For instance, if t is a scalar variable like a time, the covariance function $C(t, t')$ can be represented by a function $\Psi(\tau)$ as $C(t, t') = \Psi(t - t')$. This simple definition of stationarity does not apply to random functions that are markedly non-Gaussian.

If instead of one random function $x(t)$ one has two random functions $x^1(t)$ and $x^2(t)$, in addition to the definitions just considered, one also defines the crosscovariance

$$C^{12}(t, t') = \int dx^1 \int dx^2 (x^1 - m_2{}^1(t)) (x^2 - m_2{}^2(t')) f(x^1, x^2; t, t') \quad . \tag{5.10}$$

One then has a matrix of covariance functions

$$\mathbf{C}(t, t') = \begin{pmatrix} C^{11}(t, t') & C^{12}(t, t') \\ C^{21}(t, t') & C^{22}(t, t') \end{pmatrix} \tag{5.11}$$

---

[62]From equation (5.7), it follows that $C(t, t) = \int dx \int dx' (x - m_2(t)) (x' - m_2(t)) f(x, x'; t, t)$. Using $f(x, x'; t, t) = f(x; t) \delta(x - x')$ gives $C(t, t) = \int dx \int dx' (x - m_2(t)) (x' - m_2(t)) f(x; t) \delta(x - x') = \int dx (x - m_2(t))^2 f(x; t)$. According to equation (5.6), this is the variance $\sigma(t)^2$.

that is symmetric,

$$C^{ij}(t, t') = C^{ji}(t', t), \qquad (5.12)$$

and definite nonnegative.

The successive generalizations of this (when there are more than two random functions or when, for instance, the scalar variable t is replaced with a vector variable) are long to expose but easy to understand. They are left to the reader.

5.1.5 Covariance Functions Do Not Characterize Random Functions

Does a covariance function give a convenient description of a random function?

> In general, **NO**.

Pugachev (1965), for instance, shows an example of two random functions with completely different realizations but having exactly the same mean function and covariance function. One pseudorandom realization of each of the two random functions mentioned by Pugachev is presented in Figures 5.4–5.5.

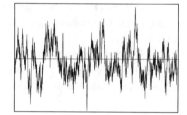

Figure 5.4. *Pseudorandom realization of a Gaussian random function with exponential covariance. The covariance function of this Gaussian random function is identical to that of the random function in Figure 5.5.*

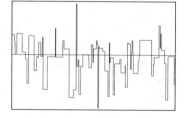

Figure 5.5. *Pseudorandom realization of a random function where the steps happen at random times (with constant probability density) and where the steps are random normally distributed (with zero mean). This (non-Gaussian) random function has the same covariance function as that in Figure 5.4, demonstrating that the mean and the covariance are not sufficient to characterize a random function.*

This should not be surprising, because this is only the equivalent, in infinite-dimensional spaces, of the well-known fact that one-dimensional random variables with completely different probability densities may well have the same mean and the same variance. The problem is that when the number of dimensions under consideration is high, it is not easy to obtain an intuitive idea of what the probability densities look like, and one may easily be misled with the assumptions made.

5.2   Solution of General Inverse Problems

In chapter 1, we introduced the notion of a (finite-dimensional) parameter manifold $\mathcal{X}$ over which probability distributions can be defined that are characterized by probability densities. It has been assumed that the notion of volume can be introduced over the manifold. When some coordinates $\{x^1, \ldots, x^n\}$ are chosen over the manifold, with this definition of volume is associated a particular probability density $\mu(\mathbf{x})$ that is homogeneous (with respect to the given definition of volume).

Given the manifold $\mathcal{X}$ and the homogeneous probability density $\mu(\mathbf{x})$, the conjunction of any two probability densities $f_1(\mathbf{x})$ and $f_2(\mathbf{x})$ has been defined as (equation (1.40))

$$\sigma(\mathbf{x}) \; = \; k \, \frac{f_1(\mathbf{x}) \, f_2(\mathbf{x})}{\mu(\mathbf{x})} \quad , \tag{5.13}$$

where k is a normalization constant. This was the basis of all the development in chapter 1 for the introduction of the general inverse problem.[63]

In section 5.1.2, we saw how actual probabilities can be computed in infinite-dimensional spaces (if the random functions are sufficiently regular). If this is the case, then equation (5.13) makes sense for every n-dimensional discretization, defining, say $\sigma_n(\mathbf{x})$. The probability $P_n(\mathfrak{A})$ of any set $\mathfrak{A}$ is computed using $\sigma_n(\mathbf{x})$, and the limit $P(\mathfrak{A}) = \lim_{n \to \infty} P_n(\mathfrak{A})$ is subsequently evaluated.

In this sense, all the results of the previous sections apply, and the only extra requirement, in the functional case, is to take the limit $n \to \infty$ of all the probabilities computed in n dimensions.

There are special circumstances where this limit can be considered from the beginning. This is the case for the Gaussian random functions, where many analytical results can be obtained, as we are about to see in the coming sections.

5.3   Introduction to Functional Least Squares

Some very general results exist concerning the solution of very special infinite-dimensional inverse problems. For instance, in a stochastic (Gaussian) context, Franklin (1970) gave the least-squares solution to inverse problems where both the model and data space can be infinite dimensional, but where data parameters are *linearly* related to model parameters. Backus (1970a,b,c) (see Appendix 6.15) gave the general Bayesian solution to problems where, if the model space can be infinite dimensional, it is assumed that the data space is finite dimensional and that the relationship between data and model parameters is linear, and where one is only interested in the prediction of a *finite* number of properties of the model (which are also *linear* functions of model parameters). These papers are historically important, but they do not provide for the infinite-dimensional problem a solution with a degree of generality comparable to that possible for finite-dimensional problems (see the first chapter of this book).

---

[63] Basically, the state of information describing measurements and a priori information was represented by a probability density $\rho(\mathbf{x})$ and the theoretical state of information by a probability density $\theta(\mathbf{x})$. The final state of information, obtained by combining these two states of information, was $\sigma(\mathbf{x}) = k \, \rho(\mathbf{x}) \, \theta(\mathbf{x}) \, / \, \mu(\mathbf{x})$.

In the following sections, we develop the theory of functional inverse problems in the special case where the 'input' random functions are Gaussian (so the 'output' random functions are not too far from Gaussian). I must again emphasize that some problems are functional but have a finite number of degrees of freedom (like when considering the random function displayed in Figure 5.2). These problems must be treated using the finite-dimensional methods of the previous chapters, and, therefore, they do not suffer from the limitation introduced by the Gaussian hypothesis.

In what follows, we assume that the functions under consideration belong to a linear space, so the sum of two functions is defined and makes physical sense (this sometimes requires a redefinition of the variables, as mentioned in section 5.1.4). The resolution of the inverse problem implies the use of differential and integral operators. Although the numerical results may be obtained after discretization, the functional language allows an indispensable compactness of the discussion. For instance, when the resolution of the forward problem involves differential equations, to obtain the differential equations giving the inverse solution is easier than when trying to formulate the inverse problem based on a discretized version of the forward problem. Once the final (functional) equation giving the solution of the inverse problem has been obtained, and in order to perform numerical computations, then (and only then) do we have to discretize everything: the original differential equation, its adjoint, and all the integral/differential equations obtained.

In writing a chapter like this one, it is difficult to choose an adequate level of mathematical rigor. Many least-squares formulas have a larger domain of validity than that for which rigorous mathematical proofs exist. In the coming pages, we shall not be preoccupied with pure mathematical aspects: common sense will guide the physicist well in this domain, with low probability of error. What matters to us here is to learn the intuitive meaning of a wide enough set of concepts and of the accompanying usual language (linear and nonlinear functionals, dual spaces, transposes of linear operators, etc.).

5.3.1 Representation of the Dual of a Linear Space

Let $\mathbb{E}$ be a linear space, eventually infinite dimensional, and let $\hat{\mathbb{E}}$ be its dual, i.e., the space of the linear real functionals over $\mathbb{E}$ (a real *functional* over $\mathbb{E}$ is an application from $\mathbb{E}$ into the real line $\mathbb{R}$). If $\mathbf{e}$ is an element of $\mathbb{E}$ and $\mathbf{f}$ is an element of $\hat{\mathbb{E}}$, the real number associated with $\mathbf{e}$ by $\mathbf{f}$ is denoted $\langle\, \mathbf{f}\,,\, \mathbf{e}\,\rangle$. It follows from Riesz's representation theorem (see, for instance, Taylor and Lay, 1980) that any linear functional $\mathbf{f}$ can be associated with an object with the same indices or variables as the elements of $\mathbb{E}$, also denoted $\mathbf{f}$, and such that if the matricial-like notation $\mathbf{f}^t\,\mathbf{e}$ represents the real (adimensional) number obtained by summing or integrating over all the (common) variables of $\mathbf{f}$ and $\mathbf{e}$, then

$$\langle\, \mathbf{f}\,,\, \mathbf{e}\,\rangle \;=\; \mathbf{f}^t\,\mathbf{e} \quad. \tag{5.14}$$

Example 5.2. *When analyzing an elastic sample in the laboratory, one may consider at every point* $\mathbf{x}$ *of the medium the mass density* $\rho(\mathbf{x})$ *and the incompressibility modulus* $\kappa(\mathbf{x})$. *As we wish to define a space of functions where the sum of two functions would conserve its meaning should one have considered, instead of mass density and incompressibility, the lightness density* $\ell = 1/\rho$ *and the compressibility modulus* $\gamma = 1/\kappa$, *we introduce the logarithmic pass density* $a(\mathbf{x}) = \log(\rho(\mathbf{x})/\rho_0)$ *and the logarithmic incompressibility* *modulus* $b(\mathbf{x}) = \log(\kappa(\mathbf{x})/\kappa_0)$ *(ρ_0 and κ_0 being two arbitrary constants). The elastic sample is then defined by the two functions*

$$\mathbf{e} = \begin{pmatrix} e^1(\mathbf{x}) \\ e^2(\mathbf{x}) \end{pmatrix} = \begin{pmatrix} a(\mathbf{x}) \\ b(\mathbf{x}) \end{pmatrix} \qquad (\mathbf{x} \in \mathcal{V}) \quad . \tag{5.15}$$

Let us denote by $\mathbb{E}$ the space of all such imaginable functions. Any element of the dual space $\hat{\mathbb{E}}$ can then be (uniquely) represented by another couple of functions (or, more generally, distributions):

$$\mathbf{f} = \begin{pmatrix} f_1(\mathbf{x}) \\ f_2(\mathbf{x}) \end{pmatrix} = \begin{pmatrix} \alpha(\mathbf{x}) \\ \beta(\mathbf{x}) \end{pmatrix} \qquad (\mathbf{x} \in \mathcal{V}) \quad , \tag{5.16}$$

such that the duality product is obtained by summing over all the variables,

$$\langle \mathbf{f} , \mathbf{e} \rangle = \mathbf{f}^t \mathbf{e} = \sum_{i=1}^{2} \int_{\mathcal{V}} dV(\mathbf{x}) \, f_i(\mathbf{x}) \, e^i(\mathbf{x}) \quad , \tag{5.17}$$

i.e.,

$$\langle \mathbf{f} , \mathbf{e} \rangle = \mathbf{f}^t \mathbf{e} = \int_{\mathcal{V}} dV(\mathbf{x}) \, \alpha(\mathbf{x}) \, a(\mathbf{x}) + \int_{\mathcal{V}} dV(\mathbf{x}) \, \beta(\mathbf{x}) \, b(\mathbf{x}) \quad . \tag{5.18}$$

5.3.2 Definition of a Covariance Operator

A *covariance operator* over $\mathbb{E}$ is, by definition, a linear symmetric definite nonnegative operator mapping $\hat{\mathbb{E}}$ into $\mathbb{E}$.

Example 5.3. *Let $\mathbf{C}$ be a covariance operator over the space $\mathbb{E}$ of Example 5.2, and let us write as*

$$\mathbf{e} = \mathbf{C}\hat{\mathbf{e}} \tag{5.19}$$

the application from $\hat{\mathbb{E}}$ into $\mathbb{E}$ defined by $\mathbf{C}$. An explicit representation of the operator $\mathbf{C}$ will be given by

$$a(\mathbf{x}) = \int_{\mathcal{V}} dV(\mathbf{x}) \, \mathbf{C}_{aa}(\mathbf{x}, \mathbf{x}') \, \alpha(\mathbf{x}) + \int_{\mathcal{V}} dV(\mathbf{x}) \, \mathbf{C}_{ab}(\mathbf{x}, \mathbf{x}') \, \beta(\mathbf{x}) \quad ,$$

$$b(\mathbf{x}) = \int_{\mathcal{V}} dV(\mathbf{x}) \, \mathbf{C}_{ba}(\mathbf{x}, \mathbf{x}') \, \alpha(\mathbf{x}) + \int_{\mathcal{V}} dV(\mathbf{x}) \, \mathbf{C}_{bb}(\mathbf{x}, \mathbf{x}') \, \beta(\mathbf{x}) \quad , \tag{5.20}$$

or, for short,

$$\begin{pmatrix} \mathbf{a} \\ \mathbf{b} \end{pmatrix} = \begin{pmatrix} \mathbf{C}_{aa} & \mathbf{C}_{ab} \\ \mathbf{C}_{ba} & \mathbf{C}_{bb} \end{pmatrix} \begin{pmatrix} \alpha \\ \beta \end{pmatrix} \quad . \tag{5.21}$$

We see that the kernel of the operator is here a 2×2 matrix of (covariance) functions.

Example 5.4. *Assume that we have N models of logarithmic density and logarithmic bulk modulus $(N \gg 1)$*

$$\begin{pmatrix} \mathbf{a}_1 \\ \mathbf{b}_1 \end{pmatrix} , \begin{pmatrix} \mathbf{a}_2 \\ \mathbf{b}_2 \end{pmatrix} , \cdots \quad . \tag{5.22}$$

Using the classical definitions of statistics, the mean model $\begin{pmatrix} \langle \mathbf{a} \rangle \\ \langle \mathbf{b} \rangle \end{pmatrix}$ *is defined by*

$$\langle a \rangle (\mathbf{x}) = \frac{1}{N} \sum_{i=1}^{N} a_i(\mathbf{x}) \quad , \quad \langle b \rangle (\mathbf{x}) = \frac{1}{N} \sum_{i=1}^{N} b_i(\mathbf{x}) \quad , \tag{5.23}$$

and the following covariance functions are defined:

$$
\begin{aligned}
C_{aa}(\mathbf{x}, \mathbf{x}') &= \frac{1}{N} \sum_{i=1}^{N} (a_i(\mathbf{x}) - \langle a \rangle (\mathbf{x}))(a_i(\mathbf{x}') - \langle a \rangle (\mathbf{x}')) \quad , \\
C_{ab}(\mathbf{x}, \mathbf{x}') &= \frac{1}{N} \sum_{i=1}^{N} (a_i(\mathbf{x}) - \langle a \rangle (\mathbf{x}))(b_i(\mathbf{x}') - \langle b \rangle (\mathbf{x}')) \quad , \\
C_{ba}(\mathbf{x}, \mathbf{x}') &= \frac{1}{N} \sum_{i=1}^{N} (b_i(\mathbf{x}) - \langle b \rangle (\mathbf{x}))(a_i(\mathbf{x}') - \langle a \rangle (\mathbf{x}')) \quad , \\
C_{bb}(\mathbf{x}, \mathbf{x}') &= \frac{1}{N} \sum_{i=1}^{N} (b_i(\mathbf{x}) - \langle b \rangle (\mathbf{x}))(b_i(\mathbf{x}') - \langle b \rangle (\mathbf{x}')) \quad .
\end{aligned}
\tag{5.24}
$$

A well-known result (e.g., Pugachev, 1965) is that the covariance functions are symmetric,

$$C_{aa}(\mathbf{x}, \mathbf{x}') = C_{aa}(\mathbf{x}', \mathbf{x}) \quad , \quad C_{ab}(\mathbf{x}, \mathbf{x}') = C_{ba}(\mathbf{x}', \mathbf{x}) \quad , \quad C_{bb}(\mathbf{x}, \mathbf{x}') = C_{bb}(\mathbf{x}', \mathbf{x}) \quad , \tag{5.25}$$

and definite nonnegative, i.e., for any functions $\phi_a(\mathbf{x})$ and $\phi_b(\mathbf{x})$, one has

$$
\begin{aligned}
&\int_{\mathcal{V}} d\mathbf{V}(\mathbf{x}) \, \phi_a(\mathbf{x}) \, C_{aa}(\mathbf{x}, \mathbf{x}') \, \phi_a(\mathbf{x}') + \int_{\mathcal{V}} d\mathbf{V}(\mathbf{x}) \, \phi_a(\mathbf{x}) \, C_{ab}(\mathbf{x}, \mathbf{x}') \, \phi_b(\mathbf{x}') \\
&+ \int_{\mathcal{V}} d\mathbf{V}(\mathbf{x}) \, \phi_b(\mathbf{x}) \, C_{ba}(\mathbf{x}, \mathbf{x}') \, \phi_a(\mathbf{x}') + \int_{\mathcal{V}} d\mathbf{V}(\mathbf{x}) \, \phi_b(\mathbf{x}) \, C_{bb}(\mathbf{x}, \mathbf{x}') \, \phi_b(\mathbf{x}') \geq 0 \quad .
\end{aligned}
\tag{5.26}
$$

These properties justify the abstract definition used for introducing covariance operators.

5.3.3 Covariance Functions and Associated Random Realizations

To have in mind what is exactly meant by a Gaussian random function with given mean and given covariance functions, let us see some examples.

Example 5.5. Exponential covariance function. *Let $\mathbb{M}$ be a linear space with functions $m(t)$, where t represents a spatial or temporal variable. A covariance operator over $\mathbb{M}$ will have as kernel a covariance function $C(t, t')$. One covariance function often encountered in practical applications is the exponential covariance function*

$$C(t, t') = \sigma^2 \exp(-|t - t'|/\tau) \quad . \tag{5.27}$$

Figure 5.6 *shows a (pseudo)random realization*[64] *of a Gaussian random function with zero expectation and such an exponential covariance. In two dimensions, the exponential covariance function is* $C(x, y, x', y') = \sigma^2 \exp(-\sqrt{(x - x')^2 + (y - y')^2}/\tau)$, *and a random realization is displayed in Figure* 5.7.

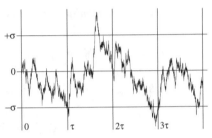

Figure 5.6. *Example of a pseudorandom realization of a Gaussian random function with zero mean and exponential covariance* $C(t, t') = \sigma^2 \exp(-|t - t'|/\tau)$. *Notice that the function has a discontinuous derivative at every point. The value of the function has been computed at* 40 000 *points (the printer device has much less resolution).*

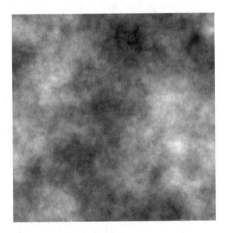

Figure 5.7. *A pseudorandom realization of a 2D Gaussian field with exponential covariance* $C(x, y, x', y') = \sigma^2 \exp(-\sqrt{(x - x')^2 + (y - y')^2}/L)$, *from Fenton (1990), with permission. The length of the side of the square is* 2.5 L .

An important classical result concerning a stationary (sufficiently regular) random function is that its realizations are differentiable n times if and only if the correlation function $\Gamma(z) = C(z, 0)$ is differentiable at $z = 0$ up to the order $2n$. For instance, for the realizations to be differentiable, the second derivative of $\Gamma(z)$ must exist at $z = 0$. This is not the case in the examples of Figures 5.6 and 5.7, so that the realizations shown are not differentiable. On the contrary, the realizations of a random function with Gaussian covariance are infinitely differentiable (they are *very* smooth), as the following example shows.

Let me mention here that, for the construction of pseudorandom realizations, like the ones displayed here, one may use different methods. One may, for instance, use the sequential realization method described in chapter 2. One may also convolve a Gaussian white noise by the appropriate filter (see section 5.3.4), a convolution that can be performed in the space domain of the Fourier domain. For details, see Fenton (1990).

---

[64]It can be demonstrated that a realization of a Gaussian random function with exponential covariance $C(t, t') = \sigma^2 \exp(-|t - t'|/\tau)$ can be generated by the algorithm $x(t + \Delta t) = \alpha\, x(t) + \sigma\sqrt{1 - \alpha^2}\,$ Gaussian$(0, 1)$, where $\alpha = \exp(-\Delta t/\tau)$.

Example 5.6. Gaussian covariance function. *Let* $\mathbb{M}$ *be a linear space with functions* $m(\mathbf{x})$, *where* $\mathbf{x}$ *represents the Cartesian coordinates of a point in a Euclidean space. Another covariance function often encountered in practical applications is the Gaussian covariance function*

$$C(\mathbf{x}, \mathbf{x}') \;=\; \sigma^2 \exp \left(-\frac{\|\mathbf{x} - \mathbf{x}'\|^2}{2\,L^2} \right) \; . \qquad (5.28)$$

The bottom of Figure 5.8 shows a pseudorandom realization of a two-dimensional Gaussian random function with zero mean and Gaussian covariance.

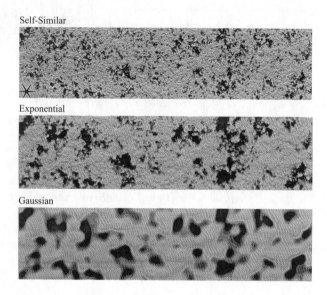

Figure 5.8. *Pseudorandom realizations of three types of two-dimensional Gaussian random fields (from Frankel and Clayton, 1986, with permission). At the top, with the Von Karman covariance function (Tatarski, 1961)* $C(x, y; x', y') = K_0(D/L)$, *where* $K_0(\cdot)$ *is the modified Bessel function of order zero and* $D^2 = (x - x')^2 + (y - y')^2$. *At the middle, with the exponential covariance function* $C(x, y; x', y') = \sigma^2 \exp(-D/L)$, *and at the bottom, with the Gaussian covariance function* $C(x, y; x', y') = \sigma^2 \exp(-D^2/(2\,L^2))$.

Example 5.7. Circular covariance function. *When working with stationary random fields in the two-dimensional Euclidean space, the* circular covariance[65] *model is often used. The covariance between two points separated by a distance* r *is*

$$\Psi(r) \;=\; \sigma^2 \left(1 - \frac{\beta + \sin\beta}{\pi} \right) \quad \text{with} \quad \beta = 2\arcsin\frac{r}{D} \qquad (5.29)$$

for $r \leq D$ *and* $\Psi(r) = 0$ *for* $r > D$.

---

[65]So called because $\Psi(r)$ is proportional to the area of intersection of two discs of diameter D whose centers are separated by the distance r.

Example 5.8. White-noise covariance function. *The covariance "function"*

$$C(t, t') \; = \; \beta \, \delta(t - t') \,, \tag{5.30}$$

where β is a finite constant and where $\delta(t)$ represents Dirac's delta distribution, corresponds to white noise (see Example 5.10). Figure 5.9 suggests what a realization of a white-noise random function looks like.

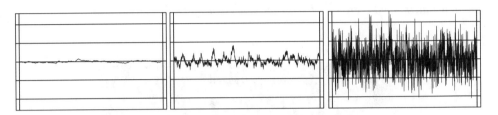

Figure 5.9. *A pseudorandom realization of a pure white-noise random function cannot be represented because the variance of the random function at any point is infinite. But white noise can be seen, for instance, as the limit of an ordinary Gaussian random function when the correlation length L tends to zero and the variance σ^2 tends to infinity, the product $L \sigma^2$ remaining constant (see Example 5.10). This figure illustrates such a limit. The three functions are pseudorandom realizations of a (stationary) Gaussian random function with zero mean value and exponential covariance function $C(z, z') = \sigma^2 \exp(-|z - z'|/L)$. The realization in the middle is the same as in Figure 5.6, represented here using a different vertical scale. At the top, the correlation length is 10 times larger (and the variance 10 times smaller), and at the bottom the correlation length is 10 times smaller (and the variance 10 times larger).*

Example 5.9. Random-walk covariance function. *Also interesting for practical applications is the covariance function*

$$C(t, t') \; = \; \beta \min(t, t') \qquad (\textit{for } t \geq 0 \textit{ and } t' \geq 0) \quad . \tag{5.31}$$

The variance at point t is $\sigma^2 = \beta \, t$. It corresponds to a (unidimensional) random walk, which, in turn, corresponds to the primitive of a white noise. See Figure 5.1 for some one-dimensional random walks.

Example 5.10. The use of white noise for modeling uncertainties. *A Gaussian random function $t \mapsto x(t)$ is named* white noise *if its covariance function is*

$$C(t, t') \; = \; w(t) \, \delta(t - t') \,, \tag{5.32}$$

where $w(t)$ is termed the intensity *of the white noise. In particular, it should be noticed that, for any t, the variance of the random variable $x(t)$ is infinite and that, for any t and t', the random variables $x(t)$ and $x(t')$ are uncorrelated. The easiest way to understand* *white noise is to consider it as a limit of a Gaussian random process where the correlation length tends to vanish. Take, for instance, a Gaussian random function with exponential covariance*

$$C(t, t') = \sigma^2 \exp\left(-\frac{|t - t'|}{T}\right) \quad . \tag{5.33}$$

In the limit when $T \to 0$, if the product $\sigma^2 T$ remains finite (and so, if the variance tends to infinity), the function $C(t, t')$ tends to the particular white-noise distribution

$$C(t, t') = 2\sigma^2 T \, \delta(t - t') \quad . \tag{5.34}$$

The adjective "white" comes from the fact that a realization of such a random function has a flat frequency spectrum (i.e., flat Fourier transform), like white light; the noun "noise" comes from the fact that a sound wave with such a spectrum really sounds like noise. The term $2\sigma^2 T$ in the last equation can be interpreted as the area under the curve $C(t, t')$:

$$\int_{-\infty}^{\infty} dt' \, C(t, t') = 2\sigma^2 T \quad \text{(for any t)} \quad . \tag{5.35}$$

It is not possible to represent a realization of a true white-noise random function because the values at any point are $+\infty$ or $-\infty$ (with probability 1), but it is possible to obtain a good intuitive feeling by a direct consideration of the concept of the limit of random functions whose correlation lengths tend to zero. Figure 5.9 shows an illustration. If instead of using an exponential covariance function we use a Gaussian covariance function,

$$C(t, t') = \sigma^2 \exp\left(-\frac{1}{2}\frac{(t - t')^2}{T^2}\right) \quad , \tag{5.36}$$

then, in the limit $T \to 0$, $\sigma^2 \to \infty$, $T\sigma^2$ constant,

$$C(t, t') = \sqrt{2\pi}\,\sigma^2 T \, \delta(t - t') \quad , \tag{5.37}$$

where, again, the factor of the delta function represents the area under the covariance function. Comparing (5.34) with (5.37), we see that the factor of $\delta(t - t')$ differs when obtaining white noise as the limit of different random functions. Assume that some data $d(t)$ consist of signal $s(t)$ plus noise $n(t)$:

$$d(t) = s(t) + n(t) \quad . \tag{5.38}$$

Usually, the d(t) data are sampled. If the correlation length of the signal is much larger than the correlation length of the noise, the last may be undersampled, and then, for all numerical purposes, its correlation length may be taken as null. The covariance function describing uncertainties in the observed data may then be approximated by that in equation (5.32), where

$$w(t) \simeq \text{(true variance of the noise)} \times \text{(true correlation length of the noise)} \quad . \tag{5.39}$$

This justifies the conventional use of white noise to represent some experimental uncertainties.

Example 5.11. Experimental estimation of a covariance function for describing exper-imental uncertainties. *Figure 5.10 shows an experimentally estimated covariance function for representing data uncertainties.*

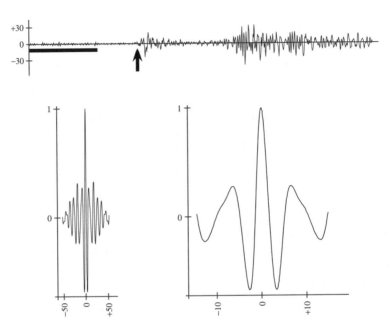

Figure 5.10. 900 s *of recording of the displacement of a point on Earth's surface south of France (top). One can see the arrival of the seismic waves produced by the big Chilean earthquake of March 3, 1985 (arrow). To estimate the noise contaminating the signal after the first wave arrival, 180 s of recording before the arrival have been selected (thick line). The normalized autocorrelation of this part of the recording is shown at the bottom in two different scales (in seconds). This can be taken directly as the correlation function for representing data uncertainties. (B. Romanowicz, pers. comm.).*

5.3.4   Realization with Prescribed Covariance

Let $\mathbf{w}$ have a Gaussian probability density with mean $\mathbf{w}_0$ and covariance $\mathbf{C}_w$:

$$f(\mathbf{w}) \; = \; \text{const. exp}\!\left(-\tfrac{1}{2}\,(\mathbf{w} - \mathbf{w}_0)^t\,\mathbf{C}_w^{-1}\,(\mathbf{w} - \mathbf{w}_0) \right) \quad . \tag{5.40}$$

If one defines

$$\mathbf{x} \; = \; \mathbf{L}\,\mathbf{w} \quad , \tag{5.41}$$

where $\mathbf{L}$ is an invertible matrix, what is the covariance matrix of $\mathbf{x}$?

The Jacobian property gives (as the Jacobian is constant) $g(\mathbf{x}) \; = \; k\,f(\mathbf{w}(\mathbf{x})) \; = \; k\,f(\mathbf{L}^{-1}\mathbf{x})$. Replacing $\mathbf{w}$ with $\mathbf{L}^{-1}\mathbf{x}$ in expression (5.40) and introducing $\mathbf{x}_0 = \mathbf{L}\,\mathbf{w}_0$

leads, after some easy rearrangements, to

$$g(\mathbf{x}) = \text{const. exp}\left(-\tfrac{1}{2}(\mathbf{x} - \mathbf{x}_0)^t \, \mathbf{C}_x^{-1} \, (\mathbf{x} - \mathbf{x}_0) \right) \tag{5.42}$$

with

$$\mathbf{x}_0 = \mathbf{L}\,\mathbf{w}_0 \quad \text{and} \quad \mathbf{C}_x = \mathbf{L}\,\mathbf{C}_w\,\mathbf{L}^t \quad . \tag{5.43}$$

This property can be used to generate a random realization of a Gaussian process with prescribed covariance. To simplify the discussion, assume that we wish to generate a random realization $\mathbf{x}$ of a Gaussian process with zero mean and covariance $\mathbf{C}_x$. It is quite easy to generate a random realization $\mathbf{w}$ of a Gaussian process with zero mean and unit covariance $\mathbf{C}_w = \mathbf{I}$ (a random realization of a Gaussian white noise). Solve the equation $\mathbf{C}_x = \mathbf{L}\mathbf{L}^t$ for $\mathbf{L}$ (using, for instance, the Cholesky decomposition of $\mathbf{C}_x$), and compute $\mathbf{x} = \mathbf{L}\mathbf{w}$. Then, $\mathbf{x}$ is a realization of a Gaussian process with zero mean and covariance $\mathbf{L}\,\mathbf{C}_w\,\mathbf{L}^t = \mathbf{L}\mathbf{L}^t = \mathbf{C}_x$.

This method has the advantage of being conceptually simple. By no means is it more efficient numerically. For a review of some efficient methods, including those working in the Fourier domain, see Fenton (1994). Frankel and Clayton (1986) show some examples of 2D random realizations of Gaussian fields (see Figure 5.8).

5.3.5 Weighting Operators and Least-Squares Norms

By definition, a covariance operator $\mathbf{C}$ is definite nonnegative and maps a linear vector space $\hat{\mathbb{E}}$ into its (anti)dual $\mathbb{E}$. If $\mathbf{C}$ is positive definite, then its inverse $\mathbf{C}^{-1}$ exists and maps $\mathbb{E}$ into $\hat{\mathbb{E}}$. $\mathbf{C}^{-1}$ is named the *weighting operator* and is often denoted by $\mathbf{W}$.

The expression

$$(\,\mathbf{e}_1\,,\,\mathbf{e}_2\,) = \mathbf{e}_1^t\,\mathbf{C}^{-1}\,\mathbf{e}_2 \tag{5.44}$$

can be shown to define a *scalar product* over $\mathbb{E}$ (see equation (6.24)). The expression

$$\|\,\mathbf{e}\,\| = (\,\mathbf{e}\,,\,\mathbf{e}\,)^{1/2} = (\mathbf{e}^t\,\mathbf{C}^{-1}\,\mathbf{e})^{1/2} \tag{5.45}$$

then defines a *norm* over $\mathbb{E}$. It is sometimes called an L_2-norm, although this terminology is improper (see Example 5.13). The name "least-squares norm" is more familiar, although a proper terminology would be "covariance-related norm."

A space $\mathbb{E}$ furnished with a scalar product is named a *Hilbert space* (in fact, it is only a *pre-Hilbert space* because it is not necessarily complete (see (6.24)), but for all interesting choices of covariance functions it will be (see, for instance, Example 5.13)).

Example 5.12. *Let $\mathbb{E}$ be the space of all functions $\mathbf{e} = \{e(t)\}$ and let $\hat{\mathbf{e}}$ be an element of $\hat{\mathbb{E}}$, dual of $\mathbb{E}$, related to an element $\mathbf{e}$ of $\mathbb{E}$ through*

$$\mathbf{e} = \mathbf{C}\,\hat{\mathbf{e}} \quad , \tag{5.46}$$

where $\mathbf{C}$ is a covariance operator over $\mathbb{E}$ with integral kernel (i.e., covariance function) $C(t, t')$:

$$e(t) = \int dt'\, C(t, t')\,\hat{e}(t') \quad . \tag{5.47}$$

By definition of the weighting operator $\mathbf{C}^{-1}$,

$$\hat{\mathbf{e}} = \mathbf{C}^{-1} \mathbf{e} \quad . \tag{5.48}$$

Formally, the integral kernel of the weighting operator $\mathbf{C}^{-1}$ *can be introduced by*

$$\hat{e}(t) = \int dt' \, C^{-1}(t, t') \, e(t') \quad . \tag{5.49}$$

This gives the formal equation

$$\int dt \int dt' \, C(t, t') \, C^{-1}(t', t'') = \delta(t - t'') \quad . \tag{5.50}$$

In fact, usual covariance functions are smooth functions, and the linear operator defined by equations (5.46)–(5.48) is a true integral operator. Its inverse is then a differential operator, and its integral representation (5.49) makes sense only if we interpret $C^{-1}(t, t')$ *as a distribution (see Example 5.13). Nevertheless, by linguistic abuse,* $C^{-1}(t, t')$ *is named the* weighting function.

Example 5.13. *Let* $C(t, t')$ *be the covariance function considered in Example 5.5:*

$$C(t, t') = \sigma^2 \exp\left(-\frac{|t - t'|}{T}\right) \quad . \tag{5.51}$$

σ^2 *is the* variance *of the random function;* T *is the* correlation length *and if the random process is Gaussian, it corresponds to the length along which successive values of any realization are correlated. The covariance operator* $\mathbf{C}$ *corresponding to the integral kernel (5.51) with any function* $\hat{e}(t)$ *associates the function*

$$e(t) = \int_{t_1}^{t_2} dt' \, C(t, t') \, \hat{e}(t') \quad . \tag{5.52}$$

As shown in Problem 7.21, we have

$$C^{-1}(t, t') = \frac{1}{2\sigma^2} \left(\frac{1}{T} \delta(t - t') - T \, \delta''(t - t')\right) \quad , \tag{5.53}$$

where the definition of the derivative of a distribution has been used.[66] For the norm of an element $\mathbf{e}$ *, this gives (see Problem 7.21), disregarding possible boundary terms,*

$$\| \mathbf{e} \|^2 = \frac{1}{2\sigma^2} \left(\frac{1}{T} \int_{t_1}^{t_2} dt \, [e(t)]^2 + T \int_{t_1}^{t_2} dt \left[\frac{de}{dt}(t)\right]^2\right) \quad . \tag{5.54}$$

This corresponds to the usual norm in the Sobolev space H^1 *(see Appendix 6.25), which equals the sum of the usual* L_2*-norm of the function and the* L_2*-norm of its derivative. See Problem 7.21 for details.*

---

[66] In particular, for Dirac's delta function, $\int dt' \, \delta^{(n)}(t - t') \, \mu(t') = (-1)^n \frac{d^n}{dt^n} \mu(t)$.

Example 5.14. *For* $0 \le t \le T$, *let* $C(t, t')$ *be the covariance function*

$$C(t, t') = \beta \min(t, t') \tag{5.55}$$

already considered in Example 5.9. The covariance operator **C** *whose kernel is this covariance function associates with any function* $\hat{e}(t)$ *the function*

$$e(t) = \int_0^L dt' \, C(t, t') \, \hat{e}(t') \quad . \tag{5.56}$$

As shown in Problem 7.22,

$$C^{-1}(t, t') = -\frac{1}{\beta} \delta''(t - t') \quad . \tag{5.57}$$

For the norm of an element $\mathbf{e} = \{e(t)\}$, *this gives (see Problem 7.22)*

$$\| \mathbf{e} \|^2 = \frac{1}{\beta} \int_0^T dt \left(\frac{de}{dt}(t) \right)^2 \quad , \tag{5.58}$$

that is, the ordinary L_2-*norm of the derivative of the function.*

Example 5.15. *The covariance function*

$$C(t, t') = w(t) \, \delta(t - t') \tag{5.59}$$

represents white noise (see Example 5.10). $w(t)$ *is termed the* intensity *of the white noise. It is easy to see that if* $e(t) = \int dt' \, C(t, t') \, \delta\hat{e}(t')$, *then* $\hat{e}(t) = \frac{1}{w(t)} e(t)$. *This corresponds to the kernel* $C^{-1}(t, t') = \frac{1}{w(t)} \delta(t - t')$. *The associated norm is the usual weighted* L_2-*norm*

$$\| \mathbf{e} \|^2 = \int dt \, \frac{e(t)^2}{w(t)} \quad . \tag{5.60}$$

5.4 Derivative and Transpose Operators in Functional Spaces

5.4.1 The Derivative Operator

When a manifold $\mathfrak{X}$ is considered that is not necessarily a linear space, endowed with some coordinates $\mathbf{x} = \{x^1, x^2, \dots\}$, with every point $\mathbf{x}_0 \in \mathfrak{X}$ one may associate a linear space $\mathbb{X}_0$, which is tangent to the manifold at the given point $\mathbf{x}_0$. A generic vector of $\mathbb{X}_0$ may be denoted $\delta\mathbf{x}$. When such a vector $\delta\mathbf{x} = \{\delta x^1, \delta x^2, \dots\}$ is small enough, it can be interpreted as the line going from point $\mathbf{x}_0 = \{x_0^1, x_0^2, \dots\}$ to point $\mathbf{x}_0 + \delta\mathbf{x} = \{x_0^1 + \delta x^1, x_0^2 + \delta x^2, \dots\}$.

Let $\mathfrak{M}$ and $\mathfrak{D}$ respectively be the model parameter manifold and the observable parameter manifold, not necessarily assumed to be linear spaces. The parameters $\{m^\alpha\}$ and

$\{d^i\}$ are interpreted as coordinates over the manifolds. Consider the (possibly nonlinear) operator $\mathbf{m} \to \mathbf{d} = \mathbf{g}(\mathbf{m})$ associating with any point $\mathbf{m}$ of the model space $\mathfrak{M}$ a point $\mathbf{d} = \mathbf{g}(\mathbf{m})$ of the data space $\mathfrak{D}$. The *tangent linear application* to $\mathbf{g}$ at a point $\mathbf{m} = \mathbf{m}_0$ is a linear operator, denoted $\mathbf{G}_0$, that maps the linear tangent space to $\mathfrak{M}$ at point $\mathbf{m}_0$ into the linear tangent space to $\mathfrak{D}$ at point $\mathbf{d}_0 = \mathbf{g}(\mathbf{m}_0)$, and is defined by the first-order development

$$\mathbf{g}(\mathbf{m}_0 + \delta\mathbf{m}) = \mathbf{g}(\mathbf{m}_0) + \mathbf{G}_0 \, \delta\mathbf{m} + \cdots \quad . \tag{5.61}$$

When applied to a scalar function $x \to g(x)$ of a one-dimensional variable, this expression defines the *tangent* (in the usual geometrical sense) to the graph $(x, g(x))$ at x_0 (see Figure 5.11), thus justifying the terminology. When $\mathfrak{M}$ and/or $\mathfrak{D}$ are functional spaces, the tangent linear application $\mathbf{G}_0$ is usually named the *Fréchet derivative* (or simply the *derivative*) of $\mathbf{g}$ at point $\mathbf{m}_0$. Note that this terminology may be misleading, because for a one-dimensional scalar function $x \to g(x)$, what is generally termed the derivative at a given point is the slope of the tangent (i.e., a number), not the tangent itself (i.e., an application).

Example 5.16. *If $\mathfrak{M}$ and $\mathfrak{D}$ are discrete,*

$$d^i = g^i(m^1, m^2, \ldots) \quad (i = 1, 2, \ldots) \quad . \tag{5.62}$$

From the definition (5.61), it follows that the kernel of the operator $\mathbf{G}_0$ is a matrix. Its elements are

$$(\mathbf{G}_0)^i{}_\alpha = \left. \frac{\partial g^i}{\partial m^\alpha} \right|_{\mathbf{m}_0} \quad . \tag{5.63}$$

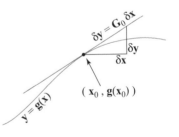

Figure 5.11. *The* tangent linear application *to the nonlinear application $y = g(x)$ at the point $(x_0, g(x_0))$ is the linear application whereby any δx is associated with $\delta y = G_0 \delta x$, where G_0 is the* slope *of the curve $y = g(x)$ at x_0. This definition generalizes to a nonlinear application $\mathbf{y} = \mathbf{g}(\mathbf{x})$ between functional spaces (see text). The kernel of the tangent linear application is then called the* Fréchet derivative *or simply the* derivative *at $\mathbf{m}_0$.*

Example 5.17. *Let $\mathfrak{M}$ be a space of functions $z \to m(z)$ for $0 \le z \le Z$, $\mathfrak{D}$ be a discrete N-dimensional manifold, and $\mathbf{m} \to \mathbf{g}(\mathbf{m})$ be the nonlinear operator from $\mathfrak{M}$ into $\mathfrak{D}$ that with the function $z \to m(z)$ associates the following element of $\mathfrak{D}$:*

$$d^i = g^i(\mathbf{m}) = \int_0^Z dz \, \beta^i(z) \, (m(z))^2 \quad (i = 1, 2, \ldots, N) \quad , \tag{5.64}$$

where $\beta^i(z)$ are given functions. If $z \rightarrow m_0(z)$ is a particular element of $\mathfrak{M}$, denoted $\mathbf{m}_0$, let us compute the derivative of $\mathbf{g}$ at the point $\mathbf{m}_0$. We have

$$
\begin{aligned}
g^i(\mathbf{m}_0 + \delta\mathbf{m}) &= \int_0^Z dz\, \beta^i(z)\, (m_0(z) + \delta m(z))^2 \\
&= \int_0^Z dz\, \beta^i(z)\, ((m_0(z))^2 + 2\, m_0(z)\, \delta m(z) + (\delta m(z))^2) \qquad (5.65) \\
&= g^i(\mathbf{m}_0) + 2 \int_0^Z dz\, \beta^i(z)\, m_0(z)\, \delta m(z) + \cdots \quad ,
\end{aligned}
$$

so, using the definition (5.61),

$$
\delta d^i = (\mathbf{G}_0\, \delta\mathbf{m})^i = 2 \int_0^Z dz\, \beta^i(z)\, m_0(z)\, \delta m(z) \quad . \qquad (5.66)
$$

In words, the derivative at the point $\mathbf{m} = \mathbf{m}_0$ of the nonlinear operator $\mathbf{m} \rightarrow \mathbf{g}(\mathbf{m})$ is the linear operator that with any function $\delta\mathbf{m}$ (i.e., any function $z \rightarrow \delta m(z)$) associates the vector given by (5.66). Introducing an integral representation of $\mathbf{G}_0$,

$$
(\mathbf{G}_0\, \delta\mathbf{m})^i = \int_0^Z dz\, G_0{}^i(z)\, \delta m(z) \quad , \qquad (5.67)
$$

gives the (integral) kernel of $\mathbf{G}_0$:

$$
G_0{}^i(z) = 2\, \beta^i(z)\, m_0(z) \quad . \qquad (5.68)
$$

Example 5.18. The Fréchet derivatives in X-ray tomography. *A useful technique for obtaining images of the interior of a human body is (computerized) X-ray tomography. As the etymology of the word indicates, tomography consists of obtaining graphics of a section of a body. Typically, X-rays are sent between a point source and a point receiver that counts the number of photons not absorbed by the medium, thus giving an indication of the integrated attenuation coefficient along that particular ray path (see Figure 5.12). Repeating the measurement for many different ray paths, conveniently sampling the medium, the bidimensional structure of the attenuation coefficient can be inferred, and so, an image of the medium be obtained. Ray paths of X-rays through an animal body can be assimilated to straight lines with an excellent approximation.*

More precisely, let ρ^i be the transmittance *along the ith ray, i.e., the* probability *of a photon being transmitted (approximately equal to the ratio between the emitted and the received intensity of the X-ray beam). It can be shown (see, for instance, Herman, 1980) that ρ^i is given by*

$$
\rho^i = \exp\left(-\int_{\mathcal{R}_i} ds\, \mu(\mathbf{x}) \right) \quad , \qquad (5.69)
$$

where $\mu(\mathbf{x})$ is the linear attenuation coefficient[67] *at point $\mathbf{x}$, $\mathcal{R}_i$ denotes the ray path, ds^i is the element of length along the ray path, and $\mathbf{x}$ denotes the current point considered in*

---

[67]The attenuation coefficient so defined is called 'linear' to distinguish it from the mass attenuation coefficient. Typical values for medical diagnostic X-rays are $\mu_{bone} = 0.480\,\text{cm}^{-1}$, $\mu_{muscle} = 0.180\,\text{cm}^{-1}$, $\mu_{blood} = 0.178\,\text{cm}^{-1}$.

Figure 5.12. *Schematic representation of an X-ray tomography experiment. Typically, a source sends a beam of X-rays, and the number of photons arriving at each of the receivers is counted. With a sufficient accuracy, photons can be assumed to propagate along straight lines. In the example in the figure, source and receivers are slowly rotating around the medium under study. Each particular measurement of the number of photons arriving at a source brings information about the integrated attenuation along a particular line across the medium. When sufficiently many integrated attenuations have been measured, with sufficiently different azimuths, the bidimensional structure of the attenuation coefficient can be inferred, thus obtaining a tomograph (i.e., an image of a section of the 3D medium).*

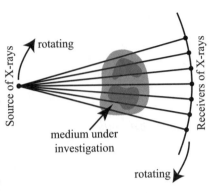

the line integral along $\mathcal{R}_i$. In a three-dimensional experience, a point $\mathbf{x}$ is represented by the Cartesian coordinates (x, y, z) or the spherical coordinates (r, θ, ϕ).

We could take here as basic parameters the transmittances $\{\rho^i\}$ and the linear attenuation coefficients $\{\mu(\mathbf{x})\}$, but, as we have in view a least-squares formulation of the inverse problem, it is better to change the variables, defining the logarithmic transmittance d^i and the logarithmic linear attenuation $m(x)$ by

$$d^i = -\log \rho^i \qquad and \qquad m(\mathbf{x}) = \log \frac{\mu(\mathbf{x})}{K} \quad , \tag{5.70}$$

where K is an arbitrary constant value of the attenuation coefficient. The reason for doing so is that positive parameters may never have Gaussian statistics (that give finite probability to negative values of the parameter). Then, rather than complicating the problem by the explicit introduction of log-normal statistics, it is better to use the logarithmic parameters just introduced.[68] The relation (5.69) between data and unknowns transforms into

$$d^i = K \int_{\mathcal{R}_i} ds \, \exp(m(\mathbf{x})) \quad , \tag{5.71}$$

an expression that, for more compactness, we can denote by $\mathbf{d} = \mathbf{g}(\mathbf{m})$, with $\mathbf{d} = \{d^i\}$ and $\mathbf{m} = \{m(\mathbf{x})\}$. Let $\mathbf{m}_0 = \{m_0(\mathbf{x})\}$ be a particular model and let us evaluate the Fréchet derivative of $\mathbf{g}$ at point $\mathbf{m}_0$. To implement here the definition given in equation (5.61), we successively write $d^i(m_0(\mathbf{x}) + \delta m(\mathbf{x})) = K \int_{\mathcal{R}_i} ds \, \exp(m_0(\mathbf{x}) + \delta m(\mathbf{x})) = K \int_{\mathcal{R}_i} ds \, \exp(m_0(\mathbf{x})) \exp(\delta m(\mathbf{x})) = K \int_{\mathcal{R}_i} ds \, \exp(m_0(\mathbf{x})) \, (1 + \delta m(\mathbf{x}) + \cdots)$, i.e.,

$$d^i(m_0(\mathbf{x}) + \delta m(\mathbf{x})) = d^i(m_0(\mathbf{x})) + K \int_{\mathcal{R}_i} ds \, \exp(m_0(\mathbf{x})) \, \delta m(\mathbf{x}) + \cdots \quad . \tag{5.72}$$

---

[68]It is here assumed that we are in the high attenuation regime, where the ρ^i are closer to zero than to one. In fact, as ρ is a probability, it takes values in the range $(0, 1)$ and one should define the new parameter $\pi = \pm \log \frac{\rho}{(1-\rho)}$.

Therefore, the tangent linear application to the nonlinear application in equation (5.71), evaluated for $\mathbf{m}_0 = \{m_0(\mathbf{x})\}$, *is the linear application that with any model perturbation* $\delta\mathbf{m} = \{\delta m(\mathbf{x})\}$ *associates the 'data perturbation'*

$$\delta d^i = K \int_{\mathcal{R}_i} ds \, \exp(m_0(\mathbf{x})) \, \delta m(\mathbf{x}) \quad . \tag{5.73}$$

Writing this formally as

$$\delta\mathbf{d} = \mathbf{G}_0 \, \delta\mathbf{m} \tag{5.74}$$

allows us to formally introduce the integral kernel (Fréchet derivative) $G_0^i(\mathbf{x})$ *of the tangent linear application* $\mathbf{G}_0$ *as*

$$\delta d^i = \int_V dV(\mathbf{x}) \, G_0^i(\mathbf{x}) \, \delta m(\mathbf{x}) \quad , \tag{5.75}$$

where V *is the volume under investigation. Introducing a delta-like function* $\Delta(\mathbf{x}; \mathcal{R}_i)$ *that is zero everywhere in the space except along the* i*th ray path by the condition that, for any function* $\psi(\mathbf{x})$,

$$\int_{\mathcal{R}_i} ds \, \psi(x) = \int_V dV(\mathbf{x}) \, \Delta(\mathbf{x}, \mathcal{R}_i) \, \psi(\mathbf{x}) \quad , \tag{5.76}$$

allows us to formally express the integral kernel $G_0^i(\mathbf{x})$ *as*

$$G_0^i(\mathbf{x}) = K \, \exp(m_0(\mathbf{x})) \, \Delta(\mathbf{x}; \mathcal{R}_i) \quad . \tag{5.77}$$

If, according to the definition at right in equation (5.70), we introduce $m_0(\mathbf{x}) = \log(\mu_0(\mathbf{x}) / K)$, *then the kernel of* $\mathbf{G}_0$ *can be written* $G^i(\mathbf{x}) = \mu_0(\mathbf{x}) \, \Delta(\mathbf{x}; \mathcal{R}_i)$.

Example 5.19. The Fréchet derivatives in travel-time tomography. *To infer the velocity structure of a medium, acoustic waves are generated by some sources, and the travel times to some receivers are measured. The main difference (and difficulty) with respect to the problem of X-ray tomography is that some media are highly heterogeneous, and the ray paths may depend on the velocity structure (see Figure 5.13).*

Here we assume that the high-frequency limit is acceptable, i.e., ray theory can be used instead of wave theory. Let $c(\mathbf{x})$ *denote the celerity of the waves at point* $\mathbf{x}$, *and let* $n(\mathbf{x})$ *be the slowness. The* i*th datum is the travel time for the* i*th ray and is given by either of the two equivalent expressions*

$$d^i = \int_{\mathcal{R}_i(\mathbf{c})} ds \, \frac{1}{c(\mathbf{x})} = \int_{\mathcal{R}_i(\mathbf{n})} ds \, n(\mathbf{x}) \quad , \tag{5.78}$$

where $\mathcal{R}_i$ *denotes the* i*th ray path (that depends on the celerity [or slowness] of the medium). We wish the formulation of the problem to be independent of arbitrarily choosing the celerity*

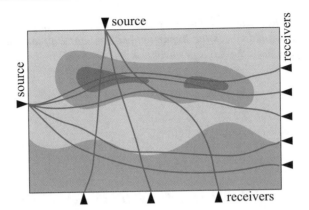

Figure 5.13. *Typical experiment of travel-time tomography. A source of acoustic waves is set at different positions, and, for each source position, some receivers are placed around the medium, which record the time of arrival of the first wavefront. The purpose of the experiment is to infer the acoustic structure of the medium. This problem is very similar to that of X-ray tomography, except that the ray paths are a priori unknown, because they depend on the actual structure of the medium.*

or the slowness, but this question is automatically solved when, for the reasons exposed in Example 5.18, one introduces the logarithmic slowness (or, equivalently, the logarithmic velocity[69])

$$m(\mathbf{x}) \;=\; \log \frac{n(\mathbf{x})}{K} \quad , \tag{5.79}$$

where K is some constant value of the slowness. Then,

$$d^i \;=\; g^i(\mathbf{m}) \;=\; K \int_{\mathcal{R}_i(\mathbf{m})} ds \; \exp(\, m(\mathbf{x})\,) \tag{5.80}$$

(written, for short, $\mathbf{d} = \mathbf{g}(\mathbf{m})$). This expression has two nonlinearities in it, one due to the exponential function and another due to the fact that the ray path depends on the function $m(\mathbf{x})$. Given a medium $\mathbf{m} = \{m(\mathbf{x})\}$, the actual ray path is obtained using Fermat's theorem (or, equivalently, the eikonal equation) and some numerical method. Here we wish to obtain the derivative of the nonlinear operator $\mathbf{g}$ at a point $\mathbf{m}_0$. We have

$$g^i(\mathbf{m}_0 + \delta\mathbf{m}) \;=\; K \int_{\mathcal{R}_i(\mathbf{m}_0 + \delta\mathbf{m})} ds \; \exp(\, m_0(\mathbf{x}) + \delta m(\mathbf{x})\,) \quad . \tag{5.81}$$

The travel time being stationary along the actual ray path (Fermat's theorem),

$$\int_{\mathcal{R}_i(\mathbf{m}_0 + \delta\mathbf{m})} ds \; \exp(\, m_0(\mathbf{x}) + \delta m(\mathbf{x})\,) = \int_{\mathcal{R}_i(\mathbf{m}_0)} ds \; \exp(\, m_0(\mathbf{x}^i) + \delta m(\mathbf{x}^i)\,) + O(\|\delta\mathbf{m}\|^2) \quad . \tag{5.82}$$

---

[69]The logarithmic velocity is just the opposite of the logarithmic slowness.

We can then make a development very similar to that in Example 5.18 to arrive at

$$g^i(\mathbf{m}_0 + \delta\mathbf{m}) = g^i(\mathbf{m}_0) + K \int_{\mathcal{R}_i(\mathbf{m}_0)} ds \, \exp(m_0(\mathbf{x})) \, \delta m(\mathbf{x}) + \cdots \quad , \tag{5.83}$$

which is very similar to equation (5.73), except that the ray path here depends on $\mathbf{m}_0$. *Therefore, the tangent linear application to the application defined by equation (5.80), when evaluated at* $\mathbf{m}_0 = \{m_0(\mathbf{x})\}$, *is the linear application that with every perturbation* $\delta m(\mathbf{x})$ *of the logarithmic slowness associates the travel-time perturbation*

$$\delta d^i = K \int_{\mathcal{R}_i(\mathbf{m}_0)} ds \, \exp(m_0(\mathbf{x})) \, \delta m(\mathbf{x}) \quad . \tag{5.84}$$

Equation (5.84) is well adapted to all numerical computations involving the derivative operator $\mathbf{G}_0$, *but for analytic developments it is sometimes useful to introduce the kernel of* $\mathbf{G}_0$ *(i.e., the Fréchet derivative). Introducing, as in the previous example, a delta-like function* $\Delta(\mathbf{x}; \mathcal{R}_i(\mathbf{m}))$ *that is zero everywhere in the space except along the* ith *ray path (this time, the ray path depends on the model* $\mathbf{m} = \{m(\mathbf{x})\}$ *) by the condition that, for any function* $\psi(\mathbf{x})$,

$$\int_{\mathcal{R}_i(\mathbf{m})} ds \, \psi(x) = \int_V dV(\mathbf{x}) \, \Delta(\mathbf{x}, \mathcal{R}_i(\mathbf{m})) \, \psi(\mathbf{x}) \quad , \tag{5.85}$$

allows us to formally express the Fréchet derivative as

$$G_0^i(\mathbf{x}) = K \, \exp(m_0(\mathbf{x})) \, \Delta(\mathbf{x}; \mathcal{R}_i(\mathbf{m}_0)) \quad , \tag{5.86}$$

or, expressing this derivative in terms of the slowness, $G_0^i(\mathbf{x}) = n_0(\mathbf{x}) \, \Delta(\mathbf{x}; \mathcal{R}_i(\mathbf{n}_0))$. *See Example 5.18 for details.*

Example 5.20. The Fréchet derivatives in the problem of inversion of acoustic waveforms. *Consider a three-dimensional unbounded medium supporting the propagation of acoustic waves, assumed to propagate according to the wave equation*

$$\frac{1}{\kappa(\mathbf{x})} \frac{\partial^2 p}{\partial t^2}(\mathbf{x}, t) - \mathrm{div}\left(\frac{1}{\rho(\mathbf{x})} \mathrm{grad} \, p(\mathbf{x}, t)\right) = S(\mathbf{x}, t) \quad , \tag{5.87}$$

where $\mathbf{x}$ *denotes a point of the medium,* t *is time,* $p(\mathbf{x}, t)$ *is the pressure perturbation at point* $\mathbf{x}$ *and time* t, $S(\mathbf{x}, t)$ *is the source function, and* $\rho(\mathbf{x})$ *and* $\kappa(\mathbf{x})$ *are respectively the mass density and bulk modulus at point* $\mathbf{x}$. *The pressure perturbation field* $p(\mathbf{x}, t)$ *is supposed to be at rest at the initial time:*

$$p(\mathbf{x}, 0) = 0 \quad , \qquad \dot{p}(\mathbf{x}, 0) = 0 \quad . \tag{5.88}$$

Assume that we have complete control of the source $S(\mathbf{x}, t)$, *so that we can send into the medium any type of wave we wish (usually the source has limited spatial extension and we send quasi-spherical waves), and that we have some sensors that are able to measure the actual value of the pressure at arbitrary points in the medium.*

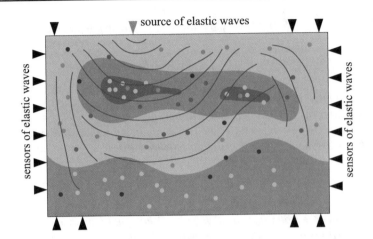

Figure 5.14. *Typical waveform acoustic experiment: Pressure waves are generated by some source, they interact with the medium, and the total pressure field is recorded by some receivers. The problem is to infer the acoustic structure of the medium.*

Assume we start an experiment at $t = 0$, sending waves into the medium by an appropriate choice of the source function $S(\mathbf{x}, t)$. From $t = 0$ to $t = T$, we record the pressure perturbation at some (finite number of) points $\mathbf{x}_r$ ($r = 1, 2, \dots$). We wish to use the results of our measurements to infer the actual value of the functions $\kappa(\mathbf{x})$ and $\rho(\mathbf{x})$ characterizing the medium (see Figure 5.14). This is, of course, an inverse problem, and will be solved in the next sections. Let us first consider the resolution of the forward problem.

Our (linear) data space $\mathbb{D}$ consists of all conceivable realizations for the results of our measurements:

$$\mathbf{d} = \{ \, p(\mathbf{x}_r, t) \quad for \quad 0 \le t \le T \quad and \quad r = 1, 2, \dots \} \quad . \tag{5.89}$$

Assume that we are only interested in the bulk modulus $\kappa(\mathbf{x})$ (the development is quite similar for the mass density). In the context of least squares, it is better to explicitly introduce the logarithmic bulk modulus

$$m(\mathbf{x}) = \log \frac{K}{\kappa(\mathbf{x})} = -\log \frac{\kappa(\mathbf{x})}{K} \quad , \tag{5.90}$$

where K is an arbitrary constant value of κ: in this way, a priori information on the bulk modulus is easily introduced using Gaussian random functions (this would not be the case for a positive parameter). Our (linear) model space $\mathbb{M}$ consists then of all possible models $m(\mathbf{x})$.

The original differential system becomes

$$\frac{\exp(m(\mathbf{x}))}{K} \frac{\partial^2 p}{\partial t^2}(\mathbf{x}, t) - \operatorname{div}\left(\frac{1}{\rho(\mathbf{x})} \operatorname{grad} p(\mathbf{x}, t) \right) = S(\mathbf{x}, t) \quad , \quad t \in [0, T] \quad ,$$

$$p(\mathbf{x}, 0) = 0 \quad ,$$

$$\dot{p}(\mathbf{x}, 0) = 0 \quad . \tag{5.91}$$

Given a model **m**, *to solve the forward problem means to compute the data vector* **d** *predicted from the knowledge of* **m**. *This is denoted by*

$$\mathbf{d} = \mathbf{g}(\mathbf{m}) \quad . \tag{5.92}$$

The most general way of solving the forward problem is to replace the differential equation in (5.91) with its finite-difference approximation (see, for instance, Alterman and Karal, 1968) and to compute numerically the approximated values of the pressure perturbation at time $t + \Delta t$ *from the knowledge of the approximated values of the pressure perturbation at times* t *and* $t - \Delta t$. *This gives the pressure perturbation field everywhere. The forward problem is solved by just picking the values obtained at the receiver locations* $\mathbf{x}_r$.

Let $\mathbf{m}_0$ *denote a particular model. We wish to compute the derivative of* **g** *at* $\mathbf{m} = \mathbf{m}_0$. *By definition,*

$$\mathbf{g}(\mathbf{m}_0 + \delta\mathbf{m}) = \mathbf{g}(\mathbf{m}_0) + \mathbf{G}_0\,\delta\mathbf{m} + O(\|\,\delta\mathbf{m}\,\|^2) \quad . \tag{5.93}$$

Let us denote

$$\delta\mathbf{p} = \mathbf{G}_0\,\delta\mathbf{m} \quad . \tag{5.94}$$

As demonstrated in Problem 7.18, $\delta p(\mathbf{x}_r, t)$ *is the value at* $\mathbf{x}_r$ *and time* t *of the field* $\delta p(\mathbf{x}, t)$, *solution of*

$$\frac{\exp(m_0(\mathbf{x}))}{K}\frac{\partial^2 \delta p}{\partial t^2}(\mathbf{x}, t) - \operatorname{div}\left(\frac{1}{\rho(\mathbf{x})}\operatorname{grad}\delta p(\mathbf{x}, t)\right)$$
$$= -\frac{\partial^2 p_0}{\partial t^2}(\mathbf{x}, t)\,\frac{\exp(m_0(\mathbf{x}))}{K}\,\delta m(\mathbf{x}) \quad , \quad t \in [0, T] \quad , \tag{5.95}$$
$$\delta p(\mathbf{x}, 0) = 0 \quad ,$$
$$\delta \dot{p}(\mathbf{x}, 0) = 0 \quad ,$$

where $p_0(\mathbf{x}, t)$ *is the field propagating in the model* $\mathbf{m}_0$:

$$\frac{\exp(m_0(\mathbf{x}))}{K}\frac{\partial^2 p_0}{\partial t^2}(\mathbf{x}, t) - \operatorname{div}\left(\frac{1}{\rho(\mathbf{x})}\operatorname{grad}p_0(\mathbf{x}, t)\right) = S(\mathbf{x}, t) \quad , \quad t \in [0, T] \quad ,$$
$$p_0(\mathbf{x}, 0) = 0 \quad ,$$
$$\dot{p}_0(\mathbf{x}, 0) = 0 \quad . \tag{5.96}$$

To compute $\delta p(\mathbf{x}, t)$, *we then essentially need to solve a forward problem (and we have already seen how to do this) using* $-\frac{\partial^2 p_0}{\partial t^2}\frac{\exp m_0}{K}\delta m$ *as source term and propagating the waves in the unperturbed medium* $m_0(\mathbf{x})$. *If we are able to compute* $\delta p(\mathbf{x}, t)$, *then from (5.94) we know the derivative operator* $\mathbf{G}_0$. *In words,* $\mathbf{G}_0$ *is the linear operator that associates the values* $\delta p(\mathbf{x}_r, t)$, *obtained by the resolution of (5.95), with the model perturbation* $\delta m(\mathbf{x})$.

Let me point out here that to estimate data $\mathbf{g}(\mathbf{m}_0 + \delta\mathbf{m})$ *using the first-order approximation*

$$\mathbf{g}(\mathbf{m}_0 + \delta\mathbf{m}) \simeq \mathbf{g}(\mathbf{m}_0) + \mathbf{G}_0\,\delta\mathbf{m} \tag{5.97}$$

corresponds to Born's approximation. *We are not using Born's approximation here. We are performing an exact evaluation of the Fréchet derivative.*

As shown in Problem 7.18, the kernel $\mathbf{G}_0(\mathbf{x}_r, t; \mathbf{x})$ *introduced by the integral representation*

$$\delta p(\mathbf{x}_r, t) = \int_{\mathcal{V}} dV(\mathbf{x}) \, \mathbf{G}_0(\mathbf{x}_r, t; \mathbf{x}) \, \delta \mathbf{m}(\mathbf{x}) \tag{5.98}$$

is given by (special case of equation (7.239))

$$\mathbf{G}_0(\mathbf{x}_r, t; \mathbf{x}) = -\frac{\exp(m_0(\mathbf{x}))}{K} \, \Gamma_0(\mathbf{x}_r, t; \mathbf{x}, 0) * \frac{\partial^2 p_0}{\partial t^2}(\mathbf{x}, t) \quad , \tag{5.99}$$

where $\Gamma_0(\mathbf{x}, t; \mathbf{x}', 0)$ *is Green's function, defined as the solution of*

$$\frac{\exp(\exp m_0(\mathbf{x}))}{K} \frac{\partial^2 \Gamma_0}{\partial t^2}(\mathbf{x}, t; \mathbf{x}', t') - \operatorname{div}\left(\frac{1}{\rho(\mathbf{x})} \operatorname{grad} \Gamma_0(\mathbf{x}, t; \mathbf{x}', t')\right)$$
$$= \delta(\mathbf{x} - \mathbf{x}')(t - t') \quad , \tag{5.100}$$

$$\Gamma_0(\mathbf{x}, t; \mathbf{x}', t') = 0 \qquad \text{for } t < t' \quad ,$$

$$\frac{\partial \Gamma_0}{\partial t}(\mathbf{x}, t; \mathbf{x}', t') = 0 \qquad \text{for } t < t' \quad .$$

5.4.2 The Transpose Operator

Let $\mathbf{G}$ represent a linear operator from a linear space $\mathbb{M}$ into a linear space $\mathbb{D}$. Its *transpose*, $\mathbf{G}^t$, is a linear operator mapping $\hat{\mathbb{D}}$ (the dual of $\mathbb{D}$) into $\hat{\mathbb{M}}$ (the dual of $\mathbb{M}$), defined by

$$\boxed{\langle \mathbf{G}^t \, \hat{\mathbf{d}} \, , \, \mathbf{m} \, \rangle_{\mathbb{M}} = \langle \, \hat{\mathbf{d}} \, , \, \mathbf{G} \mathbf{m} \, \rangle_{\mathbb{D}} \qquad \text{(for any } \hat{\mathbf{d}} \text{ and } \mathbf{m}\text{)}} \quad . \tag{5.101}$$

The reader should refer to Appendix 6.14 for (important) details.

Notice that the transpose of a linear operator is defined independently of any particular choice of scalar products over $\mathbb{M}$ and $\mathbb{D}$.

Again, let $\mathbf{G}$ represent a linear operator from a linear space $\mathbb{M}$ into a linear space $\mathbb{D}$, and assume now that $\mathbb{M}$ and $\mathbb{D}$ are scalar product vector spaces with respective scalar products

$$(\mathbf{m}_1 \, , \, \mathbf{m}_2)_{\mathbb{M}} = \langle \mathbf{C}_{\mathbb{M}}^{-1} \mathbf{m}_1 \, , \, \mathbf{m}_2 \rangle_{\mathbb{M}} \quad , \qquad (\mathbf{d}_1 \, , \, \mathbf{d}_2)_{\mathbb{M}} = \langle \mathbf{C}_{\mathbb{D}}^{-1} \mathbf{d}_1 \, , \, \mathbf{d}_2 \rangle_{\mathbb{D}} \quad . \tag{5.102}$$

The *adjoint* of $\mathbf{G}$, $\mathbf{G}^*$, is a linear operator mapping $\mathbb{D}$ into $\mathbb{M}$ defined by the equality of scalar products

$$\boxed{(\mathbf{G}^* \mathbf{d} \, , \, \mathbf{m})_{\mathbb{M}} = (\mathbf{d} \, , \, \mathbf{G} \mathbf{m})_{\mathbb{D}} \qquad \text{(for any } \mathbf{d} \text{ and } \mathbf{m}\text{)}} \quad . \tag{5.103}$$

As one can successively write

$$\begin{aligned} (\mathbf{G}^* \mathbf{d} \, , \, \mathbf{m})_{\mathbb{M}} &= (\mathbf{d} \, , \, \mathbf{G} \mathbf{m})_{\mathbb{D}} = \langle \mathbf{C}_{\mathbb{D}}^{-1} \mathbf{d} \, , \, \mathbf{G} \mathbf{m} \rangle_{\mathbb{D}} \\ &= \langle \mathbf{G}^t \, \mathbf{C}_{\mathbb{D}}^{-1} \mathbf{d} \, , \, \mathbf{m} \rangle_{\mathbb{M}} = (\mathbf{C}_{\mathbb{M}} \mathbf{G}^t \, \mathbf{C}_{\mathbb{D}}^{-1} \mathbf{d} \, , \, \mathbf{m})_{\mathbb{M}} \quad , \end{aligned} \tag{5.104}$$

one has the following relation between adjoint and transpose:

$$\boxed{\mathbf{G}^* = \mathbf{C}_\mathrm{M}\, \mathbf{G}^t\, \mathbf{C}_\mathrm{D}^{-1}} \ .$$

(5.105)

By linguistic abuse, the terms *adjoint* and *transpose* are sometimes used as synonym. This equation shows that they are not.

Let us see some examples practically illustrating the operational aspects of the definition of the transpose of a linear operator. In discrete spaces, a linear equation

$$\mathbf{d}_1 = \mathbf{G}\,\mathbf{m}_1$$

(5.106)

is explicitly written

$$(\mathbf{d}_1)^i = \sum_\alpha (\mathbf{G})^i{}_\alpha\, (\mathbf{m}_1)^\alpha \ ,$$

(5.107)

and the array $G^i{}_\alpha$ is termed the *kernel* of $\mathbf{G}$. If the indexes i and α are simple integers, then the array $G^i{}_\alpha$ is a usual (two-dimensional) matrix. A linear equation

$$\hat{\mathbf{m}}_2 = \mathbf{G}^t\, \hat{\mathbf{d}}_2$$

(5.108)

is explicitly written

$$(\mathbf{m}_2)^\alpha = \sum_i (\mathbf{G}^t)^\alpha{}_i\, (\mathbf{d}_2)^i \ ,$$

(5.109)

and the definition (5.101) gives directly

$$(\mathbf{G}^t)^\alpha{}_i = (\mathbf{G})^i{}_\alpha \ ,$$

(5.110)

which simply shows that the kernel of an operator and the kernel of the corresponding transpose operator are essentially the same, modulo a 'transposition' of the variables. In particular, if the kernel of the linear operator $\mathbf{G}$ is a matrix, the kernel of the transpose operator is simply the transpose of the matrix. If we consider more general spaces, a linear equation $\mathbf{d}_1 = \mathbf{G}\,\mathbf{m}_1$ may take, for instance, the explicit form

$$(\mathbf{d}_1)^{ij\cdots}(u, v, \dots)$$

$$= \sum_\alpha \sum_\beta \cdots \int dx \int dy \cdots\ (\mathbf{G})^{ij\cdots}{}_{\alpha\beta\dots}(u, v, \dots, x, y, \dots)\,(\mathbf{m}_1)^{\alpha\beta\cdots}(x, y, \dots) \ .$$

(5.111)

A linear equation $\hat{\mathbf{m}}_2 = \mathbf{G}^t\, \hat{\mathbf{d}}_2$ is then explicitly written

$$(\hat{\mathbf{m}}_2)^{\alpha\beta\cdots}(x, y, \dots)$$

$$= \sum_i \sum_j \cdots \int du \int dv \cdots (\mathbf{G}^t)^{\alpha\beta\cdots}{}_{ij\dots}(x, y, \dots, u, v, \dots)\,(\hat{\mathbf{d}}_2)^{ij\cdots}(u, v, \dots) \ ,$$

(5.112)

and the definition of $\mathbf{G}^t$ then gives

$$(\mathbf{G}^t)^{\alpha\beta\ldots}{}_{ij\ldots}(x, y, \ldots, u, v, \ldots) = \mathbf{G}^{ij\ldots}{}_{\alpha\beta\ldots}(u, v, \ldots, x, y, \ldots) \quad , \qquad (5.113)$$

thus generalizing the notion of "variable transposition." In words, *if the kernel of a linear operator is $\mathbf{G}^{ij\ldots}{}_{\alpha\beta\ldots}(u, v, \ldots, x, y, \ldots)$, and the application of $\mathbf{G}$ implies sums (or integrals) over the variables $\alpha, \beta, \ldots, x, y, \ldots$, then the kernel of the transpose operator is essentially the same, and the application of $\mathbf{G}^t$ implies sums over the other variables $i, j, \ldots, u, v, \ldots$.*

As a further example, consider the operator $\mathbf{G}$ mapping the space $\mathbb{M}$ into the space $\mathbb{D}$. $\mathbf{G}^t$ then maps $\hat{\mathbb{D}}$ into $\hat{\mathbb{M}}$. If $\mathbf{C}$ is a covariance operator over $\mathbb{M}$, it maps $\hat{\mathbb{M}}$ into $\mathbb{M}$, and it makes sense to consider the operator

$$\mathbf{U} = \mathbf{C}\,\mathbf{G}^t \qquad\qquad\qquad (5.114)$$

mapping $\hat{\mathbb{D}}$ into $\mathbb{M}$. Let $G^{ij\ldots\alpha\beta\ldots}(u, v, \ldots, x, y, \ldots)$ and $C^{\alpha\beta\ldots\alpha'\beta'\ldots}(x, y, \ldots, x', y', \ldots)$ be respectively the kernels of $\mathbf{G}$ and $\mathbf{C}$. As can easily be verified, the kernel of $\mathbf{U}$ is then

$$U^{\alpha\beta\ldots ij\ldots}(x, y\ldots u, v, \ldots) = \sum_{\alpha'}\sum_{\beta'}\cdots\int dx'\int dy'\ldots \qquad (5.115)$$

$$\times\; C^{\alpha\beta\ldots\alpha'\beta'\ldots}(x, y, \ldots, x', y', \ldots)\; G^{ij\ldots}{}_{\alpha'\beta'\ldots}(u, v, \ldots, x', y', \ldots) \quad .$$

In inverse problems, we always have to consider the forward equation $\mathbf{d} = \mathbf{g}(\mathbf{m})$ and the operator $\mathbf{G}_n$, the derivative of $\mathbf{g}$ at some point $\mathbf{m}_n$. In all the formulas for least-squares inversion, the transpose operator $\mathbf{G}_n^t$ appears (see next section), and, when using gradient methods of resolution, is in fact an iterative application of $\mathbf{G}_n^t$ to the residuals, which performs the data inversion (see chapter 3). The understanding of the meaning of the transpose of a linear operator is very important for functional inverse problems. In particular, it is important to understand that to compute a vector $\mathbf{G}^t\mathbf{d}$ it is *not* necessary to explicitly use the kernel of $\mathbf{G}$ (see Example 5.22 below).

Example 5.21. The derivative operator is antisymmetric. *Let $\mathbb{E}$ be a space of functions $e_1(t)$, $e_2(t)$, $\ldots$, the variable t running in the interval $(0, T)$, and let $\mathbb{F}$ be the dual of $\mathbb{E}$, i.e., a space of functions $f_1(t)$, $f_2(t)$, $\ldots$, with the variable t also running in the interval $(0, T)$, and such that for any function $\mathbf{e} \in \mathbb{E}$ and any function $\mathbf{f} \in \mathbb{F}$ the duality product*

$$\langle\, \mathbf{f}\,,\, \mathbf{e}\,\rangle_{\mathbb{E}} = \langle\, \mathbf{e}\,,\, \mathbf{f}\,\rangle_{\mathbb{F}} = \int_0^T dt\; f(t)\,e(t) = \int_0^T dt\; e(t)\,f(t) \qquad (5.116)$$

makes sense. Let $\mathbf{D}$ be the derivative operator over the functions of $\mathbb{E}$, i.e., the operator that with any function $\mathbf{e} \in \mathbb{E}$ associates its derivative:

$$e'(t) = (\mathbf{D}\,\mathbf{e})(t) = \frac{de}{dt}(t) \quad . \qquad (5.117)$$

The transpose of $\mathbf{D}$ is the operator $\mathbf{D}^t$ defined by the condition that, for any two functions (equation (5.101)),

$$\langle\, \mathbf{D}^t\mathbf{f}\,,\, \mathbf{e}\,\rangle_{\mathbb{E}} = \langle\, \mathbf{f}\,,\, \mathbf{D}\mathbf{e}\,\rangle_{\mathbb{F}} \quad . \qquad (5.118)$$

This condition gives $\int_0^T dt \ (\mathbf{D}^t \mathbf{f})(t) \, e(t) = \int_0^T dt \ f(t) \frac{de}{dt}(t)$, *and an integration by parts shows that, provided that the functions satisfy the* dual boundary conditions

$$f(T) \, e(T) = f(0) \, e(0) \quad , \tag{5.119}$$

one has $(\mathbf{D}^t \mathbf{f})(t) = -\frac{df}{dt}(t)$, *i.e., for short,*

$$\mathbf{D}^t = -\mathbf{D} \quad , \tag{5.120}$$

showing that the derivative operator is antisymmetric *(provided that the dual boundary conditions in equation (5.119) are satisfied). Assume, for instance, that the functions in* $\mathbb{E}$ *satisfy the initial boundary condition* $e(0) = 0$. *Then, the functions in* $\mathbb{F}$ *must satisfy the final boundary condition* $f(T) = 0$. *If instead of the derivative operator one considers the second derivative operator* $\mathbf{D}^2$, *one easily sees that it is symmetric,*

$$(\mathbf{D}^2)^t = \mathbf{D}^2 \quad , \tag{5.121}$$

and that, for a simple example, dual boundary conditions for the second derivative operator are initial conditions of rest for the functions in $\mathbb{E}$, $e(0) = \frac{de}{dt}(0) = 0$, *and final conditions of rest for the functions in* $\mathbb{F}$, $f(T) = \frac{df}{dt}(T) = 0$. *The existence of these dual boundary conditions for the derivative operators contrasts with discrete (finite-difference) formulations, where the transpose of a matrix is defined unconditionally, but, as shown in Appendix 6.28, a careful introduction of these matrix operators does produce boundary conditions.*

Example 5.22. The transpose operator in X-ray tomography. *We wish here to obtain the transpose of the linear operator* $\mathbf{G}_0$ *obtained in the problem of X-ray tomography (Example 5.18). We have (equation (5.73))*

$$\delta d^i = (\mathbf{G}_0 \, \delta \mathbf{m})^i = K \int_{\mathcal{R}_i} ds \ \exp(m_0(\mathbf{x})) \, \delta m(\mathbf{x}) \quad . \tag{5.122}$$

The definition of transpose (equation (5.101)),

$$\langle \, \mathbf{G}_0^t \, \delta \hat{\mathbf{d}} \, , \ \delta \mathbf{m} \, \rangle_{\mathrm{M}} = \langle \, \delta \hat{\mathbf{d}} \, , \ \mathbf{G}_0 \, \delta \mathbf{m} \, \rangle_{\mathrm{D}} \quad , \tag{5.123}$$

is here written, explicitly,

$$\int_{\mathcal{V}} dV(\mathbf{x}) \, (\mathbf{G}_0^t \, \delta \hat{\mathbf{d}})(\mathbf{x}) \, \delta m(\mathbf{x}) = K \sum_i \delta \hat{d}^i \int_{\mathcal{R}_i} ds \ \exp(m_0(\mathbf{x})) \, \delta m(\mathbf{x}) \quad . \tag{5.124}$$

We will see later (Problem 7.18) that this expression characterizes the transpose operator sufficiently well for practical computations in inversion. But let us try here to obtain a more compact representation. Using the delta-like function $\Delta(\mathbf{x}; \mathcal{R}_i)$ *introduced in equation (5.85), equation (5.124) can be rewritten*

$$\int_{\mathcal{V}} dV(\mathbf{x}) \, \delta m(\mathbf{x}) \left((\mathbf{G}_0^t \, \delta \hat{\mathbf{d}})(\mathbf{x}) - K \sum_i \delta \hat{d}^i \, \Delta(\mathbf{x}; \mathcal{R}_i) \, \exp(m_0(\mathbf{x})) \right) = 0 \quad , \tag{5.125}$$

and, this being satisfied for any $\delta\mathbf{m}$, *it follows that an expression like*

$$\delta\hat{\mathbf{m}} = \mathbf{G}_0^t \, \delta\hat{\mathbf{d}} \tag{5.126}$$

corresponds to the explicit expression

$$\delta\hat{m}(\mathbf{x}) = K \sum_i \delta\hat{d}^i \, \Delta(\mathbf{x}; \mathcal{R}_i) \, \exp(m_0(\mathbf{x})) \quad . \tag{5.127}$$

   Notice that this last result could also be obtained directly from the knowledge of the kernel of $\mathbf{G}_0$ *(given in equation (5.77)) and the rule of variable transposition.*

Example 5.23. The transpose operator in waveform inversion. *In this example, we wish to obtain the transpose of the operator* $\mathbf{G}_0$ *appearing in the problem of inversion of pressure waveforms (Example 5.19). The kernel of* $\mathbf{G}_0$ *was (equation (5.99))*

$$G_0(\mathbf{x}_r, t; \mathbf{x}) = -\frac{\exp(m_0(\mathbf{x}))}{K} \, \Gamma_0(\mathbf{x}_r, t; \mathbf{x}, 0) * \frac{\partial^2 p_0}{\partial t^2}(\mathbf{x}, t) \quad . \tag{5.128}$$

A linear equation

$$\delta\mathbf{d} = \mathbf{G}_0 \, \delta\mathbf{m} \tag{5.129}$$

is written explicitly

$$\delta d(\mathbf{x}_r, t) = \int_{\mathcal{V}} dV(\mathbf{x}) \, G_0(\mathbf{x}_r, t, \mathbf{x}) \, \delta m(\mathbf{x}) \quad , \tag{5.130}$$

while an equation (not directly related to equation (5.129))

$$\delta\hat{\mathbf{m}} = \mathbf{G}_0^t \, \delta\hat{\mathbf{d}} \tag{5.131}$$

involving $\mathbf{G}_0^t$ *is written explicitly*

$$\delta\hat{m}(\mathbf{x}) = \sum_r \int_0^T dt \, G_0^t(\mathbf{x}, \mathbf{x}_r, t) \, \delta\hat{p}(\mathbf{x}_r, t) \tag{5.132}$$

(receiver positions are assumed discrete). Using the rule that an operator and its transpose have the same kernels,

$$\delta\hat{m}(\mathbf{x}) = \sum_r \int_0^T dt \, G_0(\mathbf{x}_r, t, \mathbf{x}) \, \delta\hat{p}(\mathbf{x}_r, t) \quad . \tag{5.133}$$

   As shown in Problem 7.18, this finally gives (equation (7.255))

$$\delta\hat{m}(\mathbf{x}) = (\mathbf{G}_0^t \, \delta\hat{\mathbf{p}})(\mathbf{x}) = -\frac{\exp(m_0(\mathbf{x}))}{K} \int_0^T dt \, \dot{p}_0(\mathbf{x}, t) \, \dot{\Psi}_0(\mathbf{x}, t) \quad , \tag{5.134}$$

where $\Psi_0(\mathbf{x}, t)$ is the field defined by

$$\frac{\exp(m_0(\mathbf{x}))}{K} \frac{\partial^2 \Psi_0}{\partial t^2}(\mathbf{x}, t) - \mathrm{div}\left(\frac{1}{\rho(\mathbf{x})} \mathrm{grad}\, \Psi_0(\mathbf{x}, t)\right) = \sum_r \delta(\mathbf{x} - \mathbf{x}_r)\, \delta\hat{p}(\mathbf{x}_r, t) \quad,$$

$$\Psi_0(\mathbf{x}, T) = 0 \quad,$$

$$\dot{\Psi}_0(\mathbf{x}, T) = 0 \quad, \tag{5.135}$$

where it should be noticed that there are final *(instead of initial) conditions.*

We see thus that to evaluate the action of the operator $\mathbf{G}_0^t$ over $\delta\hat{\mathbf{p}}$, we first have to solve the propagation problem defined in equations (5.135) reversed in time, then compute the time correlation expressed in equation (5.134).

5.5   General Least-Squares Inversion

5.5.1   Linear Problems

I start by recalling some results from chapter 3 for finite-dimensional problems. Assume that the forward problem is (exactly) solved by the linear equation

$$\mathbf{d} = \mathbf{G}\mathbf{m} \quad, \tag{5.136}$$

that the results of the observations are described by a Gaussian probability with mean $\mathbf{d}_{\mathrm{obs}}$ and covariance operator $\mathbf{C_D}$, and that the a priori information is described by a Gaussian probability with mean $\mathbf{m}_{\mathrm{prior}}$ and covariance operator $\mathbf{C_M}$. Then, the posterior probability in the model space is also Gaussian, with mean

$$\begin{aligned}
\tilde{\mathbf{m}} &= \mathbf{m}_{\mathrm{prior}} + (\mathbf{G}^t\,\mathbf{C_D}^{-1}\,\mathbf{G} + \mathbf{C_M}^{-1})^{-1}\,\mathbf{G}^t\,\mathbf{C_D}^{-1}\,(\mathbf{d}_{\mathrm{obs}} - \mathbf{G}\,\mathbf{m}_{\mathrm{prior}}) \\
&= \mathbf{m}_{\mathrm{prior}} + \mathbf{C_M}\,\mathbf{G}^t\,(\mathbf{G}\,\mathbf{C_M}\,\mathbf{G}^t + \mathbf{C_D})^{-1}\,(\mathbf{d}_{\mathrm{obs}} - \mathbf{G}\,\mathbf{m}_{\mathrm{prior}})
\end{aligned} \tag{5.137}$$

and covariance operator

$$\begin{aligned}
\tilde{\mathbf{C}}_{\mathbf{M}} &= (\mathbf{G}^t\,\mathbf{C_D}^{-1}\,\mathbf{G} + \mathbf{C_M}^{-1})^{-1} \\
&= \mathbf{C_M} - \mathbf{C_M}\,\mathbf{G}^t\,(\mathbf{G}\,\mathbf{C_M}\,\mathbf{G}^t + \mathbf{C_D})^{-1}\,\mathbf{G}\,\mathbf{C_M} \quad.
\end{aligned} \tag{5.138}$$

Using the arguments of either Franklin (1970) or Backus (1970 a, b, c), it can be shown that this result remains true for infinite-dimensional problems: if priors are Gaussian and the forward equation is linear, the posterior is Gaussian, with mean and covariance operator as given above. This result shows that, when using adequate notation, the results obtained in chapter 3 are valid in a much more general context (i.e., the context of functional analysis).

The major difference between finite- and infinite-dimensional problems is that in the finite-dimensional case, probability densities can be introduced, while they cannot be introduced, as such, in the infinite-dimensional case (although the notion of a Gaussian random function is perfectly defined [see Example 5.1]).

5.5.2 The Misfit

Consider, for an n-dimensional variable $\mathbf{x} = \{x^1, \ldots, x^n\}$, the discrete n-dimensional Gaussian

$$f(\mathbf{x}) = ((2\pi)^n \det \mathbf{C})^{-1/2} \exp\left(-\tfrac{1}{2}(\mathbf{x} - \mathbf{x}_0)^t \mathbf{C}^{-1}(\mathbf{x} - \mathbf{x}_0)\right) . \tag{5.139}$$

If $\mathbf{x}$ is a random realization of $f(\mathbf{x})$, the random variable

$$\chi^2 = \|\mathbf{x} - \mathbf{x}_0\|^2 = (\mathbf{x} - \mathbf{x}_0)^t \mathbf{C}^{-1}(\mathbf{x} - \mathbf{x}_0) \tag{5.140}$$

has a χ^2 distribution (see Appendix 6.8), and its expected value is n. Clearly, when $n \to \infty$, this expected value diverges, so *the expected value of the distance between a realization $\mathbf{x}$ of a Gaussian random function and its mean $\mathbf{x}_0$ is infinite.*

Nevertheless, we have seen in section 3.3.4 that, within the context of discrete least squares, the expected value of the misfit

$$2\,S(\mathbf{m}) = (\mathbf{G}\,\mathbf{m} - \mathbf{d}_{\text{obs}})^t\, \mathbf{C}_{\text{D}}^{-1}(\mathbf{G}\,\mathbf{m} - \mathbf{d}_{\text{obs}}) + (\mathbf{m} - \mathbf{m}_{\text{prior}})^t\, \mathbf{C}_{\text{M}}^{-1}(\mathbf{m} - \mathbf{m}_{\text{prior}}) \tag{5.141}$$

equals the dimension of the data space $\mathbb{D}$. So, in the case where the data vector $\mathbf{d}$ is discrete, $\mathbf{d} = \{d^1, \ldots, d^n\}$, the expected value of the misfit S is finite, even in the case where the model space $\mathbb{M}$ is infinite dimensional.

In this case, it is clear that the solution given in equations (5.137) can be interpreted as the solution of a minimization problem over an infinite-dimensional linear space. Many functional inverse problems are in this category.

If the data space is also a functional space (i.e., it is infinite dimensional), then the misfit expression (5.141) is purely formal, but the solution in equations (5.137)–(5.138) still keeps its sense as defining the mean function and the covariance function of a Gaussian random function.

Example 5.24. Functional linear problem with discrete data. *Assume that the medium under study is described by a model function $m(\mathbf{x})$ of the space coordinates. We are told that the actual medium is a random realization of a Gaussian random function with mean $m_{\text{prior}}(\mathbf{x})$ and covariance $C_{\text{M}}(\mathbf{x}, \mathbf{x}')$. We are able to measure some linear functionals of the model,*

$$d^i = \int dV(\mathbf{x})\, G^i(\mathbf{x})\, m(\mathbf{x}) , \tag{5.142}$$

and we obtain the values d^i_{obs}, with uncertainties described by the covariance matrix C_{D}^{ij}. When using abstract notation, a model function $m(\mathbf{x})$ is denoted $\mathbf{m}$; the center of the prior Gaussian, $m_{\text{prior}}(\mathbf{x})$, is denoted $\mathbf{m}_{\text{prior}}$; and the covariance operator whose kernel is the covariance function $C_{\text{M}}(\mathbf{x}, \mathbf{x}')$ is denoted $\mathbf{C}_{\text{M}}$. The linear relation (5.142) is written, for short, $\mathbf{d} = \mathbf{G}\,\mathbf{m}$; the vector of observed values is denoted $\mathbf{d}_{\text{obs}}$; and the covariance matrix representing the uncertainties is denoted $\mathbf{C}_{\text{D}}$. The center of the posterior Gaussian is then expressed by the second of equations (5.137), and the covariance of the posterior Gaussian is expressed by the second of equations (5.138). Let us write this using explicit notation (involving the kernels). First, introducing the matrix $\mathbf{S} = \mathbf{G}\,\mathbf{C}_{\text{M}}\,\mathbf{G}^t + \mathbf{C}_{\text{D}}$, one has

$$S^{ij} = \int dV(\mathbf{x}) \int dV(\mathbf{x}')\, G^i(\mathbf{x})\, C_{\text{M}}(\mathbf{x}, \mathbf{x}')\, G^j(\mathbf{x}') + C_{\text{D}}^{ij} . \tag{5.143}$$

Then, introducing the inverse matrix

$$\mathbf{T} = \mathbf{S}^{-1} \qquad\qquad (5.144)$$

allows us to write the center of the posterior Gaussian field, $\tilde{\mathbf{m}} = \mathbf{m}_{\text{prior}} + \mathbf{C}_{\text{M}}\,\mathbf{G}^t\,(\mathbf{G}\,\mathbf{C}_{\text{M}}\,\mathbf{G}^t + \mathbf{C}_{\text{D}})^{-1}\,(\mathbf{d}_{\text{obs}} - \mathbf{G}\,\mathbf{m}_{\text{prior}})$, *as*

$$\tilde{m}(\mathbf{x}) = m_{\text{prior}}(\mathbf{x}) + \int dV(\mathbf{x}')\,C_{\text{M}}(\mathbf{x},\mathbf{x}')\,\psi(\mathbf{x}') \quad, \qquad (5.145)$$

where the function $\psi(\mathbf{x})$ *is defined as* $\psi(\mathbf{x}) = \sum_i G^i(\mathbf{x})\,\delta\hat{d}_i$, *the* weighted *data residuals* $\delta\hat{d}_i$ *are defined as* $\delta\hat{d}_i = \sum_j T_{ij}\,\delta d^j$, *and the data residuals* δd^i *are defined as* $\delta d^i = d^i_{\text{obs}} - \int dV(\mathbf{x})\,G^i(\mathbf{x})\,m_{\text{prior}}(\mathbf{x})$. *The covariance function of the posterior Gaussian field,* $\tilde{C}_{\text{M}} = C_{\text{M}} - C_{\text{M}}\,G^t\,(G\,C_{\text{M}}\,G^t + C_{\text{D}})^{-1}\,G\,C_{\text{M}}$, *is explicitly given by*

$$\tilde{C}_{\text{M}}(\mathbf{x},\mathbf{x}') = C_{\text{M}}(\mathbf{x},\mathbf{x}') - \sum_i \sum_j \Psi^i(\mathbf{x})\,T_{ij}\,\Psi^j(\mathbf{x}') \quad, \qquad (5.146)$$

where the functions $\Psi^i(\mathbf{x})$ *are defined as* $\Psi^i(\mathbf{x}) = \int dV(\mathbf{x}')\,C_{\text{M}}(\mathbf{x},\mathbf{x}')\,G^i(\mathbf{x}')$. *To understand this solution, one may carefully examine the posterior mean* $\tilde{m}(\mathbf{x})$ *and the posterior covariance* $\tilde{C}_{\text{M}}(\mathbf{x},\mathbf{x}')$, *or, much better, one may examine some (pseudo)random realizations of the (posterior) Gaussian random field whose mean function is* $\tilde{m}(\mathbf{x})$ *and whose covariance function is* $\tilde{C}_{\text{M}}(\mathbf{x},\mathbf{x}')$.

The previous example corresponds, in fact, to my preferred alternative to the Backus and Gilbert approach (see Appendix 6.16) for solving linear inverse problems.

Example 5.25. Gaussian sequential simulation. *Assume that we have the a priori information that the system under investigation is a particular realization* $m(\mathbf{x})$ *of a Gaussian random field with mean* $m_{\text{prior}}(\mathbf{x})$ *and covariance* $C_{\text{M}}(\mathbf{x},\mathbf{x}')$. *We then have been able to obtain some additional data in the form of the values of* $m(\mathbf{x})$ *at some selected points* $\{\mathbf{x}^1, \mathbf{x}^2, \dots\}$. *We can write these data as*

$$d^i = m(\mathbf{x}^i) \quad . \qquad (5.147)$$

The actual measurements have provided the observed values d^i_{obs}, *with uncertainties described by the covariance matrix* C_{D}^{ij}. *As the problem is linear, the posterior random field is also Gaussian, and we are going to evaluate its mean and covariance. The relation (5.147) can be written*

$$d^i = \int dV(\mathbf{x})\,G^i(\mathbf{x})\,m(\mathbf{x}) \qquad \text{with} \qquad G^i(\mathbf{x}) = \delta(\mathbf{x} - \mathbf{x}^i) \quad , \qquad (5.148)$$

introducing the kernel $G^i(\mathbf{x})$ *of the linear operator mapping the model into the data space. The mean of the posterior Gaussian, as given by expression (5.145), here becomes*

$$\tilde{m}(\mathbf{x}) = m_{\text{prior}}(\mathbf{x}) + \sum_i C_{\text{M}}(\mathbf{x},\mathbf{x}^i)\,\delta\hat{d}_i \quad , \qquad (5.149)$$

where the weighted data residuals $\delta \hat{d}_i$ are defined as $\delta \hat{d}_i = \sum_j T_{ij} \, \delta d^j$, and the data residuals δd^i are given by $\delta d^i = d^i_{\text{obs}} - m_{\text{prior}}(\mathbf{x}^i)$. The matrix T_{ij} is the inverse of the matrix S^{ij} introduced in equation (5.143), which here takes the simple expression $S^{ij} = C_{\text{M}}(\mathbf{x}^i, \mathbf{x}^j) + C_{\text{D}}^{ij}$. The covariance function of the posterior Gaussian field, as given by expression (5.150), becomes

$$\tilde{C}_{\text{M}}(\mathbf{x}, \mathbf{x}') = C_{\text{M}}(\mathbf{x}, \mathbf{x}') - \sum_i \sum_j C_{\text{M}}(\mathbf{x}, \mathbf{x}^i) \, T_{ij} \, C_{\text{M}}(\mathbf{x}^j, \mathbf{x}') \quad . \tag{5.150}$$

Given this mean and this covariance of the posterior Gaussian random field, one can generate some (pseudo)random realizations of it, to well apprehend the characteristics of the random field. In the field of geostatistics, an efficient technique has been developed (the Gaussian sequential simulation) to generate samples of this posterior random field (Goovaerts, 1997; Deutsch and Journel, 1998). The basic idea is to write the mean and the covariance at a particular point, say $\mathbf{x}^a$, giving $\tilde{m}(\mathbf{x}^a) = m_{\text{prior}}(\mathbf{x}^a) + \sum_i C_{\text{M}}(\mathbf{x}^a, \mathbf{x}^i) \, \delta \hat{d}_i$ and $\tilde{C}_{\text{M}}(\mathbf{x}^a, \mathbf{x}^a) = C_{\text{M}}(\mathbf{x}^a, \mathbf{x}^a) - \sum_i \sum_j C_{\text{M}}(\mathbf{x}^a, \mathbf{x}^i) \, T_{ij} \, C_{\text{M}}(\mathbf{x}^j, \mathbf{x}^a)$. A (pseudo)random value $m(\mathbf{x}^a)$ is then generated from a one-dimensional Gaussian that has this mean and this covariance. This value is then considered as one additional datum,[70] the equations (5.149)–(5.150) are rewritten with the inclusion of this new datum, and a new value of the random field is generated at some other point, say $\mathbf{x}^b$. The iteration of this procedure will produce values in as many points as we may wish. We then have a realization of the posterior random field. In the Gaussian sequential simulation technique, one is only interested in obtaining approximate samples, and, therefore, one uses the simplification, when generating a (pseudo)random value at some point $\mathbf{x}^a$, of disregarding the data points that are far from $\mathbf{x}^a$ (two points $\mathbf{x}$ and $\mathbf{x}'$ are far if the covariance $C_{\text{M}}(\mathbf{x}, \mathbf{x}')$ is negligible). This approximation keeps the size of the matrix S^{ij} small. Although the point of view used in geostatistics differs from that used here, it can be demonstrated that the mathematics is equivalent. Figure 2.5 (page 46) shows an example of Gaussian sequential simulation.

5.5.3 Nonlinear Problems

Let us now turn to the case where the equation solving the forward problem $\mathbf{d} = \mathbf{g}(\mathbf{m})$ is nonlinear. If the equation is linearizable around $\mathbf{m}_{\text{prior}}$, then, as we have seen in chapter 3, the techniques to be used are those used for linear problems. If the equation is actually nonlinear (as explained in section 3.2.1, in the context of least squares) we only consider mild nonlinearities.

For discrete problems, we have seen in chapter 3 that if the results of the observations can be described by a Gaussian probability with mean vector $\mathbf{d}_{\text{obs}}$ and covariance matrix $\mathbf{C}_{\text{D}}$, the a priori information can be described by a Gaussian probability with mean vector $\mathbf{m}_{\text{prior}}$ and covariance matrix $\mathbf{C}_{\text{M}}$, and if the equation solving the forward problem is quasi linear in the region of the model space with significant a posteriori probability, then the a posteriori probability in the model space is approximately Gaussian. The mean $\tilde{\mathbf{m}}$ of the

---

[70]If the value is generated with enough numerical precision, this is an error-free datum, i.e., the variance associated with it in the matrix C_{D}^{ij} is zero.

approximated Gaussian probability minimizes the misfit function

$$
\begin{aligned}
2\,S(\mathbf{m}) &= \|\,\mathbf{g}(\mathbf{m}) - \mathbf{d}_{\text{obs}}\,\|_{\text{D}}^{2} + \|\,\mathbf{m} - \mathbf{m}_{\text{prior}}\,\|_{\text{M}}^{2} \\
&= (\mathbf{g}(\mathbf{m}) - \mathbf{d}_{\text{obs}})^{t}\,\mathbf{C}_{\text{D}}^{-1}\,(\mathbf{g}(\mathbf{m}) - \mathbf{d}_{\text{obs}}) + (\mathbf{m} - \mathbf{m}_{\text{prior}})^{t}\,\mathbf{C}_{\text{M}}^{-1}\,(\mathbf{m} - \mathbf{m}_{\text{prior}})
\end{aligned}
\tag{5.151}
$$

and can be obtained using an iterative process (basically, a gradient-based method of minimization, like the quasi-Newton Algorithm (3.51) of the steepest descent Algorithm (3.89)). The covariance matrix of the approximated Gaussian probability is given by

$$
\begin{aligned}
\widetilde{\mathbf{C}}_{\text{M}} &= (\mathbf{G}_{\infty}^{t}\,\mathbf{C}_{\text{D}}^{-1}\,\mathbf{G}_{\infty} + \mathbf{C}_{\text{M}}^{-1})^{-1} \\
&= \mathbf{C}_{\text{M}} - \mathbf{C}_{\text{M}}\,\mathbf{G}_{\infty}^{t}\,(\mathbf{G}_{\infty}\,\mathbf{C}_{\text{M}}\,\mathbf{G}_{\infty}^{t} + \mathbf{C}_{\text{D}})^{-1}\,\mathbf{G}_{\infty}\,\mathbf{C}_{\text{M}} \quad,
\end{aligned}
\tag{5.152}
$$

where $\mathbf{G}_{\infty}$ denotes the derivative of the operator $\mathbf{g}$ at the convergence point $\widehat{\mathbf{m}} = \mathbf{m}_{\infty}$.

With the definitions and notation introduced in this chapter, almost all the equations developed in chapter 3 for discrete problems make sense for functional problems. Excluding perhaps some pathological situation, we do not need to develop the optimization algorithms again: as the examples below show, we can directly use those developed in chapter 3.

5.5.4 Comments for Functional Problems

A question arising in the practical resolution of functional least-squares problems concerns the operational significance of the resolution of a linear equation. Let, for instance, $\delta\mathbf{d}$ be a vector of data residuals. In least-squares computations, one often needs to evaluate the weighted residuals

$$
\delta\widehat{\mathbf{d}} = \mathbf{C}^{-1}\,\delta\mathbf{d} \quad.
\tag{5.153}
$$

Sometimes, the covariance operator $\mathbf{C}$ corresponds to some simple probabilistic model, and its inverse $\mathbf{C}^{-1}$ is known analytically (see Example 5.13 for the one-dimensional exponential covariance, Problem 7.23 for the three-dimensional exponential covariance, and Example 5.14 for the covariance of a one-dimensional random walk). If this is not the case, then one can rewrite the equation as

$$
\mathbf{C}\,\delta\widehat{\mathbf{d}} = \delta\mathbf{d}
\tag{5.154}
$$

and solve this equation for $\delta\widehat{\mathbf{d}}$ using an iterative algorithm. For instance, one may use the algorithm

$$
\delta\widehat{\mathbf{d}}_{n+1} = \delta\widehat{\mathbf{d}}_{n} + \mathbf{Q}\,(\delta\mathbf{d} - \mathbf{C}\,\delta\widehat{\mathbf{d}}_{n}) \quad,
\tag{5.155}
$$

where $\mathbf{Q}$ is an arbitrary operator suitably chosen to accelerate convergence. Usually, a good choice for $\mathbf{Q}$ is a diagonal operator proportional to the inverse of the variances in $\mathbf{C}$. Equation (5.155) shows that, for numerical computations, we do not explicitly need to introduce the inverse of the operator $\mathbf{C}$. Usually, a covariance operator is an integral operator, and computing $\mathbf{C}\,\delta\widehat{\mathbf{d}}_{n}$ implies some numerical method, such as, for instance, a Runge–Kutta method. Of course, any numerical method will imply a discretization of the working space (here the data space), but it is important to realize that to discretize the working space does *not* imply considering the operator $\mathbf{C}$ as a matrix and effectively building the matrix in the computer's memory (for problems other than academic this would be practically impossible).

Another problem arising in functional least squares is the computation of

$$\delta\hat{\mathbf{m}} = \mathbf{G}^t \, \delta\hat{\mathbf{d}} \quad , \tag{5.156}$$

where $\delta\hat{\mathbf{d}}$ is a weighted data vector, $\mathbf{G}$ is a linear operator from the model into the data space, and $\mathbf{G}^t$ is its transpose. We have seen in section 5.4.2 that the abstract definition of the transpose operator leads to a physical understanding of the computations to be performed: back-projection of the weighted residuals in an X-ray tomography problem, and back-propagation of the weighted residuals plus a time correlation with the current predicted field in a problem of inversion of acoustic waveforms. Again, to perform these operations numerically, some discretization of the working spaces has to be used, but the naive approach consisting of introducing a matrix representing the discretized version of $\mathbf{G}$, and interpreting (5.156) as a matrix multiplication equation, not only may destroy the physical interpretation but forces us to effectively compute the elements of the matrix $\mathbf{G}$, which is usually prohibitive because it takes too much computer time.

5.5.5 Analysis of Uncertainties

As the least-squares method provides the posterior covariance operator in the model space, $\mathbf{C}_M$, it is the analysis of $\mathbf{C}_M$ that helps us understand the actual information that the data has provided on the model $\mathbf{m}$.

There are many ways to use the posterior covariance operator. For instance, in least-squares problems, we start describing the a priori information on the model space assuming a Gaussian distribution with mean $\mathbf{m}_{\text{prior}}$ and covariance $\mathbf{C}_M$. At this level (before the inverse problem is solved), I strongly recommend generating a few pseudorandom realizations of such a Gaussian process (Figures 5.1, 5.6, and 5.7 are examples of this). One must verify that the models so obtained are actually representative of the kind of a priori information one intends to introduce. Then, one turns to the posterior Gaussian, whose center is the 'posterior model' $\tilde{\mathbf{m}}$ (given, for linear problems, by the explicit expression (5.137), or obtained, for nonlinear problems, through the minimization of the misfit in equation (5.151)) with 'posterior covariance' $\tilde{\mathbf{C}}_M$ (equations (5.138) and (5.152)). The generation of a large enough collection of models of this posterior Gaussian (and subsequent comparison with the collection obtained using the prior Gaussian) usually provides a very good understanding of the 'features' of the model that are well 'resolved' by the data (if any).

In fact, what I suggest here is to use for the Gaussian case exactly the point of view proposed in chapter 2 for a general Monte Carlo method. It is not because the probability distributions are Gaussian (or approximately Gaussian) here that the 'movie philosophy' proposed in chapter 2 loses its interest.

It remains that a direct look at the posterior covariance function itself provides useful information. To fix ideas, assume here that we examine a problem where the model parameter is a scalar function of the spatial coordinates, $\mathbf{x} \rightarrow m(\mathbf{x})$, where $\mathbf{x}$ is a point in the physical 3D space. The reader will easily generalize to the case where we need more than one function to describe the model (as is, for instance, the case in the elastodynamic waveform inversion of Problem 5.8). The kernel of the posterior covariance operator is then a single covariance function $\widetilde{C}_M(\mathbf{x}, \mathbf{x}')$. What information do we have on the true value of the parameter m at a particular point $\mathbf{x}$? As the a posteriori probability is (approximately) Gaussian, the marginal probability for the parameter $m(\mathbf{x})$ is also Gaussian. The corresponding Gaussian probability density is centered at $\widetilde{m}(\mathbf{x})$, and the standard deviation is

$$\widetilde{\sigma}(\mathbf{x}) = \sqrt{\widetilde{C}_M(\mathbf{x}, \mathbf{x})} \quad . \tag{5.157}$$

If we compute $\sigma(\mathbf{x})$ for all points $\mathbf{x}$, we may plot the estimated uncertainty $\widetilde{\sigma}(\mathbf{x})$ together with the solution $\widetilde{m}(\mathbf{x})$ (see Figure 5.17 for an example). Better than plotting covariance values is to remember that the posterior correlation between point x and point $\mathbf{x}'$ is defined as

$$\widetilde{\rho}(\mathbf{x}, \mathbf{x}') = \frac{\widetilde{C}_M(\mathbf{x}, \mathbf{x}')}{\widetilde{\sigma}(\mathbf{x}) \, \widetilde{\sigma}(\mathbf{x}')} \quad , \tag{5.158}$$

the correlation being a real number between -1 and 1. These are the correlations between the posterior uncertainties at the two points. For a few selected positions $\mathbf{x}_1, \mathbf{x}_2, \ldots$, the correlations $\widetilde{\rho}(\mathbf{x}_i, \mathbf{x})$ can be graphically represented (this is done in the example shown in Figure 5.17).

Sometimes, one may not be interested in the uncertainty on the a posteriori value of the parameter m at a given point $\mathbf{x}_0$, but rather in the uncertainty on the mean value of the parameter over a given ball[71] around $\mathbf{x}_0$,

$$\widehat{m}(\mathbf{x}_0) = \int_V dV(\mathbf{x}) \, D(\mathbf{x}_0, \mathbf{x}) \, \widetilde{m}(\mathbf{x}) \quad , \tag{5.159}$$

or, for short,

$$\widehat{\mathbf{m}} = \mathbf{D} \, \widetilde{\mathbf{m}} \quad . \tag{5.160}$$

Using the definition of covariance, it follows that

$$\widehat{\mathbf{C}}_M = \mathbf{D} \, \widetilde{\mathbf{C}}_M \, \mathbf{D}^t \quad , \tag{5.161}$$

i.e., $\widehat{C}_M(\mathbf{x}, \mathbf{x}') = \int_V dV(\mathbf{x}'') \int_V dV(\mathbf{x}''') \, D(\mathbf{x}, \mathbf{x}'') \, \widetilde{C}_M(\mathbf{x}'', \mathbf{x}''') \, D(\mathbf{x}', \mathbf{x}''')$. The uncertainty on the value $\widehat{m}(\mathbf{x}_0)$ is then $\widehat{\sigma}(\mathbf{x}_0) = (\widehat{C}_M(\mathbf{x}_0, \mathbf{x}_0))^{1/2}$.

To conclude, let me recall that the 'resolution' operator, discussed in section 3.3.2, was given by (equation (3.63))

$$\mathbf{R} = \mathbf{I} - \widetilde{\mathbf{C}}_M \, \mathbf{C}_M^{-1} \quad . \tag{5.162}$$

As discussed by Backus and Gilbert (1968), the resolution operator can be interpreted as a filter (see the comments in section 3.3.2). Instead of representing the posterior correlation $\widetilde{\rho}(\mathbf{x}, \mathbf{x}')$ (defined in equation (5.158)), one may prefer to represent the resolution $R(\mathbf{x}, \mathbf{x}')$.

---

[71] In functional inverse problems, it may well happen that the posterior variance at a given point is identical to the prior variance, and the gain of information is only seen when computing average values over well-chosen balls.

5.6   Example: X-Ray Tomography as an Inverse Problem

This problem was introduced in Examples 5.18 and 5.23. As data, we have chosen to use, instead of the transmittances $\{\rho^i\}$, the logarithmic transmittances $d^i = -\log \rho^i$, and, as parameters, instead of the linear attenuation coefficient $\{\mu(\mathbf{x})\}$, we have chosen to use the logarithmic linear attenuation $m(\mathbf{x}) = \log \frac{\mu(\mathbf{x})}{K}$, where K is an arbitrary constant value of the attenuation coefficient. We arrived at the following relation between data and parameters (equation (5.71)):

$$d^i = K \int_{\mathcal{R}_i} ds \, \exp(m(\mathbf{x})) \quad , \tag{5.163}$$

an expression that we can write $\mathbf{d} = \mathbf{g}(\mathbf{m})$, with $\mathbf{d} = \{d^i\}$ and $\mathbf{m} = \{m(\mathbf{x})\}$.

Let d^i_{obs} ($i = 1, 2, \dots$) be the observed data and C^{ij}_{D} be the elements of the covariance matrix describing experimental uncertainties. Also, let us assume that the a priori information we have on the actual model is represented by a Gaussian random field[72] with mean $m_{\text{prior}}(\mathbf{x})$ and covariance function $C_M(\mathbf{x}, \mathbf{x}')$.

We have seen that with these assumptions, the posterior distribution is (as this problem is only weakly nonlinear) approximately that of a Gaussian random field whose center is obtained through the minimization of the least-squares misfit function (equation (3.32))

$$
\begin{aligned}
2\,S(\mathbf{m}) &= \| \mathbf{g}(\mathbf{m}) - \mathbf{d}_{\text{obs}} \|_{\text{D}}^2 + \| \mathbf{m} - \mathbf{m}_{\text{prior}} \|_{\text{M}}^2 \\
&= (\mathbf{g}(\mathbf{m}) - \mathbf{d}_{\text{obs}})^t \, \mathbf{C}_{\text{D}}^{-1} \, (\mathbf{g}(\mathbf{m}) - \mathbf{d}_{\text{obs}}) + (\mathbf{m} - \mathbf{m}_{\text{prior}})^t \, \mathbf{C}_{\text{M}}^{-1} \, (\mathbf{m} - \mathbf{m}_{\text{prior}}) \quad .
\end{aligned}
\tag{5.164}
$$

The covariance function of this posterior Gaussian field is to be analyzed below.

Let us first see what a simple steepest descent algorithm would do to solve this minimization problem. We obtained, in equation (3.89),

$$\mathbf{m}_{n+1} = \mathbf{m}_n - \varepsilon_n \left(\mathbf{C}_{\text{M}} \, \mathbf{G}_n^t \, \mathbf{C}_{\text{D}}^{-1} \, (\mathbf{g}(\mathbf{m}_n) - \mathbf{d}_{\text{obs}}) + (\mathbf{m}_n - \mathbf{m}_{\text{prior}}) \right) \quad , \tag{5.165}$$

where $\mathbf{G}_n$ is the derivative operator evaluated at the current point $\mathbf{m}_n$ and $\mathbf{G}_n^t$ is its transpose. The real numbers ε_n are chosen ad hoc (as large as possible to accelerate convergence but small enough to avoid divergence).

Splitting the algorithm into its basic computations gives

$$\mathbf{d}_n = \mathbf{g}(\mathbf{m}_n) \quad , \tag{5.166}$$

$$\delta\mathbf{d}_n = \mathbf{d}_n - \mathbf{d}_{\text{obs}} \quad , \tag{5.167}$$

$$\delta\hat{\mathbf{d}}_n = \mathbf{C}_{\text{D}}^{-1} \, \delta\mathbf{d}_n \quad , \tag{5.168}$$

$$\delta\hat{\mathbf{m}}_n = \mathbf{G}_n^t \, \delta\hat{\mathbf{d}}_n \quad , \tag{5.169}$$

$$\delta\mathbf{m}_n = \mathbf{C}_{\text{M}} \, \delta\hat{\mathbf{m}}_n \quad , \tag{5.170}$$

$$\mathbf{m}_{n+1} = \mathbf{m}_n - \varepsilon_n \, (\delta\mathbf{m}_n + \mathbf{m}_n - \mathbf{m}_{\text{prior}}) \quad . \tag{5.171}$$

Let us make explicit each of these computational steps.

---

[72] Regardless of the physical goodness of this assumption, it makes, at least, mathematical sense, as we have transformed the initial positive parameter (the attenuation coefficient) into its logarithm.

Equation (5.166) corresponds to the resolution of the forward problem (equation (5.163)) corresponding to the current medium $\mathbf{m}_n$,

$$d_n^i = K \int_{\mathcal{R}_i} ds \, \exp(m_n(\mathbf{x})) \quad . \tag{5.172}$$

Equation (5.167) corresponds to the computation of the data residuals

$$\delta d_n^i = d_n^i - d_{\text{obs}}^i \quad . \tag{5.173}$$

Equation (5.168) corresponds to the computation of the weighted data residuals $(\delta\hat{\mathbf{d}}_n)_i = (\mathbf{C}_{\mathrm{D}}^{-1})_{ij} (\delta\mathbf{d}_n)^j$. This is typically done by solving the linear system

$$(\mathbf{C}_{\mathrm{D}})^{ij} (\delta\hat{\mathbf{d}}_n)_j = (\delta\mathbf{d}_n)^i \quad \mapsto \quad \delta\hat{\mathbf{d}}_n \quad . \tag{5.174}$$

Equation (5.169) corresponds to applying the operator $\mathbf{G}_n^t$ to the weighted data residuals $\delta\hat{\mathbf{d}}_n$. Using the result demonstrated in equations (5.126)–(5.127), this corresponds to $\delta\hat{m}_n(\mathbf{x}) = K \sum_i \delta\hat{d}_n^i \Delta(\mathbf{x}; \mathcal{R}_i) \exp(m_n(\mathbf{x}))$. This expression is not to be evaluated; instead it is to be used as an input to the next computation. For, using this result, equation (5.170) can be written $\delta m_n(\mathbf{x}) = \int_V dV(\mathbf{x}') C_{\mathrm{M}}(\mathbf{x}, \mathbf{x}') \delta\hat{m}_n(\mathbf{x}) = K \int_V dV(\mathbf{x}') C_{\mathrm{M}}(\mathbf{x}, \mathbf{x}') \sum_i \delta\hat{d}_n^i$ $\Delta(\mathbf{x}; \mathcal{R}_i) \exp(m_n(\mathbf{x}))$, $\delta m_n(\mathbf{x}) = K \sum_i \delta\hat{d}_n^i \int_V dV(\mathbf{x}') C_{\mathrm{M}}(\mathbf{x}, \mathbf{x}') \Delta(\mathbf{x}'; \mathcal{R}_i) \exp(m_n(\mathbf{x}'))$, i.e., using the definition of the delta-like function $\Delta(\mathbf{x}'; \mathcal{R}_i)$ (equation (5.76)),

$$\delta m_n(\mathbf{x}) = \sum_i \delta\hat{d}_n^i \alpha_i(\mathbf{x}) \quad , \tag{5.175}$$

where

$$\alpha_i(\mathbf{x}) = \int_{\mathcal{R}_i} ds(\mathbf{x}') C_{\mathrm{M}}(\mathbf{x}, \mathbf{x}') \exp(m_n(\mathbf{x}')) \quad . \tag{5.176}$$

Finally (equation (5.171)),

$$m_{n+1}(\mathbf{x}) = m_n(\mathbf{x}) - \varepsilon_n (\delta m_n(\mathbf{x}) + m_n(\mathbf{x}) - m_{\text{prior}}(\mathbf{x})) \quad . \tag{5.177}$$

The six equations (5.172)–(5.177) correspond to the actual computations to be performed at each iteration. The only nontrivial equations are (5.172) and the pair (5.175)–(5.176). Equation (5.172) corresponds to the resolution of the forward problem (estimation of the transmittances along each ray for a given model of attenuation), and we do not need to discuss it further. Equations (5.175)–(5.176) are where the inversion is made, so let us interpret them with some detail.

First, $\alpha_i(\mathbf{x})$ is a function that has negligible values everywhere in space except around the ith ray, where, typically, it has an important value on the ray itself and smaller and smaller values as the point $\mathbf{x}$ becomes more and more distant from the ith. We can call $\alpha_i(\mathbf{x})$ a "tube" around the ith ray. Then, equation (5.175) tells us that $\delta m_n(\mathbf{x})$ is made[73] as a sum

---

[73]Practically, the model $m(\mathbf{x})$ is numerically defined on a grid of points, one point per pixel of the graphic device used to plot the model.

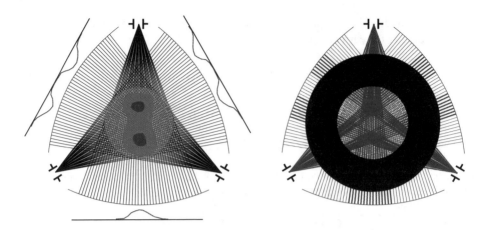

Figure 5.15. *The back-projection of data. Left: The true model and the data corresponding to three different incidences. Right: Back-projection of the three different incidences (values outside the back annulus disregarded). Iterative back-projection (of the residuals) gives the solution of the inverse problem (see text). The first iteration (back-projection of the data) gives only a rough image.*

of tubes, one for each ray, each tube being multiplied by the weighted residual $\delta \hat{d}_n^i$ associated with the given ray. This corresponds to a sort of back-projection of each weighted residual along each ray (see Figure 5.15). This notion of back-projection is usual in algebraic reconstruction techniques (see, for instance, Herman, 1980).

Now, a few words about 'preconditioning.' The steepest descent algorithm in equation (5.165) contains the steepest ascent direction $\mathbf{C}_M \, \mathbf{G}_n^t \, \mathbf{C}_D^{-1} \, (\mathbf{g}(\mathbf{m}_n) - \mathbf{d}_{obs}) + (\mathbf{m}_n - \mathbf{m}_{prior})$. This direction is only optimum for infinitesimal jumps, but in any practical algorithm, one wishes to perform as large jumps as possible to accelerate convergence. It is quite easy, in the physical sense, to apply gross corrections to this direction to obtain better ones. For instance, in the example in Figure 5.15, the X-ray being closer in the region near the sources, the 'back-projection' of the residuals along the 'tube' associated with each ray creates the geometric effect that the points near the sources receive much larger values than the points far from the sources. Preconditioning here may just amount to correcting for this effect (possibly by just putting to zero all the values outside the inner part of the ring at the right of Figure 5.15).

This amounts to replacing the steepest descent algorithm (5.165) with

$$\mathbf{m}_{n+1} \;=\; \mathbf{m}_n - \boldsymbol{\psi}_n(\delta \mathbf{m}_n) \quad , \tag{5.178}$$

where $\boldsymbol{\psi}_n(\cdot)$ is an ad hoc nonlinear function and

$$\delta \mathbf{m}_n \;=\; \mathbf{C}_M \, \mathbf{G}_n^t \, \mathbf{C}_D^{-1} \, (\mathbf{g}(\mathbf{m}_n) - \mathbf{d}_{obs}) + (\mathbf{m}_n - \mathbf{m}_{prior}) \quad . \tag{5.179}$$

With this preconditioning, the algorithm may be very rapid to converge. With a good distribution of sources and receivers, one or two iterations may be sufficient to obtain a good solution.

Assume now that the algorithm has converged into the desired minimum, $\widetilde{\mathbf{m}} = \mathbf{m}_\infty$, and let us become interested in the computation of the posterior covariance operator, as given by the second[74] of equations (5.152),

$$\widetilde{\mathbf{C}}_M = \mathbf{C}_M - \mathbf{C}_M \, \mathbf{G}'_\infty \, (\mathbf{G}_\infty \, \mathbf{C}_M \, \mathbf{G}'_\infty + \mathbf{C}_D)^{-1} \, \mathbf{G}_\infty \, \mathbf{C}_M \quad , \tag{5.180}$$

where $\mathbf{G}_\infty$ corresponds to the derivative operator evaluated at the convergence point $\mathbf{m}_\infty$. I leave as an exercise to the reader to demonstrate that the explicit expression of this equation is

$$\widetilde{C}_M(\mathbf{x}, \mathbf{x}') = C_M(\mathbf{x}, \mathbf{x}') - \sum_i \sum_j S_{ij} \, \Psi^{ij}(\mathbf{x}, \mathbf{x}') \quad , \tag{5.181}$$

where

$$\Psi^{ij}(\mathbf{x}, \mathbf{x}') = K^2 \int_{\mathcal{R}_i} ds(\mathbf{x}'') \int_{\mathcal{R}_j} ds(\mathbf{x}''') \, C_M(\mathbf{x}, \mathbf{x}'') \, \exp(m(\mathbf{x}'')) \, \exp(m(\mathbf{x}''')) \, C_M(\mathbf{x}''', \mathbf{x}')$$

$$\tag{5.182}$$

and

$$S_{ij} = C_D^{ij} + K^2 \int_{\mathcal{R}_i} ds(\mathbf{x}) \int_{\mathcal{R}_j} ds(\mathbf{x}') \, \exp(m(\mathbf{x})) \, C_M(\mathbf{x}, \mathbf{x}') \, \exp(m(\mathbf{x}')) \quad . \tag{5.183}$$

In these equations, $m(\mathbf{x})$ is the maximum likelihood point (obtained by the convergence of the iterative algorithm outlined above).

5.7 Example: Travel-Time Tomography

The inverse problem of travel-time tomography is formally very similar to the problem of X-ray tomography, the only difference being that in travel-time tomography the rays cannot be assumed to follow straight lines, but have a geometry that depends on the velocity structure of the medium. One may, for instance, replace equations (5.71) and (5.76) in Example 5.18 with equations (5.80) and (5.86) in Example 5.20.

Therefore, all equations developed in section 5.6 for the inverse problem of X-ray tomography remain valid here, except that all the integrals that were of the form

$$\int_{\mathcal{R}_i} ds \, \ldots$$

become now

$$\int_{\mathcal{R}_i(\mathbf{m}_n)} ds \, \ldots \quad ,$$

indicating that the integral along the ith ray is to be performed along the ray path predicted in the current model $m_n(\mathbf{x})$.

Figures 5.16–5.17 give an illustration of the results of a tomographic (geophysical) method. Only the mean of the posterior probability density was calculated (some random realizations of a Gaussian distribution with the posterior mean and posterior covariance should have been generated).

---

[74]We choose the second equation because, in this case where the data are discrete and the model is a function, it is easier to implement.

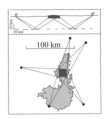

Figure 5.16. *A travel-time tomography experiment on an old volcano in France. Acoustic waves were generated at the surface of Earth. After reflection at a deep discontinuity, the travel times of the waves were observed at an array of seismic stations ($\simeq 100$ per source). The observed travel times were used to infer the velocity structure in a 3D region under the volcano (see Figure 5.17).*

Figure 5.17. *Tomographic results of a travel-time experiment on a volcanic region (Mont Dore, France), from Nercessian, Hirn, and Tarantola, 1984. Left: Horizontal sections of the velocity structure under the volcano, at respective depths of 1, 2, 3, and 4 km (the horizontal scale is that represented in Figure 5.16). Light colors indicate low velocity and dark colors indicate high velocity. Middle: The posterior covariance function $C(\mathbf{x}_0, \mathbf{x})$, for a particular given point $\mathbf{x}_0$ situated near the middle of the square, at a depth of 1 km. Light colors indicate strong correlation. One sees that the spatial resolution around $\mathbf{x}_0$ attained with this data set is of the order of 1 km. Right: A posteriori standard deviations $\sigma(\mathbf{x}) = C(\mathbf{x}, \mathbf{x})^{1/2}$ for the horizontal section 1 km deep (scale not shown). Central regions are best resolved.*

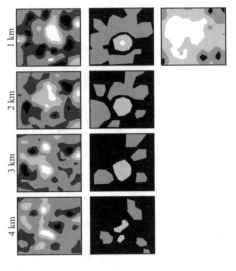

5.8   Example: Nonlinear Inversion of Elastic Waveforms

The Problem: Elastic waves, created by some controlled source, propagate on a heterogeneous isotropic linearly elastic medium, and the displacements are observed at a finite number of points (see Figure 5.18 below). Use these observations to infer the values of the mass density, the bulk modulus, and the shear modulus at every point of the medium.

5.8.1   The Forward Problem

In what follows, given a symmetric tensor t_{ij}, its *isotropic part*, denoted $\bar{t}_{ij}$, and its *deviatoric part*, denoted $\tilde{t}_{ij}$, are defined as

$$\bar{t}_{ij} = \tfrac{1}{3} t_k^k \delta_{ij} \quad , \qquad \tilde{t}_{ij} = t_{ij} - \bar{t}_{ij} = t_{ij} - \tfrac{1}{3} t_k^k \delta_{ij} \quad . \tag{5.184}$$

One clearly has $\bar{t}_k^k = t_k^k$ and $\tilde{t}_k^k = 0$.

Let $\mathcal{V}$ be the volume of the elastic medium being considered and $\mathcal{S}$ be its surface. To simplify notation, a Cartesian system of coordinates $\mathbf{x} = \{x^1, x^2, x^3\}$ is used over $\mathcal{V}$.

Denoting by $u^i(\mathbf{x}, t)$ the ith component of the displacement at space point x^i and time instant t, an elastic wavefield can be characterized by the following system of equations:

$$\rho(\mathbf{x}) \frac{\partial^2 u^i}{\partial t^2}(\mathbf{x}, t) - \frac{\partial \sigma^{ij}}{\partial x^j}(\mathbf{x}, t) = \phi^i(\mathbf{x}, t) , \qquad \mathbf{x} \in V , \ t \in (0, T) , \qquad (5.185)$$

$$\sigma^{ij}(\mathbf{x}, t) - 3\kappa(\mathbf{x}) \bar{u}^{ij}(\mathbf{x}, t) - 2\mu(\mathbf{x}) \tilde{u}^{ij}(\mathbf{x}, t) = \varphi^{ij}(\mathbf{x}, t) ,$$
$$\mathbf{x} \in V , \ t \in (0, T) , \qquad (5.186)$$

$$u_{ij}(\mathbf{x}, t) = \frac{1}{2} \left(\frac{\partial u_i}{\partial x^j}(\mathbf{x}, t) + \frac{\partial u_j}{\partial x^i}(\mathbf{x}, t) \right) , \qquad \mathbf{x} \in V , \ t \in (0, T) , \qquad (5.187)$$

$$\sigma^{ij}(\mathbf{x}, t) n_j(\mathbf{x}, t) = 0 , \qquad \mathbf{x} \in S , \ t \in (0, T) , \qquad (5.188)$$

$$u^i(\mathbf{x}, t) = 0 , \qquad \mathbf{x} \in V , \ t = 0 , \qquad (5.189)$$

$$\frac{\partial u^i}{\partial t}(\mathbf{x}, t) = 0 , \qquad \mathbf{x} \in V , \ t = 0 . \qquad (5.190)$$

Here, $\rho(\mathbf{x})$, $\kappa(\mathbf{x})$, and $\mu(\mathbf{x})$ are the parameters describing the medium (mass density, bulk modulus, and shear modulus, respectively). The sources of the waves are the volume density of force $\phi^i(\mathbf{x}, t)$ and the distribution of moments $\varphi^{ij}(\mathbf{x}, t)$. The fields $\sigma^{ij}(\mathbf{x}, t)$ and $u^{ij}(\mathbf{x}, t)$ are respectively the stress and the strain. The first of these equations expresses the fundamental dynamical equation, the second equation expresses the stress-strain relation for an isotropic, linear elastic medium (see, for instance, Landau and Lifshitz, 1986), the third equation expresses the strain as a function of the displacement, the fourth equation is a spatial boundary condition expressing that the surface of the medium is considered to be free (no tractions), and the last two equations are time-boundary (initial) conditions.

Let $G^i{}_j(\mathbf{x}, t; \mathbf{x}', t')$ be the (causal) Green's function of the problem, i.e., the displacement $u^i(\mathbf{x}, t)$ produced when the source is a point source of form $\phi^i(\mathbf{x}, t) = \delta^i_j \delta(\mathbf{x} - \mathbf{x}') \delta(t - t')$ (the reader should easily write the precise system of equations defining $G^i{}_j(\mathbf{x}, t; \mathbf{x}', t')$; see Aki and Richards (1980) for details on the linear elastic problem, Tarantola (1988) for the viscoelastic problem, and Roach (1982) or Morse and Feshbach (1953) for general notions on Green's functions).

Given the Green's function, the displacement field due to arbitrary sources $\phi^i(\mathbf{x}, t)$ and $\varphi^{ij}(\mathbf{x}, t)$ (the solution of the equations above) can be represented as (the sums over j and k are implicit)

$$u^i(\mathbf{x}, t) = \int_V dV(\mathbf{x}') \int_0^T dt' \, G^i{}_j(\mathbf{x}, t; \mathbf{x}', t') \phi^j(\mathbf{x}', t')$$
$$- \int_V dV(\mathbf{x}') \int_0^T dt' \frac{\partial G^i{}_j}{\partial x'^k}(\mathbf{x}, t; \mathbf{x}', t') \varphi^{jk}(\mathbf{x}', t') . \qquad (5.191)$$

As the values of the elastic parameters of the medium are assumed not to depend on time, it is easy to see that the Green's function is invariant by time translation:

$$G^i{}_j(\mathbf{x}, t; \mathbf{x}', t') = G^i{}_j(\mathbf{x}, t - t'; \mathbf{x}', 0) . \qquad (5.192)$$

Another important property of the Green's function is (see, for instance, Aki and Richards (1980) in the elastic context or Tarantola (1988) in the viscoelastic context) the *reciprocity*:

$$G_{ij}(\mathbf{x}, t; \mathbf{x}', t') = G_{ji}(\mathbf{x}', t; \mathbf{x}, t') \quad . \tag{5.193}$$

This means that the response at point $\mathbf{x}$ in the ith direction due to a source at point $\mathbf{x}'$ in the jth direction is identical to the response at point $\mathbf{x}'$ in the jth direction due to a source at point $\mathbf{x}$.

5.8.2 A Comment on Parameterization

The formulation above assumes that the elastic properties of the medium are continuous at every point of the medium, but it is easy to generalize the equations to the case where there are discontinuities (imposing the continuity of the displacement and the stress field across the discontinuities). If there are discontinuities, it is preferable to enter directly the geometrical properties of the discontinuity surfaces (positions, curvatures, etc.) as explicit parameters in the inverse problem. For simplicity in the exposition, this is *not* the approach followed below, where only the continuous fields $\rho(\mathbf{x})$, $\kappa(\mathbf{x})$, and $\mu(\mathbf{x})$ are considered. If such an approach is followed, many discontinuities may build up as the iterative inversion proceeds, as steep gradients, but this is not necessarily ideal.

5.8.3 Formulation of the Inverse Problem

The data of the problem are the observations of the displacement field at some discrete points:

$$\mathbf{u}_{\text{obs}} = \{u^i(\mathbf{x}_\alpha, t)_{\text{obs}}\} \quad , \quad i = \{1, 2, 3\} \quad , \quad \alpha = \{1, 2 \dots N\} \quad , \quad t \in (0, T) \quad . \tag{5.194}$$

The model parameters are the bulk modulus $\kappa(\mathbf{x})$, the shear modulus $\mu(\mathbf{x})$, and the mass density $\rho(\mathbf{x})$. As we are going to use the Gaussian assumption for the description of a priori information on the model parameters, it is better to replace these parameters with their logarithmic versions:

$$K(\mathbf{x}) = \log \frac{\kappa(\mathbf{x})}{\kappa_0} \quad , \quad M(\mathbf{x}) = \log \frac{\mu(\mathbf{x})}{\mu_0} \quad , \quad R(\mathbf{x}) = \log \frac{\rho(\mathbf{x})}{\rho_0} \quad , \tag{5.195}$$

where the denominators are arbitrary fixed values of the parameters.

Uncertainties in the data are assumed to be Gaussian, described using a general covariance function $C^{ij}(\mathbf{x}_\alpha, t; \mathbf{x}_\beta, t')$. We also introduce the weighting function $W_{ij}(\mathbf{x}_\alpha, t; \mathbf{x}_\beta, t')$, the inverse of the covariance function. The two functions[75] are related through (the sum over j is implicit)

$$\sum_{\beta=1}^{N} \int_0^T dt' \, C^{ij}(\mathbf{x}_\alpha, t; \mathbf{x}_\beta, t') \, W_{jk}(\mathbf{x}_\beta, t'; \mathbf{x}_\gamma, t'') = \delta_k^i \, \delta_{\alpha\gamma} \, \delta(t - t'') \quad . \tag{5.196}$$

---

[75] $W_{ij}(\mathbf{x}_\alpha, t; \mathbf{x}_\beta, t')$ is typically a generalized function (i.e., a distribution).

The a priori information on the model parameters is also assumed to be representable by a Gaussian distribution. More precisely, we assume that the actual elastic medium is a pseudorandom realization of a Gaussian random function (with three components) whose center and covariance are known. The centers of the Gaussian distributions are (respectively for the logarithmic bulk modulus, logarithmic shear modulus, and logarithmic mass density)

$$\mathbf{K}_{prior} = \{K(\mathbf{x})_{prior}\} \ , \ \ \mathbf{M}_{prior} = \{M(\mathbf{x})_{prior}\} \ , \ \ \mathbf{R}_{prior} = \{R(\mathbf{x})_{prior}\} \ , \ \ \mathbf{x} \in \mathcal{V} \ , \quad (5.197)$$

and the covariance functions are

$$\begin{array}{ccccc}
C_{KK}(\mathbf{x}, \mathbf{x}') & , & C_{KM}(\mathbf{x}, \mathbf{x}') & , & C_{KR}(\mathbf{x}, \mathbf{x}') & , \\
C_{MK}(\mathbf{x}, \mathbf{x}') & , & C_{MM}(\mathbf{x}, \mathbf{x}') & , & C_{MR}(\mathbf{x}, \mathbf{x}') & , \\
C_{RK}(\mathbf{x}, \mathbf{x}') & , & C_{RM}(\mathbf{x}, \mathbf{x}') & , & C_{RR}(\mathbf{x}, \mathbf{x}') & .
\end{array} \quad (5.198)$$

To simplify the discussion, let us assume that the crosscovariances are zero. We are then only left with the three covariance functions

$$C_{KK}(\mathbf{x}, \mathbf{x}') \ , \ \ C_{MM}(\mathbf{x}, \mathbf{x}') \ , \ \ C_{RR}(\mathbf{x}, \mathbf{x}') \quad (5.199)$$

and their inverses, the three weighting functions

$$W_{KK}(\mathbf{x}, \mathbf{x}') \ , \ \ W_{MM}(\mathbf{x}, \mathbf{x}') \ , \ \ W_{RR}(\mathbf{x}, \mathbf{x}') \ , \quad (5.200)$$

each related to its associated covariance function by a relation like $\int_{\mathcal{V}} dV(\mathbf{x}') C(\mathbf{x}, \mathbf{x}') W(\mathbf{x}', \mathbf{x}'') = \delta(\mathbf{x} - \mathbf{x}'')$.

The least-squares maximum likelihood model is the medium

$$\mathbf{m} = \{\mathbf{K}, \mathbf{M}, \mathbf{R}\} \quad (5.201)$$

defined through the minimization of the expression

$$2\,S(\mathbf{K}, \mathbf{M}, \mathbf{R}) = \| \mathbf{u}(\mathbf{K}, \mathbf{M}, \mathbf{R}) = \mathbf{u}_{obs} \|_u^2$$
$$+ \| \mathbf{K} - \mathbf{K}_{prior} \|_K^2 + \| \mathbf{M} - \mathbf{M}_{prior} \|_M^2 + \| \mathbf{R} - \mathbf{R}_{prior} \|_R^2 \ . \quad (5.202)$$

Here, $\mathbf{u}(\mathbf{K}, \mathbf{M}, \mathbf{R})$ denotes the displacements (at the observation points) predicted (using elastic theory) from the model $\mathbf{m} = \{\mathbf{K}, \mathbf{M}, \mathbf{R}\}$. Explicitly (the sums over i and j are implicit),

$$2\,S(\mathbf{K}, \mathbf{M}, \mathbf{R}) = \sum_{\alpha=1}^{N} \sum_{\beta=1}^{N} \int_0^T dt \int_0^T dt' \, \delta u^i(\mathbf{x}_\alpha, t) \, W_{ij}(\mathbf{x}_\alpha, t; \mathbf{x}_\beta, t') \, \delta u^j(\mathbf{x}_\beta, t')$$
$$+ \int_{\mathcal{V}} dV(\mathbf{x}) \int_{\mathcal{V}} dV(\mathbf{x}') \, \delta K(\mathbf{x}) \, W_{KK}(\mathbf{x}; \mathbf{x}') \, \delta K(\mathbf{x}')$$
$$+ \int_{\mathcal{V}} dV(\mathbf{x}) \int_{\mathcal{V}} dV(\mathbf{x}') \, \delta M(\mathbf{x}) \, W_{MM}(\mathbf{x}; \mathbf{x}') \, \delta M(\mathbf{x}')$$
$$+ \int_{\mathcal{V}} dV(\mathbf{x}) \int_{\mathcal{V}} dV(\mathbf{x}') \, \delta R(\mathbf{x}) \, W_{RR}(\mathbf{x}; \mathbf{x}') \, \delta R(\mathbf{x}') \ , \quad (5.203)$$

where (writing $\mathbf{u}_{\text{cal}}$ for $\mathbf{u}(\mathbf{K}, \mathbf{M}, \mathbf{R})$)

$$\delta u^i(\mathbf{x}_\alpha, t) = u^i(\mathbf{x}_\alpha, t)_{\text{cal}} - u^i(\mathbf{x}_\alpha, t)_{\text{obs}} \quad ,$$

$$\delta K(\mathbf{x}) = K(\mathbf{x}) - K(\mathbf{x})_{\text{prior}} = \log \frac{\kappa(\mathbf{x})}{\kappa(\mathbf{x})_{\text{prior}}} \quad ,$$

$$\delta M(\mathbf{x}) = M(\mathbf{x}) - M(\mathbf{x})_{\text{prior}} = \log \frac{\mu(\mathbf{x})}{\mu(\mathbf{x})_{\text{prior}}} \quad , \tag{5.204}$$

$$\delta R(\mathbf{x}) = R(\mathbf{x}) - R(\mathbf{x})_{\text{prior}} = \log \frac{\rho(\mathbf{x})}{\rho(\mathbf{x})_{\text{prior}}} \quad .$$

When so defined, the problem is fully nonlinear (the best model is defined without invoking any linear approximation of the basic equations, like a Born approximation). As the computed seismograms are *nonlinear* functionals of the model parameters, the functional in equations (5.202)–(5.203) is a *nonquadratic* function of the model parameters.

5.8.4 The Fréchet Derivative

When the medium parameters are $\mathbf{m} = \{\mathbf{K}, \mathbf{M}, \mathbf{R}\}$, the displacement (solution of the basic differential system) is $\mathbf{u}$. When the medium parameters are perturbed into $\mathbf{m} + \delta\mathbf{m} = \{\mathbf{K} + \delta\mathbf{K}, \mathbf{M} + \delta\mathbf{M}, \mathbf{R} + \delta\mathbf{R}\}$, the displacement becomes $\mathbf{u} + \delta\mathbf{u}$. We are interested here in evaluating the *first order* of the perturbation $\delta\mathbf{u}$.

Writing the unperturbed system (5.185)–(5.190) together with the perturbed system with the replacements[76] $u^i \mapsto u^i + \delta u^i$, $K \mapsto K + \delta K$, $M \mapsto K + \delta M$, $R \mapsto K + \delta R$, keeping only the terms that are first order, and using the unperturbed system to simplify the perturbed system, one arrives at the conclusion that the first-order perturbation δu^i is a solution of the system

$$\rho(\mathbf{x}) \frac{\partial^2 \delta u^i}{\partial t^2}(\mathbf{x}, t) - \frac{\partial \delta \sigma^{ij}}{\partial x^j}(\mathbf{x}, t) = \delta\phi^i(\mathbf{x}, t) , \qquad \mathbf{x} \in \mathcal{V} , \ t \in (0, T) , \quad (5.205)$$

$$\delta\sigma^{ij}(\mathbf{x}, t) - 3\kappa(\mathbf{x})\, \delta\bar{u}^{ij}(\mathbf{x}, t) - 2\mu(\mathbf{x})\, \delta\tilde{u}^{ij}(\mathbf{x}, t) = \varphi^{ij}(\mathbf{x}, t) ,$$
$$\mathbf{x} \in \mathcal{V} , \ t \in (0, T) , \quad (5.206)$$

$$\delta u^{ij}(\mathbf{x}, t) = \frac{1}{2} \left(\frac{\partial \delta u^i}{\partial x^j}(\mathbf{x}, t) + \frac{\partial \delta u^j}{\partial x^i}(\mathbf{x}, t) \right) , \qquad \mathbf{x} \in \mathcal{V} , \ t \in (0, T) , \quad (5.207)$$

$$\delta\sigma^{ij}(\mathbf{x}, t)\, n_j(\mathbf{x}, t) = 0 , \qquad \mathbf{x} \in \mathcal{S} , \ t \in (0, T) , \quad (5.208)$$

$$\delta u^i(\mathbf{x}, t) = 0 , \qquad \mathbf{x} \in \mathcal{V} , \ t = 0 , \quad (5.209)$$

$$\frac{\partial \delta u^i}{\partial t}(\mathbf{x}, t) = 0 , \qquad \mathbf{x} \in \mathcal{V} , \ t = 0 , \quad (5.210)$$

---

[76]One must replace any original positive parameter Ω with its logarithmic counterpart $\omega = \log(\Omega/\Omega_0)$, use the perturbation $\omega \mapsto \omega + \delta\omega$, and use the property $\Omega_0 \exp(\omega + \delta\omega) = \Omega_0 \exp\omega \exp\delta\omega = \Omega \exp\delta\omega = \Omega(1 + \delta\omega + \cdots) = \Omega + \Omega\delta\omega + \cdots$.

with the 'secondary Born sources'

$$\delta\phi^i(\mathbf{x}, t) = -\rho(\mathbf{x})\frac{\partial^2 u^i}{\partial t^2}(\mathbf{x}, t)\,\delta R(\mathbf{x}) \quad ,$$

(5.211)

$$\delta\varphi^{ij}(\mathbf{x}, t) = 3\kappa(\mathbf{x})\,\tilde{u}^{ij}(\mathbf{x}, t)\,\delta K(\mathbf{x}) + 2\mu(\mathbf{x})\,\tilde{u}^{ij}(\mathbf{x}, t)\,\delta M(\mathbf{x}) \quad .$$

We therefore see that the 'perturbation field' δu^i propagates in the unperturbed medium (i.e., the medium characterized by $\{\rho, \kappa, \mu\}$) and is excited by a double system of forces: at every point where the (logarithmic) mass density has been perturbed, there is a volume force $-\rho\,\ddot{u}^i\,\delta R$, and at every point where the (logarithmic) bulk and shear modulus have been perturbed, there is a moment source $\kappa\,u_k^k\,\delta K + 2\mu\,(u^{ij} - (1/3)\,u_k^k)\,\delta M$.

The representation theorem in equation (5.191) immediately allows us to write δu^i as

$$\delta u^i(\mathbf{x}, t) = \int_\mathcal{V} dV(\mathbf{x}')X^i(\mathbf{x}, t, \mathbf{x}')\,\delta R(\mathbf{x}') + \int_\mathcal{V} dV(\mathbf{x}')Y^i(\mathbf{x}, t, \mathbf{x}')\,\delta K(\mathbf{x}')$$

(5.212)

$$+ \int_\mathcal{V} dV(\mathbf{x}')Z^i(\mathbf{x}, t, \mathbf{x}')\,\delta M(\mathbf{x}') \quad ,$$

where (the sums over j and k are implicit)

$$X^i(\mathbf{x}, t, \mathbf{x}') = -\rho(\mathbf{x}')\int_0^T dt'\,G^i{}_j(\mathbf{x}, t; \mathbf{x}', t')\frac{\partial^2 u^j}{\partial t'^2}(\mathbf{x}', t') \quad ,$$

$$Y^i(\mathbf{x}, t, \mathbf{x}') = -3\kappa(\mathbf{x}')\int_0^T dt'\,\frac{\partial G^i{}_j}{\partial x'^k}(\mathbf{x}, t; \mathbf{x}', t')\,\tilde{u}^{jk}(\mathbf{x}', t') \quad ,$$

(5.213)

$$Z^i(\mathbf{x}, t, \mathbf{x}') = -2\mu(\mathbf{x}')\int_0^T dt'\,\frac{\partial G^i{}_j}{\partial x'^k}(\mathbf{x}, t; \mathbf{x}', t')\,\tilde{u}^{jk}(\mathbf{x}', t')$$

and where $G^i{}_j(\mathbf{x}, t; \mathbf{x}', t')$ is the Green's function of the unperturbed medium.

We have thus obtained the three integral kernels $X^i(\mathbf{x}, t, \mathbf{x}')$, $Y^i(\mathbf{x}, t, \mathbf{x}')$, and $Z^i(\mathbf{x}, t, \mathbf{x}')$ corresponding respectively to the Fréchet derivatives of the displacement field with respect to the logarithmic mass density, the logarithmic bulk modulus, and the logarithmic shear modulus.

For the resolution of the inverse problem, what we need are the Fréchet derivatives of the calculated data, i.e., the displacements at some selected points $\mathbf{x}_\alpha$ for $\alpha = 1, 2, \ldots, N$. We easily particularize to obtain

$$\delta u^i(\mathbf{x}_\alpha, t) = \int_\mathcal{V} dV(\mathbf{x})X^i(\mathbf{x}_\alpha, t, \mathbf{x})\,\delta R(\mathbf{x}) + \int_\mathcal{V} dV(\mathbf{x})Y^i(\mathbf{x}_\alpha, t, \mathbf{x})\,\delta K(\mathbf{x})$$

(5.214)

$$+ \int_\mathcal{V} dV(\mathbf{x})Z^i(\mathbf{x}_\alpha, t, \mathbf{x})\,\delta M(\mathbf{x}) \quad .$$

5.8.5   The Transpose of the Fréchet Derivative Operator

By definition, the transpose of the operator just obtained is the operator that with any $\delta \hat{u}_i(\mathbf{x}_\alpha, t)$ associates $\delta \hat{R}(\mathbf{x})$, $\delta \hat{K}(\mathbf{x})$, and $\delta \hat{M}(\mathbf{x})$ given by (the sum over i is implicit)

$$\delta \hat{R}(\mathbf{x}) = \sum_{\alpha=1}^{N} \int_{0}^{T} dt' \, X^i(\mathbf{x}_\alpha, t', \mathbf{x}) \, \delta \hat{u}_i(\mathbf{x}_\alpha, t') \quad ,$$

$$\delta \hat{K}(\mathbf{x}) = \sum_{\alpha=1}^{N} \int_{0}^{T} dt' \, Y^i(\mathbf{x}_\alpha, t', \mathbf{x}) \, \delta \hat{u}_i(\mathbf{x}_\alpha, t') \quad , \tag{5.215}$$

$$\delta \hat{M}(\mathbf{x}) = \sum_{\alpha=1}^{N} \int_{0}^{T} dt' \, Z^i(\mathbf{x}_\alpha, t', \mathbf{x}) \, \delta \hat{u}_i(\mathbf{x}_\alpha, t') \quad ,$$

where the kernels X^i, Y^i, and Z^i are those in equations (5.213). As we have seen, a linear operator and its transpose have the same kernels, the only difference arising in the variables of sum/integration, which are complementary.

This gives, after some rearranging (sum over i implicit),

$$\delta \hat{R}(\mathbf{x}) = -\int_{0}^{T} dt \left(\rho(\mathbf{x}) \frac{\partial^2 u^j}{\partial t^2}(\mathbf{x}, t) \right) \sum_{\alpha=1}^{N} \int_{0}^{T} dt' \, G^i{}_j(\mathbf{x}_\alpha, t'; \mathbf{x}, t) \, \delta \hat{u}_i(\mathbf{x}_\alpha, t') \quad ,$$

$$\delta \hat{K}(\mathbf{x}) = -\int_{0}^{T} dt \left(3 \kappa(\mathbf{x}) \, \bar{u}^{jk}(\mathbf{x}, t) \right) \sum_{\alpha=1}^{N} \int_{0}^{T} dt' \, \frac{\partial G^i{}_j}{\partial x^k}(\mathbf{x}_\alpha, t'; \mathbf{x}, t) \, \delta \hat{u}_i(\mathbf{x}_\alpha, t') \quad ,$$

$$\delta \hat{M}(\mathbf{x}) = -\int_{0}^{T} dt \left(2 \mu(\mathbf{x}) \, \tilde{u}^{jk}(\mathbf{x}, t) \right) \sum_{\alpha=1}^{N} \int_{0}^{T} dt' \, \frac{\partial G^i{}_j}{\partial x^k}(\mathbf{x}_\alpha, t'; \mathbf{x}, t) \, \delta \hat{u}_i(\mathbf{x}_\alpha, t') \quad .$$

$$\tag{5.216}$$

Defining (sum over i implicit)

$$\omega_j(\mathbf{x}, t) = \sum_{\alpha=1}^{N} \int_{0}^{T} dt' \, G^i{}_j(\mathbf{x}_\alpha, t'; \mathbf{x}, t) \, \delta \hat{u}_i(\mathbf{x}_\alpha, t') \quad , \tag{5.217}$$

this can be written[77]

$$\delta \hat{R}(\mathbf{x}) = -\rho(\mathbf{x}) \int_{0}^{T} dt \, \frac{\partial^2 u^i}{\partial t^2}(\mathbf{x}, t) \, \omega_i(\mathbf{x}, t) \quad ,$$

$$\delta \hat{K}(\mathbf{x}) = -3 \kappa(\mathbf{x}) \int_{0}^{T} dt \, \bar{u}^{ij}(\mathbf{x}, t) \, \bar{\omega}_{ij}(\mathbf{x}, t) \quad , \tag{5.218}$$

$$\delta \hat{M}(\mathbf{x}) = -2 \mu(\mathbf{x}) \int_{0}^{T} dt \, \tilde{u}^{ij}(\mathbf{x}, t) \, \tilde{\omega}_{ij}(\mathbf{x}, t) \quad ,$$

---

[77]The fact that u^{ij} is a symmetric tensor has been used as well as the fact that $\hat{u}^{ij}$ is isotropic and $\tilde{u}^{ij}$ is traceless.

where $\omega_{ij}(\mathbf{x}, t) = \frac{1}{2}\left(\frac{\partial \omega_i}{\partial x^j}(\mathbf{x}, t) + \frac{\partial \omega_j}{\partial x^i}(\mathbf{x}, t)\right)$ and, as usual, $\bar{\omega}_{ij}$ and $\tilde{\omega}$ are, respectively, the isotropic part and traceless part of ω_{ij}.

To obtain a more symmetric expression, let us integrate the first sum by parts. One has

$$\int_0^T dt \, \frac{\partial^2 u^j}{\partial t^2}(\mathbf{x}, t) \, \omega_j(\mathbf{x}, t) = \left[\frac{\partial u^j}{\partial t}(\mathbf{x}, T) \, \omega_j(\mathbf{x}, T)\right]_0^T - \int_0^T dt \, \frac{\partial u^j}{\partial t}(\mathbf{x}, t) \, \frac{\partial \omega_j}{\partial t}(\mathbf{x}, t) \quad .$$

(5.219)

As the field $u^i(\mathbf{x}, t)$ satisfies initial conditions of rest, $u^i(\mathbf{x}, 0) = 0$. We are about to see that the field $\omega_i(\mathbf{x}, t)$ satisfies *final* conditions of rest, $\omega(\mathbf{x}, T) = 0$. Therefore,

$$\int_0^T dt \, \frac{\partial^2 u^j}{\partial t^2}(\mathbf{x}, t) \, \omega_j(\mathbf{x}, t) = - \int_0^T dt \, \frac{\partial u^j}{\partial t}(\mathbf{x}, t) \, \frac{\partial \omega_j}{\partial t}(\mathbf{x}, t) \quad ,$$

(5.220)

and equations (5.218) finally become (denoting by a dot the time derivative)

$$\delta \hat{R}(\mathbf{x}) = + \rho(\mathbf{x}) \int_0^T dt \, \dot{u}^i(\mathbf{x}, t) \, \dot{\omega}_i(\mathbf{x}, t) \quad ,$$

$$\delta \hat{K}(\mathbf{x}) = - 3 \kappa(\mathbf{x}) \int_0^T dt \, \bar{u}^{ij}(\mathbf{x}, t) \, \bar{\omega}_{ij}(\mathbf{x}, t) \quad ,$$

(5.221)

$$\delta \hat{M}(\mathbf{x}) = - 2 \mu(\mathbf{x}) \int_0^T dt \, \tilde{u}^{ij}(\mathbf{x}, t) \, \tilde{\omega}_{ij}(\mathbf{x}, t) \quad .$$

The reader may easily verify that the physical dimensions of these three equations are consistent (the three give pure real numbers). We shall come back to them in a moment. Let us first obtain the interpretation of the field $\omega_i(\mathbf{x}, t)$ defined in equation (5.217).

This equation (5.217) is very close to the representation equation (5.191), except that the time variable is reversed and the source is a sum of point sources. It can be demonstrated[78] that the reversal of the time corresponds to a propagation problem with final, instead of initial, time conditions. Therefore, the field $\omega_i(\mathbf{x}, t)$ is the solution of the

---

[78]To demonstrate this, we have two routes. The direct one is to introduce the wave equation operator $\mathbf{L}$ (writing the elastic wave equation (5.185)–(5.190) formally as $\mathbf{L}\mathbf{u} = \boldsymbol{\phi}$), to define the transpose operator through $\langle \hat{\boldsymbol{\phi}}, \mathbf{L}\mathbf{u} \rangle_\phi = \langle \mathbf{L}^t \hat{\boldsymbol{\phi}}, \mathbf{u} \rangle_u$, and to verify that the operator $\mathbf{L}^t$ satisfies time-boundary conditions that are dual to those satisfied by $\mathbf{L}$ (i.e., final conditions of rest instead of initial conditions of rest), much as was done in Example 5.21. The second route is to write an equation like (5.217) first as $\omega_i(\mathbf{x}, t) = \int_V dV(\mathbf{x}) \int_0^T dt' \, G_{ji}(\mathbf{x}', t'; \mathbf{x}, t) \, \delta\hat{u}^j(\mathbf{x}', t')$, then, using the reciprocity property in equation (5.193), to obtain $\omega_i(\mathbf{x}, t) = \int_V dV(\mathbf{x}) \int_0^T dt' \, G_{ij}(\mathbf{x}, t'; \mathbf{x}', t) \, \delta\hat{u}^j(\mathbf{x}', t')$, or, using an anticausal (instead of a causal) Green's function, $\omega_i(\mathbf{x}, t) = \int_V dV(\mathbf{x}) \int_0^T dt' \, \overleftarrow{G}_{ij}(\mathbf{x}, t; \mathbf{x}', t') \, \delta\hat{u}^j(\mathbf{x}', t')$. Therefore, $\omega_i(\mathbf{x}, t)$ is the solution of a wave-propagation problem, where the sources are the $\delta\hat{u}^i(\mathbf{x}, t)$, which satisfies the final conditions of rest.

following system of equations:

$$\rho(\mathbf{x}) \frac{\partial^2 \omega^i}{\partial t^2}(\mathbf{x}, t) - \frac{\partial \sigma^{ij}}{\partial x^j}(\mathbf{x}, t) = \sum_{\alpha=1}^{N} \delta(\mathbf{x} - \mathbf{x}_\alpha) \, \delta \hat{u}^i(\mathbf{x}_\alpha, t) \,,$$
$$\mathbf{x} \in \mathcal{V}, \ t \in (0, T) \,, \qquad (5.222)$$

$$\sigma^{ij}(\mathbf{x}, t) - 3 \kappa(\mathbf{x}) \, \bar{\omega}^{ij}(\mathbf{x}, t) - 2 \mu(\mathbf{x}) \, \tilde{\omega}^{ij}(\mathbf{x}, t) = 0 \,, \quad \mathbf{x} \in \mathcal{V}, \ t \in (0, T) \,, \qquad (5.223)$$

$$\omega_{ij}(\mathbf{x}, t) = \frac{1}{2} \left(\frac{\partial \omega_i}{\partial x^j}(\mathbf{x}, t) + \frac{\partial \omega_j}{\partial x^i}(\mathbf{x}, t) \right) , \qquad \mathbf{x} \in \mathcal{V}, \ t \in (0, T) \,, \qquad (5.224)$$

$$\sigma^{ij}(\mathbf{x}, t) \, n_j(\mathbf{x}, t) = 0 \,, \qquad \mathbf{x} \in \mathcal{S}, \ t \in (0, T) \,, \qquad (5.225)$$

$$\omega^i(\mathbf{x}, t) = 0 \,, \qquad \mathbf{x} \in \mathcal{V}; \ t = T \,, \qquad (5.226)$$

$$\frac{\partial \omega^i}{\partial t}(\mathbf{x}, t) = 0 \,, \qquad \mathbf{x} \in \mathcal{V}, \ t = T \,, \qquad (5.227)$$

where the *final* conditions of rest should be noted.

We see that the sources of the field $\omega^i(\mathbf{x}, t)$ are discrete force sources, one at every point $\mathbf{x}_\alpha$ where there is a receiver, radiating the value $\delta \hat{u}^i(\mathbf{x}_\alpha, t)$.

5.8.6 A Comment on the Optimization Algorithm

Provided that the hypothesis of Gaussian uncertainties in the observed displacements and Gaussian a priori uncertainties in model parameters is realistic, the minimization of the misfit function defined in equations (5.202)–(5.203) will actually define a good model.

But there are two difficulties: (i) the relation between a wave amplitude and the elastic parameters defining a medium is essentially nonlinear, and (ii) in highly heterogeneous media the elastic wavefield may become extremely complex. Because of this, finding the minimum of the misfit function may be a difficult task.

There are no general rules, as the elastic waveform inverse problem may be applied to many problems, from the trivial to the impossible (using present-day computing power). Many problems require the combined use of a Monte Carlo search and local (steepest descent) optimization algorithms (as in the example shown in section 5.8.8 below). That I concentrate here on the description of a gradient-based algorithm does not mean that it is a panacea, but only that it will be a part of any realistic algorithm. It is only for problems where elastic waves propagate in a smooth medium presenting some diffractors (and possibly presenting multiscattering) that a gradient-based algorithm alone may produce the minimum of the misfit function.

5.8.7 The Inversion Algorithm (Steepest Descent)

Our problem here is to obtain the minimum of the misfit function defined in equation (5.202). Let us develop here the equations corresponding to a simple steepest descent algorithm. We obtained (equation (3.89))

$$\mathbf{m}_{n+1} = \mathbf{m}_n - \varepsilon_n \left(\mathbf{C}_M \, \mathbf{G}_n^t \, \mathbf{C}_D^{-1} \left(\mathbf{g}(\mathbf{m}_n) - \mathbf{d}_{\text{obs}} \right) + \left(\mathbf{m}_n - \mathbf{m}_{\text{prior}} \right) \right) \,, \qquad (5.228)$$

where $\mathbf{G}_n$ is the derivative operator evaluated at the current point $\mathbf{m}_n$ and $\mathbf{G}_m^t$ is its transpose. The real numbers ε_n are chosen ad hoc (as large as possible to accelerate convergence but small enough to avoid divergence).

Splitting the algorithm into its basic computations gives

$$\mathbf{d}_n = \mathbf{g}(\mathbf{m}_n) \quad , \tag{5.229}$$

$$\delta\mathbf{d}_n = \mathbf{d}_n - \mathbf{d}_{\text{obs}} \quad , \tag{5.230}$$

$$\delta\hat{\mathbf{d}}_n = \mathbf{C}_{\mathrm{D}}^{-1} \delta\mathbf{d}_n \quad , \tag{5.231}$$

$$\delta\hat{\mathbf{m}}_n = \mathbf{G}_n^t \delta\hat{\mathbf{d}}_n \quad , \tag{5.232}$$

$$\delta\mathbf{m}_n = \mathbf{C}_{\mathrm{M}} \delta\hat{\mathbf{m}}_n \quad , \tag{5.233}$$

$$\mathbf{m}_{n+1} = \mathbf{m}_n - \varepsilon_n (\delta\mathbf{m}_n + \mathbf{m}_n - \mathbf{m}_{\text{prior}}) \quad , \tag{5.234}$$

and, as a model $\mathbf{m}$ is made here by $\{\mathbf{R}, \mathbf{K}, \mathbf{M}\}$, we can write, with more detail,

$$\mathbf{d}_n = \mathbf{g}(\mathbf{R}_n, \mathbf{K}_n, \mathbf{M}_n) \quad , \tag{5.235}$$

$$\delta\mathbf{d}_n = \mathbf{d}_n - \mathbf{d}_{\text{obs}} \quad , \tag{5.236}$$

$$\delta\hat{\mathbf{d}}_n = \mathbf{C}_{\mathrm{D}}^{-1} \delta\mathbf{d}_n \quad , \tag{5.237}$$

$$\delta\hat{\mathbf{R}}_n = \mathbf{X}_n^t \delta\hat{\mathbf{d}}_n \quad , \quad \delta\hat{\mathbf{K}}_n = \mathbf{Y}_n^t \delta\hat{\mathbf{d}}_n \quad , \quad \delta\hat{\mathbf{M}}_n = \mathbf{Z}_n^t \delta\hat{\mathbf{d}}_n \quad , \tag{5.238}$$

$$\delta\mathbf{R}_n = \mathbf{C}_{RR} \delta\hat{\mathbf{R}}_n \quad , \quad \delta\mathbf{K}_n = \mathbf{C}_{KK} \delta\hat{\mathbf{K}}_n \quad , \quad \delta\mathbf{K}_n = \mathbf{C}_{MM} \delta\hat{\mathbf{M}}_n \quad , \tag{5.239}$$

$$\mathbf{R}_{n+1} = \mathbf{R}_n - \varepsilon_n (\delta\mathbf{R}_n + \mathbf{R}_n - \mathbf{R}_{\text{prior}}) \quad , \tag{5.240}$$

$$\mathbf{K}_{n+1} = \mathbf{K}_n - \varepsilon_n (\delta\mathbf{K}_n + \mathbf{K}_n - \mathbf{K}_{\text{prior}}) \quad , \tag{5.241}$$

$$\mathbf{M}_{n+1} = \mathbf{M}_n - \varepsilon_n (\delta\mathbf{M}_n + \mathbf{M}_n - \mathbf{M}_{\text{prior}}) \quad . \tag{5.242}$$

Here, the special structure assumed in equation (5.199) for the covariance operator has been used, as well as the transpose operators $\mathbf{X}^t$, $\mathbf{Y}^t$, and $\mathbf{Z}^t$ introduced in equation (5.215).

Let us comment on this set of equations. The algorithm is initialized at some arbitrary model, perhaps the prior model $\{\mathbf{R}_{\text{prior}}, \mathbf{K}_{\text{prior}}, \mathbf{M}_{\text{prior}}\}$ introduced in equation (5.197). Assume that a few iterations have been performed and that the current model is now $\{\mathbf{R}_n, \mathbf{K}_n, \mathbf{M}_n\}$.

Equation (5.235) corresponds to the resolution of the forward problem, i.e., the resolution of the propagation problem defined in equations (5.185)–(5.190) for the current model $\{\mathbf{R}_n, \mathbf{K}_n, \mathbf{M}_n\}$, and subsequent consideration of the displacement field $u^i(\mathbf{x}, t)$ are thus obtained at the points $\mathbf{x} = \mathbf{x}_\alpha$ where receivers were placed. This gives $u^i(\mathbf{x}_\alpha, t)_n$.

Equation (5.236) corresponds to the computation of the residuals $\delta u^i(\mathbf{x}_\alpha, t)_n = u^i(\mathbf{x}_\alpha, t)_n - u^i(\mathbf{x}_\alpha, t)_{\text{obs}}$.

Equation (5.237) corresponds to the computation of the weighted residuals. For instance, when using the weighting function $W_{ij}(\mathbf{x}_\alpha, t; \mathbf{x}_\beta, t')$ introduced in equation (5.196), this gives (sum over j implicit)

$$\delta\hat{u}_i(\mathbf{x}_\alpha, t)_n = \sum_{\beta=1}^{N} \int_0^T dt' \, W_{ij}(\mathbf{x}_\alpha, t; \mathbf{x}_\beta, t') \, \delta u^j(\mathbf{x}_\beta, t')_n \quad . \tag{5.243}$$

If the weighting function $W_{ij}(\mathbf{x}_\alpha, t; \mathbf{x}_\beta, t')$ is too difficult to obtain, one can use instead the covariance function $C^{ij}(\mathbf{x}_\alpha, t; \mathbf{x}_\beta, t')$ and solve the linear system

$$\delta u^i(\mathbf{x}_\alpha, t)_n = \sum_{\beta=1}^{N} \int_0^T dt' \, C^{ij}(\mathbf{x}_\alpha, t; \mathbf{x}_\beta, t') \, \delta \hat{u}_j(\mathbf{x}_\beta, t')_n \tag{5.244}$$

in order to obtain $\delta \hat{u}_i(\mathbf{x}_\alpha, t)_n$.

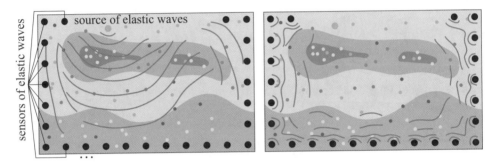

Figure 5.18. *A source generates elastic waves that propagate on a solid, and the displacements are measured at some points (left panel). These observations are used to infer the elastic structure of the medium. A steepest descent algorithm requires that each receiver act as a source, radiating the residual (calculated minus observed) displacements (right panel). The time correlation of the two propagating fields produces the updating of elastic model parameters.*

Equations (5.238) are the basic equations of the algorithm. They correspond to the computations demonstrated in equations (5.221),

$$\delta \hat{R}(\mathbf{x})_n = +\rho(\mathbf{x})_n \int_0^T dt \, \dot{u}^i(\mathbf{x}, t)_n \, \dot{\omega}_i(\mathbf{x}, t)_n \quad ,$$

$$\delta \hat{K}(\mathbf{x})_n = -3\kappa(\mathbf{x})_n \int_0^T dt \, \bar{u}^{ij}(\mathbf{x}, t)_n \, \bar{\omega}_{ij}(\mathbf{x}, t)_n \quad , \tag{5.245}$$

$$\delta \hat{M}(\mathbf{x}_n) = -2\mu(\mathbf{x})_n \int_0^T dt \, \tilde{u}^{ij}(\mathbf{x}, t)_n \, \tilde{\omega}_{ij}(\mathbf{x}, t)_n \quad ,$$

where $u^i(\mathbf{x}, t)_n$ is the field propagating in the current model $\{\mathbf{R}_n, \mathbf{K}_n, \mathbf{R}_n\}$ (excited by the actual sources and with initial conditions of rest) and $\omega^i(\mathbf{x}, t)_n$ is the field also propagating in the current medium $\{\mathbf{R}_n, \mathbf{K}_n, \mathbf{R}_n\}$, but whose fictive sources[79] are the weighted residuals $\delta \hat{u}_i(\mathbf{x}_\alpha, t)_n$, satisfying final (instead of initial) conditions of rest (see Figure 5.18). The field $\omega^i(\mathbf{x}, t)_n$ is precisely defined by the set of equations (5.222)–(5.227) (using the current medium).

---

[79] It has been considered here that the points where the displacements are observed are points inside the medium. In this case, the fictive sources are forces inside the medium. If the displacements are observed at points at the surface of the medium, it can be demonstrated that the fictive sources are surface tractions.

Equations (5.239) correspond to the application of the covariance functions introduced in equation (5.199):

$$\delta R(\mathbf{x})_n = \int_{\mathcal{V}} dV(\mathbf{x}') \, C_{RR}(\mathbf{x}, \mathbf{x}') \, \delta \hat{R}(\mathbf{x}')_n \quad,$$

$$\delta K(\mathbf{x})_n = \int_{\mathcal{V}} dV(\mathbf{x}') \, C_{KK}(\mathbf{x}, \mathbf{x}') \, \delta \hat{K}(\mathbf{x}')_n \quad, \tag{5.246}$$

$$\delta M(\mathbf{x})_n = \int_{\mathcal{V}} dV(\mathbf{x}') \, C_{MM}(\mathbf{x}, \mathbf{x}') \, \delta \hat{M}(\mathbf{x}')_n \quad.$$

Finally, equations (5.240)–(5.242) correspond to the model updatings

$$R(\mathbf{x})_{n+1} = R(\mathbf{x})_n - \varepsilon_n \left(\delta R(\mathbf{x})_n + R(\mathbf{x})_n - R(\mathbf{x})_{\text{prior}} \right) \quad,$$

$$K(\mathbf{x})_{n+1} = K(\mathbf{x})_n - \varepsilon_n \left(\delta K(\mathbf{x})_n + K(\mathbf{x})_n - K(\mathbf{x})_{\text{prior}} \right) \quad, \tag{5.247}$$

$$M(\mathbf{x})_{n+1} = M(\mathbf{x})_n - \varepsilon_n \left(\delta M(\mathbf{x})_n + M(\mathbf{x})_n - M(\mathbf{x})_{\text{prior}} \right) \quad.$$

A steepest descent algorithm, as the one presented here, never has to be used scrupulously: the steepest descent direction is only optimal for infinitesimally small steps, but nobody wants to perform infinitesimally small steps. The steepest descent direction can easily be ameliorated using some preconditioning. First, instead of the unique constant ε_n in the three equations (5.247), one may use three different constants, or even make a local optimization in the three-dimensional space defined by the three directions $\delta R(\mathbf{x})_n + R(\mathbf{x})_n - R(\mathbf{x})_{\text{prior}}$, $\delta K(\mathbf{x})_n + K(\mathbf{x})_n - K(\mathbf{x})_{\text{prior}}$, and $\delta M(\mathbf{x})_n + M(\mathbf{x})_n - M(\mathbf{x})_{\text{prior}}$. Still better, one may apply any nonlinear operator to these three directions that not only still gives a direction of descent for the misfit function but defines a much better direction (for a finite jump).

As a final comment, although I have chosen here to work with mass density, bulk modulus, and shear modulus, the special geometry of the problem at hand may suggest the use of other parameters.

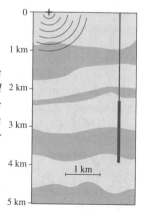

Figure 5.19. *One source is used to excite elastic waves that propagate into a geological medium. Inside a well, situated at a distance of 2 km from the source, three-component receivers were installed between 2 km and 4 km in depth. The medium a priori is known to present strong vertical gradients and milder horizontal ones. One is interested in imaging both of them.*

5.8.8 A Geological Example

As an example of the use of the methods just discussed, I choose to present here a result obtained by my team when working in a geophysical context.

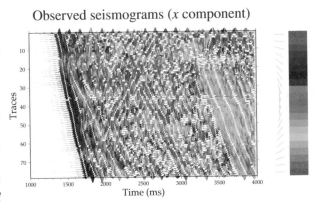

Figure 5.20. *Observed seismograms. Horizontal component displayed (vertical component also used in the inversion). Observe the complexity of the elastic field, with refractions, reflections, and reverberations. The color code displays the instantaneous polarization of the wave. See Charara, Barnes, and Tarantola (1996, 2000) and Barnes, Charara, and Tarantola (1998).*

The problem was to use waveform data measured inside a well (Figure 5.19) to infer the elastic structure of the medium. The data were three-component seismograms obtained at different positions between 2 and 4 km in depth. Of the three components of data recorded, the horizontal component inside the plane of Figure 5.19 is represented in Figure 5.20.

The inverse problem was essentially set following the methods presented earlier in this section, except that the parameters chosen here were the velocity of the longitudinal waves, the velocity of the transverse waves, the mass density, and the attenuation (too important in this example to be neglected).

The numerical simulation of the propagation of elastic waves was performed using a finite-difference approximation to the viscoelastic wave equation. It was assumed (quite unrealistically) that the medium presented a symmetry of revolution around the source (so a two-dimensional simulation could be used, instead of the more expensive three-dimensional simulation).

An extensive Monte Carlo search of the model space was performed (see Barnes, Charara, and Tarantola, 1998) until a model that was able to fit most of the arrival times of the packets of elastic energy was found. From this point on, a few tens of iterations of a preconditioned steepest descent algorithm were able to provide a model whose predicted seismograms fit the observations well (see Charara, Barnes, and Tarantola, 1996, 2000).

The obtained model is presented in Figure 5.21 (velocities are in m/s, mass density is in g/cm^3, and attenuation is in logarithmic units), the seismograms predicted from this model (using the viscoelastic wave equation) are presented in Figure 5.22, and the residual seismograms are presented in Figure 5.23.

In this problem, the estimation of uncertainties was basically obtained during the first phase of a Monte Carlo search. No special effort was made to compute the covariance operator during the gradient-based optimization phase.

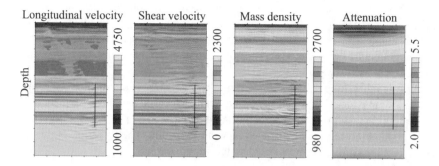

Figure 5.21. *The best model of seismic velocities, mass density, and attenuation.*

Calculated seismograms (*x* component)

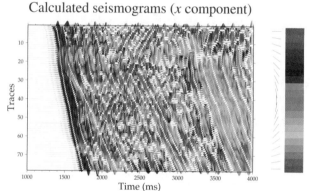

Figure 5.22. *Cal-culated seismograms from the model in Figure 5.21, using the viscoelastic wave equation. Compare with the observed seismograms in Figure 5.20.*

Residual seismograms (*x* component)

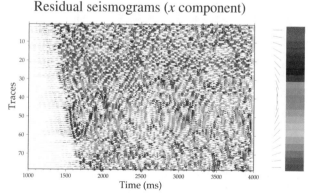

Figure 5.23. *Resid-ual seismograms (difference between the observations in Figure 5.20 and the calcula-tions in Figure 5.22). Note that most of the remaining energy is incoherent.*

In spite of some serious limitations (the actual three-dimensionality of the problem was only approximately taken into account and only one source location was available), this example shows that quite complex seismic waveforms can be fitted. As the initial models available here predicted seismograms that were quite different from those actually observed, gradient-based methods could not be applied from the start, and the Monte Carlo initial search phase was essential.

Among other things, this example suggested that there is not much to be gained when changing a simple preconditioned steepest descent algorithm to a more sophisticated quasi-Newton algorithm: with a good preconditioning (suggested by physical intuition), the extra computational effort required by a quasi-Newton algorithm is better spent (in this example) on extra iterations of the steepest descent algorithm.

Chapter 6

Appendices

6.1 Volumetric Probability and Probability Density

A probability distribution $\mathcal{A} \to P(\mathcal{A})$ over a manifold can be represented by a *volumetric probability* $F(\mathbf{x})$, defined through

$$P(\mathcal{A}) = \int_{\mathcal{A}} dV(\mathbf{x}) \, F(\mathbf{x}) \quad , \tag{6.1}$$

or by a *probability density* $f(\mathbf{x})$, defined through

$$P(\mathcal{A}) = \int_{\mathcal{A}} d\mathbf{x} \, f(\mathbf{x}) \quad , \tag{6.2}$$

where $d\mathbf{x} = dx^1 \, dx^2 \ldots$. While, under a change of variables, a probability density behaves as a density (i.e., its value at a point gets multiplied by the Jacobian of the transformation), a volumetric probability is a scalar (i.e., its value at a point remains invariant: it is defined independently of any coordinate system).

Defining the *volume density* through

$$V(\mathcal{A}) = \int_{\mathcal{A}} d\mathbf{x} \, v(\mathbf{x}) \tag{6.3}$$

and considering the expression $V(\mathcal{A}) = \int_{\mathcal{A}} dV(\mathbf{x})$, we obtain

$$dV(\mathbf{x}) = v(\mathbf{x}) \, d\mathbf{x} \quad . \tag{6.4}$$

It follows that the relation between volumetric probability and probability density is

$$f(\mathbf{x}) = v(\mathbf{x}) \, F(\mathbf{x}) \quad . \tag{6.5}$$

While the homogeneous probability distribution (the one assigning equal probabilities to equal volumes of the space) is, in general, *not* represented by a constant probability density, it is always represented by a constant volumetric probability.

Although I prefer, in my own work, to use volumetric probabilities, I have chosen in this text to use probability densities (for pedagogical reasons).

159

6.2 Homogeneous Probability Distributions

This appendix is reproduced from Mosegaard and Tarantola, 2002.

In some parameter spaces, there is an obvious definition of distance between points, and therefore of volume. For instance, in the 3D Euclidean space, the distance between two points is just the Euclidean distance (which is invariant under translations and rotations). Should we choose to parameterize the position of a point by its Cartesian coordinates $\{x, y, z\}$, the volume element in the space would be $dV(x, y, z) = dx\, dy\, dz$, while if we chose to use geographical coordinates, the volume element would be $dV(r, \theta, \varphi) = r^2 \sin\theta dr\, d\vartheta\, d\varphi$.

Definition. *The* homogeneous probability distribution *is the probability distribution that assigns to each region of the space a probability proportional to the volume of the region.*

Then, which probability density represents such a homogeneous probability distribution? Let us give the answer in three steps.

- If we use Cartesian coordinates $\{x, y, z\}$, as we have $dV(x, y, z) = dx\, dy\, dz$, the probability density representing the homogeneous probability distribution is constant: $f(x, y, z) = k$.

- If we use geographical coordinates $\{r, \theta, \varphi\}$, as we have $dV(r, \theta, \varphi) = r^2 \sin\theta\, dr\, d\theta\, d\varphi$, the probability density representing the homogeneous probability distribution is $g(r, \theta, \varphi) = k\, r^2 \sin\theta$.

- Finally, if we use an arbitrary system of coordinates $\{u, v, w\}$, in which the volume element of the space is $dV(u, v, w) = v(u, v, w)\, du\, dv\, dw$, the homogeneous probability distribution is represented by the probability density $h(u, v, w) = k\, v(u, v, w)$.

This is obviously true, since if we calculate the probability of a region $\mathcal{A}$ of the space, with volume $V(\mathcal{A})$, we get a number proportional to $V(\mathcal{A})$.

From these observations, we can arrive at conclusions that are of general validity. First, the homogeneous probability distribution over some space is represented by a constant probability density *only* if the space is flat (in which case rectilinear systems of coordinates exist) and if we use Cartesian (or rectilinear) coordinates. The other conclusions can be stated as rules.

Rule 6.1. *The probability density representing the homogeneous probability distribution is easily obtained if the expression of the volume element $dV(u_1, u_2, \dots) = v(u_1, u_2, \dots)\, du_1 du_2 \dots$ of the space is known, as it is then given by $h(u_1, u_2, \dots) = k\, v(u_1, u_2, \dots)$, where k is a proportionality constant (that may have physical dimensions).*

Rule 6.2. *If there is a metric $g_{ij}(u_1, u_2, \dots)$ in the space, then the volume element is given by $dV(u_1, u_2, \dots) = \sqrt{\det \mathbf{g}(u_1, u_2, \dots)}\, du_1 du_2 \dots$, i.e., we have $v(u_1, u_2, \dots) = \sqrt{\det \mathbf{g}(u_1, u_2, \dots)}$. The probability density representing the homogeneous probability distribution is, then, $h(u_1, u_2, \dots) = k\sqrt{\det \mathbf{g}(u_1, u_2, \dots)}$.*

Rule 6.3. *If the expression of the probability density representing the homogeneous probability distribution is known in one system of coordinates, then it is known in any other system of coordinates through the Jacobian rule (equation (1.18)).*

Indeed, in the expression above, $g(r, \theta, \varphi) = k\, r^2 \sin\theta$, we recognize the Jacobian between the geographical and the Cartesian coordinates (where the probability density is constant).

For short, when we say *the homogeneous probability density*, we mean *the probability density representing the homogeneous probability distribution. One should remember that, in general, the homogeneous probability density is* not *constant.*

Let us now examine positive parameters, like a temperature, a period, or a seismic wave propagation velocity. One of the properties of the parameters we have in mind is that they occur in pairs of mutually reciprocal parameters:

Period	$T = 1/\nu$	,	Frequency	$\nu = 1/T$;
Resistivity	$\rho = 1/\sigma$	,	Conductivity	$\sigma = 1/\rho$;
Temperature	$T = 1/(k\beta)$	,	Thermodynamic parameter	$\beta = 1/(kT)$;
Mass density	$\rho = 1/\ell$	,	Lightness	$\ell = 1/\rho$;
Compressibility	$\gamma = 1/\kappa$	,	Bulk modulus (incompress.)	$\kappa = 1/\gamma$;
Wave velocity	$c = 1/n$	,	Wave slowness	$n = 1/c$.

When working with physical theories, one may freely choose one of these parameters or its reciprocal.

Sometimes, these pairs of equivalent parameters come from a definition, like when we define frequency ν as a function of the period T by $\nu = 1/T$. Sometimes, these parameters arise when analyzing an idealized physical system. For instance, Hooke's law, relating stress σ_{ij} to strain ε_{ij}, can be expressed as $\sigma_{ij} = c_{ij}{}^{k\ell} \varepsilon_{k\ell}$, thus introducing the stiffness tensor $c_{ijk\ell}$, or as $\varepsilon_{ij} = d_{ij}{}^{k\ell} \sigma_{k\ell}$, thus introducing the compliance tensor $d_{ijk\ell}$, the inverse of the stiffness tensor. Then, the respective eigenvalues of these two tensors belong to the class of scalars analyzed here.

Let us take, as an example, the pair conductivity-resistivity (this may be thermal, electric, etc.). Assume we have two samples in the laboratory, S_1 and S_2, whose resistivities are respectively ρ_1 and ρ_2. Correspondingly, their conductivities are $\sigma_1 = 1/\rho_1$ and $\sigma_2 = 1/\rho_2$. How should we define the *distance* between the electrical properties of the two samples? As we have $|\rho_2 - \rho_1| \neq |\sigma_2 - \sigma_1|$, choosing one of the two expressions as the distance would be arbitrary. Consider the following definition of distance between the two samples:

$$D(S_1, S_2) = \left| \log \frac{\rho_2}{\rho_1} \right| = \left| \log \frac{\sigma_2}{\sigma_1} \right| . \tag{6.6}$$

This definition (i) treats symmetrically the two equivalent parameters ρ and σ and, more importantly, (ii) has an *invariance of scale* (what matters is how many *octaves* we have between the two values, not the plain difference between the values). In fact, it is the only definition of distance between the two samples S_1 and S_2 that has an invariance of scale and is additive (i.e., $D(S_1, S_2) + D(S_2, S_3) = D(S_1, S_3)$).

Example 6.1. *We have just considered two samples. Can we define the mean sample? The mean sample should be defined as the sample that is equidistant from the two given samples. Using the distance in equation (6.6), one easily finds[80] that the resistivity of the mean sample is*

$$\rho = \sqrt{\rho_1 \rho_2} \quad . \tag{6.7}$$

Equivalently, the conductivity of the mean sample is

$$\sigma = \sqrt{\sigma_1 \sigma_2} \quad . \tag{6.8}$$

Note that (i) *these two expressions are formally identical, and* (ii) *the conductivity of the mean sample equals the inverse of the resistivity of the mean sample (while the arithmetic mean of the resistivities is not the inverse of the arithmetic mean of the conductivities).*

Associated with the distance $D(x_1, x_2) = |\log(x_2/x_1)|$ is the distance element (differential form of the distance)

$$dL(x) = dx/x \quad . \tag{6.9}$$

This being a one-dimensional volume, we can apply now Rule 6.1 above to get the expression of the homogeneous probability density for such a positive parameter:

$$f(x) = k/x \quad . \tag{6.10}$$

Defining the reciprocal parameter $y = 1/x$ and using the Jacobian rule, we arrive at the homogeneous probability density for y :

$$g(y) = k/y \quad . \tag{6.11}$$

These two probability densities have the same form: the two reciprocal parameters are treated symmetrically. Introducing the logarithmic parameters

$$x^* = \log(x/x_0) \quad , \qquad y^* = \log(y/y_0) \quad , \tag{6.12}$$

where x_0 and y_0 are arbitrary positive constants, and using the Jacobian rule, we arrive at the homogeneous probability densities

$$f'(x^*) = k \quad , \qquad g'(y^*) = k \quad . \tag{6.13}$$

This shows that the logarithm of a positive parameter (of the type considered above) is a Cartesian parameter. In fact, it is the consideration of equations (6.13), together with the Jacobian rule, that allows full understanding of the (homogeneous) probability densities (6.10)–(6.11).

The association of the probability density $f(u) = k/u$ with positive parameters was first made by Jeffreys (1939). To honor him, I propose to use the term *Jeffreys parameters* for

---

[80]Using, for instance, the resistivity, the equidistance condition is written $\log(\rho_1/\rho) = \log(\rho/\rho_2)$, i.e., $\rho^2 = \rho_1 \rho_2$.

all the parameters of the type considered above. The $1/u$ probability density was advocated by Jaynes (1968), and a nontrivial use of it was made by Rietsch (1977) in the context of inverse problems.

Rule 6.4. *The homogeneous probability density for a Jeffreys quantity u is $f(u) = k/u$.*

Rule 6.5. *The homogeneous probability density for a* Cartesian *parameter u (like the logarithm of a Jeffreys parameter, an actual Cartesian coordinate in an Euclidean space, or the Newtonian time coordinate) is $f(u) = k$. The homogeneous probability density for an angle describing the position of a point in a circle is also constant.*

If a parameter u is a Jeffreys parameter with the homogeneous probability density $f(u) = k/u$, then its inverse, its square, and, in general, any power of the parameter is also a Jeffreys parameter, as can easily be seen using the Jacobian rule.

Rule 6.6. *Any power of a Jeffreys quantity (including its inverse) is a Jeffreys quantity.*

It is important to recognize when we do *not* face a Jeffreys parameter. Among the many parameters used in the literature to describe an isotropic linear elastic medium, we find parameters like Lamé's coefficients λ and μ, the bulk modulus κ, the Poisson ratio σ, etc. A simple inspection of the theoretical range of variation of these parameters shows that the first Lamé parameter λ and the Poisson ratio σ may take negative values, so they are certainly not Jeffreys parameters. In contrast, Hooke's law $\sigma_{ij} = c_{ijk\ell}\, \varepsilon^{k\ell}$, defining a linearity between stress σ_{ij} and strain ε_{ij}, defines the positive definite stiffness tensor $c_{ijk\ell}$ or, if we write $\varepsilon_{ij} = d_{ijk\ell}\, \sigma^{k\ell}$, defines its inverse, the compliance tensor $d_{ijk\ell}$. The two reciprocal tensors $c_{ijk\ell}$ and $d_{ijk\ell}$ are Jeffreys tensors. This is a notion whose development is beyond the scope of this book, but we can give the following rule.

Rule 6.7. *The eigenvalues of a Jeffreys tensor are Jeffreys quantities.*[81]

As the two (different) eigenvalues of the stiffness tensor $c_{ijk\ell}$ are $\lambda_\kappa = 3\kappa$ (with multiplicity 1) and $\lambda_\mu = 2\mu$ (with multiplicity 5), we see that the incompressibility modulus κ and the shear modulus μ are Jeffreys parameters[82] (as is any parameter proportional to them, or any power of them, including the inverses). If for some reason, instead of working with κ and μ, we wish to work with other elastic parameters, like for instance the Young modulus Y and the Poisson ratio σ, or the two elastic wave velocities, then the homogeneous probability distribution must be found using the Jacobian of the transformation (see Appendix 6.3).

---

[81]This solves the complete problem for isotropic tensors only. It is beyond the scope of this text to propose rules valid for general anisotropic tensors: the necessary mathematics has not yet been developed.

[82]The definition of the elastic constants was made before the tensorial structure of the theory was understood. Seismologists today should not use, at a theoretical level, parameters like the first Lamé coefficient λ or the Poisson ratio σ. Instead, they should use κ and μ (and their inverses). In fact, our suggestion is to use the true eigenvalues of the stiffness tensor, $\lambda_\kappa = 3\kappa$ and $\lambda_\mu = 2\mu$, which we propose to call the *eigen-bulk-modulus* and the *eigen-shear-modulus*, respectively.

Some probability densities have conspicuous 'dispersion parameters,' like the σ's in the normal probability density $f(x) = k \, \exp\left(-\frac{(x-x_0)^2}{2\sigma^2}\right)$, in the log-normal probability $g(X) = \frac{k}{X} \, \exp\left(-\frac{(\log X/X_0)^2}{2\sigma^2}\right)$, or in the Fischer probability density (Fischer, 1953) $h(\vartheta, \varphi) = k \, \sin\theta \, \exp\left(\cos\theta \, / \, \sigma^2\right)$. A consistent probability model requires that when the dispersion parameter σ tends to infinity, the probability density tends to the homogeneous probability distribution. For instance, in the three examples just given, $f(x) \to k$, $g(X) \to k/X$, and $h(\theta, \varphi) \to k \, \sin\theta$, which are the respective homogeneous probability densities for a Cartesian quantity, a Jeffreys quantity, and the geographical coordinates on the surface of the sphere. We can state the following rule.

Rule 6.8. *If a probability density has some dispersion parameters, then, in the limit where the dispersion parameters tend to infinity, the probability density must tend to the homogeneous one.*

As an example, using the normal probability density $f(x) = k \, \exp\left(-\frac{(x-x_0)^2}{2\sigma^2}\right)$ for a Jeffreys parameter is not consistent. Note that it would assign a finite probability to negative values of a parameter that, by definition, is positive. More technically, this would violate the condition that all probability densities be absolutely continuous with respect to the homogeneous probability density. Using the log-normal probability density for a Jeffreys parameter is perfectly acceptable.

There is a problem of terminology in the Bayesian literature. The homogeneous probability distribution is a very special distribution. When the problem of selecting a 'prior' probability distribution arises in the absence of any information, except the fundamental symmetries of the problem, one may select as a prior probability distribution the homogeneous distribution. But enthusiastic Bayesians do not call it 'homogeneous,' but 'noninformative.' I cannot recommend using this terminology. The homogeneous probability distribution is as informative as any other distribution, it is just the homogeneous one.

In general, each time we consider an abstract parameter space, each point being represented by some parameters $\mathbf{x} = \{x^1, x^2, \ldots, x^n\}$, we will start by solving the (sometimes nontrivial) problem of defining a distance between points that respects the necessary symmetries of the problem. Only exceptionally will this distance be a quadratic expression of the parameters (coordinates) being used (i.e., only exceptionally will our parameters correspond to Cartesian coordinates in the space). From this distance, a volume element $dV(\mathbf{x}) = v(\mathbf{x}) \, d\mathbf{x}$ will be deduced, from which the expression $f(\mathbf{x}) = k \, v(\mathbf{x})$ of the homogeneous probability density will follow. Sometimes, we can directly define the volume element without the need of a distance. We emphasize the need of defining a distance — or a volume element — in the parameter space, from which the notion of homogeneity will follow. With this point of view, we slightly depart from the original work by Jeffreys and Jaynes.

6.3 Homogeneous Distribution for Elastic Parameters

Consider an ideally elastic, homogeneous (although perhaps anisotropic) medium. Hooke's law relating stress σ^{ij} to strain $\varepsilon^{k\ell}$ can be written

$$\sigma^{ij} = c^{ij}{}_{k\ell} \, \varepsilon^{k\ell} \quad , \tag{6.14}$$

where $c^{ij}{}_{k\ell}$ is the tensor of elastic *stiffnesses*. Alternatively, one can write

$$\varepsilon^{ij} = s^{ij}{}_{k\ell}\,\sigma^{k\ell} \quad, \tag{6.15}$$

where $c^{ij}{}_{k\ell}$ is the tensor of elastic *compliances*. The stiffness and compliance tensors are mutually inverse, and both are positive definite. In our language, they are *Jeffreys tensors*. Because of the different symmetries among their components, there are only 21 degrees of freedom to represent an ideally elastic medium. There are many possible choices for the 21 quantities needed to represent an elastic medium. But whatever our choice for these quantities, there is one general answer to the problem of obtaining the associated homogeneous probability density.

In the 21-dimensional abstract space where each point represents one elastic medium, there is only one choice of distance between two points (i.e., between two elastic media) that has all the necessary invariances (it must satisfy the axioms of a distance, the expression has to be the same when using stiffness or compliance, there must be invariance of scale). The distance between the elastic medium $\mathcal{M}_1$ (represented by the stiffness tensor $\mathbf{c}_1$ or the compliance tensor $\mathbf{s}_1$) and the elastic medium $\mathcal{M}_2$ (represented by the stiffness tensor $\mathbf{c}_2$ or the compliance tensor $\mathbf{s}_2$) is

$$D(\mathcal{M}_1, \mathcal{M}_2) = \| \log(\mathbf{c}_2\,\mathbf{c}_1^{-1}) \| = \| \log(\mathbf{s}_2\,\mathbf{s}_1^{-1}) \| \quad. \tag{6.16}$$

One should remember here that the logarithm of a tensor can be defined using a Taylor development. Equivalently, the logarithm of a (positive definite) tensor has the same eigendirection as the original tensor but its eigenvalues are the logarithm of the original eigenvalues. The norm of a tensor is defined as usual.[83]

Once a definition of distance (a metric) has been introduced in a manifold, it is just a matter of simple (although lengthy) computations to deduce an expression for the volume element of the space (using the given coordinates). And once the volume element has been expressed, the expression for the homogeneous probability density immediately follows, as explained in Appendix 6.2.

Rather than developing the general theory here, let us concentrate on the much simpler example where the elastic medium is *isotropic*. Then, only two quantities are required, for instance, the incompressibility modulus and the shear modulus. These are a pair of Jeffreys parameters (as they are eigenvalues of the stiffness tensor $c^{ij}{}_{k\ell}$). It is quite elementary to obtain the homogeneous probability density for these two parameters. Then, the expression of the homogeneous probability density for other sets of elastic parameters, like the set {Young's modulus, Poisson ratio} or the set {Longitudinal wave velocity, Tranverse wave velocity} is obtained using the Jacobian rule.

Let us develop this elementary theory.

6.3.1 Incompressibility Modulus and Shear Modulus

The Cartesian parameters of elastic theory are the logarithm of the incompressibility modulus and the logarithm of the shear modulus,

$$\kappa^* = \log(\kappa/\kappa_0) \quad, \quad \mu^* = \log(\mu/\mu_0) \quad, \tag{6.17}$$

---

[83] For instance, the squared norm of a tensor $\psi^{ij}{}_{k\ell}$ is $\| \boldsymbol{\psi} \|^2 = g_{ip}\,g_{jq}\,g^{kr}\,g^{\ell s}\,\psi^{ij}{}_{k\ell}\,\psi^{pq}{}_{rs}$, where g_{ij} are the components of the metric tensor in the given coordinates.

where κ_0 and μ_0 are two arbitrary constants. The homogeneous probability density is constant for these parameters (a constant that we set arbitrarily to one):

$$f_{\kappa^*\mu^*}(\kappa^*, \mu^*) = 1 \quad . \tag{6.18}$$

As is often the case for homogeneous 'probability' densities, $f_{\kappa^*\mu^*}(\kappa^*, \mu^*)$ is not normalizable. Using the Jacobian rule, it is easy to transform this probability density into the equivalent one for the positive parameters themselves:

$$f_{\kappa\mu}(\kappa, \mu) = 1/(\kappa\,\mu) \quad . \tag{6.19}$$

This $1/x$ form of the probability density remains invariant if we take any power of κ and of μ. In particular, if instead of using the incompressibility κ we use the compressibility $\gamma = 1/\kappa$, the Jacobian rule simply gives $f_{\gamma\mu}(\gamma, \mu) = 1/(\gamma\,\mu)$.

Associated with the probability density (6.18) is the Euclidean definition of distance

$$ds^2 = (d\kappa^*)^2 + (d\mu^*)^2 \quad , \tag{6.20}$$

which corresponds, in the variables (κ, μ), to

$$ds^2 = (d\kappa/\kappa)^2 + (d\mu/\mu)^2 \quad , \tag{6.21}$$

i.e., to the metric

$$\begin{pmatrix} g_{\kappa\kappa} & g_{\kappa\mu} \\ g_{\mu\kappa} & g_{\mu\mu} \end{pmatrix} = \begin{pmatrix} 1/\kappa^2 & 0 \\ 0 & 1/\mu^2 \end{pmatrix} \quad . \tag{6.22}$$

6.3.2 Young Modulus and Poisson Ratio

The Young modulus Y and the Poisson ratio σ can be expressed as a function of the incompressibility modulus and the shear modulus as

$$Y = \frac{9\kappa\mu}{3\kappa + \mu} \quad , \qquad \sigma = \frac{1}{2}\frac{3\kappa - 2\mu}{3\kappa + \mu} \tag{6.23}$$

or, reciprocally,

$$\kappa = \frac{Y}{3(1 - 2\sigma)} \quad , \qquad \mu = \frac{Y}{2(1 + \sigma)} \quad . \tag{6.24}$$

The absolute value of the Jacobian of the transformation is easily computed,

$$J = \frac{Y}{2(1 + \sigma)^2(1 - 2\sigma)^2} \quad , \tag{6.25}$$

and the Jacobian rule transforms the probability density (6.19) into

$$f_{Y\sigma}(Y, \sigma) = \frac{1}{\kappa\mu}J = \frac{3}{Y(1 + \sigma)(1 - 2\sigma)} \quad , \tag{6.26}$$

which is the probability density representing the homogeneous probability distribution for elastic parameters using the variables (Y, σ). This probability density is the product of the probability density $1/Y$ for the Young modulus and the probability density

$$g(\sigma) = \frac{3}{Y(1+\sigma)(1-2\sigma)} \tag{6.27}$$

for the Poisson ratio. This probability density is represented in Figure 6.1. From the definition of σ it can be demonstrated that its values must range in the interval $-1 < \sigma < 1/2$, and we see that the homogeneous probability density is singular at these points. Although most rocks have positive values of the Poisson ratio, there are materials where σ is negative (e.g., Yeganeh-Haeri et al., 1992).

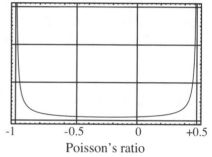

Figure 6.1. *The homogeneous probability density for the Poisson ratio, as deduced from the condition that the incompressibility and the shear modulus are Jeffreys parameters.*

-1	-0.5	0	+0.5

Poisson's ratio

It may be surprising that the probability density in Figure 6.1 corresponds to a homogeneous distribution. If we have many samples of elastic materials, and if their logarithmic incompressibility modulus κ^* and their logarithmic shear modulus μ^* have a constant probability density (which *is* the definition of homogeneous distribution of elastic materials), then σ will be distributed according to the $g(\sigma)$ of the figure.

To be complete, let me mention that with a change of variables $x^i \rightleftharpoons x^I$, a metric g_{ij} changes to

$$g_{IJ} = \Lambda_I{}^i \Lambda_J{}^j g_{ij} = \frac{\partial x^i}{\partial x^I} \frac{\partial x^j}{\partial x^J} g_{ij} \quad . \tag{6.28}$$

The metric (6.21) then transforms into

$$\begin{pmatrix} g_{YY} & g_{Y\sigma} \\ g_{\sigma Y} & g_{\sigma\sigma} \end{pmatrix} = \begin{pmatrix} \frac{2}{Y^2} & \frac{2}{(1-2\sigma)Y} - \frac{1}{(1+\sigma)Y} \\ \frac{2}{(1-2\sigma)Y} - \frac{1}{(1+\sigma)Y} & \frac{4}{(1-2\sigma)^2} + \frac{1}{(1+\sigma)^2} \end{pmatrix} \quad . \tag{6.29}$$

The surface element is

$$dS_{Y\sigma}(Y, \sigma) = \sqrt{\det g}\, dY\, d\sigma = \frac{3\, dY\, d\sigma}{Y(1+\sigma)(1-2\sigma)} \quad , \tag{6.30}$$

a result from which expression (6.26) can be inferred.

Although the Poisson ratio has historical interest, it is not a simple parameter, as shown by its theoretical bounds $-1 < \sigma < 1/2$ or the form of the homogeneous probability density

(Figure 6.1). In fact, the Poisson ratio σ depends only on the ratio κ/μ (incompressibility modulus over shear modulus), as we have

$$\frac{1+\sigma}{1-2\sigma} = \frac{3}{2}\frac{\kappa}{\mu} \quad . \tag{6.31}$$

The ratio $J = \kappa/\mu$ of two Jeffreys parameters being a Jeffreys parameter, a useful pair of Jeffreys parameters may be $\{\kappa, J\}$. The ratio $J = \kappa/\mu$ has an easy to grasp physical interpretation (as the ratio between the incompressibility and the shear modulus) and should be preferred, in theoretical developments, to the Poisson ratio, as it has simpler theoretical properties. As the name of the nearest metro station to the university of the author is *Jussieu*, we accordingly call J the *Jussieu ratio*.

6.3.3 Longitudinal and Transverse Wave Velocities

Equation (6.19) gives the probability density representing the homogeneous probability distribution of elastic media, when parameterized by the incompressibility modulus and the shear modulus:

$$f_{\kappa\mu}(\kappa, \mu) = 1/(\kappa\,\mu) \quad . \tag{6.32}$$

Should we have been interested, in addition, in the mass density ρ, then we would have arrived (as ρ is another Jeffreys parameter) at the probability density

$$f_{\kappa\mu\rho}(\kappa, \mu, \rho) = 1/(\kappa\,\mu\,\rho) \quad . \tag{6.33}$$

This is the starting point for this section.

What about the probability density representing the homogeneous probability distribution of elastic materials when we use as parameters the mass density and the two wave velocities? The longitudinal wave velocity α and the shear wave velocity β are related to the incompressibility modulus κ and the shear modulus μ through

$$\alpha = \sqrt{(\kappa + 4\mu/3)/\rho} \quad , \quad \beta = \sqrt{\mu/\rho} \,, \tag{6.34}$$

and a direct use of the Jacobian rule transforms the probability density (6.33) into

$$f_{\alpha\beta\rho}(\alpha, \beta, \rho) = \frac{1}{\rho\,\alpha\,\beta\left(\frac{3}{4} - \frac{\beta^2}{\alpha^2}\right)} \quad , \tag{6.35}$$

which is the answer to our question.

That this function becomes singular for $\alpha = 2\beta/\sqrt{3}$ is due to the fact that the boundary $\alpha = 2\beta/\sqrt{3}$ cannot be crossed: the fundamental inequalities $\kappa > 0$, $\mu > 0$ impose that the two velocities are linked by the inequality constraint

$$\alpha > 2\beta/\sqrt{3} \quad . \tag{6.36}$$

Let us focus for a moment on the homogeneous probability density for the two wave velocities (α, β) existing in an elastic solid (disregard here the mass density ρ). We have

$$f_{\alpha\beta}(\alpha, \beta) = \frac{1}{\alpha\,\beta\left(\frac{3}{4} - \frac{\beta^2}{\alpha^2}\right)} \quad . \tag{6.37}$$

This is displayed in Figure 6.2.

Figure 6.2. *The joint homogeneous probabil-
ity density for the velocities* (α, β) *of the longitudinal
and transverse waves propagating in an elastic solid.
Contrary to the incompressibility modulus and the shear
modulus, which are independent parameters, the longitudinal wave velocity and the transversal wave velocity
are not independent (see text for an explanation). The
scales for the velocities are unimportant: it is possible
to multiply the two velocity scales by any factor without modifying the form of the probability (which is itself
defined up to a multiplicative constant).*

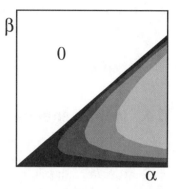

Let us demonstrate that the marginal probability density for both α and β is of the
form $1/x$. For we have to compute

$$f_\alpha(\alpha) = \int_0^{\sqrt{3}\alpha/2} d\beta \; f(\alpha, \beta) \tag{6.38}$$

and

$$f_\beta(\beta) = \int_{2\beta/\sqrt{3}}^{+\infty} d\alpha \; f(\alpha, \beta) \tag{6.39}$$

(the bounds of integration can easily be understood by a look at Figure 6.2). These integrals
can be evaluated as

$$f_\alpha(\alpha) = \lim_{\varepsilon \to 0} \int_{\sqrt{\varepsilon}\sqrt{3}\alpha/2}^{\sqrt{1-\varepsilon}\sqrt{3}\alpha/2} d\beta \; f(\alpha, \beta) = \lim_{\varepsilon \to 0} \left(\frac{4}{3} \log \frac{1-\varepsilon}{\varepsilon} \right) \frac{1}{\alpha} \tag{6.40}$$

and

$$f_\beta(\beta) = \lim_{\varepsilon \to 0} \int_{\sqrt{1+\varepsilon}\,2\beta/\sqrt{3}}^{2\beta/(\sqrt{\varepsilon}\sqrt{3})} d\alpha \; f(\alpha, \beta) = \lim_{\varepsilon \to 0} \left(\frac{2}{3} \log \frac{1/\varepsilon - 1}{\varepsilon} \right) \frac{1}{\beta} \quad . \tag{6.41}$$

The numerical factors tend to infinity, but this is only one more manifestation of the fact that
the homogeneous probability densities are usually improper (not normalizable). Dropping
these numerical factors gives

$$f_\alpha(\alpha) = 1/\alpha \tag{6.42}$$

and

$$f_\beta(\beta) = 1/\beta \quad . \tag{6.43}$$

It is interesting to note that we have here an example where two parameters look like Jeffreys
parameters, but are not, because they are not independent (the homogeneous joint probability
density is not the product of the homogeneous marginal probability densities).

It is also worth knowing that using slownesses instead of velocities ($n = 1/\alpha, \eta = 1/\beta$) leads, as one would expect, to

$$f_{n\eta\rho}(n, \eta, \rho) = \frac{1}{\rho \, n \, \eta \left(\frac{3}{4} - \frac{n^2}{\eta^2}\right)} \quad . \tag{6.44}$$

6.4 Homogeneous Distribution for Second-Rank Tensors

The usual definition of the norm of a tensor provides the only natural definition of distance in the space of all possible tensors. This shows that, when using a Cartesian system of coordinates, the components of a tensor are the Cartesian coordinates in the 6D space of symmetric tensors. The homogeneous distribution is then represented by a constant (nonnormalizable) probability density:

$$f(\sigma_{xx}, \sigma_{yy}, \sigma_{zz}, \sigma_{xy}, \sigma_{yz}, \sigma_{zx}) = k \quad . \tag{6.45}$$

Instead of using the components, we may use the three eigenvalues $\{\lambda_1, \lambda_2, \lambda_3\}$ of the tensor and the three Euler angles $\{\psi, \theta, \varphi\}$ defining the orientation of the eigendirections in the space. As the Jacobian of the transformation

$$\{\sigma_{xx}, \sigma_{yy}, \sigma_{zz}, \sigma_{xy}, \sigma_{yz}, \sigma_{zx}\} \rightleftharpoons \{\lambda_1, \lambda_2, \lambda_3, \psi, \theta, \varphi\} \tag{6.46}$$

is

$$\left| \frac{\partial(\sigma_{xx}, \sigma_{yy}, \sigma_{zz}, \sigma_{xy}, \sigma_{yz}, \sigma_{zx})}{\partial(\lambda_1, \lambda_2, \lambda_3, \psi, \theta, \varphi)} \right| = (\lambda_1 - \lambda_2)(\lambda_2 - \lambda_3)(\lambda_3 - \lambda_1) \sin\theta \quad , \tag{6.47}$$

the homogeneous probability density (6.45) transforms into

$$g(\lambda_1, \lambda_2, \lambda_3, \psi, \theta, \varphi) = k (\lambda_1 - \lambda_2)(\lambda_2 - \lambda_3)(\lambda_3 - \lambda_1) \sin\theta \quad . \tag{6.48}$$

Although this is not obvious, this probability density is isotropic in spatial directions (i.e., the 3D referentials defined by the three Euler angles are isotropically distributed). In this sense, we recover isotropy as a special case of homogeneity.

Rule 6.8, imposing that any probability density of the variables $\{\lambda_1, \lambda_2, \lambda_3, \psi, \theta, \varphi\}$ has to tend to the homogeneous probability density (6.48) when the dispersion parameters tend to infinity imposes a strong constraint on the form of acceptable probability densities that is, generally, overlooked.

For instance, a Gaussian model for the variables $\{\sigma_{xx}, \sigma_{yy}, \sigma_{zz}, \sigma_{xy}, \sigma_{yz}, \sigma_{zx}\}$ is consistent (as the limit of a Gaussian is constant). This induces, via the Jacobian rule, a probability density for the variables $\{\lambda_1, \lambda_2, \lambda_3, \psi, \theta, \varphi\}$, a probability density that is not simple, but consistent. A Gaussian model for the parameters $\{\lambda_1, \lambda_2, \lambda_3, \psi, \theta, \varphi\}$ would not be consistent.

6.5 Central Estimators and Estimators of Dispersion

Probability densities are, in general, defined over manifolds, and the definitions of the 'center' and of the 'dispersion' of the probability distribution are technically difficult. In the case where the probability density is defined over a *linear space* (i.e., over a space where sum and multiplication by a scalar make clear sense), the theory is very simple (as shown below).

One-Dimensional Case

Given a normalized one-dimensional probability density function $f(x)$, defined over a variable x for which the expression $|x_2 - x_1|$ is an acceptable definition of distance, consider the expression

$$s_p(m) = \left(\int_{-\infty}^{+\infty} dx \, |x - m|^p \, f(x) \right)^{1/p} . \tag{6.49}$$

For given p, the value of m that makes s_p minimum is termed the *center of $f(x)$ in the ℓ_p-norm sense* and is denoted by m_p. The value m_1 is termed the *median*, m_2 is the *mean* (or *mathematical expectation*), and m_∞ is the *midrange*. The following properties hold.

- *Median* (minimum ℓ_1-norm):

$$\int_{-\infty}^{+\infty} dx \, |x - m_1| \, f(x) \quad \text{minimum} \quad \Leftrightarrow \quad \int_{-\infty}^{m_1} dx \, f(x) = \int_{m_1}^{+\infty} dx \, f(x) = \frac{1}{2} . \tag{6.50}$$

- *Mean* (minimum ℓ_2-norm):

$$\int_{-\infty}^{+\infty} dx \, (x - m_2)^2 \, f(x) \quad \text{minimum} \quad \Leftrightarrow \quad m_2 = \int_{-\infty}^{+\infty} dx \, x \, f(x) . \tag{6.51}$$

- *Midrange* (minimum ℓ_∞-norm):

$$\lim_{p \to \infty} \int_{-\infty}^{+\infty} dx \, |x - m_\infty|^p \, f(x) \quad \text{minimum} \quad \Leftrightarrow \quad m_\infty = \frac{x_{\max} + x_{\min}}{2} , \tag{6.52}$$

where $x_{\max}$ (resp., $x_{\min}$) is the maximum (resp., minimum) value of x for which $f(x) \neq 0$.

The value of $s_p(m)$ at the minimum is termed the *dispersion of $f(\mathbf{x})$ in the ℓ_p-norm sense* and is denoted σ_p:

$$\sigma_p = s_p(m_p) . \tag{6.53}$$

The value σ_1 is termed the *mean deviation*, σ_2 is the *standard deviation*, and σ_∞ is the *half-range*. The following properties hold.

- *Mean deviation* (minimum ℓ_1-norm):

$$\sigma_1 = \int_{-\infty}^{+\infty} dx \, |x - m_1| \, f(x) \quad \Leftrightarrow \quad \sigma_1 = \int_{m_1}^{+\infty} dx \, x \, f(x) - \int_{-\infty}^{m_1} dx \, x \, f(x) . \tag{6.54}$$

- *Standard deviation* (minimum ℓ_2-norm):

$$\sigma_2{}^2 = \int_{-\infty}^{+\infty} dx\,(x - m_2)^2\,f(x) \quad \Leftrightarrow \quad \sigma_2{}^2 = \int_{-\infty}^{+\infty} dx\,x^2\,f(x) - m_2{}^2 \quad .$$

$$(6.55)$$

- *Half-range* (minimum ℓ_∞-norm):

$$\sigma_\infty = \lim_{p \to \infty} \left(\int_{-\infty}^{+\infty} dx\,|x - m_\infty|^p\,f(x) \right)^{1/p} \quad \Leftrightarrow \quad \sigma_\infty = \frac{x_{\sup} - x_{\inf}}{2} \quad .$$

$$(6.56)$$

Multidimensional Case

Given a probability density function $f(\mathbf{x})$ defined for the vector variable $\mathbf{x}$ (element of a linear space where the Euclidean norm of a vector makes sense), consider the operator $\mathbf{C}(\mathbf{m})$ defined by its components,

$$C^{ij}(\mathbf{m}) = \int d\mathbf{x}\,(x^i - m^i)\,(x^j - m^j)\,f(\mathbf{x}) \quad . \tag{6.57}$$

The vector $\mathbf{m}$ that minimizes the determinant of $\mathbf{C}(\mathbf{m})$ is termed the *mean* (or mathematical expectation) of $\mathbf{x}$ *in the ℓ_2-norm sense*. It is given by

$$\mathbf{m}_2 = \int d\mathbf{x}\,\mathbf{x}\,f(\mathbf{x}) \quad . \tag{6.58}$$

The value at $\mathbf{m} = \mathbf{m}_2$ of the operator in equation (6.57) is termed the *covariance* of $\mathbf{x}$ *in the ℓ_2-norm sense*, and is simply denoted by $\mathbf{C}$:

$$\mathbf{C} = \mathbf{C}(\mathbf{m}_2) \quad . \tag{6.59}$$

The diagonal elements of $\mathbf{C}$ clearly equal the variances (square of standard deviations) previously defined:

$$C^{ii} = (\sigma^i)^2 \quad . \tag{6.60}$$

The covariance operator in the ℓ_2-norm sense (or *ordinary* covariance operator) has the following properties (see, for instance, Pugachev, 1965):

1. $\mathbf{C}$ is symmetric:

$$C^{ij} = C^{ji} \quad . \tag{6.61}$$

2. $\mathbf{C}$ is definite nonnegative: for any vector $\mathbf{x}$,

$$\mathbf{x}^t\,\mathbf{C}^{-1}\,\mathbf{x} \geq 0 \quad . \tag{6.62}$$

3. If $\mathbf{C}$ is positive definite, then, for any vector $\mathbf{x}$, the quantity

$$\| \mathbf{x} \| = (\mathbf{x}^t \mathbf{C}^{-1} \mathbf{x})^{1/2} \tag{6.63}$$

has the properties of a norm. It is termed the *weighted ℓ_2-norm* of the vector $\mathbf{x}$.

4. The *correlation coefficients* ρ^{ij} defined by

$$\rho^{ij} = \frac{C^{ij}}{\sigma^i \sigma^j} \tag{6.64}$$

have the property

$$-1 \leq \rho^{ij} \leq +1 \quad . \tag{6.65}$$

5. The probability density

$$f(\mathbf{x}) = ((2\pi)^N \det \mathbf{C})^{-1/2} \exp\left(-\frac{1}{2} (\mathbf{x} - \mathbf{x}_0)^t \mathbf{C}^{-1} (\mathbf{x} - \mathbf{x}_0) \right) \quad , \tag{6.66}$$

where N is the dimension of the vector $\mathbf{x}$, is normalized, with a mean value $\mathbf{x}_0$ and covariance operator $\mathbf{C}$ (e.g., Dubes, 1968). From the results of problem (6.26), it follows that among all the probability densities with given ℓ_2-norm covariance operator, the Gaussian function has minimum information content (i.e., it has maximum 'spreading').

Intuitive estimation of the numerical value of a covariance is not easy, while it is very easy to intuitively estimate the value of a correlation. In 2D, when the two standard deviations and the correlation are estimated, equation (6.64) can be used to estimate the covariance. See Figures 6.3–6.4.

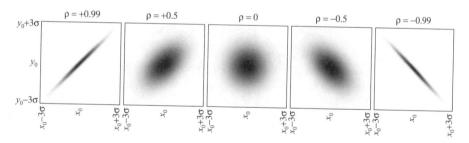

Figure 6.3. *A two-dimensional Gaussian plotted using three standard deviations in each of the two axes for different values of the coefficient of correlation.*

Two variables for which $\rho = 0$ are called *uncorrelated*. The reader should keep in mind that while the notion of *independent variables* is general (see section 1.2.8), the notion of *uncorrelated variables* is not, and should only be used if the probability density under consideration is approximately Gaussian. While (independent) $\Rightarrow$ (uncorrelated), (uncorrelated) $\not\Rightarrow$ (independent), as illustrated in Figure 6.5.

The discussion of the multidimensional spaces has been limited to the ℓ_2-norm case. It is not easy to generalize these concepts to the general ℓ_p-norm case.

Figure 6.4. *The gross intuitive es-timation of the mean values and of the stan-dard deviations of a two-dimensional prob-ability density is easy, but this is not so for the covariance. The covariance is better es-timated by first estimating the correlation, then using equation (6.64).*

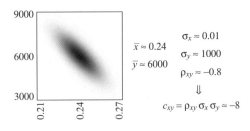

$$\sigma_x \approx 0.01$$
$$\bar{x} \approx 0.24 \qquad \sigma_y \approx 1000$$
$$\bar{y} \approx 6000 \qquad \rho_{xy} \approx -0.8$$
$$\Downarrow$$
$$c_{xy} = \rho_{xy}\,\sigma_x\,\sigma_y \approx -8$$

Figure 6.5. *The correlation ρ is a meaningful param-eter only if the probability density under consideration is not too far from a Gaussian. A circular probability density, for instance, has zero correlation, but the variables are far from being inde-pendent.*

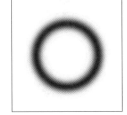

6.6   Generalized Gaussian

As shown in problem (6.26), among all the normalized probability densities $f(x)$ with fixed ℓ_p-norm estimator of dispersion,

$$\int_{-\infty}^{+\infty} dx\, |x - x_0|^p\, f(x) \;=\; (\sigma_p)^p \quad, \tag{6.67}$$

the one with *minimum information content* (i.e., with maximum 'spreading') is given by

$$f_p(x) \;=\; \frac{p^{1-1/p}}{2\,\sigma_p\,\Gamma(1/p)} \, \exp\left(-\frac{1}{p}\frac{|x-x_0|^p}{(\sigma_p)^p}\right) \quad, \tag{6.68}$$

where $\Gamma(\cdot)$ denotes the gamma function.

Figure 6.6 shows some examples with p respectively equal to 1, 1.5, 2, 3, and 10. For $p = 1$,

$$f_1(x) \;=\; \frac{1}{2\,\sigma_1} \, \exp\left(-\frac{|x-x_0|}{\sigma_1}\right) \quad, \tag{6.69}$$

and $f_1(x)$ is a symmetric exponential, centered at $x = x_0$ with *mean deviation* equal to σ_1. For $p = 2$,

$$f_2(x) \;=\; \frac{1}{\sqrt{2\,\pi}\,\sigma_2} \, \exp\left(-\frac{1}{2}\frac{(x-x_0)^2}{\sigma_2{}^2}\right) \quad, \tag{6.70}$$

and $f_2(x)$ is a Gaussian function, centered at $x = x_0$ with *standard deviation* equal to σ_2. For $p \to \infty$,

$$f_\infty(x) \;=\; \begin{cases} 1/(2\,\sigma_\infty) & \text{for } x_0 - \sigma_\infty \le x \le x_0 + \sigma_\infty \quad, \\ 0 & \text{otherwise} \quad, \end{cases} \tag{6.71}$$

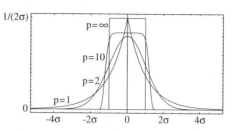

Figure 6.6. *Generalized Gaussians of order p (centered at zero). The value* $p = 1$ *gives a double exponential,* $p = 2$ *gives an ordinary Gaussian, and* $p = \infty$ *gives a boxcar function. The parameter* σ *of the figure is the* σ_p *of the text.*

and $f_\infty(x)$ is a box function, centered at $x = x_0$ with *midrange* equal to σ_∞. Problem (6.32) shows that $f_p(x)$ is normalized to unity.

The function $f_p(x)$ defined in equation (6.68) can be termed a *generalized Gaussian*, because it generates a family of well-behaved functions containing the Gaussian function as a particular case. Symmetric exponentials, Gaussian functions, and boxcar functions are often used to model error distribution. The definition of a generalized Gaussian slightly widens the possibility of choice.

6.7 Log-Normal Probability Density

The log-normal probability density is defined by

$$f(x) = \frac{1}{(2\pi)^{1/2} s} \frac{1}{x} \exp\left(-\frac{1}{2 s^2}\left(\log \frac{x}{x_0}\right)^2\right) \quad . \tag{6.72}$$

Figure 6.7 shows some examples of this probability density.

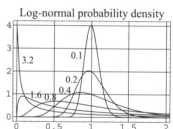

Figure 6.7. *The log-normal probability density (equation (6.72)). Note that when the dispersion parameter s tends to* ∞ *, the probability density tends to the function* $1/x$ *, the homogeneous probability density for a Jeffreys quantity.*

The log-normal probability density is so called because the *logarithm* of the variable has a normal (Gaussian) probability density. For the change of variables

$$x^* = \beta \log(x/\gamma) \quad , \quad x = \gamma \exp(x^*/\beta) \tag{6.73}$$

transforms $f(x)$ into

$$f^*(x^*) = \frac{1}{(2\pi)^{1/2} \sigma} \exp\left(-\frac{1}{2}\frac{(x^* - x_0^*)^2}{\sigma^2}\right) \quad , \tag{6.74}$$

with $\sigma = s \beta$, $s = \sigma/\beta$, and

$$x_0^* = \beta \log(x_0/\gamma) \quad , \quad x_0 = \gamma \exp(x_0^*/\beta) \quad . \tag{6.75}$$

In (6.73), the constant β is often $\log_e 10$, which corresponds to defining $x^* = \log_{10}(x/\gamma)$. The constant γ often corresponds to the physical unit used for x (see Example 1.30). Alternatively, the particular choice

$$x^* = \frac{1}{s} \log\left(\frac{x}{x_0}\right) \quad , \quad x = x_0 \exp(s\, x^*) \tag{6.76}$$

leads to a Gaussian density with zero mean and unit standard deviation:

$$f^*(x^*) = \frac{1}{(2\pi)^{1/2}} \exp\left(-\frac{(x^*)^2}{2}\right) . \tag{6.77}$$

Figure 6.7 suggests that, for given x_0, when the dispersion s is very small, the log-normal probability density tends to a Gaussian function. This is indeed the case. For, when $s \to 0$, $f(x)$ takes significant value only in the vicinity of x_0, and $f(x) = \frac{1}{(2\pi)^{1/2} s}\frac{1}{x_0} \exp(-\frac{1}{2s^2}(-1 + \frac{x}{x_0} - \cdots)^2)$, i.e.,

$$f(x) \simeq \frac{1}{(2\pi)^{1/2}(s\, x_0)} \exp\left(-\frac{1}{2}\frac{(x - x_0)^2}{(s\, x_0)^2}\right) . \tag{6.78}$$

If, for given x_0, the dispersion s is very large, the log-normal probability density tends to a log-uniform probability density (i.e., proportional to $1/x$; see section 1.2.4). For any x not too close to the origin, the argument of the exponential in (6.72) can be taken as null, thus showing that, at the limit $s \to \infty$,

$$f(x) \simeq \frac{1}{(2\pi)^{1/2} s}\frac{1}{x} . \tag{6.79}$$

The convergence of (6.72) into (6.79) is not a uniform convergence, in the sense that while the function (6.79) tends to infinity when x tends to 0, the log-normal (6.72) takes the value 0 at the origin. But for values of x of the same order of magnitude as x_0, the approximation (6.79) is adequate (for instance, for the values $x_0 = 1$, $s = 10$, the log-normal function and the log-uniform function are indistinguishable in Figure 6.7).

As suggested in section 1.2.4, the log-normal probability density is often adequate to represent probability distributions for variables that by definition are constrained to be positive. The reader will easily verify that if a variable x has a log-normal distribution, the variable $y = 1/x$ has the same distribution.

The function

$$f(x) = \frac{p^{1-1/p}}{2s\,\Gamma(1/p)}\frac{1}{x} \exp\left(-\frac{1}{p\, s^p}\left|\log\frac{x}{x_0}\right|^p\right) \tag{6.80}$$

transforms, under the change of variables (6.73), into the generalized Gaussian

$$f^*(x^*) = \frac{p^{1-1/p}}{2\sigma\,\Gamma(1/p)} \exp\left(-\frac{1}{p}\frac{|x^* - x_0^*|^p}{\sigma^p}\right) , \tag{6.81}$$

where σ and x_0^* are given by (6). This suggests that 6.80 can be referred to as the generalized log-normal in the ℓ_p-norm sense.

6.8 Chi-Squared Probability Density

The probability density

$$f_\nu(u) = \frac{1}{2^{\nu/2}\,\Gamma(\nu/2)}\, u^{\frac{\nu}{2}-1}\, e^{-\frac{u}{2}} \tag{6.82}$$

is called the χ^2 probability density with parameter ν (one usually says with ν degrees of freedom). Sometimes, the variable u in equation (6.82) is denoted χ^2 (leading to ambiguous notations).

Figure 6.8 displays the χ^2 probability density for some selected values of ν. Note that, for $\nu = 1$, the value at the origin is infinite, and that for $\nu = 2$, one has the Laplace probability density (exponential law). For large values of ν, the χ^2 probability density can be roughly approximated (near its maximum) by a Gaussian probability density with mean value ν and standard deviation $\sqrt{2\nu}$.

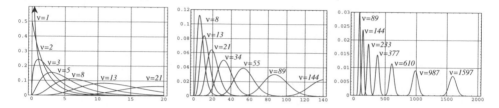

Figure 6.8. *The χ^2 probability density for some selected values of ν.*

First Property

Let $\mathbf{y} = \{y^1, \ldots, y^p\}$ be a p-dimensional Gaussian random vector with mean value $\mathbf{m}$ and covariance matrix $\mathbf{C}$. With each random realization $\mathbf{y}_0$ of the vector $\mathbf{y}$ associate the number

$$\chi^2 = (\mathbf{y}_0 - \mathbf{m})^t\, \mathbf{C}^{-1}\, (\mathbf{y}_0 - \mathbf{m}) \quad . \tag{6.83}$$

Then, this random variable is distributed according to the χ^2 probability density with p degrees of freedom (see Rao, 1973, or Afifi and Azen, 1979).

Second Property

Let $\mathbf{y}$ be a p-dimensional Gaussian random vector with unspecified mean and with covariance matrix $\mathbf{C}$. Let $\mathbf{A}$ be a $p \times q$ matrix, with $p \geq q$, the matrix $\mathbf{A}$ having full rank (so that, given $\mathbf{y}$, the system $\mathbf{y} = \mathbf{A}\mathbf{x}$ is not underdetermined for $\mathbf{x}$). With each random realization $\mathbf{y}_0$ of $\mathbf{y}$ associate the vector $\mathbf{x}_0$ defined by the minimization of

$$\chi^2(\mathbf{x}) = (\mathbf{A}\mathbf{x} - \mathbf{y}_0)^t\, \mathbf{C}^{-1}\, (\mathbf{A}\mathbf{x} - \mathbf{y}_0) \quad . \tag{6.84}$$

Then, the random variable $\chi^2(\mathbf{x}_0)$ is distributed according to the χ^2 probability density with

$$\nu = \dim(\mathbf{y}) - \operatorname{rank}(\mathbf{A}) = p - q \tag{6.85}$$

degrees of freedom (Rao, 1973).

We may note that, as the solution of the minimization problem is (this is a special case of the first of equations (3.37))

$$\mathbf{x}_0 = (\mathbf{A}^t \, \mathbf{C}^{-1} \, \mathbf{A})^{-1} \, \mathbf{A}^t \, \mathbf{C}^{-1} \, \mathbf{y}_0 \quad , \tag{6.86}$$

we obtain, after some easy simplifications,

$$\chi^2(\mathbf{x}_0) = \mathbf{y}_0^t \, (\mathbf{C}^{-1} - \mathbf{C}^{-1} \, \mathbf{A} \, (\mathbf{A}^t \, \mathbf{C}^{-1} \, \mathbf{A})^{-1} \, \mathbf{A}^t \, \mathbf{C}^{-1}) \, \mathbf{y}_0 \quad . \tag{6.87}$$

Third Property

Setting

$$\mathbf{x} = \mathbf{m} \quad , \quad \mathbf{y} = \begin{pmatrix} \mathbf{d} \\ \mathbf{m} \end{pmatrix} \quad , \quad \mathbf{y}_0 = \begin{pmatrix} \mathbf{d}_{\text{obs}} \\ \mathbf{m}_{\text{prior}} \end{pmatrix} \quad , \quad \mathbf{C} = \begin{pmatrix} \mathbf{C}_{\text{D}} & \mathbf{0} \\ \mathbf{0} & \mathbf{C}_{\text{M}} \end{pmatrix} \quad , \quad \mathbf{A} = \begin{pmatrix} \mathbf{G} \\ \mathbf{I} \end{pmatrix} \tag{6.88}$$

in the equations above, the theorem translates as follows.

Let $\mathbf{y}$ be an n-dimensional Gaussian vector with unspecified mean and with covariance matrix $\mathbf{C}_{\text{M}}$. Let $\mathbf{m}$ be an m-dimensional Gaussian vector with unspecified mean and with covariance matrix $\mathbf{C}_{\text{M}}$. Let $\mathbf{G}$ be an $n \times m$ matrix (of arbitrary rank). With each random realization $\mathbf{d}_{\text{obs}}$ of $\mathbf{d}$ and each random realization $\mathbf{m}_{\text{prior}}$ of $\mathbf{m}$ associate the vector $\mathbf{m}_0$ defined by the minimization of

$$\chi^2(\mathbf{m}) = (\mathbf{G} \, \mathbf{m} - \mathbf{d}_{\text{obs}})^t \, \mathbf{C}_{\text{D}}^{-1} \, (\mathbf{G} \, \mathbf{m} - \mathbf{d}_{\text{obs}}) + (\mathbf{m} - \mathbf{m}_{\text{prior}})^t \, \mathbf{C}_{\text{M}}^{-1} \, (\mathbf{m} - \mathbf{m}_{\text{prior}}) \quad . \tag{6.89}$$

Then, the random variable $\chi^2(\mathbf{m}_0)$ is distributed according to the χ^2 probability density with

$$\nu = \dim(\mathbf{d}) = n \tag{6.90}$$

degrees of freedom.

As above, we may note that, as the solution of the minimization problem is (first of equations (3.37))

$$\mathbf{m}_0 = \left(\mathbf{G}^t \, \mathbf{C}_{\text{D}}^{-1} \, \mathbf{G} + \mathbf{C}_{\text{M}}^{-1} \right)^{-1} \left(\mathbf{G}^t \, \mathbf{C}_{\text{D}}^{-1} \, \mathbf{d}_{\text{obs}} + \mathbf{C}_{\text{M}}^{-1} \, \mathbf{m}_{\text{prior}} \right) \quad , \tag{6.91}$$

one obtains, after some simplifications,

$$\chi^2(\mathbf{m}_0) = (\mathbf{G} \, \mathbf{m}_{\text{prior}} - \mathbf{d}_{\text{obs}})^t \, (\mathbf{G} \, \mathbf{C}_{\text{M}} \, \mathbf{G}^t + \mathbf{C})^{-1} \, (\mathbf{G} \, \mathbf{m}_{\text{prior}} - \mathbf{d}_{\text{obs}}) \quad . \tag{6.92}$$

Fourth Property

Let x be a random variable that may take the values $\{x_1, x_2, \ldots, x_n\}$ with the probabilities p_i. The probabilities p_i are assumed to be known a priori and are normed: $\sum_{i=1}^{n} p_i = 1$.

A large number of realizations of the variable x have given the experimental frequencies f_i. As we must satisfy the constraint $\sum_{i=1}^{n} f_i = 1$, the number of values f_i that are independent (i.e., the degrees of freedom) is $n - 1$.

Then, the random variable

$$\chi^2 = \sum_{i=1}^{n} \frac{(f_i - p_i)^2}{p_i} \qquad (6.93)$$

is distributed according to the χ^2 probability density with $n - 1$ degrees of freedom. If the probability distribution p_i is not given a priori, but k parameters of the law are estimated from the experimental frequencies (for instance, the mean and the variance), then the random variable defined by equation (6.93) is distributed according to the χ^2 probability density with $n - k - 1$ degrees of freedom.

This is, of course, the basis of the well-known *goodness-of-fit test*: one knows that the variable χ^2 in the sum (6.93) has a chi-squared distribution, so if the actually obtained value of χ^2 is too large (with respect to the values one may expect to randomly get from the chi-squared distribution), the theoretical probabilities p_i are not consistent with the experimental frequencies f_i.

Fifth Property

If x and y are two independent random variables distributed according to the χ^2 law with, respectively, n_x and n_y degrees of freedom, then the sum $z = x + y$ is a random variable distributed according to the χ^2 law with $n_z = n_x + n_y$ degrees of freedom.

6.9 Monte Carlo Method of Numerical Integration

Consider an s-dimensional manifold $\mathfrak{M}$ with coordinates $\{m^1, \ldots, m^s\}$. For a point of the manifold, we use the notation $\mathbf{m} \in \mathfrak{M}$.

Let $\phi(\mathbf{m})$ be an arbitrary scalar function defined over $\mathfrak{M}$ and assume that we need to evaluate the sum

$$I = \int_{\mathfrak{M}} d\mathbf{m} \, \phi(\mathbf{m}) = \underbrace{\int dm^1 \cdots \int dm^s \, \phi(m^1, \ldots, m^s)}_{\mathfrak{M}} \ . \qquad (6.94)$$

If $\mathfrak{M}$ has finite volume, the simplest method of evaluating I numerically is to define a regular grid of points in $\mathfrak{M}$, to compute $\phi(\mathbf{m})$ at each point of the grid, and to approximate the integral in equation (6.94) by a discrete sum. But as the number of points in a regular grid is a rapidly increasing function of the dimension of the space (N proportional to a constant raised to the power s), the method becomes impractical for large-dimensional spaces (say $s \geq 4$). The Monte Carlo method of numerical integration consists of replacing the regular grid of points with a pseudorandom grid generated by a computer code based on a pseudorandom-number generator. Although it is not possible to give any general rule for the number of points needed for an accurate evaluation of the sum (because this number is very much dependent on the form of $\phi(\mathbf{m})$), it turns out *in practical applications* that, for well-behaved functions, $\phi(\mathbf{m})$ can be smaller, by some orders of magnitude, than the number of points needed in a regular grid.

Let $p(\mathbf{m})$ be an arbitrary normed ($\int d\mathbf{m} \, p(\mathbf{m}) = 1$) probability density over $\mathfrak{M}$ that we choose to use to generate pseudorandom points over $\mathfrak{M}$ (the homogeneous probability density is the simplest choice,[84] but using a probability density that samples preferentially the regions of the space where $\phi(\mathbf{m})$ has significant values improves the efficiency of the algorithm).

Defining $\psi(\mathbf{m}) = \frac{\phi(\mathbf{m})}{p(\mathbf{m})}$, the sum we wish to evaluate can be written

$$I = \int_{\mathfrak{M}} d\mathbf{m} \, p(\mathbf{m}) \, \psi(\mathbf{m}) \quad . \tag{6.95}$$

Let $\mathbf{m}_1, \ldots, \mathbf{m}_N$ be a suite of N points collectively independent and randomly distributed over $\mathfrak{M}$ with a probability density $p(\mathbf{m})$. Defining

$$\psi_n = \psi(\mathbf{m}_n) \quad , \quad I_N = \frac{1}{N} \sum_{n=1}^{N} \psi_n \quad , \quad V_N = \frac{N}{N+1} \left(\frac{1}{N} \sum_{n=1}^{N} \psi_n^2 - I_N^2 \right) \, , \tag{6.96}$$

it can easily be seen that the mathematical expectation of I_N is

$$\langle I_N \rangle = \int_{\mathfrak{M}} d\mathbf{m} \, p(\mathbf{m}) \, \psi(\mathbf{m}) = I \quad , \tag{6.97}$$

so that I_N is an unbiased estimate of I. Using the central limit theorem, it can be shown (see, for instance, Bakhvalov, 1977) that, for large N, the probability that the relative error $|I_N - I|/|I|$ is bounded as

$$\frac{|I_N - I|}{|I|} \leq \frac{k\sqrt{V}}{|I|\sqrt{N}} \quad , \tag{6.98}$$

where V is the (unknown) variance of $\psi(\mathbf{m})$, is asymptotically equal to

$$P(k) = 1 - \frac{2}{\sqrt{2\pi}} \int_{k}^{+\infty} dt \, \exp\left(-\frac{t^2}{2} \right) \quad . \tag{6.99}$$

For large N, a useful estimate of the right-hand side of equation (6.98) is

$$\frac{k\sqrt{V}}{|I|\sqrt{N}} \simeq \frac{k\sqrt{V_N}}{|I_N|\sqrt{N}} \quad , \tag{6.100}$$

where I_N and V_N are defined in equations (6.96).

This method of numerical integration is used as follows: first, one selects the value of the confidence level, $P(k)$, at which the bound equation (6.98) is required to hold (for instance, $P(k) = 0.99$). The corresponding value of k is easily deduced using equation (6.99) and the error-function tables ($k \simeq 3$ for $P(k) = 0.99$). A pseudorandom-number generator is then used to obtain the points $\mathbf{m}_1, \mathbf{m}_2, \ldots$ distributed with the probability $p(\mathbf{m})$, and, for each new point, the right-hand side of equation (6.98) is estimated using equation (6.100). The computations are stopped when this number equals the relative accuracy desired (for instance, 10^{-3}). The typical statement that can then be made is as follows: The value of I can be estimated by I_N, with a probability of $P(k)$ (e.g., 99%) for the relative error being smaller than ϵ (e.g., 10^{-3}).

For more details, see, for instance, Hammersley and Handscomb (1964).

---

[84]This requires, in fact, that the manifold $\mathfrak{M}$ have finite volume.

6.10 Sequential Random Realization

Let us write here the details of the decomposition of a joint probability density as a product of one-dimensional marginals and conditionals.

Let us apply a first time the partition of a joint probability density as the product of a marginal and a conditional. Defining

$$f_1(x^1) = \int dx^2 \ldots \int dx^n \, f_n(x^1, x^2, \ldots, x^n) \tag{6.101}$$

and

$$f_{n-1|1}(x^2, \ldots, x^n | x^1) = \frac{f_n(x^1, x^2, \ldots, x^n)}{\int dx^2 \ldots \int dx^n \, f_n(x^1, x^2, \ldots, x^n)} = \frac{f_n(x^1, x^2, \ldots, x^n)}{f_1(x^1)} \tag{6.102}$$

gives

$$f_n(x^1, x^2, \ldots, x^n) = f_1(x^1) \, f_{n-1|1}(x^2, \ldots, x^n | x^1) \quad . \tag{6.103}$$

Let us apply the partition again. Defining

$$f_{1|1}(x^2 | x^1) = \int dx^3 \ldots \int dx^n \, f_{n-1|1}(x^2, \ldots, x^n | x^1) \tag{6.104}$$

and

$$f_{n-2|2}(x^3, \ldots, x^n | x^1, x^2) = \frac{f_{n-1|1}(x^2, \ldots, x^n | x^1)}{\int dx^3 \ldots \int dx^n \, f_{n-1|1}(x^2, \ldots, x^n | x^1)}$$

$$= \frac{f_{n-1|1}(x^2, \ldots, x^n | x^1)}{f_{1|1}(x^2 | x^1)} \tag{6.105}$$

gives

$$f_{n-1|1}(x^2, \ldots, x^n | x^1) = f_{1|1}(x^2 | x^1) \, f_{n-2|2}(x^3, \ldots, x^n | x^1, x^2) \quad , \tag{6.106}$$

and, with this, equation (6.103) can be written

$$f_n(x^1, x^2, \ldots, x^n) = f_1(x^1) \, f_{1|1}(x^2 | x^1) \, f_{n-2|2}(x^3, \ldots, x^n | x^1, x^2) \quad . \tag{6.107}$$

Continuing this procedure, one arrives at

$$f_n(x^1, x^2, \ldots, x^n) = f_1(x^1) \, f_{1|1}(x^2 | x^1) \, f_{1|2}(x^3 | x^1, x^2) \, f_{1|3}(x^4 | x^1, x^2, x^3)$$

$$\times \cdots \times f_{1|n-1}(x^n | x^1, x^2, \ldots, x^{n-1}) \quad , \tag{6.108}$$

expressing the joint probability density as a product of different conditional probability densities.

This immediately suggests a method for generating a random point that only uses one-dimensional random generations. One first generates a random value for x^1 using the unconditional (marginal) density $f_1(x^1)$. This gives some value, say x_0^1. Given this value, one then generates a random value for x^2 using the conditional probability density $f_{1|1}(x^2 | x_0^1)$. This gives some value, say x_0^2. Then, one generates a random value for x^3 using $f_{1|2}(x^3 | x_0^1, x_0^2)$, and so on until one generates a random value for x^n using $f_{1|n-1}(x^n | x_0^1, x_0^2, \ldots, x_0^{n-1})$.

6.11   Cascaded Metropolis Algorithm

Let $\mu(\mathbf{x})$ be the homogeneous probability density and $f_1(\mathbf{x})$, $f_2(\mathbf{x})$, ..., $f_p(\mathbf{x})$ be p probability densities. Our goal here is to develop a Metropolis random walk that samples the conjunction

$$h(\mathbf{x}) \;=\; k\,\mu(\mathbf{x}) \frac{f_1(\mathbf{x})}{\mu(\mathbf{x})} \frac{f_2(\mathbf{x})}{\mu(\mathbf{x})} \;\cdots\; \frac{f_p(\mathbf{x})}{\mu(\mathbf{x})} \quad . \tag{6.109}$$

We need to define the likelihood functions (or volumetric probabilities)

$$\varphi_i(\mathbf{x}) \;=\; f_i(\mathbf{x}) \,/\, \mu(\mathbf{x}) \quad . \tag{6.110}$$

Assume that some random rules define a random walk that samples a probability density $f_1(\mathbf{x})$. At a given step, the random walker is at point $\mathbf{x}_i$, and the application of the rules would lead to a transition to point $\mathbf{x}_j$. When all such proposed transitions $\mathbf{x}_i \rightarrow \mathbf{x}_j$ are accepted, the random walker will sample the probability density $f_1(\mathbf{x})$. Instead of always accepting the proposed transition $\mathbf{x}_i \rightarrow \mathbf{x}_j$, we reject it sometimes by using the following rules (to decide if the random walker is allowed to move to $\mathbf{x}_j$ or if it must stay at $\mathbf{x}_i$):

(a) If $\varphi_2(\mathbf{x}_j) \geq \varphi_2(\mathbf{x}_i)$, then go to step (c).

(b) If $\varphi_2(\mathbf{x}_j) < \varphi_2(\mathbf{x}_i)$, then decide randomly to go to step (c) or to reject the proposed move, with the following probability of going to step (c):

$$P \;=\; \varphi_2(\mathbf{x}_j) \,/\, \varphi_2(\mathbf{x}_i) \quad . \tag{6.111}$$

(c) If $\varphi_3(\mathbf{x}_j) \geq \varphi_3(\mathbf{x}_i)$, then go to step (e).

(d) If $\varphi_3(\mathbf{x}_j) < \varphi_3(\mathbf{x}_i)$, then decide randomly to go to step (e) or to reject the proposed move, with the following probability of going to step (e):

$$P \;=\; \varphi_3(\mathbf{x}_j) \,/\, \varphi_3(\mathbf{x}_i) \quad . \tag{6.112}$$

(e) If ..., then go to ....

(f) If ..., then decide randomly ....

(w) If $\varphi_{p-1}(\mathbf{x}_j) \geq \varphi_{p-1}(\mathbf{x}_i)$, then go to step (y).

(x) If $\varphi_{p-1}(\mathbf{x}_j) < \varphi_{p-1}(\mathbf{x}_i)$, then decide randomly to go to step (y) or to reject the proposed move, with the following probability of going to step (y):

$$P \;=\; \varphi_{p-1}(\mathbf{x}_j) \,/\, \varphi_{p-1}(\mathbf{x}_i) \quad . \tag{6.113}$$

(y) If $\varphi_p(\mathbf{x}_j) \geq \varphi_p(\mathbf{x}_i)$, then accept the proposed transition to $\mathbf{x}_j$.

(z) If $\varphi_p(\mathbf{x}_j) < \varphi_p(\mathbf{x}_i)$, then decide randomly to move to $\mathbf{x}_j$ or to stay at $\mathbf{x}_i$, with the following probability of accepting the move to $\mathbf{x}_j$:

$$P_{i \rightarrow j} \;=\; \varphi_p(\mathbf{x}_j) \,/\, \varphi_p(\mathbf{x}_i) \quad . \tag{6.114}$$

Then, the random walk samples the conjunction $h(\mathbf{x})$ (equation (6.109)) of the probability densities $f_1(\mathbf{x})$, $f_2(\mathbf{x})$, ..., $f_p(\mathbf{x})$. This property immediately results from the validity of the rule for the conjunction of two probability densities.

6.12 Distance and Norm

Let $\mathfrak{M}$ be a space of points, with points denoted $\mathcal{P}_1, \mathcal{P}_2, \dots$. A *distance* over $\mathfrak{E}$ associates a real number with any pair of points $(\mathcal{P}_1, \mathcal{P}_2)$ of $\mathfrak{E}$. This distance is denoted $D(\mathcal{P}_1, \mathcal{P}_2)$ and has the properties

$$
\begin{aligned}
D(\mathcal{P}_1, \mathcal{P}_2) &= 0 \quad \Leftrightarrow \quad \mathcal{P}_1 = \mathcal{P}_2 & \text{for any } \mathcal{P}_1 \text{ and } \mathcal{P}_2 \quad, \\
D(\mathcal{P}_1, \mathcal{P}_2) &= D(\mathcal{P}_2, \mathcal{P}_1) & \text{for any } \mathcal{P}_1 \text{ and } \mathcal{P}_2 \quad, \\
D(\mathcal{P}_1, \mathcal{P}_3) &\leq D(\mathcal{P}_1, \mathcal{P}_2) + D(\mathcal{P}_2, \mathcal{P}_3) & \text{for any } \mathcal{P}_1, \mathcal{P}_2, \text{ and } \mathcal{P}_3 \quad,
\end{aligned}
\tag{6.115}
$$

the last property being called the *triangular inequality* (the length of one side of a triangle is less than or equal to the sum of the lengths of the other two sides). When such a distance has been defined, $\mathfrak{M}$ is termed a *metric space*.

Let $\mathbb{V}$ be a linear space. A *norm* over $\mathbb{V}$ associates a positive real number with any element $\mathbf{v}$ of $\mathbb{V}$. This norm is denoted $\|\mathbf{v}\|$ and has the properties (parallel to those in equations (6.115))

$$
\begin{aligned}
\|\mathbf{v}\| &= 0 \quad \Leftrightarrow \quad \mathbf{v} = 0 & \text{for any } \mathbf{v} \quad, \\
\|\lambda \mathbf{v}\| &= |\lambda| \, \|\mathbf{v}\| & \text{for any } \mathbf{v} \text{ and any real } \lambda \quad, \\
\|\mathbf{v}_1 + \mathbf{v}_2\| &\leq \|\mathbf{v}_1\| + \|\mathbf{v}_2\| & \text{for any } \mathbf{v}_1 \text{ and } \mathbf{v}_2 \quad.
\end{aligned}
\tag{6.116}
$$

A linear space furnished with a norm is termed a *normed linear space*.

Let $(\mathbf{v}_1, \mathbf{v}_2)$ denote a scalar product (see section 3.1.3). From $0 \leq (\mathbf{v}_1 - \lambda \mathbf{v}_2, \mathbf{v}_1 - \lambda \mathbf{v}_2) = (\mathbf{v}_1, \mathbf{v}_1) - 2\lambda(\mathbf{v}_1, \mathbf{v}_2) + \lambda^2(\mathbf{v}_2, \mathbf{v}_2)$, there follows, taking $\lambda = (\mathbf{v}_1, \mathbf{v}_2)/(\mathbf{v}_2, \mathbf{v}_2)$,

$$
|(\mathbf{v}_1, \mathbf{v}_2)| \leq (\mathbf{v}_1, \mathbf{v}_1)^{1/2} (\mathbf{v}_2, \mathbf{v}_2)^{1/2} \qquad \text{(Cauchy–Schwarz inequality)} \quad.
\tag{6.117}
$$

The expression

$$
\|\mathbf{v}\| = (\mathbf{v}, \mathbf{v})^{1/2}
\tag{6.118}
$$

defines a norm. Only the triangular property needs to be proved. We successively have $\|\mathbf{v}_1 + \mathbf{v}_2\|^2 = (\mathbf{v}_1 + \mathbf{v}_2, \mathbf{v}_1 + \mathbf{v}_2) = (\mathbf{v}_1, \mathbf{v}_1) + 2(\mathbf{v}_1, \mathbf{v}_2) + (\mathbf{v}_2, \mathbf{v}_2) = \|\mathbf{v}_1\|^2 + 2(\mathbf{v}_1, \mathbf{v}_2) + \|\mathbf{v}_2\|^2 \leq \|\mathbf{v}_1\|^2 + 2|(\mathbf{v}_1, \mathbf{v}_2)| + \|\mathbf{v}_2\|^2$. Using the Cauchy–Schwarz inequality (equation (6.117)) gives $\|\mathbf{v}_1 + \mathbf{v}_2\|^2 \leq \|\mathbf{v}_1\|^2 + 2\|\mathbf{v}_1\| \, \|\mathbf{v}_2\| + \|\mathbf{v}_2\|^2 = (\|\mathbf{v}_1\| + \|\mathbf{v}_2\|)^2$, from which the triangular inequality follows.

6.13 The Different Meanings of the Word Kernel

There are different mathematical meanings for the term 'kernel.' Let us recall the two main ones.

a) *A kernel may be a subspace.* Let $\mathbb{M}$ and $\mathbb{D}$ be two linear spaces and $\mathbf{G}$ be a linear operator mapping vectors of $\mathbb{M}$ into vectors of $\mathbb{D}$: $\mathbf{m} \mapsto \mathbf{d} = \mathbf{G} \mathbf{m}$. The linear subspace $\mathbb{M}_0 \subset \mathbb{M}$ of elements $\mathbf{m}$ such that

$$
\mathbf{G} \mathbf{m} = 0
\tag{6.119}
$$

is termed the *kernel* of $\mathbf{G}$ (or the *null space* of $\mathbf{G}_0$). If the kernel of a linear operator is not reduced to the zero element, the operator is not invertible.

b) *A kernel may be the representation of a linear operator.* Let $\mathbb{M}$ and $\mathbb{D}$ be two linear spaces and $\mathbf{G}$ be a linear operator mapping $\mathbb{M}$ into $\mathbb{D}$. According to whether $\mathbb{M}$ or $\mathbb{D}$ is a discrete or a continuous space, the abstract linear equation

$$\mathbf{d} = \mathbf{G}\,\mathbf{m} \tag{6.120}$$

may take one of the following explicit representations:

$$d^i = \sum_\alpha \mathbf{G}^{i\alpha}\,\mathbf{m}^\alpha \qquad (\alpha \in \mathbf{I}_\mathrm{M},\ i \in \mathbf{I}_\mathrm{D}) \quad,$$

$$d(\mathbf{y}) = \sum_\alpha \mathbf{G}^\alpha(\mathbf{y})\,\mathbf{m}^\alpha \qquad (\mathbf{y} \in \mathbf{V}_y,\ \alpha \in \mathbf{I}_\mathrm{M}) \quad,$$

$$\tag{6.121}$$

$$d^i = \int_{\mathbf{V}_x} d\mathbf{x}\,\mathbf{G}^i(\mathbf{x})\,\mathbf{m}(\mathbf{x}) \qquad (i \in \mathbf{I}_\mathrm{D},\ \mathbf{x} \in \mathbf{V}_x) \quad,$$

$$d(\mathbf{y}) = \int_{\mathbf{V}_x} d\mathbf{x}\,\mathbf{G}(\mathbf{y},\mathbf{x})\,\mathbf{m}(\mathbf{x}) \qquad (\mathbf{y} \in \mathbf{V}_y,\ \mathbf{x} \in \mathbf{V}_x) \quad.$$

The matrix $\mathbf{G}^{i\alpha}$, the arrays of functions $\mathbf{G}^\alpha(\mathbf{y})$ or $\mathbf{G}^i(\mathbf{x})$, and the function $\mathbf{G}(\mathbf{y},\mathbf{x})$ are termed the *kernel* of the linear operator $\mathbf{G}$.

6.14   Transpose and Adjoint of a Differential Operator

Let $\mathbb{M}$ and $\mathbb{D}$ represent two linear spaces and $\mathbb{M}^*$ and $\mathbb{D}^*$ be their respective duals. The duality product of $\hat{\mathbf{d}}_1 \in \mathbb{D}^*$ and $\mathbf{d}_2 \in \mathbb{D}$ (resp., of $\hat{\mathbf{m}}_1 \in \mathbb{M}^*$ and $\mathbf{m}_2 \in \mathbb{M}$) is denoted $\langle\, \hat{\mathbf{d}}_1\,,\ \mathbf{d}_2\,\rangle_\mathrm{D}$ (resp., $\langle\, \hat{\mathbf{m}}_1\,,\ \mathbf{m}_2\,\rangle_\mathrm{M}$).

Let $\mathbf{G}$ be a linear operator mapping $\mathbb{M}$ into $\mathbb{D}$. If the spaces $\mathbb{M}$ and $\mathbb{D}$ are finite-dimensional discrete spaces, or if $\mathbf{G}$ is an ordinary *integral operator* between functional spaces, the definitions of transpose and adjoint are as in section 3.1.2 (page 59) and section 3.1.4 (page 61). If $\mathbf{G}$ is a *differential operator* between functional spaces, these definitions need some generalization.

Let $\mathbf{G}$ be a differential operator mapping the functional space $\mathbb{M}$ into the functional space $\mathbb{D}$, $\mathbf{x} = (x^1, x^2, x^3, \ldots)$ be the (common) variables of these functional spaces, $\mathcal{V}$ be the (generalized) volume under consideration, $\mathcal{S}$ be its boundary, and $n^i(\mathbf{x})$ be the (contravariant) components of the outward normal unit vector on $\mathcal{S}$. The *formal transpose* of $\mathbf{G}$, $\mathbf{G}^t$, is the unique operator mapping $\mathbb{D}^*$ into $\mathbb{M}^*$ such that the difference $\langle\, \hat{\mathbf{d}}\,,\ \mathbf{Gm}\,\rangle_\mathrm{D} - \langle\, \mathbf{G}^t\hat{\mathbf{d}}\,,\ \mathbf{m}\,\rangle_\mathrm{M}$ equals the volume integral of the divergence of a certain bilinear (vector) form $\mathbf{P}(\hat{\mathbf{d}},\mathbf{m})$,

$$\langle\, \hat{\mathbf{d}}\,,\ \mathbf{Gm}\,\rangle_\mathrm{D} - \langle\, \mathbf{G}^t\hat{\mathbf{d}}\,,\ \mathbf{m}\,\rangle_\mathrm{M} = \int_\mathcal{V} dV(\mathbf{x})\,(\,\nabla_i\,P^i(\hat{\mathbf{d}},\mathbf{m})\,)(\mathbf{x}) \quad, \tag{6.122}$$

where an implicit sum over i is assumed. Using Green's theorem, this difference can be written as the boundary integral of the flux of the bilinear form:

$$\langle\, \hat{\mathbf{d}}\,,\ \mathbf{Gm}\,\rangle_\mathrm{D} - \langle\, \mathbf{G}^t\hat{\mathbf{d}}\,,\ \mathbf{m}\,\rangle_\mathrm{M} = \int_\mathcal{S} dS(\mathbf{x})\,n_i(\mathbf{x})\,P^i(\hat{\mathbf{d}},\mathbf{m})(\mathbf{x}) \quad. \tag{6.123}$$

Although $\mathbf{G}^t$ is uniquely defined by (6.122) or (6.123), the vector $\mathbf{P}[\cdot, \cdot]$ is defined except for the possible addition of a divergence-free vector.

Example 6.2. The transpose of the gradient operator is the negative of the divergence operator. *Let $\mathcal{V}$ denote a volume in the physical (Euclidean) 3D space, bounded by a surface $\mathcal{S}$, and let (x, y, z) denote Cartesian coordinates. We consider a space of functions $m(x, y, z)$ defined inside (and at the surface of) $\mathcal{V}$. The gradient operator*

$$\nabla = \begin{pmatrix} \partial/\partial x \\ \partial/\partial y \\ \partial/\partial z \end{pmatrix} \tag{6.124}$$

associates with a scalar function $m(x, y, z)$ its gradient

$$\mathbf{d} = \nabla \mathbf{m} = \begin{pmatrix} \partial m/\partial x \\ \partial m/\partial y \\ \partial m/\partial z \end{pmatrix} . \tag{6.125}$$

Let us verify that the transpose of the gradient operator equals the divergence operator with reversed sign:

$$\nabla^t = \begin{bmatrix} \partial/\partial x \; \partial/\partial y \; \partial/\partial z \end{bmatrix} = -\nabla . \tag{6.126}$$

For any $\hat{\mathbf{d}} \in \mathbb{D}^*$ and $\mathbf{m} \in \mathbb{M}$, we have

$$\langle \hat{\mathbf{d}}, \nabla \mathbf{m} \rangle_\mathrm{D} - \langle \nabla^t \hat{\mathbf{d}}, \mathbf{m} \rangle_\mathrm{M}$$

$$= \int_\mathcal{V} dV(\mathbf{x}) \, \hat{d}^i(\mathbf{x}) \, \frac{\partial m}{\partial x^i}(\mathbf{x}) + \int_\mathcal{V} dV(\mathbf{x}) \, \frac{\partial \hat{d}^i}{\partial x^i}(\mathbf{x}) \, m(\mathbf{x}) \tag{6.127}$$

$$= \int_\mathcal{V} dV(\mathbf{x}) \, \frac{\partial}{\partial x^i}(\hat{d}^i(\mathbf{x}) \, m(\mathbf{x})) = \int_\mathcal{S} dS(\mathbf{x}) \, n_i(\mathbf{x}) \, \hat{d}^i(\mathbf{x}) \, m(\mathbf{x}) ,$$

and the components of the bilinear form are

$$P^i(\hat{\mathbf{d}}, \mathbf{m})(\mathbf{x}) = \hat{d}^i(\mathbf{x}) \, m(\mathbf{x}) . \tag{6.128}$$

Example 6.3. Demonstration of $(\mathbf{A}\,\mathbf{B})^t = \mathbf{B}^t\,\mathbf{A}^t$. *Let $\mathbf{B} : \mathbb{E} \to \mathbb{F}$ and $\mathbf{A} : \mathbb{F} \to \mathbb{G}$. Then, $\mathbf{A}\,\mathbf{B} : \mathbb{E} \to \mathbb{G}$. From*

$$\langle \hat{\mathbf{g}}, \mathbf{A}\mathbf{f} \rangle_G - \langle \mathbf{A}^t \hat{\mathbf{g}}, \mathbf{f} \rangle_F = \int dV \, D_i \, a^i ,$$

$$\langle \hat{\mathbf{f}}, \mathbf{B}\mathbf{e} \rangle_F - \langle \mathbf{A}^t \hat{\mathbf{f}}, \mathbf{e} \rangle_E = \int dV \, D_i \, b^i , \tag{6.129}$$

it follows, setting $\mathbf{f} = \mathbf{B}\mathbf{e}$ and $\hat{\mathbf{f}} = \mathbf{A}^t \hat{\mathbf{g}}$, that

$$\langle \hat{\mathbf{g}}, \mathbf{A}\mathbf{B}\mathbf{e} \rangle_G - \langle \mathbf{A}^t \hat{\mathbf{g}}, \mathbf{B}\mathbf{e} \rangle_F = \int dV \, D_i \, a^i ,$$

$$\langle \mathbf{A}^t \hat{\mathbf{g}}, \mathbf{B}\mathbf{e} \rangle_F - \langle \mathbf{B}^t \mathbf{A}^t \hat{\mathbf{e}}, \mathbf{e} \rangle_E = \int dV \, D_i \, b^i , \tag{6.130}$$

and

$$\langle\, \hat{\mathbf{g}}\,,\ \mathbf{A}\,\mathbf{B}\,\mathbf{e}\,\rangle_G - \langle\, \mathbf{B}^t\,\mathbf{A}^t\,\hat{\mathbf{e}}\,,\ \mathbf{e}\,\rangle_{eE}\ =\ \int dV\, D_i\,(a^i + b^i)\quad,\tag{6.131}$$

i.e.,

$$(\mathbf{A}\,\mathbf{B})^t\ =\ \mathbf{B}^t\,\mathbf{A}^t\quad.\tag{6.132}$$

If the right-hand sides of (6.122)–(6.123) vanish for any $\mathbf{m}$ and $\hat{\mathbf{d}}$,

$$\int_{\mathcal{V}} dV(\mathbf{x})\,(\,\nabla_i P^i(\hat{\mathbf{d}},\mathbf{m})\,)(\mathbf{x})\ =\ \int_S dS(\mathbf{x})\, n_i(\mathbf{x})\, P^i(\hat{\mathbf{d}},\mathbf{m})(\mathbf{x})\ =\ 0\quad,\tag{6.133}$$

then the *formal transpose* is simply termed the *transpose*, and we have

$$\langle\,\hat{\mathbf{d}}\,,\ \mathbf{G}\mathbf{m}\,\rangle_{\mathrm{D}}\ =\ \langle\,\mathbf{G}^t\hat{\mathbf{d}}\,,\ \mathbf{m}\,\rangle_{\mathrm{M}}\quad.\tag{6.134}$$

In that case, we can harmlessly use all the equations involving transpositions as if we were dealing with discrete spaces.

Usually, the domains of definition of $\mathbf{G}$ and $\mathbf{G}^t$ are restricted so as to satisfy (6.133). It is then said that these domains of definition satisfy *dual boundary conditions*. See below for an example.

In the special case where a linear operator $\mathbf{W}$ maps a space $\mathbb{E}$ into its dual $\mathbb{E}^*$, then $\mathbf{W}^t$ also maps $\mathbb{E}$ into $\hat{\mathbb{E}}$. In that case, it may happen that $\mathbf{W}^t = \mathbf{W}$, and the operator $\mathbf{W}$ is *symmetric*. Let us come to this definition with some care.

Assume that a linear operator $\mathbf{W}$ maps $\mathbb{E}_0 \subset \mathbb{E}$ into $\mathbb{Y}$, and define its transpose $\mathbf{W}^t$ as mapping $\mathbb{Y}_1^* \subset \mathbb{Y}^*$ into $\mathbb{E}^*$. If the subspaces $\mathbb{E}_0$ and $\mathbb{Y}_1^*$ satisfy *dual boundary conditions* (see above), then

$$\forall\,\mathbf{e}_0 \in \mathbb{E}_0\quad,\qquad \forall\,\hat{\mathbf{y}}_1 \in \mathbb{Y}_1^*\quad,\qquad \langle\,\hat{\mathbf{y}}_1\,,\ \mathbf{W}\,\mathbf{e}_0\,\rangle_Y\ =\ \langle\,\mathbf{W}^t\,\hat{\mathbf{y}}_1\,,\ \mathbf{e}_0\,\rangle_E\quad.\tag{6.135}$$

If $\mathbb{Y} = \mathbb{E}^*$ and we identify the bidual of $\mathbb{E}$ to $\mathbb{E}$, then $\mathbf{W} : \mathbb{E}_0 \to \mathbb{E}^*$ and $\mathbf{W}^t : \mathbb{E}_1 \to \mathbb{E}^*$, and, by definition of transpose,

$$\forall\,\mathbf{e}_0 \in \mathbb{E}_0\quad,\qquad \forall\,\mathbf{e}_1 \in \mathbb{E}_1\quad,\qquad \langle\,\mathbf{e}_1\,,\ \mathbf{W}\,\mathbf{e}_0\,\rangle_{E^*}\ =\ \langle\,\mathbf{W}^t\,\mathbf{e}_1\,,\ \mathbf{e}_0\,\rangle_E\quad.\tag{6.136}$$

Now, if

$$\forall\,\mathbf{e}_0 \in \mathbb{E}_0 \cap \mathbb{E}_1\quad,\qquad \forall\,\mathbf{e}' \in \mathbb{E}_0 \cap \mathbb{E}_1\quad,\qquad \langle\,\mathbf{W}\,\mathbf{e}\,,\ \mathbf{e}'\,\rangle_E\ =\ \langle\,\mathbf{W}^t\,\mathbf{e}\,,\ \mathbf{e}'\,\rangle_E\quad,\tag{6.137}$$

the operator $\mathbf{W}$ is termed *symmetric*, the notation

$$\mathbf{W}\ =\ \mathbf{W}^t\tag{6.138}$$

is used, and the identities

$$\langle\,\mathbf{e}\,,\ \mathbf{W}^t\mathbf{e}'\,\rangle_{E^*}\ =\ \langle\,\mathbf{W}\,\mathbf{e}\,,\ \mathbf{e}'\,\rangle_E\ =\ \langle\,\mathbf{W}^t\,\mathbf{e}\,,\ \mathbf{e}'\,\rangle_E\ =\ \langle\,\mathbf{e}\,,\ \mathbf{W}\mathbf{e}'\,\rangle_{E^*}\tag{6.139}$$

hold $\forall\,\mathbf{e}, \mathbf{e}' \in \mathbb{E}_0 \cap \mathbb{E}_1$.

In defining the transpose of a linear operator, it is not assumed that the linear spaces under consideration have a scalar product. If they do, then it is possible to define the adjoint of a linear operator.

Let $\mathbb{M}$ and $\mathbb{D}$ represent two scalar product linear spaces. The scalar product of $\mathbf{d}_1$ and $\mathbf{d}_2$ (resp., of $\mathbf{m}_1$ and $\mathbf{m}_2$) is denoted $(\mathbf{d}_1, \mathbf{d}_2)_\mathrm{D}$ (resp., $(\mathbf{m}_1, \mathbf{m}_2)_\mathrm{M}$).

Using the same notation as above, the *formal adjoint* of $\mathbf{G}$, $\mathbf{G}^*$, is the unique operator mapping $\mathbb{D}$ into $\mathbb{M}$ such that

$$(\mathbf{d}, \mathbf{G\,m})_\mathrm{D} - (\mathbf{G}^*\mathbf{d}, \mathbf{m})_\mathrm{M} = \int_{\mathcal{V}} dV(\mathbf{x})\,(\nabla_i P^i(\mathbf{d}, \mathbf{m}))(\mathbf{x}) \quad , \qquad (6.140)$$

or, using Green's theorem,

$$(\mathbf{d}, \mathbf{G\,m})_\mathrm{D} - (\mathbf{G}^*\mathbf{d}, \mathbf{m})_\mathrm{M} = \int_{S} dS(\mathbf{x})\, n_i(\mathbf{x})\, P^i(\mathbf{d}, \mathbf{m})(\mathbf{x}) \quad . \qquad (6.141)$$

Again, although $\mathbf{G}^*$ is uniquely defined by (6.140) or (6.141), the vector $\mathbf{P}(\cdot, \cdot)$ is defined for the addition of a divergence-free vector.

Let $\mathbf{C}_\mathrm{M}$ and $\mathbf{C}_\mathrm{D}$ be the covariance operators defining the scalar products over $\mathbb{M}$ and $\mathbb{D}$, respectively (and, thus, the natural isomorphisms between $\mathbb{M}$ and $\mathbb{D}$ and their respective duals):

$$(\mathbf{d}_1, \mathbf{d}_2)_\mathrm{D} = \langle\, \mathbf{C}_\mathrm{D}^{-1}\mathbf{d}_1, \mathbf{d}_2 \,\rangle_\mathrm{D} \quad , \qquad (6.142)$$

$$(\mathbf{m}_1, \mathbf{m}_2)_\mathrm{M} = \langle\, \mathbf{C}_\mathrm{M}^{-1}\mathbf{m}_1, \mathbf{m}_2 \,\rangle_\mathrm{M} \quad . \qquad (6.143)$$

Using, for instance, (6.122), we obtain

$$(\mathbf{d}, \mathbf{G\,m})_\mathrm{D} - (\mathbf{C}_\mathrm{M}\,\mathbf{G}^t\,\mathbf{C}_\mathrm{D}^{-1}\mathbf{d}, \mathbf{m})_\mathrm{M} = \int_{\mathcal{V}} dV(\mathbf{x})\,\frac{DP^i[d, m]}{DX^i}(\mathbf{x}) \quad , \qquad (6.144)$$

or, for short,

$$\mathbf{G}^* = \mathbf{C}_\mathrm{M}\,\mathbf{G}^t\,\mathbf{C}_\mathrm{D}^{-1} \quad . \qquad (6.145)$$

The reader will easily give sense to and demonstrate the property

$$(\mathbf{A}\,\mathbf{B})^* = \mathbf{B}^*\,\mathbf{A}^* \quad . \qquad (6.146)$$

If the right-hand side of (6.140)–(6.141) vanishes for any $\mathbf{m}$ and $\mathbf{d}$,

$$\int_{\mathcal{V}} dV(\mathbf{x})\,(\nabla_i P^i(\mathbf{d}, \mathbf{m}))(\mathbf{x}) = \int_{S} dS(\mathbf{x})\, n_i(\mathbf{x})\, P^i(\mathbf{d}, \mathbf{m})(\mathbf{x}) = 0 \quad , \qquad (6.147)$$

then the *formal adjoint* is simply termed the *adjoint*, and we have

$$(\mathbf{d}, \mathbf{G\,m})_\mathrm{D} = (\mathbf{G}^*\mathbf{d}, \mathbf{m})_\mathrm{M} \quad . \qquad (6.148)$$

In the special case where a linear operator $\mathbf{L}$ maps a space E into itself, then the adjoint operator $\mathbf{L}^*$ also maps E into itself. In that case, it may happen that $\mathbf{L}^* = \mathbf{L}$, and the operator $\mathbf{L}$ is *self-adjoint*. Let us replace duality products with scalar products in the equations above.

Assume that a linear operator $\mathbf{W}$ maps $\mathbb{E}_0 \subset \mathbb{E}$ into $\mathbb{Y}$, and define its adjoint $\mathbf{W}^*$ as mapping $\mathbb{Y}_1 \subset \mathbb{Y}$ into $\mathbb{E}$. If the subspaces $\mathbb{E}_0$ and $\mathbb{Y}_1$ satisfy *dual boundary conditions* (see above), then

$$\forall \mathbf{e}_0 \in \mathbb{E}_0 \quad , \quad \forall \mathbf{y}_1 \in \mathbb{Y}_1 \quad , \quad (\mathbf{y}_1, \mathbf{W}\mathbf{e}_0)_Y = (\mathbf{W}^* \mathbf{y}_1, \mathbf{e}_0)_E \quad . \tag{6.149}$$

If $\mathbb{Y} = \mathbb{E}$, then $\mathbf{W} : \mathbb{E}_0 \to \mathbb{E}$ and $\mathbf{W}^* : \mathbb{E}_1 \to \mathbb{E}$, and, by definition of adjoint,

$$\forall \mathbf{e}_0 \in \mathbb{E}_0 \quad , \quad \forall \mathbf{e}_1 \in \mathbb{E}_1 \quad , \quad (\mathbf{e}_1, \mathbf{W}\mathbf{e}_0)_E = (\mathbf{W}^* \mathbf{e}_1, \mathbf{e}_0)_E \quad . \tag{6.150}$$

Now, if

$$\forall \mathbf{e}_0 \in \mathbb{E}_0 \cap \mathbb{E}_1 \quad , \quad \forall \mathbf{e}' \in \mathbb{E}_0 \cap \mathbb{E}_1 \quad , \quad (\mathbf{W}\mathbf{e}, \mathbf{e}')_E = (\mathbf{W}^* \mathbf{e}, \mathbf{e}')_E \quad , \tag{6.151}$$

the operator $\mathbf{W}$ is termed *self-adjoint*, the notation

$$\mathbf{W} = \mathbf{W}^* \tag{6.152}$$

is used, and the identities

$$(\mathbf{e}, \mathbf{W}^* \mathbf{e}')_E = (\mathbf{W}\mathbf{e}, \mathbf{e}')_E = (\mathbf{W}^* \mathbf{e}, \mathbf{e}')_E = (\mathbf{e}, \mathbf{W}\mathbf{e}')_E \tag{6.153}$$

hold $\forall \ \mathbf{e}, \mathbf{e}' \in \mathbb{E}_0 \cap \mathbb{E}_1$.

It is easy to see that identities (6.153) define a scalar product, denoted $\mathbf{W}(\mathbf{e}, \mathbf{e}')$:

$$\mathbf{W}(\mathbf{e}, \mathbf{e}') = (\mathbf{e}, \mathbf{W}^* \mathbf{e}')_E = (\mathbf{W}\mathbf{e}, \mathbf{e}')_E = (\mathbf{W}^* \mathbf{e}, \mathbf{e}')_E = (\mathbf{e}, \mathbf{W}\mathbf{e}')_E \quad . \tag{6.154}$$

If $\mathbf{W}$ is the operator defining the original scalar product over $\mathbb{E}$,

$$\mathbf{W}(\mathbf{e}, \mathbf{e}') = (\mathbf{e}, \mathbf{e}')_E \quad , \tag{6.155}$$

it is named the *weighting* operator over $\mathbb{E}$, or the *inverse of the covariance* operator over $\mathbb{E}$, and the following notation is used:

$$\mathbf{W}(\mathbf{e}, \mathbf{e}') = (\mathbf{e}, \mathbf{e}')_E = \mathbf{e}^t \mathbf{W} \mathbf{e}' \quad . \tag{6.156}$$

Example 6.4. The transpose of the elastodynamics operator. *Let $\mathbf{L}$ denote the elastic wave equation operator that with a displacement field $\mu(\mathbf{x}, t)$ associates its source $\phi(\mathbf{x}, t)$ (volume density of external forces):*

$$\mathbf{L}\mu = \phi \quad . \tag{6.157}$$

Explicitly, using tensor notation,

$$\rho(\mathbf{x}) \frac{\partial^2 u^i}{\partial t^2}(\mathbf{x}, t) - \frac{\partial}{\partial x^j}\left(c^{ijkl}(\mathbf{x}) \frac{\partial u_k}{\partial x^l}(\mathbf{x}, t) \right) = \phi^i(\mathbf{x}, t) \quad . \tag{6.158}$$

Let $\boldsymbol{\mu} \in \mathbb{U}$ and $\boldsymbol{\phi} \in \mathbb{F}$. *Each source vector $\boldsymbol{\phi}$ can be considered as a linear form over the space of displacements:*

$$\langle \boldsymbol{\phi} , \boldsymbol{\mu} \rangle = \int_V dV(\mathbf{x}) \, \phi^i(\mathbf{x}, t) \, u_i(\mathbf{x}, t) = \int_V dV(\mathbf{x}) \, \phi_i(\mathbf{x}, t) \, u^i(\mathbf{x}, t) \quad . \tag{6.159}$$

The physical dimension of $\langle \boldsymbol{\phi} , \boldsymbol{\mu} \rangle_U$ is an action. *Equation (6.139) allows us to identify $\mathbb{F}$ as $\mathbb{U}^*$, the dual of $\mathbb{U}$. Furthermore, identifying the bidual of $\mathbb{U}$ with $\mathbb{U}$, we see that $\mathbf{L}$ and $\mathbf{L}^t$ both map $\mathbb{U} = \mathbb{F}^*$ into $\mathbb{F} = \mathbb{U}^*$. We will now check under which conditions the wave equation operator is symmetric:*

$$\mathbf{L}^t = \mathbf{L} \quad . \tag{6.160}$$

For any $\hat{\boldsymbol{\phi}} \in \mathbb{F}^* = \mathbb{U}$ and any $\mathbf{u} \in \mathbb{U} = \mathbb{F}^*$,

$$
\begin{aligned}
\langle \hat{\boldsymbol{\phi}} , \mathbf{L}\boldsymbol{\mu} \rangle_F - \langle \mathbf{L}^t \hat{\boldsymbol{\phi}} , \boldsymbol{\mu} \rangle_U &= \int_V dV(\mathbf{x}) \int_{t_0}^{t_1} dt \, \hat{\phi}^i \left\{ \rho \frac{\partial^2 u^i}{\partial t^2} - \frac{\partial}{\partial x^j} \left(c^{ijkl} \frac{\partial u_k}{\partial x^l} \right) \right\} \\
&\quad - \int_V dV(\mathbf{x}) \int_{t_0}^{t_1} dt \left\{ \rho \frac{\partial^2 \phi^i}{\partial t^2} - \frac{\partial}{\partial x^j} \left(c_{ijkl} \frac{\partial \hat{\phi}_k}{\partial \mathbf{x}_l} \right) \right\} u^i \\
&= \int_V dV(\mathbf{x}) \int_{t_0}^{t_1} dt \, \frac{\partial}{\partial t} \left\{ \rho \left(\hat{\phi}_i \frac{\partial u^i}{\partial t} - \frac{\partial \hat{\phi}_i}{\partial t} u^i \right) \right\} \\
&\quad - \int_V dV(\mathbf{x}) \int_{t_0}^{t_1} dt \, \frac{\partial}{\partial x^j} \left\{ c^{ijkl} \left(\hat{\phi}^i \frac{\partial u_k}{\partial x^l} - \frac{\partial \hat{\phi}_k}{\partial \mathbf{x}^l} u^i \right) \right\} \quad ,
\end{aligned}
\tag{6.161}
$$

and the (time-space) components of the bilinear form are

$$P^t(\hat{\phi}, u)(\mathbf{x}, t) = \rho(\mathbf{x}) \left(\hat{\phi}_i(\mathbf{x}, t) \frac{\partial u^i}{\partial t}(\mathbf{x}, t) - \frac{\partial \hat{\phi}_i}{\partial t}(\mathbf{x}, t) u^i(\mathbf{x}, t) \right) \quad , \tag{6.162}$$

$$P^j(\hat{\phi}, u)(\mathbf{x}, t) = c^{ijkl}(\mathbf{x}) \left(\frac{\partial \hat{\phi}_k}{\partial x^l}(\mathbf{x}, t) u^i(\mathbf{x}, t) - \hat{\phi}^i(\mathbf{x}, t) \frac{\partial u_k}{\partial x^l}(\mathbf{x}, t) \right) \quad . \tag{6.163}$$

Using Green's theorem, we obtain

$$
\begin{aligned}
&\langle \hat{\boldsymbol{\phi}} , \mathbf{L}\mathbf{u} \rangle_F - \langle \mathbf{L}^t \hat{\boldsymbol{\phi}} , \mathbf{u} \rangle_U \\
&= \int_V dV(\mathbf{x}) \, \rho \left(\hat{\phi}_i \frac{\partial u^i}{\partial t} - \frac{\partial \hat{\phi}_i}{\partial t} u^i \right) \Bigg|_{t_0}^{t_1} - \int_{t_0}^{t_1} dt \int_S dS(\mathbf{x}) \, n_j \, c^{ijkl} \left(\hat{\phi}^i \frac{\partial u^k}{\partial \partial x^l} - \frac{\partial \hat{\phi}_k}{\partial x^l} u^i \right) \quad .
\end{aligned}
\tag{6.164}
$$

Assume, for instance, that the fields $\mu(\mathbf{x}, t)$ *and* $\hat{\phi}(\mathbf{x}, t)$ *satisfy the boundary conditions*

$$u^i(\mathbf{x}, t_0) = 0 \quad , \quad \frac{\partial u^i}{\partial t}(\mathbf{x}, t_0) = 0 \quad ,$$

$$\hat{\phi}^i(\mathbf{x}, t_1) = 0 \quad , \quad \frac{\partial \hat{\phi}^i}{\partial t}(\mathbf{x}, t_1) = 0 \quad , \tag{6.165}$$

$$n_j(\mathbf{x}) \, c^{ijkl}(\mathbf{x}) \frac{\partial u_k}{\partial x^\ell}(\mathbf{x}, t) = 0 \quad , \quad n_j(\mathbf{x}) \, c^{ijkl}(\mathbf{x}) \frac{\partial \hat{\phi}_k}{\partial x^\ell}(\mathbf{x}, t) = 0 \quad .$$

As expression (6.164) *then vanishes, these are dual boundary conditions. If we only consider fields* $\mu(\mathbf{x}, t)$ *and* $\hat{\phi}(\mathbf{x}, t)$ *satisfying these conditions, then*

$$\langle \, \hat{\phi} \, , \, \mathbf{L} \mathbf{u} \, \rangle_F = \langle \, \mathbf{L}^t \, \hat{\phi} \, , \, \mathbf{u} \, \rangle_U \quad , \tag{6.166}$$

and the symmetry property (6.160) *holds. As we saw in section* 5.8.7, *when solving usual inverse problems, we are actually faced with wave fields satisfying dual boundary conditions.*

Notice that the usual textbooks (Morse and Feshbach, 1953; Courant and Hilbert, 1966; Dautray and Lions, 1984) talk about the "self-adjointness" of the wave equation operator. This assumes that a scalar product can be defined over $\mathbb{U}$, which is not necessarily the case (for instance, $\mathbb{U}$ may be a general Banach space). Only the symmetry of the wave equation operator is needed, not the self-adjointness.

6.15   The Bayesian Viewpoint of Backus (1970)

In his paper "Inference from inadequate and inaccurate data," Backus (1970a,b,c) made the first effort to formalize the probabilistic approach to inverse problems. Although indigestible, the paper is historically important. The essentials of the theory are as follows.

The infinite-dimensional linear model space $\mathbb{M}$ is assumed to be a Hilbert space with scalar product denoted by $(\cdot, \cdot)$. The true (unknown) model is denoted by $\mathbf{m}_{\text{true}}$. n measurements of physical properties of $\mathbf{m}_{\text{true}}$ give the n real quantities $d^1, d^2, \ldots, d^n$, which are assumed linearly related with $\mathbf{m}_{\text{true}}$. Then, there exist n vectors $\mathbf{g}^1, \mathbf{g}^2, \ldots, \mathbf{g}^n$ of $\mathbb{M}$, called 'data vectors,' such that if $\mathbf{m}$ was $\mathbf{m}_{\text{true}}$, we could predict the n real quantities $d^1, d^2, \ldots, d^n$ by the scalar products

$$d^1 = (\mathbf{g}^1, \mathbf{m}) \quad , \quad d^2 = (\mathbf{g}^2, \mathbf{m}) \quad , \quad d^n = (\mathbf{g}^n, \mathbf{m}) \quad . \tag{6.167}$$

(In fact, introducing $\mathbf{g}^1, \mathbf{g}^2, \ldots, \mathbf{g}^n$ as elements of the dual of $\mathbb{M}$, and considering duality products instead of scalar products, allows us to drop the assumption that $\mathbb{M}$ is a Hilbert space, which seems unnecessary.)

The n actual measurements give the values $d^1_{\text{obs}}, d^2_{\text{obs}}, \ldots, d^n_{\text{obs}}$, as well as an estimation of the probabilistic distribution of experimental errors.

Backus is not interested in the description of $\mathbf{m}_{\text{true}}$ itself, but only in the prediction of m numerical properties of $\mathbf{m}_{\text{true}}$, which are also assumed to be linear functions of the model:

$$\delta^1 = (\boldsymbol{\gamma}^1, \mathbf{m}) \quad , \quad \delta^2 = (\boldsymbol{\gamma}^2, \mathbf{m}) \quad , \quad \cdots \quad , \quad \delta^m = (\boldsymbol{\gamma}^m, \mathbf{m}) \quad , \tag{6.168}$$

where $\boldsymbol{\gamma}^1, \boldsymbol{\gamma}^2, \ldots, \boldsymbol{\gamma}^m$ are the 'prediction vectors.'

The following finite-dimensional spaces are introduced:

- $\mathbb{G}$: The n-dimensional linear space generated by the data vectors $\mathbf{g}^1, \mathbf{g}^2, \ldots, \mathbf{g}^n$. It is a subspace of $\mathbb{M}$.

- $\mathbb{D}$: The n-dimensional linear space where the measurements $d^1, \ldots, d^n$ may take their values (in fact, $\Re^n$).

- Γ : The m-dimensional linear space generated by the prediction vectors $\gamma^1, \gamma^2, \ldots, \gamma^m$. It is a subspace of $\mathbb{M}$.

- Δ : The m-dimensional linear space where the predictions $\delta^1, \ldots, \delta^m$ may take their values (in fact, $\Re^m$).

- $\mathbb{S}$: An arbitrary *finite-dimensional* subspace of $\mathbb{M}$ containing $\mathbb{G}$ and Γ as subspaces.

The n measurements only give information on the projection of $\mathbf{m}_{\text{true}}$ over $\mathbb{G}$. As we are only interested in the m predictions, we only need information on the projection of $\mathbf{m}_{\text{true}}$ over Γ. We can then drop the infinite-dimensional model space $\mathbb{M}$ from our attention and only consider the finite-dimensional space $\mathbb{S}$, which contains the projections of $\mathbf{m}_{\text{true}}$ over both $\mathbb{G}$ and Γ. As $\mathbb{S}$ is finite dimensional, the standard Bayesian inference can be used.

For instance, let us denote by $\mathbf{s}$ a generic element of $\mathbb{S}$. We can introduce the probability density $\rho_{\text{meas}}(\mathbf{d} \mid \mathbf{s})$ over $\mathbb{D}$ representing the density of probability of obtaining $\mathbf{d}$ as the result of our measurements if (the projection of) the true model is $\mathbf{s}$. This probability density is practically obtained from the knowledge of the data vectors $\mathbf{g}^1, \mathbf{g}^2, \ldots, \mathbf{g}^m$ and the error statistics of our measuring instruments (see, for instance, chapter 1). The a priori information over $\mathbb{S}$ is described using a probability density $\rho_{\text{prior}}(\mathbf{s})$. The Bayes rule then gives the posterior probability density over $\mathbb{S}$,

$$\rho_{\text{post}}(\mathbf{s} \mid \mathbf{d}_{\text{obs}}) = k\,\rho_{\text{meas}}(\mathbf{d}_{\text{obs}} \mid \mathbf{s})\,\rho_{\text{prior}}(\mathbf{s}) \quad, \tag{6.169}$$

where k is a normalization constant.

Once this posterior probability has been defined over $\mathbb{S}$, the corresponding probability over Γ is obtained as a marginal probability. As we know how to associate the m-dimensional vector $\delta^1, \ldots, \delta^m$ with any elements of Γ (equation (6.168)), it is then easy to deduce the corresponding posterior probability over Δ, the space of predictions.

The conceptually important result proved by Backus (Theorem 29 on page 54 of his paper) is that if M_{prior} is a *cylinder measure* over $\mathbb{M}$, and for each choice of the finite-dimensional space $\mathbf{S}$ the a priori probability over $\mathbb{S}$ is the corresponding marginal probability of the cylinder measure M_{prior}, then all the posterior probability distributions over $\mathbb{S}$ obtained from the Bayesian solution are marginal distributions of a single cylinder measure M_{post} over $\mathbb{M}$. The results are then independent of the particular choice of $\mathbb{S}$. The simplest results are obtained for the smallest $\mathbb{S}$, i.e., $\mathbb{S} = \mathbb{G} + \Gamma$.

This theory can be applied to linearized inverse problems, but does not generalize to true nonlinear problems.

6.16 The Method of Backus and Gilbert

In a famous paper, Backus and Gilbert (1970) gave a conceptually simple philosophy for dealing with linear, essentially underdetermined, problems.

Assume that a model is described by a *function*, $m(r)$, and that we consider a finite amount of *discrete data*, $d^1, d^2, \ldots, d^n$, which are linear functionals of $m(r)$ through kernels $G^i(r)$:

$$d^i = \int dr \, G^i(r) \, m(r) \quad . \tag{6.170}$$

The kernels $G^i(r)$ are assumed regular enough for equation (6.170) to be an ordinary integral equation.

The true (unknown) model is denoted $m_{\text{true}}(r)$. The observed data are denoted d^i_{obs} and are assumed error free. Then,

$$d^i_{\text{obs}} = \int dr \, G^i(r) \, m_{\text{true}}(r) \quad . \tag{6.171}$$

The problem is to obtain a good estimator of $m_{\text{true}}(r)$ at a given point $r = r_0$. Let us denote this estimator by $m_{\text{est}}(r_0)$. As the forward problem is linear, Backus and Gilbert impose that the value $m_{\text{est}}(r_0)$ be a *linear function of the observed data*, i.e., they assume the form

$$m_{\text{est}}(r_0) = \sum_i Q^i(r_0) \, d^i_{\text{obs}} \quad , \tag{6.172}$$

where, at a given point r_0, $Q^i(r_0)$ are some constants. The problem now is to obtain the best constants $Q^i(r_0)$. Inserting (6.171) into (6.172) gives

$$m_{\text{est}}(r_0) = \sum_i Q^i(r_0) \int dr \, G^i(r) \, m_{\text{true}}(r) \quad , \tag{6.173}$$

and defining

$$R(r_0, r) = \sum_i Q^i(r_0) \, G^i(r) \tag{6.174}$$

gives

$$m_{\text{est}}(r_0) = \int dr \, R(r_0, r) \, m_{\text{true}}(r) \quad . \tag{6.175}$$

This last equation shows that the estimate at $r = r_0$ will be a filtered version of the true value, with filter $R(r_0, r)$. This filter is called the *resolving kernel*. We are only able to see the true world through this filter. The sharper the filter is around r_0, the better our estimate (see Figure 7.34). One can arbitrarily choose the coefficients $Q^i(r_0)$. The *deltaness criterion* consists of choosing these coefficients in such a way that the resulting resolving kernel is the closest to a delta function,

$$R(r_0, r) \simeq \delta(r_0 - r) \quad , \tag{6.176}$$

in which case

$$m_{\text{est}}(r_0) \simeq m_{\text{true}}(r_0) \quad . \tag{6.177}$$

Using, for instance, a least-squares deltaness criterion

$$\int dr \left(R(r_0, r) - \delta(r_0 - r) \right)^2 \qquad \text{minimum} \qquad (6.178)$$

gives (see Problem 7.20)

$$Q^i(r_0) = \sum_i (\mathbf{S}^{-1})_{ij} G^i(r_0) \quad , \qquad (6.179)$$

where

$$S^{ij} = \int dr \, G^i(r) \, G^j(r) \quad . \qquad (6.180)$$

This gives

$$m_{\text{est}}(r_0) = \sum_i \sum_j G^i(r_0) \, (\mathbf{S}^{-1})_{ij} \, d_{\text{obs}}^j \quad , \qquad (6.181)$$

which corresponds to the Backus and Gilbert solution for the estimation problem (the reader will easily verify that the matrix $\mathbf{S}$ is regular if the $G^j(r)$ are linearly independent, i.e., each datum depends differently on model parameters). Using (6.179), the resolving kernel is given by

$$R(r_0, r) = \sum_i \sum_j G^i(r_0) \, (\mathbf{S}^{-1})_{ij} \, G^j(r) \quad . \qquad (6.182)$$

It is easy to see that the data predicted from $m_{\text{est}}(r)$ exactly verify the observations:

$$d_{\text{cal}}^i = \int dr \, G^i(r) \, m_{\text{est}}(r) = \sum_j \sum_k S^{ij} \, (\mathbf{S}^{-1})_{jk} \, d_{\text{obs}}^k = d_{\text{obs}}^i \quad . \qquad (6.183)$$

Using more compact notation, all previous equations can be rewritten as follows:

$$\begin{aligned}
\mathbf{d} &= \mathbf{G}\,\mathbf{m} \quad , \\
\mathbf{d}_{\text{obs}} &= \mathbf{G}\,\mathbf{m}_{\text{true}} \quad , \\
\mathbf{m}_{\text{est}} &= \mathbf{Q}^t \,\mathbf{d}_{\text{obs}} \quad , \\
\mathbf{R} &= \mathbf{Q}^t \,\mathbf{G} \quad , \\
\mathbf{m}_{\text{est}} &= \mathbf{R}\,\mathbf{m}_{\text{true}} \quad , \\
\mathbf{R} &\simeq \mathbf{I} \quad , \\
\mathbf{m}_{\text{est}} &\simeq \mathbf{m}_{\text{true}} \quad , \qquad\qquad (6.184) \\
\| \mathbf{R} - \mathbf{I} \|^2 & \qquad \text{minimum} \quad , \\
\mathbf{Q} &= (\mathbf{G}\,\mathbf{G}^t)^{-1}\,\mathbf{G} \quad ,
\end{aligned}$$

$$\mathbf{m}_{\text{est}} = \mathbf{G}^t \, (\mathbf{G}\,\mathbf{G}^t)^{-1}\,\mathbf{d}_{\text{obs}} \quad ,$$

and

$$\mathbf{R} = \mathbf{G}^t \, (\mathbf{G}\,\mathbf{G}^t)^{-1}\,\mathbf{G} \quad .$$

Of course, there is no reason for the estimate $\mathbf{m}_{est}$ to equal the true model $\mathbf{m}_{true}$, which is generally not attainable with a finite amount of data. But as Backus and Gilbert assume exact data, it is easy to show that the true model is necessarily of the form

$$\boxed{\mathbf{m} = \mathbf{m}_{est} + (\mathbf{I} - \mathbf{R})\,\mathbf{m}_0} \quad , \tag{6.185}$$

where $\mathbf{m}_0$ is an arbitrary model. For it is sufficient to verify, using the last of equations (6.184), that $(\mathbf{I} - \mathbf{R})\,\mathbf{m}_0$ belongs to the null space of $\mathbf{G}$, i.e., it is such that

$$\mathbf{G}\left((\mathbf{I} - \mathbf{R})\,\mathbf{m}_0\right) = \mathbf{0} \quad . \tag{6.186}$$

Then,

$$\mathbf{G}\,\mathbf{m} = \mathbf{G}\,\mathbf{m}_{est} = \mathbf{G}\,\mathbf{m}_{true} = \mathbf{d}_{obs} \quad . \tag{6.187}$$

For Backus and Gilbert, the solution $\mathbf{m}_{est} = \mathbf{G}^t\,(\mathbf{G}\,\mathbf{G}^t)^{-1}\,\mathbf{d}_{obs}$ gives a *particular solution* of the inverse problem, while $\mathbf{m} = \mathbf{m}_{est} + (\mathbf{I} - \mathbf{R})\,\mathbf{m}_0$ gives the *general solution*.

Let us make the comparison between Backus and Gilbert's philosophy and the probabilistic approach. For Backus and Gilbert, $\mathbf{m}_{est}$ is the best estimate of $\mathbf{m}_{true}$ and turns out to be a filtered version of it. From a probabilistic point of view, we have some a priori information on $\mathbf{m}_{true}$ described through the a priori model $\mathbf{m}_{prior}$ and the a priori covariance operator $\mathbf{C}_M$. Observations $\mathbf{d}_{obs}$ have estimated errors described by $\mathbf{C}_D$. If a priori information and observational errors are adequately described using the Gaussian hypothesis, then the posterior probability in the model space is also Gaussian, with mathematical expectation (third of equations (3.37))

$$\tilde{\mathbf{m}} = \mathbf{m}_{prior} + \mathbf{C}_M\,\mathbf{G}^t\left(\mathbf{G}\,\mathbf{C}_M\,\mathbf{G}^t + \mathbf{C}_D\right)^{-1}\left(\mathbf{d}_{obs} - \mathbf{G}\,\mathbf{m}_{prior}\right) \tag{6.188}$$

and posterior covariance operator (second of equations (3.38))

$$\tilde{\mathbf{C}}_M = \mathbf{C}_M - \mathbf{C}_M\,\mathbf{G}^t\left(\mathbf{G}\,\mathbf{C}_M\,\mathbf{G}^t + \mathbf{C}_D\right)^{-1}\mathbf{G}\,\mathbf{C}_M \quad . \tag{6.189}$$

Backus and Gilbert do not use a priori information in the model space. This corresponds in a probabilistic context to the particular assumption of white noise (infinite variance and null correlations):

$$\mathbf{C}_M \simeq k\,\mathbf{I} \qquad (k \to \infty) \quad . \tag{6.190}$$

Equations (6.188)–(6.189) give then, respectively,

$$\tilde{\mathbf{m}} = \mathbf{G}^t\,(\mathbf{G}\,\mathbf{G}^t)^{-1}\,\mathbf{d}_{obs} + (\mathbf{I} - \mathbf{R})\,\mathbf{m}_{prior} \quad ,$$
$$\tilde{\mathbf{C}}_M = (\mathbf{I} - \mathbf{R})\,\mathbf{C}_M \quad . \tag{6.191}$$

The first of these equations is identical to equation (6.185), where $\mathbf{m}_{prior}$ replaces the arbitrary $\mathbf{m}_0$. Note that if $\mathbf{R} \simeq \mathbf{I}$, in the Backus and Gilbert context $\mathbf{m}_{est} \simeq \mathbf{m}_{true}$, while in the probabilistic context, $\tilde{\mathbf{C}}_M \simeq \mathbf{0}$ (no error in the a posteriori solution), which means the same thing. I believe that the probabilistic approach is richer than the mathematical approach of Backus and Gilbert, but they probably feel the opposite way.

Please also read Example 5.24.

6.17 Disjunction and Conjunction of Probabilities

6.17.1 Conjunction of Probabilities

Let $\mathcal{X}$ be the finite-dimensional manifold where we assume our probability distributions are defined. In section 1.2.6, the conjunction $P_1 \wedge P_2$ of two probability distributions P_1 and P_2 was introduced. Let us find an explicit expression for this operation using probability densities. Let $f_1(\mathbf{x})$, $f_2(\mathbf{x})$, and $\mu(\mathbf{x})$ be the probability densities defined by the condition that, for any $\mathcal{A} \subset \mathcal{X}$,

$$P_1(\mathcal{A}) = \int_{\mathcal{A}} d\mathbf{x}\, f_1(\mathbf{x}) \quad , \quad P_2(\mathcal{A}) = \int_{\mathcal{A}} d\mathbf{x}\, f_2(\mathbf{x}) \quad , \quad M(\mathcal{A}) = \int_{\mathcal{A}} d\mathbf{x}\, \mu(\mathbf{x}) \quad ,$$
(6.192)

where M is the homogeneous probability distribution. We seek the probability density $(f_1 \wedge f_2)(\mathbf{x})$ defined by the condition that, for any $\mathcal{A}$,

$$(P_1 \wedge P_2)(\mathcal{A}) = \int_{\mathcal{A}} d\mathbf{x}\, (f_1 \wedge f_2)(\mathbf{x}) \quad .$$
(6.193)

One of the conditions defining $P_1 \wedge P_2$ is that, for any $\mathcal{A}$,

$$P_1(\mathcal{A}) = 0 \quad \Rightarrow \quad (P_1 \wedge P_2)(\mathcal{A}) = 0 \quad .$$
(6.194)

In mathematical terminology, this condition means that the probability $(P_1 \wedge P_2)$ is absolutely continuous with respect to the probability P_1. The Radon–Nikodym theorem (e.g., Taylor, 1966) then states that there exists a unique positive function $\phi_1(\mathbf{x})$ such that, for any $\mathcal{A} \subseteq \mathcal{X}$, $(P_1 \wedge P_2)(\mathcal{A}) = \int_{\mathcal{A}} d\mathbf{x}\, f_1(\mathbf{x})\, \phi_1(\mathbf{x})$. Of course, this function ϕ_1 may depend on $\mathbf{f}_1$, $\mathbf{f}_2$, and μ, so we may, more explicitly, use the notation $\phi(\mathbf{x}; \mathbf{f}_1, \mathbf{f}_2, \mu)$ for it.[85] Then, for any $\mathcal{A}$,

$$(P_1 \wedge P_2)(\mathcal{A}) = \int_{\mathcal{A}} d\mathbf{x}\, f_1(\mathbf{x})\, \phi_1(\mathbf{x}; \mathbf{f}_1, \mathbf{f}_2, \mu) \quad .$$
(6.195)

Similarly, the condition $P_2(\mathcal{A}) = 0 \Rightarrow (P_1 \wedge P_2)(\mathcal{A}) = 0$ implies that there exists a positive function $\phi_2(\mathbf{x}; \mathbf{f}_2, \mathbf{f}_1, \mu)$ such that, for any $\mathcal{A}$,

$$(P_1 \wedge P_2)(\mathcal{A}) = \int_{\mathcal{A}} d\mathbf{x}\, f_2(\mathbf{x})\, \phi_2(\mathbf{x}; \mathbf{f}_2, \mathbf{f}_1, \mu) \quad .$$
(6.196)

As the conjunction operation is commutative, one can permute $\mathbf{f}_1$ and $\mathbf{f}_2$ in these equations, from which it follows that $\phi_2 = \phi_1$. Denoting this unique function by ϕ allows us to write

$$(P_1 \wedge P_2)(\mathcal{A}) = \int_{\mathcal{A}} d\mathbf{x}\, f_1(\mathbf{x})\, \phi(\mathbf{x}; \mathbf{f}_1, \mathbf{f}_2, \mu) = \int_{\mathcal{A}} d\mathbf{x}\, f_2(\mathbf{x})\, \phi(\mathbf{x}; \mathbf{f}_2, \mathbf{f}_1, \mu) \quad .$$
(6.197)

---

[85]To understand this possible dependence of ϕ_1 on both $\mathbf{f}_1$ and $\mathbf{f}_2$, consider one possible solution to the condition (6.194), $(P_1 \wedge P_2)(\mathcal{A}) = \int_{\mathcal{A}} d\mathbf{x}\, \min(f_1(\mathbf{x}), f_2(\mathbf{x}))$, in which case $\phi_1(\mathbf{x}; \mathbf{f}_1, \mathbf{f}_2, \mu) = \min(f_1(\mathbf{x}), f_2(\mathbf{x})) / f_1(\mathbf{x})$.

To introduce more symmetric notation, rather than using the function ϕ, let us introduce the function ω defined, for any probability densities $\mathbf{f}$ and $\mathbf{g}$, through

$$\phi(\mathbf{x}; \mathbf{f}, \mathbf{g}, \mu) = \frac{g(\mathbf{x})}{\omega(\mathbf{x}; \mathbf{f}, \mathbf{g}, \mu)} \quad . \tag{6.198}$$

Then,

$$(P_1 \wedge P_2)(\mathcal{A}) = \int_{\mathcal{A}} d\mathbf{x} \, \frac{f_1(\mathbf{x}) \, f_2(\mathbf{x})}{\omega(\mathbf{x}; \mathbf{f}_1, \mathbf{f}_2, \mu)} = \int_{\mathcal{A}} d\mathbf{x} \, \frac{f_1(\mathbf{x}) \, f_2(\mathbf{x})}{\omega(\mathbf{x}; \mathbf{f}_2, \mathbf{f}_1, \mu)} \quad , \tag{6.199}$$

demonstrating that the function ω is symmetric in $\mathbf{f}_1$ and $\mathbf{f}_2$,

$$\omega(\mathbf{x}; \mathbf{f}_1, \mathbf{f}_2, \mu) = \omega(\mathbf{x}; \mathbf{f}_2, \mathbf{f}_1, \mu) \quad . \tag{6.200}$$

The condition that, for any probability distribution P, $P \wedge M = M \wedge P = P$, imposes that, for any $\mathbf{f}$,

$$\omega(\mathbf{x}; \mathbf{f}, \mu, \mu) = \omega(\mathbf{x}; \mu, \mathbf{f}, \mu) = \mu(\mathbf{x}) \quad . \tag{6.201}$$

The simplest solution is obtained when taking $\omega(\mathbf{x}; \mathbf{f}_1, \mathbf{f}_2, \mu)$ independent of $\mathbf{f}_1$ and $\mathbf{f}_2$, $\omega(\mathbf{x}; \mathbf{f}_1, \mathbf{f}_2, \mu) = \nu \, \mu(\mathbf{x})$, where ν is a constant. Then, the probability density $(f_1 \wedge f_2)(\mathbf{x})$ representing the conjunction $P_1 \wedge P_2$ is

$$(f_1 \wedge f_2)(\mathbf{x}) = \frac{1}{\nu} \frac{f_1(\mathbf{x}) \, f_2(\mathbf{x})}{\mu(\mathbf{x})} \quad , \tag{6.202}$$

where ν is the normalization constant $\nu = \int_{\mathcal{X}} d\mathbf{x} \, \frac{f_1(\mathbf{x}) \, f_2(\mathbf{x})}{\mu(\mathbf{x})}$.

6.17.2 Disjunction of Probabilities

Let us now turn our attention to the disjunction $P_1 \vee P_2$, using the notation introduced above. We seek here the probability density $(f_1 \vee f_2)(\mathbf{x})$ defined by the condition that, for any $\mathcal{A}$,

$$(P_1 \vee P_2)(\mathcal{A}) = \int_{\mathcal{A}} d\mathbf{x} \, (f_1 \vee f_2)(\mathbf{x}) \quad . \tag{6.203}$$

One of the conditions defining $P_1 \vee P_2$ is that, for any $\mathcal{A}$,

$$P_1(\mathcal{A}) \neq 0 \quad \Rightarrow \quad (P_1 \vee P_2)(\mathcal{A}) \neq 0 \quad . \tag{6.204}$$

Using a line of reasoning quite similar to that used above, we arrive at the expression

$$(P_1 \vee P_2)(\mathcal{A}) = k \int_{\mathcal{A}} d\mathbf{x} \, (f_1(\mathbf{x}) + f_2(\mathbf{x}) + \omega(\mathbf{x}; \mathbf{f}_1, \mathbf{f}_2, \mu)) \quad , \tag{6.205}$$

where k is a constant and ω is an arbitrary nonnegative function symmetric in $\mathbf{f}_1$ and $\mathbf{f}_2$:

$$\omega(\mathbf{x}; \mathbf{f}_1, \mathbf{f}_2, \mu) = \omega(\mathbf{x}; \mathbf{f}_2, \mathbf{f}_1, \mu) \quad . \tag{6.206}$$

The simplest solution is obtained when taking $\omega(\mathbf{x}; \mathbf{f}, \boldsymbol{\mu}, \boldsymbol{\mu})$ equal to zero. In that case, the measure density $(f_1 \vee f_2)(\mathbf{x})$ representing the disjunction $P_1 \vee P_2$ is

$$(f_1 \vee f_2)(\mathbf{x}) = \tfrac{1}{2}(f_1(\mathbf{x}) + f_2(\mathbf{x})) \quad . \tag{6.207}$$

6.18 Partition of Data into Subsets

In a linear least-squares problem, if a data set can be divided into two subsets such that the covariances between the different subsets are zero,

$$\mathbf{d}_{\mathrm{obs}} = \begin{pmatrix} \mathbf{d}_1 \\ \mathbf{d}_2 \\ \mathbf{d}_3 \\ \vdots \end{pmatrix} \quad , \qquad \mathbf{C}_{\mathrm{D}} = \begin{pmatrix} \mathbf{K}_1 & \mathbf{0} & \mathbf{0} & \cdots \\ \mathbf{0} & \mathbf{K}_2 & \mathbf{0} & \cdots \\ \mathbf{0} & \mathbf{0} & \mathbf{K}_3 & \cdots \\ \vdots & \vdots & \vdots & \ddots \end{pmatrix} \quad , \tag{6.208}$$

then solving one global inverse problem is equivalent to solving a series of smaller problems, introducing the data sets one by one, and using the posterior solution of each partial problem (model and covariance matrix) as prior information for the next.

We need to introduce the partitioned matrix

$$\mathbf{G} = \begin{pmatrix} \mathbf{G}_1 \\ \mathbf{G}_2 \\ \mathbf{G}_3 \\ \vdots \end{pmatrix} \quad , \tag{6.209}$$

and, for more clarity in the notation, let us modify the notation introduced in chapter 3, writing

$$\mathbf{m}_0 \equiv \mathbf{m}_{\mathrm{prior}} \quad , \qquad \mathbf{C}_{\mathrm{M}} \equiv \mathbf{C}_0 \quad . \tag{6.210}$$

Then, as demonstrated below, solving the global linear problem using equations (3.37)–(3.38) of chapter 3 is equivalent to solving a series of partial problems according to the algorithm

$$\boxed{\begin{aligned} \mathbf{m}_{i+1} &= \mathbf{m}_i + \mathbf{C}_i\,\mathbf{G}_{i+1}^t\,(\mathbf{K}_{i+1} + \mathbf{G}_{i+1}\,\mathbf{C}_i\,\mathbf{G}_{i+1}^t)^{-1}\,(\mathbf{d}_{i+1} - \mathbf{G}_{i+1}\,\mathbf{m}_i) \quad , \\ \mathbf{C}_{i+1} &= \mathbf{C}_i - \mathbf{C}_i\,\mathbf{G}_{i+1}^t\,(\mathbf{K}_{i+1} + \mathbf{G}_{i+1}\,\mathbf{C}_i\,\mathbf{G}_{i+1}^t)^{-1}\,\mathbf{G}_{i+1}\,\mathbf{C}_i \quad . \end{aligned}} \tag{6.211}$$

Equivalently, using the matrix identities given in Appendix 6.30,

$$\begin{aligned} \mathbf{m}_{i+1} &= \mathbf{m}_i + (\mathbf{G}_{i+1}^t\,\mathbf{K}_{i+1}^{-1}\,\mathbf{G}_{i+1} + \mathbf{C}_i^{-1})^{-1}\,\mathbf{G}_{i+1}^t\,\mathbf{K}_{i+1}^{-1}\,(\mathbf{d}_{i+1} - \mathbf{G}_{i+1}\,\mathbf{m}_i) \quad , \\ \mathbf{C}_{i+1} &= (\mathbf{G}_{i+1}^t\,\mathbf{K}_{i+1}^{-1}\,\mathbf{G}_{i+1} + \mathbf{C}_i^{-1})^{-1} \quad . \end{aligned} \tag{6.212}$$

When the last data set has been taken into account in this way, we get exactly the same solution (model vector and covariance matrix) as we would have obtained using equations (3.37)–(3.38) (see demonstration below).

We see that, at each step, to integrate the data set $\mathbf{d}_{i+1}$, the previous model $\mathbf{m}_i$ and the previous covariance matrix $\mathbf{C}_i$ must be used. This way of solving a linear least-squares problem is reminiscent of the Kalman filter (Kalman, 1960), which applies to a slightly different problem.

In the form proposed in equations (6.211), the matrix to be inverted at each step has the dimension of the data subset being integrated. If all the data have independent uncertainties (i.e., if the original matrix $\mathbf{C}_D$ is diagonal), then we can integrate the data one by one, and there is no matrix to be inverted any more, as the term $\mathbf{K}_{i+1} + \mathbf{G}_{i+1}\,\mathbf{C}_i\,\mathbf{G}_{i+1}^t$ degenerates into a scalar.

The reader is invited to verify that the following very simple computer code solves the least-squares linear inverse problem in the special case where data uncertainties are uncorrelated.

```
subroutine oneone(nd,nm,d0,cd,m0,cm,g,q)
dimension d0(nd),cd(nd),m0(nm),cm(nm,nm),g(nd,nm),q(nm)
   do k = 1,nd
   v = d0(k)
         do i = 1,nm
         v = v - g(k,i)*m0(i)
         q(i) = 0.
            do j = 1,nm
            q(i) = q(i) + cm(i,j)*g(k,j)
            end do
         end do
      a = cd(k)
         do i = 1,nm
         a = a + g(k,i)*q(i)
         end do
         do i = 1,nm
         m0(i) = m0(i) + q(i)*v/a
            do j = 1,nm
            cm(i,j) = cm(i,j) - q(i)*q(j)/a
            end do
         end do
      end do
return
end
```

Let us now move to the demonstration. It is done when only two data subsets are considered, but the generalization is obvious.

The least-squares solution is (see section 3.2.2)

$$\tilde{\mathbf{m}} = \mathbf{m}_{\text{prior}} + \left(\mathbf{G}^t\,\mathbf{C}_D^{-1}\,\mathbf{G} + \mathbf{C}_M^{-1}\right)^{-1}\mathbf{G}^t\,\mathbf{C}_D^{-1}(\mathbf{d}_{\text{obs}} - \mathbf{G}\,\mathbf{m}_{\text{prior}}) \quad ,$$

$$\tilde{\mathbf{C}}_M = \left(\mathbf{G}^t\,\mathbf{C}_D^{-1}\,\mathbf{G} + \mathbf{C}_M^{-1}\right)^{-1} \quad .$$

(6.213)

Introducing

$$\begin{pmatrix} \mathbf{d}_1 \\ \mathbf{d}_2 \end{pmatrix} = \begin{pmatrix} \mathbf{G}_1 \\ \mathbf{G}_2 \end{pmatrix} \mathbf{m} \quad , \tag{6.214}$$

we have

$$\tilde{\mathbf{m}} = \mathbf{m}_{\text{prior}} + \left[\begin{bmatrix} \mathbf{G}_1^t & \mathbf{G}_2^t \end{bmatrix} \begin{pmatrix} \mathbf{C}_1^{-1} & 0 \\ 0 & \mathbf{C}_2^{-1} \end{pmatrix} \begin{pmatrix} \mathbf{G}_1 \\ \mathbf{G}_2 \end{pmatrix} + \mathbf{C}_M^{-1} \right]^{-1} \begin{bmatrix} \mathbf{G}_1^t & \mathbf{G}_2^t \end{bmatrix}$$

$$\times \begin{pmatrix} \mathbf{C}_1^{-1} & 0 \\ 0 & \mathbf{C}_2^{-1} \end{pmatrix} \left[\begin{pmatrix} \mathbf{d}_1 \\ \mathbf{d}_2 \end{pmatrix} - \begin{pmatrix} \mathbf{G}_1 \\ \mathbf{G}_2 \end{pmatrix} \mathbf{m}_{\text{prior}} \right]$$

$$= \mathbf{m}_{\text{prior}} + (\mathbf{S}_0 + \mathbf{S}_1 + \mathbf{S}_2)^{-1} \left(\mathbf{G}_1^t \, \mathbf{C}_1^{-1} (\mathbf{d}_1 - \mathbf{G}_1 \, \mathbf{m}_{\text{prior}}) + \mathbf{G}_2^t \, \mathbf{C}_2^{-1} (\mathbf{d}_2 - \mathbf{G}_2 \, \mathbf{m}_{\text{prior}}) \right),$$

$$\tag{6.215}$$

where

$$\mathbf{S}_0 = \mathbf{C}_M^{-1} \quad , \qquad \mathbf{S}_1 = \mathbf{G}_1^t \, \mathbf{C}_1^{-1} \, \mathbf{G}_1 \quad , \qquad \mathbf{S}_2 = \mathbf{G}_2^t \, \mathbf{C}_2^{-1} \, \mathbf{G}_2 \quad . \tag{6.216}$$

The following identity holds:

$$\begin{aligned} (\mathbf{S}_0 + \mathbf{S}_1 + \mathbf{S}_2)^{-1} &= (\mathbf{S}_0 + \mathbf{S}_1 + \mathbf{S}_2)^{-1} \left((\mathbf{S}_0 + \mathbf{S}_1 + \mathbf{S}_2) - \mathbf{S}_2 \right) (\mathbf{S}_0 + \mathbf{S}_1)^{-1} \\ &= (\mathbf{S}_0 + \mathbf{S}_1 + \mathbf{S}_2)^{-1} (\mathbf{S}_0 + \mathbf{S}_1 + \mathbf{S}_2) \left(\mathbf{I} - (\mathbf{S}_0 + \mathbf{S}_1 + \mathbf{S}_2)^{-1} \mathbf{S}_2 \right) (\mathbf{S}_0 + \mathbf{S}_1)^{-1} \\ &= \left(\mathbf{I} - (\mathbf{S}_0 + \mathbf{S}_1 + \mathbf{S}_2)^{-1} \mathbf{S}_2 \right) (\mathbf{S}_0 + \mathbf{S}_1)^{-1} \\ &= (\mathbf{S}_0 + \mathbf{S}_1)^{-1} - (\mathbf{S}_0 + \mathbf{S}_1 + \mathbf{S}_2)^{-1} \mathbf{S}_2 \, (\mathbf{S}_0 + \mathbf{S}_1)^{-1} \quad . \end{aligned} \tag{6.217}$$

This gives

$$\tilde{\mathbf{m}} = \mathbf{m}_A + \left(\mathbf{G}_2^t \, \mathbf{C}_2^{-1} \, \mathbf{G}_2 + \mathbf{C}_A^{-1} \right)^{-1} \mathbf{G}_2^t \, \mathbf{C}_2^{-1} (\mathbf{d}_2 - \mathbf{G}_2 \, \mathbf{m}_A) \tag{6.218}$$

and

$$\tilde{\mathbf{C}}_M = \left(\mathbf{G}_2^t \, \mathbf{C}_2^{-1} \, \mathbf{G}_2 + \mathbf{C}_A^{-1} \right)^{-1} \quad , \tag{6.219}$$

where

$$\mathbf{m}_A = \mathbf{m}_{\text{prior}} + \left(\mathbf{G}_1^t \, \mathbf{C}_1^{-1} \, \mathbf{G}_1 + \mathbf{C}_M^{-1} \right)^{-1} \mathbf{G}_1^t \, \mathbf{C}_1^{-1} (\mathbf{d}_1 - \mathbf{G}_1 \, \mathbf{m}_{\text{prior}}) \tag{6.220}$$

and

$$\mathbf{C}_A = \left(\mathbf{G}_1^t \, \mathbf{C}_1^{-1} \, \mathbf{G}_1 + \mathbf{C}_M^{-1} \right)^{-1} \quad . \tag{6.221}$$

6.19 Marginalizing in Linear Least Squares

Let $\mathbf{d} = \mathbf{G}\,\mathbf{m}$ be the equation solving the forward problem, and let $\mathbf{d}_{\text{obs}}$, $\mathbf{C}_{\text{D}}$, $\mathbf{m}_{\text{prior}}$, and $\mathbf{C}_{\text{M}}$ be as usual in least-squares problems. The solution in the least-squares sense is given, for instance, by (see section 3.2.2)

$$\tilde{\mathbf{m}} = \mathbf{m}_{\text{prior}} + \left(\mathbf{G}^t\,\mathbf{C}_{\text{D}}^{-1}\,\mathbf{G} + \mathbf{C}_{\text{M}}^{-1}\right)^{-1}\mathbf{G}^t\,\mathbf{C}_{\text{D}}^{-1}(\mathbf{d}_{\text{obs}} - \mathbf{G}\,\mathbf{m}_{\text{prior}}) \quad, \tag{6.222}$$

and the posterior covariance operator is given, for instance, by

$$\tilde{\mathbf{C}}_{\text{M}} = \left(\mathbf{G}^t\,\mathbf{C}_{\text{D}}^{-1}\,\mathbf{G} + \mathbf{C}_{\text{M}}^{-1}\right)^{-1} \quad. \tag{6.223}$$

Assume that we can partition the model vector into

$$\mathbf{m} = \begin{pmatrix} \mathbf{m}_1 \\ \mathbf{m}_2 \end{pmatrix} \quad, \qquad \text{with} \qquad \mathbf{C}_{\text{M}} = \begin{pmatrix} \mathbf{C}_1 & 0 \\ 0 & \mathbf{C}_2 \end{pmatrix} \quad, \tag{6.224}$$

and that we are only interested in $\mathbf{m}_1$. Let us derive the corresponding formulas.

Introducing

$$\Delta\mathbf{d} = \mathbf{d}_{\text{obs}} - \mathbf{G}\,\mathbf{m}_{\text{prior}} \quad, \qquad \Delta\mathbf{m} = \mathbf{m} - \mathbf{m}_{\text{prior}} \quad, \qquad \mathbf{G} = \begin{bmatrix} \mathbf{G}_1 & \mathbf{G}_2 \end{bmatrix} \quad, \tag{6.225}$$

we obtain

$$\begin{pmatrix} \Delta\mathbf{m}_1 \\ \Delta\mathbf{m}_2 \end{pmatrix} = \begin{pmatrix} \mathbf{S}_{11} & \mathbf{S}_{12} \\ \mathbf{S}_{21} & \mathbf{S}_{22} \end{pmatrix}^{-1} \begin{pmatrix} \mathbf{G}_1^t \\ \mathbf{G}_2^t \end{pmatrix} \mathbf{C}_{\text{D}}^{-1}\,\Delta\mathbf{d} \quad, \tag{6.226}$$

where

$$\mathbf{S}_{11} = \mathbf{G}_1^t\,\mathbf{C}_{\text{D}}^{-1}\,\mathbf{G}_1 + \mathbf{C}_1^{-1} \;, \quad \mathbf{S}_{22} = \mathbf{G}_2^t\,\mathbf{C}_{\text{D}}^{-1}\,\mathbf{G}_2 + \mathbf{C}_2^{-1} \;, \quad \mathbf{S}_{12} = \mathbf{G}_1^t\,\mathbf{C}_{\text{D}}^{-1}\,\mathbf{G}_2 \;, \quad \mathbf{S}_{12} = \mathbf{S}_{12}^t \;. \tag{6.227}$$

Using an inversion per block (see Appendix 6.31) gives

$$\begin{pmatrix} \Delta\mathbf{m}_1 \\ \Delta\mathbf{m}_2 \end{pmatrix} = \begin{pmatrix} \mathbf{A} & \mathbf{B} \\ \mathbf{C} & \mathbf{D} \end{pmatrix} \begin{pmatrix} \mathbf{G}_1^t \\ \mathbf{G}_2^t \end{pmatrix} \mathbf{C}_{\text{D}}^{-1}\,\Delta\mathbf{d} \quad, \tag{6.228}$$

where

$$\begin{aligned} \mathbf{A} &= (\mathbf{S}_{11} - \mathbf{S}_{12}\,\mathbf{S}_{22}^{-1}\,\mathbf{S}_{12}^t)^{-1} \;, \qquad & \mathbf{D} &= (\mathbf{S}_{22} - \mathbf{S}_{12}^t\,\mathbf{S}_{11}^{-1}\,\mathbf{S}_{12})^{-1} \;, \\ \mathbf{B} &= -\mathbf{A}\,\mathbf{S}_{12}\,\mathbf{S}_{22}^{-1} \;, \qquad & \mathbf{C} &= \mathbf{B}^t \;. \end{aligned} \tag{6.229}$$

For $\Delta\mathbf{m}_1$ we obtain

$$\Delta\mathbf{m}_1 = (\mathbf{A}\,\mathbf{G}_1^t + \mathbf{B}\,\mathbf{G}_2^t)\,\mathbf{C}_{\text{D}}^{-1}\,\Delta\mathbf{d} \quad. \tag{6.230}$$

This gives

$$\Delta\mathbf{m}_1 = (\mathbf{S}_{11} - \mathbf{G}_1^t\,\mathbf{T}_{22}\mathbf{G}_1)^{-1}\mathbf{G}_1^t\,(\mathbf{C}_{\text{D}}^{-1} - \mathbf{T}_{22})\,\Delta\mathbf{d} \quad, \tag{6.231}$$

where

$$T_{22} = C_D^{-1} G_2 S_{22}^{-1} G_2' C_D^{-1} \quad . \tag{6.232}$$

The a posteriori covariance operator for m_1 clearly equals A .

The previous formulas allow us to take into consideration model parameters that, although not interesting in themselves, are poorly known, and they introduce uncertainties that cannot be ignored.

6.20   Relative Information of Two Gaussians

Let $f_1(x)$ and $f_0(x)$ represent two normalized probability density functions. The relative information on f_1 with respect to f_0 is defined by

$$I(f_1; f_0) = \int dx \, f_1(x) \, \log \frac{f_1(x)}{f_0(x)} \quad . \tag{6.233}$$

Let us demonstrate that if f_1 and f_0 are Gaussian probability densities with mathematical expectations respectively equal to x_1 and x_0 and covariance operators respectively equal to C_1 and C_0 , then

$$I(f_1; f_0) = \log \frac{\det^{1/2} C_0}{\det^{1/2} C_1} + \frac{1}{2} (x_1 - x_0)^t \, C_0^{-1} (x_1 - x_0) + \frac{1}{2} \operatorname{trace}(C_1 \, C_0^{-1} - I) \quad . \tag{6.234}$$

By definition,

$$f_1(x) = \frac{1}{(2\pi)^{n/2} \det^{1/2} C_1} \exp\left(-\frac{1}{2} (x - x_1)^t \, C_1^{-1} (x - x_1) \right) \tag{6.235}$$

and

$$f_0(x) = \frac{1}{(2\pi)^{n/2} \det^{1/2} C_0} \exp\left(-\frac{1}{2} (x - x_0)^t \, C_0^{-1} (x - x_0) \right) \quad . \tag{6.236}$$

Inserting these expressions in equation (6.233) gives

$$I(f_1; f_0) = \log \left(\frac{\det^{1/2} C_0}{\det^{1/2} C_1} \right) - \frac{1}{2} E_1\left((x - x_1)^t \, C_1^{-1} (x - x_1) \right)$$

$$+ \frac{1}{2} E_1\left((x - x_0)^t \, C_0^{-1} (x - x_0) \right), \tag{6.237}$$

where $E_1(\cdot)$ denotes the mathematical expectation with respect to f_1 :

$$E_1(\Psi(x)) \equiv \int dx \, f_1(x) \, \Psi(x) \quad . \tag{6.238}$$

From the definition of covariance operator, and using the linearity of the mathematical expectation, we obtain $C_1 = E_1((x - x_1)(x - x_1)^t) = E_1(x x^t - 2 x_1 x^t + x_1 x_1^t) =$

$E_1(\mathbf{x}\,\mathbf{x}^t) - 2\,\mathbf{x}_1\,E_1(\mathbf{x}^t) + \mathbf{x}_1\,\mathbf{x}_1{}^t = E_1(\mathbf{x}\,\mathbf{x}^t) - \mathbf{x}_1\,\mathbf{x}_1{}^t$, whence, using a tensor notation, we deduce

$$E_1(x^\alpha x^\beta) = C_1{}^{\alpha\beta} + x_1{}^\alpha x_1{}^\beta \quad . \tag{6.239}$$

We have $E_1((\mathbf{x} - \mathbf{x}_1)^t\,C_1^{-1}\,(\mathbf{x} - \mathbf{x}_1)) = \cdots = E_1(\mathbf{x}^t\,C_1^{-1}\,\mathbf{x}) - \mathbf{x}_1^t\,C_1^{-1}\,\mathbf{x}_1 = (C_1^{-1})^{\alpha\beta}\,E_1(x^\alpha x^\beta)$ $- (C_1^{-1})^{\alpha\beta}\,x_1{}^\alpha x_1{}^\beta$, whence, using equation (6.239), we deduce

$$E_1\left((\mathbf{x} - \mathbf{x}_1)^t\,C_1^{-1}\,(\mathbf{x} - \mathbf{x}_1)\right) = (C_1^{-1})^{\alpha\beta}\,C_1{}^{\alpha\beta} = \text{trace }\mathbf{I} \quad . \tag{6.240}$$

We also have $E_1((\mathbf{x} - \mathbf{x}_0)^t\,C_0^{-1}\,(\mathbf{x} - \mathbf{x}_0)) = \cdots = E_1(\mathbf{x}^t\,C_0^{-1}\,\mathbf{x}) - 2\,\mathbf{x}_0{}^t\,C_0^{-1}\,\mathbf{x}_1 + \mathbf{x}_0{}^t\,C_0^{-1}\,\mathbf{x}_0$ $= (C_0^{-1})^{\alpha\beta}\,E_1(x^\alpha x^\beta) - 2\,(C_0^{-1})^{\alpha\beta}\,x_0{}^\alpha x_1{}^\beta + (C_0^{-1})^{\alpha\beta}\,x_0{}^\alpha x_0{}^\beta$, whence, using equation (6.239), we deduce

$$E_1\left((\mathbf{x} - \mathbf{x}_0)^t\,C_0^{-1}\,(\mathbf{x} - \mathbf{x}_0)\right) = (C_0^{-1})^{\alpha\beta}\,(C_0^{-1})^{\alpha\beta} + (C_0^{-1})^{\alpha\beta}\,(x_1{}^\alpha - x_0{}^\alpha)(x_1{}^\beta - x_0{}^\beta)$$
$$= \text{trace}\,(C_0^{-1}\,C_1) + (\mathbf{x}_1 - \mathbf{x}_0)^t\,C_0^{-1}\,(\mathbf{x}_1 - \mathbf{x}_0) \quad . \tag{6.241}$$

Inserting (6.240) and (6.241) into (6.237), result (6.234) follows.

Notice that the factor $\det^{1/2}\mathbf{C}$ represents the (hyper)volume of the hyperellipsoid representing the covariance operator $\mathbf{C}$.

6.21   Convolution of Two Gaussians

Let us evaluate the sum

$$I = \int d\mathbf{d}\,\exp\left(-\frac{1}{2}\left((\mathbf{d} - \mathbf{d}_0)^t\,C_\mathbf{d}^{-1}\,(\mathbf{d} - \mathbf{d}_0) + (\mathbf{d} - \mathbf{g}(\mathbf{m}))^t\,C_T^{-1}\,(\mathbf{d} - \mathbf{g}(\mathbf{m}))\right)\right) \quad . \tag{6.242}$$

The separation of the quadratic terms from the linear terms leads to

$$I = \int d\mathbf{d}\,\exp\left(-\frac{1}{2}\,(\mathbf{d}^t\,\mathbf{A}\,\mathbf{d} - 2\,\mathbf{b}^t\,\mathbf{d} + c)\right) \quad , \tag{6.243}$$

where

$$\mathbf{A} = C_\mathbf{d}^{-1} + C_T^{-1} \quad ,$$
$$\mathbf{b}^t = \mathbf{d}_0{}^t\,C_\mathbf{d}^{-1} + \mathbf{g}(\mathbf{m})^t\,C_T^{-1} \quad , \tag{6.244}$$
$$c = \mathbf{d}_0{}^t\,C_\mathbf{d}^{-1}\,\mathbf{d}_0 + \mathbf{g}(\mathbf{m})^t\,C_T^{-1}\,\mathbf{g}(\mathbf{m}) \quad .$$

Since $\mathbf{A}$ is positive definite, it follows that

$$I = \int d\mathbf{d}\,\exp\left(-\frac{1}{2}\left((\mathbf{d} - \mathbf{A}^{-1}\,\mathbf{b})^t\,\mathbf{A}\,(\mathbf{d} - \mathbf{A}^{-1}\,\mathbf{b}) + (c - \mathbf{b}^t\,\mathbf{A}^{-1}\,\mathbf{b})\right)\right)$$
$$= \exp\left(-\frac{1}{2}\,(c - \mathbf{b}^t\,\mathbf{A}^{-1}\,\mathbf{b})\right)\int d\mathbf{d}\,\exp\left(-\frac{1}{2}\,(\mathbf{d} - \mathbf{A}^{-1}\,\mathbf{b})^t\,\mathbf{A}\,(\mathbf{d} - \mathbf{A}^{-1}\,\mathbf{b})\right)$$
$$= (2\pi)^{n/2}\,(\det\mathbf{A})^{-1/2}\,\exp\left(-\frac{1}{2}\,(c - \mathbf{b}^t\,\mathbf{A}^{-1}\,\mathbf{b})\right) \quad . \tag{6.245}$$

By substitution we obtain

$$c - \mathbf{b}^t \mathbf{A}^{-1} \mathbf{b} = \mathbf{d}_0^t \left(\mathbf{C}_d^{-1} - \mathbf{C}_d^{-1}(\mathbf{C}_d^{-1} + \mathbf{C}_T^{-1})^{-1} \mathbf{C}_d^{-1} \right) \mathbf{d}_0$$

$$+ \mathbf{g}(\mathbf{m})^t \left(\mathbf{C}_T^{-1} - \mathbf{C}_T^{-1}(\mathbf{C}_d^{-1} + \mathbf{C}_T^{-1})^{-1} \mathbf{C}_T^{-1} \right) \mathbf{g}(\mathbf{m}) \qquad (6.246)$$

$$- 2 \mathbf{g}(\mathbf{m})^t \mathbf{C}_T^{-1} (\mathbf{C}_d^{-1} + \mathbf{C}_T^{-1})^{-1} \mathbf{C}_d^{-1} \mathbf{d}_0 \quad .$$

Thus, by using the two identities demonstrated in Appendix 6.30, we get

$$c - \mathbf{b}^t \mathbf{A}^{-1} \mathbf{b} = \mathbf{d}_0^t (\mathbf{C}_d + \mathbf{C}_T)^{-1} \mathbf{d}_0 + \mathbf{g}(\mathbf{m})^t (\mathbf{C}_d + \mathbf{C}_T)^{-1} \mathbf{g}(\mathbf{m}) - 2 \mathbf{g}(\mathbf{m})^t (\mathbf{C}_d + \mathbf{C}_T)^{-1} \mathbf{d}_0$$

$$= (\mathbf{d}_0 - \mathbf{g}(\mathbf{m}))^t (\mathbf{C}_d + \mathbf{C}_T)^{-1} (\mathbf{d}_0 - \mathbf{g}(\mathbf{m})) \quad . \qquad (6.247)$$

Finally,

$$I = (2\pi)^{n/2} \det (\mathbf{C}_d^{-1} + \mathbf{C}_T^{-1})^{-1/2} \exp \left(-\frac{1}{2} (\mathbf{d}_0 - \mathbf{g}(\mathbf{m}))^t (\mathbf{C}_d + \mathbf{C}_T)^{-1} (\mathbf{d}_0 - \mathbf{g}(\mathbf{m})) \right).$$

$$(6.248)$$

6.22 Gradient-Based Optimization Algorithms

There are two fundamentally different methods of optimization, those based on the local computation of the function to be optimized and those based on a random search. We are interested here in the first class of methods. The advantage of the gradient-based method is that, when it works, it may be very efficient. The disadvantage is that the local properties of the function to be optimized may be of little interest — if the function is complex enough.

Gradient methods may have increasing levels of sophistication. The simplest methods just use the local direction of steepest ascent (or descent). Or this direction may be preconditioned using different operators (ranging from ad hoc fixed operators to variable metric operators, passing by the Newton methods).

Some of the results given below are applicable when an ℓ_p-norm is used in the model space, and some results are only valid for least-squares (ℓ_2-norm) problems. My own experience is that while sophisticated methods may work well for least-squares problems, the simpler (steepest descent) methods are preferable for general ℓ_p-norm problems: although many techniques are presented as valid for nonquadratic optimization problems, ℓ_p-norm optimization problems may be so strongly nonquadratic that they may fail. Figure 6.9 shows a function used as an example by Gill, Murray, and Wright (1981). It is obviously a nonquadratic function, but it is not strongly nonquadratic.

There are many good books on gradient methods. English books (Walsh, 1975; Fletcher, 1980; Powell, 1981; Scales, 1985) are excellent for their empirical taste. French books (Céa, 1971; Ciarlet, 1982) are good at generality.

6.22.1 Gradient, Hessian, Steepest Ascent, Curvature

Gradient and Hessian of a Scalar Function

Let $\mathbf{m} \mapsto \psi(\mathbf{m})$ be a nonlinear form (i.e., an application into $\Re$) over a finite-dimensional linear space $\mathbb{M}$ (this linear space is not (yet) assumed to be normed).

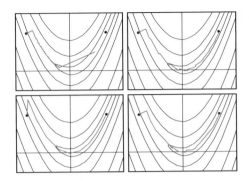

Figure 6.9. *The* Rosenbrock function $S(x, y) = 100(y - x^2)^2 + (1 - x)^2$, *often used to test optimization algorithms. It has a unique minimum at point* $(x, y) = (1, 1)$. *Gill, Murray, and Wright (1981) show the minimization paths obtained with different algorithms, all initialized at* $(x, y) = (-1.2, 1.0)$. *Top left: Steepest descent algorithm. Note that the algorithm would have failed in the vicinity of the point* $(x, y) = (-0.3, 0.1)$ *but for the fact that the linear search found, by chance, the second minimum along the search direction. Several hundred iterations were performed close to the new point without any perceptible change. Top right: Conjugate directions algorithm. Although the method is not intended for problems with such a small number of parameters, the figure is useful in illustrating the cyclic nature of the algorithm. Bottom left: Modified Newton's algorithm. The method follows the base of the valley in an almost optimal number of steps. Bottom right: Variable metric algorithm. Like Newton's method, the algorithm makes good progress at points remote from the solution.*

A series development of $\psi(\mathbf{m})$ around a point $\mathbf{m}_0$ can be written, using different notation, as

$$
\begin{aligned}
\psi(\mathbf{m}_0 + \delta\mathbf{m}) &= \psi(\mathbf{m}_0) + \langle\, \hat{\boldsymbol{\gamma}}_0\,,\, \delta\mathbf{m}\,\rangle + \tfrac{1}{2}\,\langle\, \hat{\mathbf{H}}_0\,\delta\mathbf{m}\,,\, \delta\mathbf{m}\,\rangle + O(3) \\[4pt]
&= \psi(\mathbf{m}_0) + \hat{\boldsymbol{\gamma}}_0^t\,\delta\mathbf{m} + \tfrac{1}{2}\,(\hat{\mathbf{H}}_0\,\delta\mathbf{m})^t\,\delta\mathbf{m} + O(3) \qquad\qquad (6.249)\\[4pt]
&= \psi(\mathbf{m}_0) + (\hat{\boldsymbol{\gamma}}_0)_\alpha\,\delta m^\alpha + \tfrac{1}{2}\,(\hat{\mathbf{H}}_0)_{\alpha\beta}\,\delta m^\alpha\,\delta m^\beta + O(3)\quad,
\end{aligned}
$$

where $\hat{\boldsymbol{\gamma}}_0$ is the *gradient* of ψ at point $\mathbf{m}_0$ and $\hat{\mathbf{H}}_0$ is the *Hessian*:

$$
(\hat{\boldsymbol{\gamma}}_0)_\alpha = \left(\frac{\partial\psi}{\partial m^\alpha}\right)_{\mathbf{m}_0} \quad, \qquad (\hat{\mathbf{H}}_0)_{\alpha\beta} = \left(\frac{\partial\hat{\gamma}_\alpha}{\partial m^\beta}\right)_{\mathbf{m}_0} = \left(\frac{\partial^2\psi}{\partial m^\alpha\,\partial m^\beta}\right)_{\mathbf{m}_0} \quad. \qquad (6.250)
$$

Here, $O(3)$ denotes a term that tends to zero more rapidly than the second-order term when $\delta\mathbf{m} \to 0$ (this is independent of the particular norm being used). In the third of expressions (6.249), the convention of implicit sum over repeated indices is used.

These equations show that the gradient $\hat{\boldsymbol{\gamma}}$ at a given point defines a linear application from $\mathbb{M}$ into the real line $\mathfrak{R}$: the gradient is a form, i.e., an element of $\mathbb{M}^*$, the dual of $\mathbb{M}$. The Hessian $\hat{\mathbf{H}}$ is a linear operator mapping $\mathbb{M}$ into $\mathbb{M}^*$.

Gradient of the Misfit Function

In nonlinear least squares, the function of interest is the misfit function given by (equation (3.46))

$$2\,S(\mathbf{m}) = \|\,\mathbf{g}(\mathbf{m}) - \mathbf{d}_{\text{obs}}\,\|_{\text{D}}^2 + \|\,\mathbf{m} - \mathbf{m}_{\text{prior}}\,\|_{\text{M}}^2$$

$$= (\mathbf{g}(\mathbf{m}) - \mathbf{d}_{\text{obs}})^t\,\mathbf{C}_{\text{D}}^{-1}\,(\mathbf{g}(\mathbf{m}) - \mathbf{d}_{\text{obs}}) + (\mathbf{m} - \mathbf{m}_{\text{prior}})^t\,\mathbf{C}_{\text{M}}^{-1}\,(\mathbf{m} - \mathbf{m}_{\text{prior}}) \quad . \tag{6.251}$$

The gradient of S at a point $\mathbf{m}_0$ is (equation (3.88))

$$\hat{\boldsymbol{\gamma}}_0 = \mathbf{G}_0^t\,\mathbf{C}_{\text{D}}^{-1}\,(\mathbf{g}(\mathbf{m}_0) - \mathbf{d}_{\text{obs}}) + \mathbf{C}_{\text{M}}^{-1}\,(\mathbf{m}_0 - \mathbf{m}_{\text{prior}}) \quad . \tag{6.252}$$

Here, $\mathbf{G}_0$ stands for the matrix of partial derivatives

$$(\mathbf{G}_0)^i{}_\alpha = \left(\frac{\partial g^i}{\partial m^\alpha}\right)_{\mathbf{m}_0} \quad . \tag{6.253}$$

In ℓ_p-norm problems, where the misfit function is (equation (4.26))

$$S(\mathbf{m}) = \frac{1}{s}\sum_i \frac{|g^i(\mathbf{m}) - d_{\text{obs}}^i|^s}{(\sigma_{\text{D}}^i)^s} + \frac{1}{r}\sum_\alpha \frac{|m^\alpha - m_{\text{prior}}^\alpha|^r}{(\sigma_{\text{M}}^\alpha)^r} \quad , \tag{6.254}$$

the gradient of S at a point $\mathbf{m}_0$ is (equation (4.27))

$$(\hat{\boldsymbol{\gamma}}_0)_\alpha = \sum_i \frac{1}{\sigma_{\text{D}}^i}\,(\mathbf{G}_0)^i{}_\alpha \left(\frac{g^i(\mathbf{m}_0) - d_{\text{obs}}^i}{\sigma_{\text{D}}^i}\right)^{\{s-1\}} + \sum_\alpha \frac{1}{\sigma_{\text{M}}^\alpha}\left(\frac{m_0^\alpha - m_{\text{prior}}^\alpha}{\sigma_{\text{M}}^\alpha}\right)^{\{r-1\}} \quad . \tag{6.255}$$

Hessian of the Least-Squares Misfit Function

The Hessian associated with the least-squares misfit function (6.251) is (equation (3.91))

$$(\hat{\mathbf{H}}_0)_{\alpha\beta} = (\mathbf{G}_0)^i{}_\alpha\,(\mathbf{C}_{\text{D}}^{-1})_{ij}\,(\mathbf{G}_0)^j{}_\beta + (\mathbf{C}_{\text{M}}^{-1})_{\alpha\beta} + \left(\frac{\partial G^i{}_\alpha}{\partial m^\beta}\right)_{\mathbf{m}_0}(\mathbf{C}_{\text{D}}^{-1})_{ij}\,(g^j(\mathbf{m}_0) - d_{\text{obs}}^j) \quad , \tag{6.256}$$

or, with the usual approximation of dropping the second derivatives (equation (3.92)),

$$\hat{\mathbf{H}}_0 \simeq \mathbf{G}_0^t\,\mathbf{C}_{\text{D}}^{-1}\,\mathbf{G}_0 + \mathbf{C}_{\text{M}}^{-1} \quad . \tag{6.257}$$

Note that in equation (6.256) the implicit sum convention is used.

Gradient Versus Steepest Ascent Vector

Some elementary texts tend to use the terms *gradient* and *steepest ascent vector* as almost synonymous. In fact, they correspond to two very different concepts (and, numerically, to two very different quantities).

Let $\mathbb{M}$ be a linear space, with vectors denoted $\mathbf{m}, \mathbf{m}', \ldots$, and $\mathbf{m} \mapsto \psi(\mathbf{m})$ be a (scalar) function to be maximized. The gradient, as introduced above, is a linear form. In fact, it can be understood as the linear[86] function over $\mathbb{M}$ that is tangent to the function $\psi(\mathbf{m})$ (at the point $\mathbf{m}_0$ where the gradient is evaluated). The gradient is defined irrespective of any possible definition of norm over $\mathbb{M}$.

To define the direction of steepest ascent (at the given point $\mathbf{m}_0$), one must define a norm (for instance an ℓ_1- or an ℓ_2-norm). When a norm is given, one may consider an infinitesimally small circle (defined according to the given norm) around the point $\mathbf{m}_0$. The direction of steepest ascent is defined by the point on the circle where the function $\psi(\mathbf{m})$ reaches the maximum value.

Changing the norm changes the circle, and the direction of steepest ascent, as suggested in Figure 6.10.

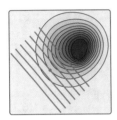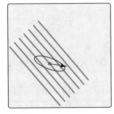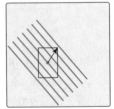

Figure 6.10. *In the first panel, the level lines of a real function $S(\mathbf{m})$ in a 2D space and the mille-feuilles representing the gradient of S at a given point are shown. The three other panels represent a small circle (defined using different definitions of norm) around the given point and the respective direction of steepest ascent. This direction is defined by the point on the small circle where the gradient takes its maximum value. In the second panel, the circle is defined in the ℓ_1-norm sense. In the third panel, it is defined using an ℓ_2-norm (with covariances). In the last panel, it is defined using an ℓ_∞-norm. Notice that, while the gradient of the function $S(\mathbf{m})$ at a given point is defined irrespective of any definition of norm, the direction of steepest ascent fundamentally depends on the norm being used.*

Given a function $\mathbf{m} \mapsto \psi(\mathbf{m})$, the evaluation of the gradient just involves taking partial derivatives (equation at left in (6.250)). To evaluate the direction of steepest ascent, one may take an infinitesimal circle, as explained above, and seek the maximum of the function $\psi(\mathbf{m})$ or, equivalently, take a finite circle and seek the maximum of the linear tangent application (as defined by the gradient).

The result of this computation is simple enough. Let $\hat{\boldsymbol{\gamma}}_0$ denote the gradient of the function $\psi(\mathbf{m})$ at a point $\mathbf{m}_0$. By definition, it is an element of $\mathbb{M}^*$, the dual of $\mathbb{M}$. The steepest ascent vector, say $\boldsymbol{\gamma}_0$, is the vector of $\mathbb{M}$ that is related to $\hat{\boldsymbol{\gamma}}_0$ through the usual duality relation between vectors and forms. Before demonstrating this result, let us make it more explicit.

---

[86]To speak properly, an *affine* function.

Given a function $\mathbf{m} \mapsto \psi(\mathbf{m})$, the gradient at a given point is the form whose components are

$$\hat{\gamma}_\alpha = \frac{\partial \psi}{\partial m^\alpha} \quad . \tag{6.258}$$

- When using an ℓ_p-norm over $\mathbb{M}$,

$$\| \mathbf{m} \| = \left(\sum_\alpha \frac{| m^\alpha |^p}{(\sigma^\alpha)^p} \right)^{1/p} \quad , \tag{6.259}$$

it makes sense to consider the ℓ_q-norm over $\mathbb{M}^*$,

$$\| \hat{\mathbf{m}} \| = \left(\sum_\alpha \frac{| \hat{m}_\alpha |^q}{(\hat{\sigma}_\alpha)^q} \right)^{1/q} \quad , \tag{6.260}$$

where

$$\frac{1}{p} + \frac{1}{q} = 1 \quad , \quad \hat{\sigma}_\alpha = \frac{1}{\sigma^\alpha} \quad . \tag{6.261}$$

The vector of $\mathbb{M}$ associated with the gradient $\hat{\boldsymbol{\gamma}}$ by the basic duality between $\mathbb{M}^*$ and $\mathbb{M}$ is (see equations (4.13) and (4.16) in the main text)

$$\gamma^\alpha = \frac{1}{\hat{\sigma}_\alpha} \, \mathrm{sg}(\hat{\gamma}_\alpha) \left| \frac{\hat{\gamma}_\alpha}{\hat{\sigma}_\alpha} \right|^{q-1} = \frac{1}{\hat{\sigma}_\alpha} \left(\frac{\hat{\gamma}_\alpha}{\hat{\sigma}_\alpha} \right)^{\{q-1\}} \quad , \tag{6.262}$$

and the norms of $\hat{\boldsymbol{\gamma}}$ and $\boldsymbol{\gamma}$ are related as (equation (4.12))

$$\| \boldsymbol{\gamma} \|^p = \| \hat{\boldsymbol{\gamma}} \|^q \quad . \tag{6.263}$$

In this situation, the vector $\boldsymbol{\gamma}$ in equation (6.262) is the steepest ascent vector associated with the gradient $\hat{\boldsymbol{\gamma}}$ in equation (6.258).

- When using an ℓ_2-norm over $\mathbb{M}$,

$$\| \mathbf{m} \| = \left(\mathbf{m}^t \, \mathbf{C}^{-1} \, \mathbf{m} \right)^{1/2} \quad , \tag{6.264}$$

($\mathbf{C}$ being a covariance matrix), it makes sense to consider the ℓ_2-norm over $\mathbb{M}^*$,

$$\| \hat{\mathbf{m}} \| = \left(\hat{\mathbf{m}}^t \, \hat{\mathbf{C}}^{-1} \, \hat{\mathbf{m}} \right)^{1/2} \quad , \tag{6.265}$$

where

$$\hat{\mathbf{C}} = \mathbf{C}^{-1} \quad . \tag{6.266}$$

The vector of $\mathbb{M}$ associated with the gradient $\hat{\boldsymbol{\gamma}}$ by the basic duality between $\mathbb{M}^*$ and $\mathbb{M}$ is (see equation (3.18) in the main text)

$$\boldsymbol{\gamma} = \mathbf{C}\hat{\boldsymbol{\gamma}} \quad , \tag{6.267}$$

and the norms of $\hat{\boldsymbol{\gamma}}$ and $\boldsymbol{\gamma}$ are obviously related as

$$\| \boldsymbol{\gamma} \| = \| \hat{\boldsymbol{\gamma}} \| \quad . \tag{6.268}$$

In this situation, the vector $\boldsymbol{\gamma}$ in equation (6.267) is the steepest ascent vector associated with the gradient $\hat{\boldsymbol{\gamma}}$ in equation (6.258).

The demonstration is very similar for the ℓ_p- and ℓ_2-norm cases, but, as the ℓ_2-norm case is not a special case of the ℓ_p-norm case (because of the covariances considered in ℓ_2), let us make two separate demonstrations, starting with the ℓ_p case.

At a given point $\mathbf{m}_0$, we seek the direction $\mathbf{d}_0$ such that in the limit $\varepsilon \to 0$ we obtain the minimum of $\psi(\mathbf{m}_0 + \varepsilon\, \mathbf{d}_0)$. As

$$\psi(\mathbf{m}_0 + \varepsilon\, \mathbf{d}_0) \; = \; \psi(\mathbf{m}_0) + \varepsilon\, \langle\, \hat{\boldsymbol{\gamma}}_0\, ,\, \mathbf{d}_0\, \rangle + \cdots \quad , \tag{6.269}$$

where $\hat{\boldsymbol{\gamma}}_0$ is the gradient of $\psi(\mathbf{m})$ at point $\mathbf{m}_0$, we see that we can formulate the problem as follows:

$$\begin{cases} \text{find the } \mathbf{d}_0 \text{ that maximizes } \langle\, \hat{\boldsymbol{\gamma}}_0\, ,\, \mathbf{d}_0\, \rangle \\ \text{under the constraint } \| \mathbf{d}_0 \| \; = \; \text{const.} \; = \; k \quad . \end{cases} \tag{6.270}$$

Introducing the Lagrange parameters (see Appendix 6.29), this problem of constrained maximization can be transformed into a problem of unconstrained maximization:

$$\begin{cases} \text{find the } \mathbf{d}_0 \text{ and the } \lambda \text{ that maximize the function} \\ \Psi(\mathbf{d}_0, \lambda) \; = \; \langle\, \hat{\boldsymbol{\gamma}}_0\, ,\, \mathbf{d}_0\, \rangle - \frac{\lambda}{p} \left(\| \mathbf{d}_0 \|^p - K^p \right) \quad . \end{cases} \tag{6.271}$$

More explicitly,

$$\Psi(\mathbf{d}_0, \lambda) \; = \; \sum_\alpha (\hat{\boldsymbol{\gamma}}_0)_\alpha\, (\mathbf{d}_0)^\alpha - \frac{\lambda}{p} \left(\sum_\alpha \frac{|\, (\mathbf{d}_0)^\alpha\, |^p}{(\sigma^\alpha)^p} - K^p \right) \quad . \tag{6.272}$$

The condition $\partial\Psi/\partial\lambda = 0$ gives $\| \mathbf{d}_0 \| = K$, as it should. The condition $\partial\Psi/\partial(\mathbf{d}_0)^\alpha = 0$ gives

$$\frac{(\hat{\boldsymbol{\gamma}}_0)_\alpha}{\hat{\sigma}_\alpha} \; = \; \lambda \left(\frac{(\mathbf{d}_0)^\alpha}{\sigma^\alpha} \right)^{\{p-1\}} \quad , \tag{6.273}$$

where $\hat{\sigma}_\alpha = 1/\sigma^\alpha$ and where the symbol $u^{\{r\}}$ has been defined in equation (4.14). Using the relations (4.16) gives

$$\frac{(\mathbf{d}_0)^\alpha}{\sigma^\alpha} \; = \; \frac{1}{\lambda} \left(\frac{(\hat{\boldsymbol{\gamma}}_0)_\alpha}{\hat{\sigma}_\alpha} \right)^{\{q-1\}} \quad , \tag{6.274}$$

where $1/p + 1/q = 1$. The value of λ is so far arbitrary. Taking the value $\lambda = 1$ makes $\mathbf{d}_0$ the element dual to $\boldsymbol{\gamma}_0$ (equation (4.16)), and then (equation (4.12))

$$\| \mathbf{d}_0 \|^p \; = \; \| \hat{\boldsymbol{\gamma}}_0 \|^q \quad . \tag{6.275}$$

When passing from the ℓ_p to the ℓ_2 case, the unconstrained optimization problem in equation (6.271) becomes

$$\begin{cases} \text{find the } \mathbf{d}_0 \text{ and the } \lambda \text{ that maximize the function} \\ \Psi(\mathbf{d}_0, \lambda) \; = \; \langle\, \hat{\boldsymbol{\gamma}}_0\, ,\, \mathbf{d}_0\, \rangle - \frac{\lambda}{2} \left(\| \mathbf{d}_0 \|^2 - K^2 \right) \quad . \end{cases} \tag{6.276}$$

Here, explicitly,

$$\Psi(\mathbf{d}_0, \lambda) = \sum_\alpha (\hat{\boldsymbol{\gamma}}_0)_\alpha (\mathbf{d}_0)^\alpha - \frac{\lambda}{2} \left(\sum_\alpha \sum_\beta (\mathbf{d}_0)^\alpha (\mathbf{C}^{-1})_{\alpha\beta} (\mathbf{d}_0)^\beta - K^2 \right) . \quad (6.277)$$

The condition $\partial \Psi / \partial \lambda = 0$ gives $\| \mathbf{d}_0 \| = K$, as it should. The condition $\partial \Psi / \partial (\mathbf{d}_0)^\alpha = 0$ gives

$$\mathbf{d}_0 = \frac{1}{\lambda} \mathbf{C} \hat{\boldsymbol{\gamma}}_0 , \quad (6.278)$$

the value of λ being arbitrary. Taking the value $\lambda = 1$ makes $\mathbf{d}_0$ the element dual to $\boldsymbol{\gamma}_0$ (equation (3.18)), and then

$$\| \mathbf{d}_0 \| = \| \hat{\boldsymbol{\gamma}}_0 \| . \quad (6.279)$$

This concludes the demonstrations.

Norm of the Gradient

Let $\mathbf{m} \mapsto \psi(\mathbf{m})$ be a nonlinear form over a linear space $\mathbb{M}$ and $\hat{\boldsymbol{\gamma}}_0$ be the gradient of ψ at a point $\mathbf{m}_0$. By definition of gradient, for any $\delta \mathbf{m}$,

$$\psi(\mathbf{m}_0 + \delta \mathbf{m}) = \psi(\mathbf{m}_0) + \langle \hat{\boldsymbol{\gamma}}_0 , \delta \mathbf{m} \rangle + O(2) . \quad (6.280)$$

Let $\boldsymbol{\gamma}_0$ be the dual of $\hat{\boldsymbol{\gamma}}_0$ (we have seen that $\boldsymbol{\gamma}_0$ represents the steepest ascent vector for ψ at $\mathbf{m}_0$). From the equation above, it follows that

$$\psi\left(\mathbf{m}_0 + \frac{1}{\| \boldsymbol{\gamma}_0 \|} \boldsymbol{\gamma}_0 \right) = \psi(\mathbf{m}_0) + \frac{1}{\| \boldsymbol{\gamma}_0 \|} \langle \hat{\boldsymbol{\gamma}}_0 , \boldsymbol{\gamma}_0 \rangle + O(2) , \quad (6.281)$$

and, using the property $\langle \hat{\boldsymbol{\gamma}}_0 , \boldsymbol{\gamma}_0 \rangle = \| \hat{\boldsymbol{\gamma}}_0 \| \| \boldsymbol{\gamma}_0 \|$ (equation (4.12)),

$$\psi\left(\mathbf{m}_0 + \frac{1}{\| \boldsymbol{\gamma}_0 \|} \boldsymbol{\gamma}_0 \right) = \psi(\mathbf{m}_0) + \| \hat{\boldsymbol{\gamma}}_0 \| + O(2) . \quad (6.282)$$

This shows that the norm of $\boldsymbol{\gamma}_0$ represents the variation of the linear application tangent to $\psi(\mathbf{m})$ at $\mathbf{m}_0$ per unit norm-length of variation of $\mathbf{m}$ in the direction of steepest ascent, i.e., the norm of $\boldsymbol{\gamma}_0$ represents the *slope* of ψ at $\mathbf{m}_0$.

Curvature

Let $\mathbb{M}$ be a finite-dimensional linear space and $\mathbf{m} \mapsto \psi(\mathbf{m})$ be a nonlinear form over $\mathbb{M}$. We have seen that the gradient of ψ at a point $\mathbf{m}$ (which we have denoted $\hat{\boldsymbol{\gamma}}_0$) is a form over $\mathbb{M}$, i.e., it is an element of $\mathbb{M}^*$ dual to $\mathbb{M}$. When introducing a norm over $\mathbb{M}$, we have seen that we have a bijection between $\mathbb{M}$ and $\mathbb{M}^*$. Also, the element $\boldsymbol{\gamma}_0$ of $\mathbb{M}$ associated with the form $\hat{\boldsymbol{\gamma}}_0$ corresponds to the direction of ascent of ψ at point $\mathbf{m}_0$.

When using an ℓ_2-norm associated with a covariance operator $\mathbf{C}$ through the expression $\| \mathbf{m} \|^2 = \langle \mathbf{C}^{-1} \mathbf{m}, \mathbf{m} \rangle = \mathbf{m}^t \mathbf{C}^{-1} \mathbf{m}$, the relation between $\hat{\boldsymbol{\gamma}}_0$ and $\boldsymbol{\gamma}_0$ is $\boldsymbol{\gamma}_0 = \mathbf{C} \hat{\boldsymbol{\gamma}}_0$, so, in ℓ_2-norm problems, the terminology is as follows:

$$\hat{\boldsymbol{\gamma}}_0 = \left(\frac{\partial \psi}{\partial \mathbf{m}} \right)_{\mathbf{m}_0} \quad : \quad \text{gradient of } \psi \text{ at } \mathbf{m}_0 \quad , \tag{6.283}$$

$$\boldsymbol{\gamma}_0 = \mathbf{C} \hat{\boldsymbol{\gamma}}_0 \quad : \quad \text{steepest ascent vector for } \psi \text{ at } \mathbf{m}_0 \quad .$$

The Hessian of $\psi(\mathbf{m})$ at a given point $\mathbf{m}_0$ is a linear operator $\hat{\mathbf{H}}_0$ mapping $\mathbb{M}$ into $\mathbb{M}^*$ (see, for instance, the first of equations (6.249)). The *curvature* operator $\mathbf{H}_0$ is defined as

$$(\mathbf{H}_0)^{\alpha}{}_{\beta} = \left(\frac{\partial \gamma^{\alpha}}{\partial m^{\beta}} \right)_{\mathbf{m}_0} \quad , \tag{6.284}$$

which should be compared to that defining the Hessian:

$$(\hat{\mathbf{H}}_0)_{\alpha\beta} = \left(\frac{\partial \hat{\gamma}_{\alpha}}{\partial m^{\beta}} \right)_{\mathbf{m}_0} \tag{6.285}$$

(while the Hessian is the derivative of the gradient, the curvature is the derivative of the steepest ascent vector).

In least squares, where the relation between the primal space $\mathbb{M}$ and the dual space $\mathbb{M}^*$ is linear, it follows directly from $\boldsymbol{\gamma} = \mathbf{C} \hat{\boldsymbol{\gamma}}$ that the relation between curvature and Hessian is $\mathbf{H}_0 = \mathbf{C} \hat{\mathbf{H}}_0$. Therefore, in ℓ_2-norm problems, the terminology is as follows:

$$\hat{\mathbf{H}}_0 = \left(\frac{\partial^2 \psi}{\partial^2 \mathbf{m}} \right)_{\mathbf{m}_0} = \left(\frac{\partial \hat{\boldsymbol{\gamma}}}{\partial \mathbf{m}} \right)_{\mathbf{m}_0} \quad : \quad \text{Hessian of } \psi \text{ at } \mathbf{m}_0 \quad , \tag{6.286}$$

$$\mathbf{H}_0 = \mathbf{C} \hat{\mathbf{H}}_0 = \left(\frac{\partial \boldsymbol{\gamma}}{\partial \mathbf{m}} \right)_{\mathbf{m}_0} \quad : \quad \text{curvature of } \psi \text{ at } \mathbf{m}_0 \quad .$$

$\boldsymbol{\gamma}_0$ is sometimes abusively called the gradient, while $\mathbf{H}_0$ is sometimes called the Hessian. This terminology can be misleading because although these elements are isomorphic, they are by no means identical.

For instance, while in least-squares problems the Hessian operator is (approximately) given by (equation (6.257))

$$\hat{\mathbf{H}} \simeq \mathbf{C}_{\mathrm{M}}^{-1} + \mathbf{G}^t \mathbf{C}_{\mathrm{D}}^{-1} \mathbf{G} \quad , \tag{6.287}$$

the curvature operator is

$$\mathbf{H} = \mathbf{I} + \mathbf{C}_{\mathrm{M}} \mathbf{G}^t \mathbf{C}_{\mathrm{D}}^{-1} \mathbf{G} \quad . \tag{6.288}$$

Introducing the adjoint of $\mathbf{G}$, one has $\mathbf{H} = \mathbf{I} + \mathbf{G}^* \mathbf{G}$. Note that while the Hessian is symmetric, the curvature is self-adjoint.

6.22.2   Newton Method for ℓ_p-Norms

Let $\mathbb{M}$ be a linear space and $\psi(\mathbf{m})$ be a nonlinear form over $\mathbb{M}$. Introducing the gradient and Hessian of ψ, we can write the two series developments

$$\psi(\mathbf{m}) = \psi(\mathbf{m}_0) + \langle \hat{\boldsymbol{\gamma}}_0, \mathbf{m} - \mathbf{m}_0 \rangle + \tfrac{1}{2} \langle \hat{\mathbf{H}}_0 (\mathbf{m} - \mathbf{m}_0), \mathbf{m} - \mathbf{m}_0 \rangle + \cdots \quad ,$$

$$\hat{\boldsymbol{\gamma}}(\mathbf{m}) = \hat{\boldsymbol{\gamma}}_0 + \hat{\mathbf{H}}_0 (\mathbf{m} - \mathbf{m}_0) + \cdots \quad . \tag{6.289}$$

Locating the point $\mathbf{m}$ where ψ has an optimum (maximum or minimum) is equivalent to locating a point where the gradient $\hat{\boldsymbol{\gamma}}(\mathbf{m})$ vanishes. Imposing $\hat{\boldsymbol{\gamma}}(\mathbf{m}) = \mathbf{0}$ gives, using the second of equations (6.289),

$$\mathbf{m} \approx \mathbf{m}_0 - \hat{\mathbf{H}}_0^{-1} \hat{\boldsymbol{\gamma}}_0 \quad . \tag{6.290}$$

Should the function ψ be strictly quadratic, the Newton method would converge in only one iteration. If the function ψ is not too far from being quadratic, the Newton algorithm

$$\mathbf{m}_{n+1} \approx \mathbf{m}_n - \hat{\mathbf{H}}_n^{-1} \hat{\boldsymbol{\gamma}}_n \tag{6.291}$$

converges well.

But it converges well *only* when the function ψ is not too far from being quadratic. The typical functions encountered when working with inverse problems that are based on ℓ_p-norms are usually not close to being quadratic. A minor modification of the Newton algorithm provides a much better convergence.

To see this, examine a one-dimensional minimization problem where the function to be minimized is $S(x) = |x|^p$. Instead of trying the Newton algorithm $x_{n+1} = x_n - S_n'/S_n''$, one may try the algorithm $x_{n+1} = x_n - k\, S_n'/S_n''$, where k is a constant to be determined. An easy computation shows that the choice $k = p - 1$ makes this iterative algorithm converge in only one iteration, whatever the value of p is. Therefore, when dealing with ℓ_p-norm optimization problems, the right version of the Newton algorithm is

$$\boxed{\mathbf{m}_{n+1} \approx \mathbf{m}_n - (p-1)\,\hat{\mathbf{H}}_n^{-1}\,\hat{\boldsymbol{\gamma}}_n \quad .} \tag{6.292}$$

When $p = 2$, we obtain the standard version of the algorithm. In the two limit cases $p \to 1$ and $p \to \infty$, the algorithm becomes undetermined.

6.22.3 Minimization Along a Given Direction

Let us face now the problem of one-dimensional optimization. Although in inverse problems one is typically faced with the need to find the *minimum* of a misfit function $S(\mathbf{m})$, let us examine here the problem of finding the *maximum* of a function $\psi(\mathbf{m})$ (this simplifies language and notation, and the adaptations are trivial).

So, let us consider a finite-dimensional linear space $\mathbb{M}$ and a nonlinear form $\psi(\mathbf{m})$ defined over $\mathbb{M}$. We assume we are given a particular point $\mathbf{m}_0$ and a direction $\boldsymbol{\phi}_0$ at $\mathbf{m}_0$. This direction may be the steepest ascent direction for ψ, or any other direction. We wish to perform a jump along the direction defined by $\boldsymbol{\phi}_0$ (i.e., we wish to move from point $\mathbf{m}_0$ to a point $\mathbf{m} = \mathbf{m}_0 + \mu\,\boldsymbol{\phi}_0$) such that the new value of ψ takes its minimum value along that direction. In other words, which value should we give to the scalar μ in order to obtain the minimum value of $\psi(\mathbf{m}_0 + \mu\,\boldsymbol{\phi}_0)$?

Usual methods for obtaining adequate values for μ are, for instance, trial and error and parabolic interpolation (the value of ψ is computed for three values of μ, a parabola is fitted to these three points, and the value of μ giving the minimum of the parabola is chosen). These methods, although robust, need the computation of ψ at some points, and, for large-dimensional problems, this can be too expensive.

If the functional ψ is sufficiently well behaved, a useful approximation for μ can be obtained by considering not ψ but its second-order approximation around $\mathbf{m}_0$. By definition of gradient and Hessian, we can write

$$\psi(\mathbf{m}_0 + \mu\,\boldsymbol{\phi}_0) = \psi(\mathbf{m}_0) + \mu\,\langle\,\hat{\boldsymbol{\gamma}}_0\,,\,\boldsymbol{\phi}_0\,\rangle + \frac{\mu^2}{2}\,\langle\,\hat{\mathbf{H}}_0\,\boldsymbol{\phi}_0\,,\,\boldsymbol{\phi}_0\,\rangle + O(3) \quad . \qquad (6.293)$$

The condition $\partial\psi/\partial\mu = 0$ then gives

$$\mu \simeq \frac{\langle\,\hat{\boldsymbol{\gamma}}_0\,,\,\boldsymbol{\phi}_0\,\rangle}{\langle\,\hat{\mathbf{H}}_0\,\boldsymbol{\phi}_0\,,\,\boldsymbol{\phi}_0\,\rangle} \quad . \qquad (6.294)$$

This, in fact, results from an application of a one-dimensional version of the Newton algorithm. But we have seen above that when the function to be optimized is defined via the use of an ℓ_p-norm, it is better to use a modified version of the Newton algorithm. Then, a better value for the scalar μ is (see equation (6.292))

$$\mu \simeq (p-1)\frac{\langle\,\hat{\boldsymbol{\gamma}}_0\,,\,\boldsymbol{\phi}_0\,\rangle}{\langle\,\hat{\mathbf{H}}_0\,\boldsymbol{\phi}_0\,,\,\boldsymbol{\phi}_0\,\rangle} \quad . \qquad (6.295)$$

When working with ℓ_2-norms, $p - 1 = 1$, and the usual expression is found.

Obviously, this only gives a reasonable estimate of the step length to be used. One has to check that, with such a step length, the function ψ actually takes better values. If not, the value of μ has to be diminished until the condition is met.

6.22.4 Steepest Descent

Let $\mathbb{M}$ be the model space and $S(\mathbf{m})$ be the misfit function (whose minimum is sought). The model space is assumed to be a linear space and is assumed to be endowed with a norm $\mathbf{m} \mapsto \|\,\mathbf{m}\,\|$ (typically an ℓ_p-norm). The steepest descent algorithm corresponds to taking the steepest descent direction as the direction of search.

The gradient of the misfit function, an element of $\mathbb{M}^*$, is

$$\hat{\gamma}_\alpha(\mathbf{m}) = \frac{\partial S}{\partial m^\alpha}(\mathbf{m}) \quad , \qquad (6.296)$$

and, as we have seen, the steepest ascent vector is the vector $\boldsymbol{\gamma}$ of $\mathbb{M}$ that is related to $\hat{\boldsymbol{\gamma}}$ via the duality relation.

The steepest descent algorithm is, then,

$$\mathbf{m}_{n+1} = \mathbf{m}_n - \mu_n\,\boldsymbol{\gamma}_n \qquad (6.297)$$

(the minus sign is because we want a direction of descent). The right value of the scalar μ_n is to be obtained by linear search. As we have seen (equation (6.295)), for a smooth enough misfit function, a reasonable value may be

$$\mu_n \simeq (p-1)\frac{\langle\,\hat{\boldsymbol{\gamma}}_n\,,\,\boldsymbol{\gamma}_n\,\rangle}{\langle\,\hat{\mathbf{H}}_n\,\boldsymbol{\gamma}_n\,,\,\boldsymbol{\gamma}_n\,\rangle} \quad , \qquad (6.298)$$

where $\hat{\mathbf{H}}$ is the Hessian of the misfit function,

$$\hat{H}_{\alpha\beta}(\mathbf{m}) = \frac{\partial \hat{\gamma}_\alpha}{\partial m^\beta}(\mathbf{m}) = \frac{\partial^2 S}{\partial m^\alpha m^\beta}(\mathbf{m}) \quad , \tag{6.299}$$

and where the factor $(p-1)$ is there because it is assumed that the misfit function has been derived using an ℓ_p-norm.

Steepest Descent in ℓ_p-Norm Inverse Problems

In the ℓ_p-norm formulation of inverse problems, the misfit function is (equation (6.254))

$$S(\mathbf{m}) = \frac{1}{s} \sum_i \frac{|g^i(\mathbf{m}) - d^i_{\text{obs}}|^s}{(\sigma^i_D)^s} + \frac{1}{r} \sum_\alpha \frac{|m^\alpha - m^\alpha_{\text{prior}}|^r}{(\sigma^\alpha_M)^r} \quad , \tag{6.300}$$

where an ℓ_s-norm is used in the data space and an ℓ_r-norm is used in the model space. The gradient of S is (equation (6.255))

$$\hat{\gamma}_\alpha(\mathbf{m}) = \sum_i \frac{1}{\sigma^i_D} G^i_\alpha(\mathbf{m}) \left(\frac{g^i(\mathbf{m}) - d^i_{\text{obs}}}{\sigma^i_D} \right)^{\{s-1\}} + \sum_\alpha \frac{1}{\sigma^\alpha_M} \left(\frac{m^\alpha - m^\alpha_{\text{prior}}}{\sigma^\alpha_M} \right)^{\{r-1\}} \quad , \tag{6.301}$$

where the G^i_α are the partial derivatives $G^i_\alpha = \partial g^i / \partial m^\alpha$.

The steepest ascent vector is (see equation (6.262))

$$\gamma^\alpha(\mathbf{m}) = \sigma^\alpha \left(\sigma^\alpha \hat{\gamma}_\alpha(\mathbf{m}) \right)^{\{q-1\}} \quad , \tag{6.302}$$

where the parameter q is related to the parameter r (defining an ℓ_r-norm in the model space) via

$$\frac{1}{r} + \frac{1}{q} = 1 \quad . \tag{6.303}$$

The steepest descent algorithm is then

$$\mathbf{m}_{n+1} = \mathbf{m}_n - \mu_n \gamma(\mathbf{m}_n) \quad , \tag{6.304}$$

where the real number μ_n is to be obtained by linear search (or one may try to use the approximation proposed by equation (6.298)).

Steepest Descent in Least-Squares Inverse Problems

In the least-squares formulation of inverse problems, the misfit function S is given by (equation (6.251))

$$2 S(\mathbf{m}) = (\mathbf{g}(\mathbf{m}) - \mathbf{d}_{\text{obs}})^t \mathbf{C}_D^{-1} (\mathbf{g}(\mathbf{m}) - \mathbf{d}_{\text{obs}}) + (\mathbf{m} - \mathbf{m}_{\text{prior}})^t \mathbf{C}_M^{-1} (\mathbf{m} - \mathbf{m}_{\text{prior}}) \quad , \tag{6.305}$$

the gradient of S at a point $\mathbf{m}$ being (equation (6.252))

$$\hat{\gamma}(\mathbf{m}) \;=\; \mathbf{G}^t(\mathbf{m})\, \mathbf{C}_\mathrm{D}^{-1}\, (\mathbf{g}(\mathbf{m}) - \mathbf{d}_\mathrm{obs}) + \mathbf{C}_\mathrm{M}^{-1}\, (\mathbf{m} - \mathbf{m}_\mathrm{prior}) \quad , \tag{6.306}$$

where $\mathbf{G}$ stands for the matrix of partial derivatives $G^i{}_\alpha = \partial g^i / \partial m^\alpha$.

In this ℓ_2-norm formulation, the steepest ascent vector, $\boldsymbol{\gamma}$, is related to the gradient $\hat{\boldsymbol{\gamma}}$ via (equation (6.262)) $\boldsymbol{\gamma} = \mathbf{C}_\mathrm{M}\,\hat{\boldsymbol{\gamma}}$. Therefore,

$$\boldsymbol{\gamma}(\mathbf{m}) \;=\; \mathbf{C}_\mathrm{M}\, \mathbf{G}^t(\mathbf{m})\, \mathbf{C}_\mathrm{D}^{-1}\, (\mathbf{g}(\mathbf{m}) - \mathbf{d}_\mathrm{obs}) + (\mathbf{m} - \mathbf{m}_\mathrm{prior}) \quad . \tag{6.307}$$

The steepest descent algorithm $\mathbf{m}_{n+1} = \mathbf{m}_n - \mu_n\, \boldsymbol{\gamma}(\mathbf{m}_n)$ here becomes

$$\mathbf{m}_{n+1} \;=\; \mathbf{m}_n - \mu_n \left(\mathbf{C}_\mathrm{M}\, \mathbf{G}_n^t\, \mathbf{C}_\mathrm{D}^{-1}\, (\mathbf{g}_n - \mathbf{d}_\mathrm{obs}) + (\mathbf{m}_n - \mathbf{m}_\mathrm{prior}) \right) \quad , \tag{6.308}$$

where, for short, the notation $\mathbf{g}_n = \mathbf{g}(\mathbf{m}_n)$ and $\mathbf{G}_n = \mathbf{G}(\mathbf{m}_n)$ has been used.

The real number μ_n is to be obtained by linear search. Using, instead, the approximation proposed in equation (6.298), one has (at $p = 2$) $\mu_n \simeq \langle \hat{\boldsymbol{\gamma}}_n\,,\, \boldsymbol{\gamma}_n \rangle / \langle \hat{\mathbf{H}}_n\, \boldsymbol{\gamma}_n\,,\, \boldsymbol{\gamma}_n \rangle$. These duality products being in the model space, one has $\langle\, \mathbf{a}\,,\,\mathbf{b}\,\rangle = \mathbf{a}^t\, \mathbf{C}_\mathrm{M}^{-1}\, \mathbf{b}$. Using the approximation for the Hessian proposed in equation (6.257), one finally obtains

$$\mu_n \;\simeq\; \frac{\boldsymbol{\gamma}_n^t\, \mathbf{C}_\mathrm{M}^{-1}\, \boldsymbol{\gamma}_n}{\boldsymbol{\gamma}_n^t\, \mathbf{C}_\mathrm{M}^{-1}\, \boldsymbol{\gamma}_n + \mathbf{b}_n^t\, \mathbf{C}_\mathrm{D}^{-1}\, \mathbf{b}_n} \quad , \qquad \text{where} \qquad \mathbf{b}_n \;=\; \mathbf{G}_n\, \boldsymbol{\gamma}_n \quad . \tag{6.309}$$

6.22.5   Preconditioned Steepest Descent

The steepest descent direction is an optimal *local* descent direction. As the property is local, one should only use the direction of steepest descent if one wants to perform infinitesimally small jumps. In practical algorithms of optimization, one wishes, on the contrary, to perform as large jumps as possible, to achieve convergence in as small a number of iterations as possible. Instead of using the direction of steepest descent, one could, for instance, use the direction that for a given (finite) size of the jump gives the smallest value of the misfit function.

Instead, what one usually does is to compute the direction of steepest ascent (as accurately as possible), then to use physical common sense to modify this direction at will. It is impossible to give any rule valid for all inverse problems. For instance, in the problem of X-ray tomography, the direction of steepest descent tends to give too much weight to perturbations in the model that are close to the source. Then one may (brutally) damp these perturbations in the steepest descent vector. Doing this, one passes from the steepest descent vector $\boldsymbol{\gamma}_n$ to another direction (hopefully better for finite jumps)

$$\boldsymbol{\phi}_n \;=\; \mathbf{f}(\boldsymbol{\gamma}_n) \quad , \tag{6.310}$$

and one says that the steepest descent direction has been *preconditioned*. The function $\mathbf{f}(\cdot)$ can be any nonlinear function, with the sole requirement that the direction it generates is still a direction of descent for the misfit S.

Sometimes, one only considers linear preconditioning operators,[87] in which case

$$\boldsymbol{\phi}_n = \mathbf{F}_n \, \boldsymbol{\gamma}_n \quad , \tag{6.311}$$

where the linear operator may depend on $\mathbf{m}_n$ (hence the index in $\mathbf{F}_n$).

As there is not much to be added for the ℓ_p-norm inverse problem, let us directly analyze the least-squares problem.

Whatever the preconditioning one chooses to do (linear or nonlinear), one starts by computing the steepest ascent vector (equation (6.307))

$$\boldsymbol{\gamma}_n = \mathbf{C}_\mathrm{M} \, \mathbf{G}_n^t \, \mathbf{C}_\mathrm{D}^{-1} \, (\mathbf{g}_n - \mathbf{d}_\mathrm{obs}) + (\mathbf{m} - \mathbf{m}_\mathrm{prior}) \tag{6.312}$$

and obtains from it a (preconditioned) direction of search $\boldsymbol{\phi}_n = \mathbf{f}(\boldsymbol{\gamma}_n)$, and the preconditioned steepest descent algorithm is

$$\mathbf{m}_{n+1} = \mathbf{m}_n - \mu_n \, \boldsymbol{\phi}_n \quad , \tag{6.313}$$

where the real number μ_n is to be obtained by linear search. Using the approximation suggested in equation (6.295), one obtains

$$\mu_n \simeq \frac{\boldsymbol{\gamma}_n^t \, \mathbf{C}_\mathrm{M}^{-1} \, \boldsymbol{\phi}_n}{\boldsymbol{\phi}_n^t \, \mathbf{C}_\mathrm{M}^{-1} \, \boldsymbol{\phi}_n + \mathbf{b}_n^t \, \mathbf{C}_\mathrm{D}^{-1} \, \mathbf{b}_n} \quad , \qquad \text{where} \qquad \mathbf{b}_n = \mathbf{G}_n \, \boldsymbol{\phi}_n \quad . \tag{6.314}$$

When limiting the choice to linear preconditioning operators, $\boldsymbol{\phi}_n = \mathbf{F}_n \, \boldsymbol{\gamma}_n$, it is good to have in mind the quasi-Newton algorithm (see section 6.22.6 below), which can be written $\mathbf{m}_{n+1} = \mathbf{m}_n - (\mathbf{I} + \mathbf{C}_\mathrm{M} \, \mathbf{G}_n^t \, \mathbf{C}_\mathrm{D}^{-1} \, \mathbf{G}_n)^{-1} \, \boldsymbol{\gamma}_n$. This suggests using as linear preconditioning operator any operator $\mathbf{F}_n$ that can be an approximation of the operator appearing in the quasi-Newton algorithm, i.e.,

$$\mathbf{F}_n \approx \left(\mathbf{I} + \mathbf{C}_\mathrm{M} \, \mathbf{G}_n^t \, \mathbf{C}_\mathrm{D}^{-1} \, \mathbf{G}_n \right)^{-1} \quad . \tag{6.315}$$

Even very crude approximations may work well.

In Algorithm (6.313), the starting point $\mathbf{m}_0$ is arbitrary, the simplest choice being $\mathbf{m}_0 = \mathbf{m}_\mathrm{prior}$. The use of different starting points may help to check the existence of secondary minima.

6.22.6 Quasi-Newton Method

To obtain the quasi-Newton algorithm for the iterative resolution of the least-squares inverse problem, we just need to collect here some of the expressions written above.

The misfit function S is given by (equation (6.251))

$$2 \, S(\mathbf{m}) = (\mathbf{g}(\mathbf{m}) - \mathbf{d}_\mathrm{obs})^t \, \mathbf{C}_\mathrm{D}^{-1} \, (\mathbf{g}(\mathbf{m}) - \mathbf{d}_\mathrm{obs}) + (\mathbf{m} - \mathbf{m}_\mathrm{prior})^t \, \mathbf{C}_\mathrm{M}^{-1} \, (\mathbf{m} - \mathbf{m}_\mathrm{prior}) \quad , \tag{6.316}$$

---

[87]The name *preconditioning operator* takes its source in the theory of the resolution of linear systems: letting $\mathbf{A}\mathbf{x} = \mathbf{y}$ represent a system to be solved, and $\mathbf{F} \simeq \mathbf{A}^{-1}$, the system $\mathbf{F}\mathbf{A}\mathbf{x} = \mathbf{F}\mathbf{y}$ is equivalent to the first, but if $\mathbf{F}$ is astutely chosen, it has a lower *condition number* (see Problem 7.8), so its numerical resolution is more stable: the system has been preconditioned.

its gradient is (equation (6.252))

$$\hat{\gamma}(\mathbf{m}) \;=\; \mathbf{G}^t(\mathbf{m})\,\mathbf{C}_{\mathrm{D}}^{-1}\,(\mathbf{g}(\mathbf{m}) - \mathbf{d}_{\mathrm{obs}}) + \mathbf{C}_{\mathrm{M}}^{-1}\,(\mathbf{m} - \mathbf{m}_{\mathrm{prior}})\quad, \tag{6.317}$$

and the (usual approximation of the) Hessian is (equation (6.257))

$$\hat{\mathbf{H}}(\mathbf{m}) \;\simeq\; \mathbf{C}_{\mathrm{M}}^{-1} + \mathbf{G}^t(\mathbf{m})\,\mathbf{C}_{\mathrm{D}}^{-1}\,\mathbf{G}(\mathbf{m})\quad. \tag{6.318}$$

In these equations, $G^i{}_\alpha = \partial g^i/\partial m^\alpha$.

The Newton algorithm (equation (6.291)) then gives[88]

$$\mathbf{m}_{n+1} \;=\; \mathbf{m}_n - \nu_n\left(\mathbf{C}_{\mathrm{M}}^{-1} + \mathbf{G}_n^t\,\mathbf{C}_{\mathrm{D}}^{-1}\,\mathbf{G}_n\right)^{-1}\left(\mathbf{G}_n^t\,\mathbf{C}_{\mathrm{D}}^{-1}\,(\mathbf{g}_n - \mathbf{d}_{\mathrm{obs}}) + \mathbf{C}_{\mathrm{M}}^{-1}\,(\mathbf{m}_n - \mathbf{m}_{\mathrm{prior}})\right)\quad, \tag{6.319}$$

where $\mathbf{G}_n = \mathbf{G}(\mathbf{m}_n)$, $\mathbf{g}_n = \mathbf{g}(\mathbf{m}_n)$, and ν_n is a real number to be obtained by linear search (but that, in most common situations, is of the order of one).

This expression uses the gradient and the Hessian of the misfit. An equivalent algorithm is obtained when using the steepest ascent vector and the curvature:

$$\mathbf{m}_{n+1} \;=\; \mathbf{m}_n - \nu_n\left(\mathbf{I} + \mathbf{G}_n^*\,\mathbf{G}_n\right)^{-1}\left(\mathbf{G}_n^*\,(\mathbf{g}_n - \mathbf{d}_{\mathrm{obs}}) + (\mathbf{m}_n - \mathbf{m}_{\mathrm{prior}})\right)\quad, \tag{6.320}$$

where $\mathbf{G}_n^*$, the adjoint of $\mathbf{G}_n$, is (see section 3.1.4)

$$\mathbf{G}_n^* \;=\; \mathbf{C}_{\mathrm{M}}\,\mathbf{G}_n^t\,\mathbf{C}_{\mathrm{D}}^{-1}\quad. \tag{6.321}$$

As usual, the starting point $\mathbf{m}_0$ is arbitrary, the simplest choice is $\mathbf{m}_0 = \mathbf{m}_{\mathrm{prior}}$, but the use of different starting points helps to check the existence of secondary minima.

6.22.7 Conjugate Directions

The conjugate directions method for the maximization of a function Σ is based on the following idea: Let $\mathbf{m}_0$ be the starting point and $\boldsymbol{\gamma}_0$ be the steepest ascent vector at $\mathbf{m}_0$. The point $\mathbf{m}_1$ is defined as in the steepest ascent method. Let $\boldsymbol{\gamma}_1$ be the steepest ascent vector at $\mathbf{m}_1$. The point $\mathbf{m}_2$ is not defined as the point maximizing Σ in the direction given by $\boldsymbol{\gamma}_1$, but as the point maximizing Σ *in the subspace generated by $\boldsymbol{\gamma}_0$ and $\boldsymbol{\gamma}_1$.* The point $\mathbf{m}_3$ is defined as the point maximizing Σ in the subspace generated by $\boldsymbol{\gamma}_0$, $\boldsymbol{\gamma}_1$, and $\boldsymbol{\gamma}_3$, and so on until convergence. To precondition the conjugate directions method means, as above, to replace the $\boldsymbol{\gamma}_n$ with some other direction of ascent (hopefully better for the finite jumps intended).

It can be shown (see, for instance, Fletcher, 1980) that this method generally converges at the same rate as the variable metric method (it has quadratic convergence for linear problems). The miracle is that the computations needed to perform this method are not more difficult than those needed in using the steepest ascent method.

For least-squares problems, the choice between conjugate directions and variable metric has to be made by considering that the first needs less computer memory, but the second gives a direct approximation to the posterior covariance operator.

---

[88]These formulas were first derived by Rodgers (1976) and rediscovered by Tarantola and Valette (1982b).

Conjugate Directions in Normed Spaces

Here, we wish to minimize the misfit function $S(\mathbf{m})$ defined in equation (6.300). The gradient of $S(\mathbf{m})$ (given in equation (6.301)) is denoted by $\hat{\boldsymbol{\gamma}}(\mathbf{m})$ and the Hessian is denoted by $\hat{\mathbf{H}}(\mathbf{m})$. The steepest ascent vector $\boldsymbol{\gamma}(\mathbf{m})$ (given in equation (6.302)) is the vector dual to the gradient.

The general algorithm for the method of conjugate directions is (see Céa, 1971; Walsh, 1975; Fletcher, 1980; Powell, 1981; Ciarlet, 1982; or Scales, 1985)

$$\boldsymbol{\lambda}_n = \mathbf{F}_0 \boldsymbol{\gamma}_n \quad ,$$

$$\boldsymbol{\phi}_n = \boldsymbol{\lambda}_n + \alpha_n \boldsymbol{\phi}_{n-1} \qquad (\alpha_n \text{ defined below}) \quad , \tag{6.322}$$

$$\mathbf{m}_{n+1} = \mathbf{m}_n - \mu_n \boldsymbol{\phi}_n \qquad (\text{obtain } \mu_n \text{ by linear search}) \ .$$

The algorithm is initialized at an arbitrary point $\mathbf{m}_0$, $\mathbf{F}_0$ is an arbitrary preconditioning operator, and the vector $\boldsymbol{\phi}_n$ is initialized at $\boldsymbol{\phi}_0 = \boldsymbol{\lambda}_0$.

Different expressions can be obtained for α_n:

$$\alpha_n = \frac{\langle \hat{\boldsymbol{\gamma}}_n , \boldsymbol{\lambda}_n \rangle}{\langle \hat{\boldsymbol{\gamma}}_{n-1} , \boldsymbol{\lambda}_{n-1} \rangle} \qquad (\text{Fletcher and Reeves, 1964}) \quad ,$$

$$\alpha_n = \frac{\langle \hat{\boldsymbol{\gamma}}_n - \hat{\boldsymbol{\gamma}}_{n-1} , \boldsymbol{\lambda}_n \rangle}{\langle \hat{\boldsymbol{\gamma}}_{n-1} , \boldsymbol{\lambda}_{n-1} \rangle} \qquad (\text{Polak and Ribière, 1969}) \quad , \tag{6.323}$$

$$\alpha_n = \frac{\langle \hat{\boldsymbol{\gamma}}_n - \hat{\boldsymbol{\gamma}}_{n-1} , \boldsymbol{\lambda}_n \rangle}{\langle \hat{\boldsymbol{\gamma}}_n - \hat{\boldsymbol{\gamma}}_{n-1} , \boldsymbol{\phi}_{n-1} \rangle} \qquad (\text{Hestenes and Stiefel, 1952}) \quad .$$

Although these expressions are perfectly equivalent for quadratic functions, they are not for general problems. The Fletcher–Reeves formula is still probably the most widely used. Nevertheless, Powell (1977) has suggested that in some situations, the Polak–Ribière formula may give superior results.[89]

For quadratic problems, it can be shown that if N iterations are effectively performed, the actual minimum is attained, where N is the dimension of the model space $\mathbb{M}$.

It remains to fix a value for the real numbers μ_n. Ideally, one should use a linear search. If not, using the approximation given in equation (6.295) gives

$$\mu_n \simeq (p - 1) \frac{\langle \hat{\boldsymbol{\gamma}}_n , \boldsymbol{\phi}_n \rangle}{\langle \hat{\mathbf{H}}_n \boldsymbol{\phi}_n , \boldsymbol{\phi}_n \rangle} \quad . \tag{6.324}$$

Conjugate Directions for Least Squares

The formulas for least-squares problems are very similar to those just written, except that the misfit function is now (equation (6.305))

$$2 S(\mathbf{m}) = (\mathbf{g}(\mathbf{m}) - \mathbf{d}_{\text{obs}})^t \, \mathbf{C}_{\text{D}}^{-1} \, (\mathbf{g}(\mathbf{m}) - \mathbf{d}_{\text{obs}}) + (\mathbf{m} - \mathbf{m}_{\text{prior}})^t \, \mathbf{C}_{\text{M}}^{-1} \, (\mathbf{m} - \mathbf{m}_{\text{prior}}) \quad , \tag{6.325}$$

---

[89]For instance, in nonquadratic minimizations, it may happen that $\boldsymbol{\phi}_n$ becomes almost orthogonal to the steepest descent vector $\boldsymbol{\lambda}_n$. In that case, $\mathbf{m}_{n+1} \simeq \mathbf{m}_n$, $\hat{\boldsymbol{\gamma}}_{n+1} \simeq \hat{\boldsymbol{\gamma}}_n$, and $\boldsymbol{\lambda}_{n+1} \simeq \boldsymbol{\lambda}_n$. The Fletcher–Reeves formula then gives $\alpha_{n+1} \simeq 1$ and $\boldsymbol{\phi}_{n+1} \simeq \boldsymbol{\lambda}_{n+1} + \boldsymbol{\phi}_n$, while the Polak–Ribière formula gives $\alpha_n \simeq 0$ and $\boldsymbol{\phi}_{n+1} \simeq \boldsymbol{\lambda}_{n+1}$. This shows that in critical situations, when a small advance can be made, the Polak–Ribière method is more robust because it has a tendency to take the steepest descent direction as a direction of search.

the gradient is (equation (6.306))

$$\hat{\boldsymbol{\gamma}}(\mathbf{m}) \;=\; \mathbf{G}^t(\mathbf{m})\,\mathbf{C}_\mathrm{D}^{-1}\,(\mathbf{g}(\mathbf{m}) - \mathbf{d}_\mathrm{obs}) + \mathbf{C}_\mathrm{M}^{-1}\,(\mathbf{m} - \mathbf{m}_\mathrm{prior}) \quad, \tag{6.326}$$

and the (approximate) Hessian is (equation (6.257))

$$\hat{\mathbf{H}}_0 \;\simeq\; \mathbf{G}_0^t\,\mathbf{C}_\mathrm{D}^{-1}\,\mathbf{G}_0 + \mathbf{C}_\mathrm{M}^{-1} \quad. \tag{6.327}$$

Finally, the steepest ascent vector is (equation (6.307))

$$\boldsymbol{\gamma}(\mathbf{m}) \;=\; \mathbf{C}_\mathrm{M}\,\mathbf{G}^t(\mathbf{m})\,\mathbf{C}_\mathrm{D}^{-1}\,(\mathbf{g}(\mathbf{m}) - \mathbf{d}_\mathrm{obs}) + (\mathbf{m} - \mathbf{m}_\mathrm{prior}) \quad. \tag{6.328}$$

This gives the algorithm

$$\boldsymbol{\gamma}_n \;=\; \mathbf{C}_\mathrm{M}\,\mathbf{G}_n^t\,\mathbf{C}_\mathrm{D}^{-1}\,(\mathbf{g}_n - \mathbf{d}_\mathrm{obs}) + (\mathbf{m}_n - \mathbf{m}_\mathrm{prior}) \quad,$$

$$\boldsymbol{\lambda}_n \;=\; \mathbf{F}_0\,\boldsymbol{\gamma}_n \quad,$$

$$\boldsymbol{\phi}_n \;=\; \boldsymbol{\lambda}_n + \alpha_n\,\boldsymbol{\phi}_{n-1} \qquad \big(\alpha_n \text{ defined below}\big) \quad, \tag{6.329}$$

$$\mathbf{m}_{n+1} \;=\; \mathbf{m}_n - \mu_n\,\boldsymbol{\phi}_n \qquad (\text{obtain } \mu_n \text{ by linear search}) \quad.$$

The starting point $\mathbf{m}_0$ is arbitrary, the simplest choice being $\mathbf{m}_0 = \mathbf{m}_\mathrm{prior}$, although the use of different starting points helps to check the existence of secondary solutions.

The simplest choice for the preconditioning operator $\mathbf{F}_0$ is $\mathbf{F}_0 = \mathbf{I}$. Usually, some approximation of the initial curvature

$$\mathbf{F}_0 \;\simeq\; \big(\mathbf{I} + \mathbf{C}_\mathrm{M}\,\mathbf{G}_0^t\,\mathbf{C}_\mathrm{D}^{-1}\,\mathbf{G}_0\big)^{-1} \tag{6.330}$$

gives good results. The vector $\boldsymbol{\phi}_n$ is initialized at $\boldsymbol{\phi}_0 = \boldsymbol{\lambda}_0$.

Using, for instance, the Polak–Ribière formula gives

$$\alpha_n \;=\; \frac{\omega_n - \boldsymbol{\gamma}_{n-1}{}^t\,\mathbf{C}_\mathrm{M}^{-1}\,\boldsymbol{\lambda}_n}{\omega_{n-1}} \quad, \tag{6.331}$$

where

$$\omega_n \;=\; \boldsymbol{\gamma}_n{}^t\,\mathbf{C}_\mathrm{M}^{-1}\,\boldsymbol{\lambda}_n \quad. \tag{6.332}$$

The value μ_n has to be obtained by linear search. Alternatively, a linearization of $\mathbf{g}(\mathbf{m})$ around $\mathbf{g}(\mathbf{m}_n)$ gives

$$\mu_n \;\simeq\; \frac{\boldsymbol{\gamma}_n{}^t\,\mathbf{C}_\mathrm{M}^{-1}\,\boldsymbol{\phi}_n}{\boldsymbol{\phi}_n{}^t\,\mathbf{C}_\mathrm{M}^{-1}\,\boldsymbol{\phi}_n + \mathbf{b}_n{}^t\,\mathbf{C}_\mathrm{D}^{-1}\,\mathbf{b}_n} \quad, \qquad \text{where} \qquad \mathbf{b}_n \;=\; \mathbf{G}_n\,\boldsymbol{\phi}_n \quad. \tag{6.333}$$

Notice that the numerator can be written

$$\boldsymbol{\gamma}_n{}^t\,\mathbf{C}_\mathrm{M}^{-1}\,\boldsymbol{\phi}_n \;=\; \boldsymbol{\gamma}_n{}^{-1}\,\mathbf{C}_\mathrm{M}^{-1}\,\boldsymbol{\lambda}_n - \alpha_n\,\boldsymbol{\gamma}_n{}^t\,\mathbf{C}_\mathrm{M}^{-1}\,\boldsymbol{\phi}_{n-1} \quad. \tag{6.334}$$

If the linear searches are accurate, the steepest ascent vector $\boldsymbol{\gamma}_n$ is approximately orthogonal to the previous search direction $\boldsymbol{\phi}_{n-1}$,

$$\boldsymbol{\gamma}_n{}^t\,\mathbf{C}_\mathrm{M}^{-1}\,\boldsymbol{\phi}_{n-1} \;\simeq\; 0 \quad, \tag{6.335}$$

and the following simplification can be used:

$$\mu_n \;\simeq\; \frac{\omega_n}{\boldsymbol{\phi}_n{}^t\,\mathbf{C}_\mathrm{M}^{-1}\,\boldsymbol{\phi}_n + \mathbf{b}_n{}^t\,\mathbf{C}_\mathrm{D}^{-1}\,\mathbf{b}_n} \quad, \qquad \text{where} \qquad \mathbf{b}_n \;=\; \mathbf{G}_n\,\boldsymbol{\phi}_n \quad. \tag{6.336}$$

6.22.8 Variable Metric Methods

Variable Metric in General

Let $\mathbb{M}$ be a finite-dimensional linear space, $S(\mathbf{m})$ be the real function to be minimized, $\hat{\boldsymbol{\gamma}}(\mathbf{m})$ be the gradient of S, and $\hat{\mathbf{H}}(\mathbf{m})$ be the Hessian of S.

The central idea of variable metric methods is to allow the preconditioning operator (discussed in the previous sections) to vary from iteration to iteration and to find an updating formula

$$\hat{\mathbf{F}}_{n+1} = \hat{\mathbf{F}}_n + \delta\hat{\mathbf{F}}_n \tag{6.337}$$

such that, as the iterations proceed, the preconditioning operator tends to the inverse Hessian,

$$\hat{\mathbf{F}}_n \to \hat{\mathbf{H}}_n^{-1} \quad , \tag{6.338}$$

so that a variable metric method will start behaving like a steepest descent method, but will finish behaving like a Newton method (with its rapid termination). The name *variable metric* makes sense, as the inverse Hessian can be interpreted as a metric over the space $\mathbb{M}$.

In the previous sections, the preconditioning operator was denoted by $\mathbf{F}$. If I denote it here with a hat, $\hat{\mathbf{F}}$, it is because the preconditioning operator was assumed above to be of the same type as the inverse of the curvature operator, while here it is assumed to be of the same type as the inverse of the Hessian operator (to fix ideas, in the context of least squares, where Hessian and curvature are given by the two equations (6.287)–(6.288), one has $\hat{\mathbf{F}} = \mathbf{F}\, \mathbf{C}_{\mathbf{M}}$).

Let $\hat{\mathbf{F}}_0$ be an arbitrary symmetric positive definite operator, hopefully a good approximation of the inverse of the initial Hessian

$$\hat{\mathbf{F}}_0 \simeq \hat{\mathbf{H}}_0^{-1} \quad . \tag{6.339}$$

The general structure of a variable metric method is (see Céa, 1971; Walsh, 1975; Fletcher, 1980; Powell, 1981; Ciarlet, 1982; Scales, 1985)

$$
\begin{aligned}
\boldsymbol{\phi}_n &= \hat{\mathbf{F}}_n\, \hat{\boldsymbol{\gamma}}_n \quad , \\
\mathbf{m}_{n+1} &= \mathbf{m}_n - \mu_n\, \boldsymbol{\phi}_n \qquad \text{(obtain } \mu_n \text{ by linear search)} \quad , \\
\hat{\mathbf{F}}_{n+1} &= \hat{\mathbf{F}}_n + \delta\hat{\mathbf{F}}_n \qquad (\delta\hat{\mathbf{F}}_n \text{ defined below}) \quad .
\end{aligned}
\tag{6.340}
$$

In numerical applications, the kernels of the operators $\hat{\mathbf{F}}_1, \hat{\mathbf{F}}_2, \dots$ do not need to be explicitly computed. All that is needed is the possibility of computing the result of the action of the operators on arbitrary vectors. In the following, $\hat{\mathbf{F}}_n(\cdot)$ represents the result of the application of $\hat{\mathbf{F}}_n$ over a generic vector represented by "$\cdot$." Also, the following notations are used:

$$
\begin{aligned}
\delta\mathbf{m}_n &= \mathbf{m}_{n+1} - \mathbf{m}_n \quad , \\
\delta\hat{\boldsymbol{\gamma}}_n &= \hat{\boldsymbol{\gamma}}_{n+1} - \hat{\boldsymbol{\gamma}}_n \quad , \\
\mathbf{v}_n &= \hat{\mathbf{F}}_n\, \delta\hat{\boldsymbol{\gamma}}_n \quad , \\
\mathbf{u}_n &= \delta\mathbf{m}_n - \mathbf{v}_n \quad .
\end{aligned}
\tag{6.341}
$$

Many different formulas exist for the updating of $\hat{\mathbf{F}}_n$, for instance, the symmetric rank-one formula (due to Davidon, 1959)

$$\hat{\mathbf{F}}_{n+1}(\cdot) = \hat{\mathbf{F}}_n(\cdot) + \frac{\langle \, \cdot \, , \, \mathbf{u}_n \, \rangle}{\langle \, \delta\hat{\boldsymbol{\gamma}}_n \, , \, \mathbf{u}_n \, \rangle} \mathbf{u}_n \quad , \tag{6.342}$$

the DFP formula (due to Davidon, 1959, and to Fletcher and Powell, 1963)

$$\hat{\mathbf{F}}_{n+1}(\cdot) = \hat{\mathbf{F}}_n(\cdot) + \frac{\langle \, \cdot \, , \, \delta\mathbf{m}_n \, \rangle}{\langle \, \delta\hat{\boldsymbol{\gamma}}_n \, , \, \delta\mathbf{m}_n \, \rangle} \delta\mathbf{m}_n - \frac{\langle \, \cdot \, , \, \mathbf{v}_n \, \rangle}{\langle \, \delta\hat{\boldsymbol{\gamma}}_n \, , \, \mathbf{v}_n \, \rangle} \mathbf{v}_n \quad , \tag{6.343}$$

and the BFGS formula (due to Broyden, 1967, Fletcher, 1980, Goldfarb, 1976, and Shannon, 1948)

$$\hat{\mathbf{F}}_{n+1}(\cdot) = \hat{\mathbf{F}}_n(\cdot) + \frac{\beta_n \langle \, \cdot \, , \, \delta\mathbf{m}_n \, \rangle \delta\mathbf{m}_n - \langle \, \delta\hat{\boldsymbol{\gamma}}_n \, , \, \hat{\mathbf{F}}_n(\cdot) \, \rangle \delta\mathbf{m}_n - \langle \, \cdot \, , \, \delta\mathbf{m}_n \, \rangle \mathbf{v}_n}{\langle \, \delta\hat{\boldsymbol{\gamma}}_n \, , \, \delta\mathbf{m}_n \, \rangle} \, , \tag{6.344}$$

where

$$\beta_n = 1 + \frac{\langle \, \delta\hat{\boldsymbol{\gamma}}_n \, , \, \mathbf{v}_n \, \rangle}{\langle \, \delta\hat{\boldsymbol{\gamma}}_n \, , \, \delta\mathbf{m}_n \, \rangle} \quad . \tag{6.345}$$

Although it is easy to see that all the operators $\hat{\mathbf{F}}_n$ thus defined are positive definite, they may become numerically singular. It seems that the BFGS formula has a greater tendency to keep the definiteness of $\hat{\mathbf{F}}_n$. Maybe this is the reason why it is today the most widely used updating formula.

Let N denote the dimension of the space $\mathbb{M}$. When using these formulas for quadratic problems, it can be shown that if N iterations are effectively performed, the actual minimum is attained, and

$$\hat{\mathbf{F}}_N = \hat{\mathbf{H}}^{-1} \quad , \tag{6.346}$$

where $\hat{\mathbf{H}}$ is the (constant) Hessian of $\hat{\mathbf{F}}$. For large-dimensional problems, good approximations of the minimum are sometimes obtained after a few iterations. Of course, the better $\hat{\mathbf{F}}_0$ approximates $\hat{\mathbf{H}}_0^{-1}$, the better the convergence is in general.

As previously indicated, to operate with $\hat{\mathbf{F}}_n$, we do not need to build the kernel of the operator, we only need to store in the computer's memory some vectors and scalars. For instance, letting $\hat{\boldsymbol{\chi}}$ be an arbitrary vector, we have, for the rank-one formula,

$$\hat{\mathbf{F}}_n \hat{\boldsymbol{\chi}} = \hat{\mathbf{F}}_0 \hat{\boldsymbol{\chi}} + \sum_{k=0}^{n-1} \frac{\langle \, \hat{\boldsymbol{\chi}} \, , \, \mathbf{u}_k \, \rangle}{v_k} \mathbf{u}_k \quad , \tag{6.347}$$

where v_k are the real numbers

$$v_k = \langle \, \delta\hat{\boldsymbol{\gamma}}_k \, , \, \mathbf{u}_k \, \rangle \quad , \tag{6.348}$$

which shows that, in order to operate with $\hat{\mathbf{F}}_n$, we only have to store in the computer's memory the vectors $\mathbf{u}_0, \dots, \mathbf{u}_n$ and the scalars $v_0, \dots, v_n$.

For small-sized problems, it is nevertheless possible explicitly to compute the kernels of $\hat{\mathbf{F}}_n$. Using the notation $\langle\,\hat{\chi}\,,\,\mathbf{u}\,\rangle = \hat{\chi}^t\,\mathbf{u}$ for the duality product, the formulas (6.342)–(6.344) can be written

$$\hat{\mathbf{F}}_{n+1} = \hat{\mathbf{F}}_n + \frac{\mathbf{u}_n\,\mathbf{u}_n{}^t}{\delta\hat{\boldsymbol{\gamma}}_n^t\,\mathbf{u}_n} \quad, \tag{6.349}$$

$$\hat{\mathbf{F}}_{n+1} = \hat{\mathbf{F}}_n + \frac{\delta\mathbf{m}_n\,\delta\mathbf{m}_n{}^t}{\delta\hat{\boldsymbol{\gamma}}^t\,\delta\mathbf{m}_n} - \frac{\mathbf{v}_n\,\mathbf{v}_n{}^t}{\delta\hat{\boldsymbol{\gamma}}_n^t\,\mathbf{v}_n} \quad, \tag{6.350}$$

$$\hat{\mathbf{F}}_{n+1} = \left(\mathbf{I} - \frac{\delta\mathbf{m}_n\,\delta\hat{\boldsymbol{\gamma}}_n^t}{\delta\hat{\boldsymbol{\gamma}}_n^t\,\delta\mathbf{m}_n}\right)\hat{\mathbf{F}}_n\left(\mathbf{I} - \frac{\delta\mathbf{m}_n\,\delta\hat{\boldsymbol{\gamma}}_n^t}{\delta\hat{\boldsymbol{\gamma}}_n^t\,\delta\mathbf{m}_n}\right)^t + \frac{\delta\mathbf{m}_n\,\delta\mathbf{m}_n{}^t}{\delta\hat{\boldsymbol{\gamma}}^t\,\delta\mathbf{m}_n} \quad. \tag{6.351}$$

Using the approximation (6.295) for μ_n gives

$$\mu_n \simeq (p-1)\,\frac{\langle\,\hat{\boldsymbol{\gamma}}_n\,,\,\boldsymbol{\phi}_n\,\rangle}{\langle\,\hat{\mathbf{H}}_n\,\boldsymbol{\phi}_n\,,\,\boldsymbol{\phi}_n\,\rangle} \quad. \tag{6.352}$$

Variable Metric for Least Squares

To obtain the formulas of the variable metric method for least squares, we only need to particularize the formulas of the previous section (although it is better here to use steepest descent vector and curvature instead of gradient and Hessian).

The preconditioning operator $\mathbf{F}$ is initialized as an arbitrary approximation of the initial curvature $\mathbf{H}_0^{-1}$,

$$\mathbf{F}_0 \simeq \mathbf{H}_0^{-1} \simeq \left(\mathbf{I} + \mathbf{C}_M\,\mathbf{G}_0{}^t\,\mathbf{C}_D^{-1}\,\mathbf{G}_0\right)^{-1}, \tag{6.353}$$

and the variable metric method updates it in such a way that, at least for linear functions $\mathbf{g}(\mathbf{m})$ (i.e., for quadratic misfit functions $S(\mathbf{m})$),

$$\mathbf{F}_n \quad\rightarrow\quad \mathbf{H}_n^{-1} \quad. \tag{6.354}$$

There are two advantages in this. First, as already noted, although the method starts out by behaving like a preconditioned steepest descent, it ends up by behaving like the Newton method, with its rapid termination. Second, as we have seen in chapter 3, the final value of the inverse Hessian equals the posterior covariance operator $\widetilde{\mathbf{C}}_M$, so *the variable metric method directly provides the posterior covariance operator* without any need to invert a matrix (see below).

We obtain the following algorithm (initialized at an arbitrary $\mathbf{m}_0$)

$$\begin{aligned}
\boldsymbol{\gamma}_n &= \mathbf{C}_M\,\mathbf{G}_n{}^t\,\mathbf{C}_D^{-1}\big(\mathbf{g}(\mathbf{m}_n) - \mathbf{d}_{\text{obs}}\big) + (\mathbf{m}_n - \mathbf{m}_{\text{prior}}) \quad, \\
\boldsymbol{\phi}_n &= \mathbf{F}_n\,\boldsymbol{\gamma}_n \quad, \\
\mathbf{m}_{n+1} &= \mathbf{m}_n - \mu_n\,\boldsymbol{\phi}_n \qquad \text{(obtain } \mu_n \text{ by linear search)} \quad, \\
\mathbf{F}_{n+1} &= \mathbf{F}_n + \delta\mathbf{F}_n \qquad (\delta\mathbf{F}_n \text{ defined below}) \quad.
\end{aligned} \tag{6.355}$$

Then the rank-one formula gives

$$\mathbf{F}_{n+1} = \mathbf{F}_n + \frac{\mathbf{u}_n \, \mathbf{u}_n^t \, \mathbf{C}_M^{-1}}{\mathbf{u}_n^t \, \mathbf{C}_M^{-1} \, \delta\boldsymbol{\gamma}_n^t} \quad , \tag{6.356}$$

the DFP formula gives

$$\mathbf{F}_{n+1} = \mathbf{F}_n + \frac{\delta\mathbf{m}_n \, \delta\mathbf{m}_n^t \, \mathbf{C}_M^{-1}}{\delta\mathbf{m}_n^t \, \mathbf{C}_M^{-1} \, \delta\boldsymbol{\gamma}_n} - \frac{\mathbf{v}_n \, \mathbf{v}_n^t \, \mathbf{C}_M^{-1}}{\mathbf{v}_n^t \, \mathbf{C}_M^{-1} \, \delta\boldsymbol{\gamma}_n} \quad , \tag{6.357}$$

and the BFGS formula gives

$$\mathbf{F}_{n+1} = \left(\mathbf{I} - \frac{\delta\mathbf{m}_n \, \delta\boldsymbol{\gamma}_n^t \, \mathbf{C}_M^{-1}}{\delta\boldsymbol{\gamma}_n^t \, \mathbf{C}_M^{-1} \, \delta\mathbf{m}_n} \right) \mathbf{F}_n \left(\mathbf{I} - \frac{\delta\mathbf{m}_n \, \delta\boldsymbol{\gamma}_n^t \, \mathbf{C}_M^{-1}}{\delta\boldsymbol{\gamma}_n^t \, \mathbf{C}_M^{-1} \, \delta\mathbf{m}_n} \right) + \frac{\delta\mathbf{m}_n \, \delta\mathbf{m}_n^t \, \mathbf{C}_M^{-1}}{\delta\boldsymbol{\gamma}_n^t \, \mathbf{C}_M^{-1} \, \delta\mathbf{m}_n} \quad . \tag{6.358}$$

In these equations, $\delta\mathbf{m}_n = \mathbf{m}_{n+1} - \mathbf{m}_n$, $\delta\boldsymbol{\gamma}_n = \boldsymbol{\gamma}_{n+1} - \boldsymbol{\gamma}_n$, $\mathbf{v}_n = \mathbf{F}_n \, \delta\boldsymbol{\gamma}_n$, and $\mathbf{u}_n = \delta\mathbf{m}_n - \mathbf{v}_n$. As mentioned above, the kernels (matrices) representing the operators $\mathbf{F}_n$ should only be explicitly computed for small-sized problems. For large-sized problems, it should be noticed that all that we need is to be able to compute the result of the action of $\mathbf{F}_n$ on an arbitrary model vector $\mathbf{f}$. Using, for instance, the rank-one formula, we have

$$\mathbf{F}_n \, \mathbf{f} = \mathbf{F}_0 \, \mathbf{f} + \sum_{k=0}^{n-1} \frac{\mathbf{u}_k^{\,t} \, \mathbf{C}_M^{-1} \, \mathbf{f}}{\nu_k} \, \mathbf{u}_k \quad , \tag{6.359}$$

where ν_k are the real numbers

$$\nu_k = \mathbf{u}_k^{\,t} \, \mathbf{C}_M^{-1} \, \delta\boldsymbol{\gamma}_k \quad . \tag{6.360}$$

This shows that, in order to operate with $\mathbf{F}_n$, we only have to store in the computer memory the vectors $\mathbf{u}_0, \dots, \mathbf{u}_n$ and the scalars $\nu_0, \dots, \nu_n$.

This remark also applies for the estimation of the posterior covariance operator in large-sized problems. We have seen that one interesting property of the variable metric method is that it allows, at least in principle, an inexpensive estimate of the posterior covariance operator: the property

$$\mathbf{F}_n \to \left(\mathbf{I} + \mathbf{C}_M \, \mathbf{G}_n^{\,t} \, \mathbf{C}_D^{-1} \, \mathbf{G}_n \right)^{-1} \tag{6.361}$$

gives

$$\widetilde{\mathbf{C}}_M \simeq \mathbf{F}_K \, \mathbf{C}_M \quad , \tag{6.362}$$

where the index K represents the value of n for which iterations are stopped. Using, for instance, the rank-one formula, gives (using equation (6.356)),

$$\left(\widetilde{\mathbf{C}}_M \right)^{\alpha\beta} \approx \left(\mathbf{F}_K \, \mathbf{C}_M \right)^{\alpha\beta} = \left(\mathbf{F}_0 \mathbf{C}_M \right)^{\alpha\beta} + \sum_{n=0}^{K-1} \frac{u_n^{\,\alpha} \, u_n^{\,\beta}}{\phi_n} \quad . \tag{6.363}$$

For large-sized problems, good approximations of the solution are usually obtained after a few iterations. Unfortunately, very little is known about the accuracy of the covariance operator obtained after a few iterations of a variable metric method.

It remains to fix a value to the real number μ_n. Ideally, we should perform a linear search. Alternatively, a linearization of $\mathbf{g}(\mathbf{m})$ around $\mathbf{g}(\mathbf{m}_n)$ gives

$$\mu_n \simeq \frac{\delta_n{}^t \, \mathbf{C}_{\mathrm{M}}^{-1} \, \boldsymbol{\phi}_n}{\boldsymbol{\phi}_n{}^t \, \mathbf{C}_{\mathrm{M}}^{-1} \, \boldsymbol{\phi}_n + \mathbf{b}_n{}^t \, \mathbf{C}_{\mathrm{D}}^{-1} \, \mathbf{b}_n} \quad , \qquad \text{where} \qquad \mathbf{b}_n = \mathbf{G}_n \, \boldsymbol{\phi}_n \quad . \tag{6.364}$$

6.23 Elements of Linear Programming

6.23.1 Simplex Method of Linear Programming

Let the following be given:

$$\begin{aligned} \mathbf{M} \quad &: \quad (n \times m) \text{ rectangular matrix,} \\ \hat{\boldsymbol{\chi}} \quad &: \quad (m \times 1) \text{ column matrix,} \\ \mathbf{y} \quad &: \quad (n \times 1) \text{ column matrix.} \end{aligned} \tag{6.365}$$

We wish to obtain an $(m \times 1)$ column matrix $\mathbf{x}$ to solve the problem

$$\text{minimize} \quad \hat{\boldsymbol{\chi}}^t \, \mathbf{x} \tag{6.366}$$

subject to the constraints

$$\mathbf{M}\mathbf{x} = \mathbf{y} \quad , \qquad \mathbf{x} \geq \mathbf{0} \quad , \tag{6.367}$$

where by $\mathbf{x} \geq \mathbf{0}$ we mean that all the components of $\mathbf{x}$ are nonnegative. In the application we will consider, the matrix $\mathbf{M}$ will have the following properties:

$$\begin{aligned} m > n \quad &\text{(more columns than rows)} \quad , \\ \mathbf{M} \text{ is full rank} \quad &\text{(rows are linearly independent)} \quad , \\ &\text{there is no row with all elements but one null} \quad . \end{aligned} \tag{6.368}$$

The first of conditions (6.367) defines a hyperplane with dimension $m - n$. The first two conditions of (6.368) assure that this hyperplane is not empty and is not reduced to a single point. The third of conditions (6.368) ensures that the hyperplane is not parallel to one of the coordinate axes. The set of constraints (6.367) then always has an infinity of solutions. The set of solutions can easily be seen to constitute a nonempty convex set (convex means that the straight segment joining two arbitrary points of the set belongs to the set). The problem then reduces to one of obtaining the point $\mathbf{x}$ of the convex set that minimizes the scalar function $\hat{\boldsymbol{\chi}}^t \, \mathbf{x}$ defined by (6.366). As $\hat{\boldsymbol{\chi}}^t \, \mathbf{x}$ is a linear function of $\mathbf{x}$, the minimum of $\hat{\boldsymbol{\chi}}^t \, \mathbf{x}$ is attained at a vertex of the convex set (or at an edge, if the solution is not unique).

The idea of the *simplex method* for obtaining the solution is to start at an arbitrary vertex of the convex polyhedron and to follow the steepest descending edge to the next vertex. It is clear that the minimum will be attained in a finite number of steps. It can be shown that the vector $\mathbf{x}$ solution of (6.366)–(6.367) has at most n components different from zero. We can then limit ourselves to looking for the solution in a subspace with n eventually nonnull components.

We start by arbitrarily choosing n components (called the *basic* components) among the m components of $\mathbf{x}$ (remember that $m > n$). Setting all the other components to zero and using the first of equations (6.367) allows us to compute the values of the basic components. This gives a vertex of the polyhedron. If, by chance, we have the minimum of $\hat{\chi}^t \mathbf{x}$, we stop the computations; if not, we drop one of the basic components and replace it with another component. This gives a new vertex connected to the old one by an edge. The simplex method chooses as a new vertex the one for which the corresponding edge has maximum (descending) slope. Let us see how this can be done.

Assume that one has chosen the starting n components of $\mathbf{x}$ arbitrarily. For simplicity, assume that the components of $\mathbf{x}$ are classed in such a way that the basic components appear in the first n places of the column matrix representing $\mathbf{x}$:

$$
\mathbf{x} = \begin{pmatrix} \mathbf{x}^B \\ \mathbf{x}^N \end{pmatrix} = \begin{pmatrix} x^1 \\ x^2 \\ \dots \\ x^n \\ x^{n+1} \\ \dots \\ x^m \end{pmatrix} = \begin{pmatrix} \text{basic} \\ \text{components} \\ \\ \text{other} \\ \text{components} \end{pmatrix} . \tag{6.369}
$$

The matrix $\mathbf{M}$ can then be written in partitioned form:

$$
\mathbf{M} = \begin{pmatrix} \mathbf{M}^B & \mathbf{M}^N \end{pmatrix} . \tag{6.370}
$$

As we have assumed that $\mathbf{M}$ has full rank, we can always choose our basic components in such a way that $\mathbf{M}^B$ is regular. The equation $\mathbf{M}\mathbf{x} = \mathbf{y}$ can now be rewritten

$$
\begin{pmatrix} \mathbf{M}^B & \mathbf{M}^N \end{pmatrix} \begin{pmatrix} \mathbf{x}^B \\ \mathbf{x}^N \end{pmatrix} = \mathbf{y} , \tag{6.371}
$$

i.e., $\mathbf{M}^B \mathbf{x}^B = \mathbf{y} - \mathbf{M}^N \mathbf{x}^N$, which gives

$$
\mathbf{x}^B = (\mathbf{M}^B)^{-1}(\mathbf{y} - \mathbf{M}^N \mathbf{x}^N) . \tag{6.372}
$$

As we have not yet used the positivity constraint $\mathbf{x} \geq \mathbf{0}$, we have no reason to have $\mathbf{x}_B \geq \mathbf{0}$. We then have to change the choice of basic components until this condition is fulfilled. When it is, we can pass to the study of $\hat{\chi}^t \mathbf{x}$. We have

$$
\hat{\chi}^t \mathbf{x} = \begin{pmatrix} \hat{\chi}^B \\ \hat{\chi}^N \end{pmatrix}^t \begin{pmatrix} \mathbf{x}^B \\ \mathbf{x}^B \end{pmatrix} = (\hat{\chi}^B)^t \mathbf{x}^N + (\hat{\chi}^N)^t \mathbf{x}^N \tag{6.373}
$$

and, using (6.372)

$$
\hat{\chi}^t \mathbf{x} = (\hat{\chi}^B)^t (\mathbf{M}^B)^{-1} \mathbf{y} + (\gamma^N)^t \mathbf{x}^N , \tag{6.374}
$$

where

$$(\boldsymbol{\gamma}^N)^t = (\hat{\boldsymbol{\chi}}^N)^t - (\hat{\boldsymbol{\chi}}^B)^t \, (\mathbf{M}^B)^{-1} \, \mathbf{M}^N \quad . \tag{6.375}$$

If all the components of $\boldsymbol{\gamma}^N$ are greater than or equal to 0, $\hat{\boldsymbol{\chi}}^t \mathbf{x}$ is clearly minimized for $\mathbf{x}^N = 0$, and then the solution obtained using (6.372),

$$\mathbf{x} = \begin{pmatrix} \mathbf{x}^B \\ \mathbf{x}^N \end{pmatrix} = \begin{pmatrix} (\mathbf{M}^B)^{-1}\,\mathbf{y} \\ \mathbf{0} \end{pmatrix} \quad , \tag{6.376}$$

is the solution of the problem. If some of the components of $\boldsymbol{\gamma}^N$ are negative, one chooses the most negative. Denote as x^k the corresponding component of $\mathbf{x}$ (which is not in the basic components). If one gives increasingly positive values to x^k and computes the corresponding values for $\mathbf{x}^B$ using (6.372), one of the basic components, say x^j, will vanish. It can be seen that replacing x^j with x^k in the basic components corresponds to a move from one vertex to the neighboring vertex whose direction is the steepest one.

Iterating this procedure, one ends at a vertex for which all the components of $\boldsymbol{\gamma}^N$ are positive, and the solution is attained.

For important questions concerning implementation, the reader may refer to Cuer (1984) or to the references there quoted. For algorithmic questions, see references in the footnote.[90]

6.23.2 Dual Problems in Linear Programming

Let $\mathbb{X}$ and $\mathbb{Y}$ be two abstract linear spaces and $\mathbf{M}$ be a linear operator mapping $\mathbb{X}$ into $\mathbb{Y}$. Assume that the dimension of $\mathbb{X}$ is greater than the dimension of $\mathbb{Y}$ and that $\mathbf{M}$ is full rank. Let $\mathbf{y}$ denote a given vector of $\mathbb{Y}$ and $\hat{\boldsymbol{\chi}}$ be a given form over $\mathbb{X}$, i.e., an element of $\mathbb{X}^*$, dual of $\mathbb{X}$. The problem of obtaining the vector $\mathbf{x}$ of $\mathbb{X}$ satisfying the conditions

$$\text{minimize} \quad \hat{\boldsymbol{\chi}}^t \mathbf{x} \quad \text{subject to the constraints} \quad \mathbf{M}\mathbf{x} = \mathbf{y} \quad , \quad \mathbf{x} \geq \mathbf{0} \quad , \tag{6.377}$$

is called the *standard problem* of linear programming. As usual, $\mathbf{x} \geq \mathbf{0}$ means that all the components of $\mathbf{x}$ are not less than 0.

Let $\mathbf{M}$, $\mathbf{y}$, and $\hat{\boldsymbol{\chi}}$ be the same as above. As $\mathbf{M}$ maps $\mathbb{X}$ into $\mathbb{Y}$, by definition of the transpose operator (see section 3.1.2), $\mathbf{M}^t$ maps $\mathbb{Y}^*$, dual of $\mathbb{Y}$, into $\mathbb{X}^*$, dual of $\mathbb{X}$. The element $\mathbf{y}$ of $\mathbb{Y}$ defines a linear form over $\mathbb{Y}^*$. The problem of obtaining the vector $\hat{\boldsymbol{v}}$ of $\mathbb{Y}^*$ satisfying the conditions

$$\text{maximize} \quad \mathbf{y}^t \, \hat{\boldsymbol{v}} \quad \text{subject to the constraints} \quad \mathbf{M}^t \, \hat{\boldsymbol{v}} \leq \hat{\boldsymbol{\chi}} \tag{6.378}$$

is called the *canonical* problem of linear programming.

The inputs of the two problems are the same ($\mathbf{M}$, $\mathbf{y}$, and $\hat{\boldsymbol{\chi}}$). While the unknown of the first (standard) problem is an element of $\mathbb{X}$, the unknown of the second (canonical)

---

[90]Klee and Minty, 1972; Jeroslow, 1973; Hacijan, 1979; Gacs and Lovasc, 1981; Kônig and Pallaschke, 1981; Cottle and Dantzig, 1968; Bartels, 1971; Mangasarian, 1981, 1983; Ciarlet and Thomas, 1982; Censor and Elfving, 1982; Cimino, 1938; Magnanti, 1976; Nazareth, 1984; McCall, 1982; Ecker and Kupferschmid, 1985; Karmarkar, 1984; de Ghellinck and Vial, 1985, 1986; Nash and Sofer, 1996.

problem is an element of $\mathbb{Y}^*$, dual of $\mathbb{Y}$. Any of the two problems is termed the *dual* of the other problem, then called *primal*. This definition is useful because the solution of one problem also gives the solution to the associated dual problem.

The basic duality theorem (Dantzig, 1963; Gass, 1975) is as follows: if feasible solutions to both the primal and dual problems exist, there exists an optimum solution to both problems, and

$$\text{minimum of } \hat{\chi}^t\,\mathbf{x} \;=\; \text{maximum of } \mathbf{y}^t\,\hat{\upsilon} \quad. \tag{6.379}$$

The solution of the standard problem (6.377) can be written (see Appendix 6.23.1)

$$\mathbf{x}_{\text{sol}} \;=\; \begin{pmatrix} \mathbf{x}^B \\ \mathbf{x}^L \end{pmatrix} \;=\; \begin{pmatrix} (\mathbf{M}^B)^{-1}\mathbf{y} \\ \mathbf{0} \end{pmatrix} \quad, \tag{6.380}$$

where $\mathbf{x}^B$ is the column matrix of *basic components* and $\mathbf{M}^B$ is the *basic submatrix* of $\mathbf{M}$. Let

$$\hat{\chi} \;=\; \begin{pmatrix} \hat{\chi}^B \\ \hat{\chi}^N \end{pmatrix} \tag{6.381}$$

represent the partition of $\hat{\chi}$ into basic and nonbasic components. It can be shown (see, for instance, Gass, 1975) that the solution of the dual problem (6.381) is then given by

$$\hat{\upsilon}_{\text{sol}} \;=\; (\mathbf{M}^B)^{-t}\,\hat{\chi}^B \quad, \tag{6.382}$$

where $(\mathbf{M}^B)^{-t}$ denotes the transpose of the inverse of $\mathbf{M}^B$. The property (6.379) is then easily verified:

$$\hat{\chi}^t\,\mathbf{x}_{\text{sol}} \;=\; (\hat{\chi}^B)^t\,\mathbf{x}^B \;=\; (\hat{\chi}^B)^t\,(\mathbf{M}_B)^{-1}\,\mathbf{y} \;=\; \mathbf{y}^t\,(\mathbf{M}^B)^{-t}\,\hat{\chi}^B \;=\; \mathbf{y}^t\,\hat{\upsilon}_{\text{sol}} \quad. \tag{6.383}$$

The standard problem has been arbitrarily assumed to be a minimization problem. The dual of a maximization problem can be obtained simply by changing the signs of $\mathbf{M}$, $\mathbf{y}$, and $\hat{\chi}$. This gives the primal problem

$$\text{maximize} \quad \hat{\chi}^t\,\mathbf{x} \quad \text{subject to the constraints} \quad \mathbf{M}\mathbf{x} = \mathbf{y} \quad, \quad \mathbf{x} \geq \mathbf{0} \quad, \tag{6.384}$$

and the dual problem

$$\text{minimize} \quad \mathbf{y}^t\,\hat{\upsilon} \quad \text{subject to the constraints} \quad \mathbf{M}^t\,\hat{\upsilon} \geq \hat{\chi} \quad. \tag{6.385}$$

6.23.3 Slack Variables in Linear Programming

The most general form of a linear programming problem is

$$\text{minimize (resp., maximize)} \quad \mathbf{a}^t\,\mathbf{b} \tag{6.386}$$

under the constraints

$$\mathbf{L}_1\,\mathbf{b} = \mathbf{c}_1 \quad, \qquad \mathbf{L}_2\,\mathbf{b} \geq \mathbf{c}_2 \quad, \qquad \mathbf{L}_3\,\mathbf{b} \leq \mathbf{c}_3 \quad, \tag{6.387}$$

where $\mathbf{b}$ is the unknown vector, $\mathbf{a}$, $\mathbf{c}_1$, $\mathbf{c}_2$, and $\mathbf{c}_3$ are given vectors, and $\mathbf{L}_1$, $\mathbf{L}_2$, and $\mathbf{L}_3$ are given linear operators.

The particular choice $\mathbf{a} = \hat{\chi}$, $\mathbf{b} = \mathbf{x}$, $\mathbf{L}_1 = \mathbf{M}$, $\mathbf{c}_1 = \mathbf{y}$, $\mathbf{L}_2 = \mathbf{I}$, $\mathbf{c}_2 = \mathbf{0}$, $\mathbf{L}_3 = \mathbf{0}$, and $\mathbf{c}_3 = \mathbf{0}$ leads to the problem

$$\text{minimize (resp., maximize)} \quad \hat{\chi}^t \mathbf{x} \quad \text{subject to the constraints} \quad \mathbf{M}\mathbf{x} = \mathbf{y} \quad , \quad \mathbf{x} \geq \mathbf{0}, \tag{6.388}$$

which is the standard problem of Appendix 6.23.2.

The particular choice $\mathbf{a} = -\mathbf{y}$, $\mathbf{b} = \hat{v}$, $\mathbf{L}_1 = \mathbf{0}$, $\mathbf{c}_1 = \mathbf{0}$, $\mathbf{L}_2 = \mathbf{0}$, $\mathbf{c}_2 = \mathbf{0}$, $\mathbf{L}_3 = \mathbf{M}^t$, and $\mathbf{c}_3 = \hat{\chi}$ leads to the problem

$$\text{maximize (resp., minimize)} \quad \mathbf{y}^t \hat{v} \quad \text{subject to the constraints} \quad \mathbf{M}^t \hat{v} \leq \hat{\chi} \quad , \tag{6.389}$$

which is the canonical problem of Appendix 6.23.2.

We see that the standard and canonical problems are special cases of (6.386)–(6.387). We will now see that the reciprocal is also true.

Problem (6.386)–(6.387) can be written

$$\text{minimize (resp., maximize)} \quad \mathbf{a}^t \mathbf{b} \tag{6.390}$$

under the constraints

$$\mathbf{L}_1 \mathbf{b} \leq \mathbf{c}_1 \quad , \quad -\mathbf{L}_1 \mathbf{b} \leq -\mathbf{c}_1 \quad , \quad -\mathbf{L}_2 \mathbf{b} \leq -\mathbf{c}_2 \quad , \quad \mathbf{L}_3 \mathbf{b} \leq \mathbf{c}_3 , \tag{6.391}$$

and using

$$\mathbf{a} = -\mathbf{y} , \ \mathbf{b} = \hat{v} , \ \mathbf{M}^t = \begin{pmatrix} \mathbf{L}_1 \\ -\mathbf{L}_1 \\ -\mathbf{L}_2 \\ \mathbf{L}_3 \end{pmatrix} , \ \hat{\chi} = \begin{pmatrix} \mathbf{c}_1 \\ -\mathbf{c}_1 \\ -\mathbf{c}_2 \\ \mathbf{c}_3 \end{pmatrix}$$

leads to the canonical form (6.389).

Introducing the *slack variables* $\mathbf{b}' \geq \mathbf{0}$, $\mathbf{b}'' \geq \mathbf{0}$, $\mathbf{c}_2' \geq \mathbf{0}$, and $\mathbf{c}_3' \geq \mathbf{0}$, and writing $\mathbf{b} = \mathbf{b}' - \mathbf{b}''$, the problem (6.386)–(6.387) becomes

$$\text{minimize (resp., maximize)} \quad \mathbf{a}^t (\mathbf{b}' - \mathbf{b}'') \tag{6.392}$$

under the constraints

$$\mathbf{L}_1 (\mathbf{b}' - \mathbf{b}'') = \mathbf{c}_1 \quad , \quad \mathbf{L}_2 (\mathbf{b}' - \mathbf{b}'') - \mathbf{c}_2' = \mathbf{c}_2 \quad , \quad \mathbf{L}_3 (\mathbf{b}' - \mathbf{b}'') + \mathbf{c}_3' = \mathbf{c}_3 \quad ,$$
$$\mathbf{b}' \geq \mathbf{0} \quad , \quad \mathbf{b}'' \geq \mathbf{0} \quad , \quad \mathbf{c}_2' \geq \mathbf{0} \quad , \quad \mathbf{c}_3' \geq \mathbf{0}, \tag{6.393}$$

and using

$$\mathbf{x} = \begin{pmatrix} \mathbf{b}' \\ \mathbf{b}'' \\ \mathbf{c}_2' \\ \mathbf{c}_3' \end{pmatrix} , \ \hat{\chi} = \begin{pmatrix} \mathbf{a} \\ -\mathbf{a} \\ 0 \\ 0 \end{pmatrix} , \ \mathbf{y} = \begin{pmatrix} \mathbf{c}_1 \\ \mathbf{c}_2 \\ \mathbf{c}_3 \end{pmatrix} , \ \text{and} \ \mathbf{M} = \begin{pmatrix} \mathbf{L}_1 & -\mathbf{L}_1 & 0 & 0 \\ \mathbf{L}_2 & -\mathbf{L}_2 & -\mathbf{I} & 0 \\ \mathbf{L}_3 & -\mathbf{L}_3 & 0 & \mathbf{I} \end{pmatrix}$$

leads to the standard form (6.388).

6.23.4 ℓ_1-Norm Minimization Using Linear Programming

Below, the index i is assumed to belong to some set $\mathbf{I_D}$ and the index α is assumed to belong to some set $\mathbf{I_M}$. Assume the constants $\{d^i_{\text{obs}}\}$, $\{\sigma^i_D\}$, $\{m^\alpha_{\text{prior}}\}$, $\{\sigma^\alpha_M\}$, $\{G^{i\alpha}\}$ are given, and consider the problem of obtaining the unknowns $\{u^i\}$, $\{v^i\}$, $\{a^\alpha\}$, $\{b^\alpha\}$, solution of the following problem of constrained minimization:

$$\text{minimize } S = \sum_i \frac{u^i + v^i}{\sigma^i_D} + \sum_\alpha \frac{a^\alpha + b^\alpha}{\sigma^\alpha_M}$$

$$\text{subject to } \begin{cases} \mathbf{G}\,(\mathbf{a} - \mathbf{b}) - (\mathbf{u} - \mathbf{v}) = \mathbf{d}_{\text{obs}} - \mathbf{G}\,\mathbf{m}_{\text{prior}} \quad, \\ \mathbf{u, v, a, b} \geq \mathbf{0} \quad. \end{cases} \tag{6.394}$$

Defining

$$w^i_D = 1/\sigma^i_D \quad, \qquad w^\alpha_M = 1/\sigma^\alpha_M \quad, \tag{6.395}$$

and setting

$$\mathbf{x} = \begin{pmatrix} \mathbf{a} \\ \mathbf{u} \\ \mathbf{b} \\ \mathbf{v} \end{pmatrix} \quad, \quad \hat{\boldsymbol{\chi}} = \begin{pmatrix} \mathbf{w_M} \\ \mathbf{w_D} \\ \mathbf{w_M} \\ \mathbf{w_D} \end{pmatrix} \quad, \quad \mathbf{M} = \begin{pmatrix} \mathbf{G} & -\mathbf{I} & -\mathbf{G} & \mathbf{I} \end{pmatrix} \quad, \quad \mathbf{y} = \mathbf{d}_{\text{obs}} - \mathbf{G}\,\mathbf{m}_{\text{prior}} \quad, \tag{6.396}$$

equations (6.394) can be written

$$\text{minimize } \quad \hat{\boldsymbol{\chi}}^t \mathbf{x} \quad \text{subject to} \quad \mathbf{M}\,\mathbf{x} = \mathbf{y} \quad, \quad \mathbf{x} \geq \mathbf{0} \quad, \tag{6.397}$$

which corresponds to the standard form of the linear programming problem.

   To see the equivalence between this problem and the unconstrained ℓ_1-norm minimization problem, let us define $\mathbf{m}$ by

$$\mathbf{a} - \mathbf{b} = \mathbf{m} - \mathbf{m}_{\text{prior}} \quad. \tag{6.398}$$

The second of conditions (6.394) then becomes

$$\mathbf{u} - \mathbf{v} = \mathbf{G}\,\mathbf{m} - \mathbf{d}_{\text{obs}} \quad. \tag{6.399}$$

It can be shown that in each vertex of the convex polyhedron defined by (6.394), and due to the particular structure of the linear system there, for any α, both a^α and b^α cannot be $\neq 0$ simultaneously, and for any i, both u^i and v^i cannot be $\neq 0$ simultaneously. Then, at each vertex,

$$u^i + v^i = |u^i - v^i| \quad, \qquad a^\alpha + b^\alpha = |a^\alpha - b^\alpha| \quad, \tag{6.400}$$

so that the S appearing in the first of equations (6.394) can be written

$$S = \sum_i \frac{\left| (\mathbf{G}\,\mathbf{m})^i - d^i_{\text{obs}} \right|}{\sigma^i_D} + \sum_\alpha \frac{\left| m^\alpha - m^\alpha_{\text{prior}} \right|}{\sigma^\alpha_M} \quad, \tag{6.401}$$

which corresponds to the standard cost function for the ℓ_1-norm criterion for the resolution of linear problems. We have thus seen that the minimization of the S appearing in the first of equations (6.394) under the constraints expressed there (what constitutes a linear programming problem) is equivalent to the usual unconstrained ℓ_1-norm minimization.

In numerical analysis, ℓ_1-norm minimization problems have been studied in approximation theory (Barrodale and Young, 1996; Barrodale, 1970; Barrodale and Roberts, 1973, 1974; Bartels, Conn, and Sinclair, 1978; Armstrong and Golfrey, 1979; Watson, 1980).

A useful package of routines has been developed by Cuer and Bayer (1980a,b). They apply to the general problem

$$\text{minimize} \sum_{A \in I_x} \frac{|x^A - x^A_{\text{prior}}|}{\sigma^A} \quad \text{under the constraints} \quad \begin{cases} \mathbf{A}\mathbf{x} = \mathbf{b} \quad , \\ \mathbf{0} \leq \mathbf{x} \leq \mathbf{x}_{\text{max}} \quad , \end{cases} \tag{6.402}$$

where $\mathbf{x}_{\text{prior}}$, $\mathbf{x}_{\text{max}}$, $\mathbf{A}$, and $\mathbf{b}$ are given, and where some of the constants σ^A may be infinite. These algorithms allow the resolution of the standard linear programming problem.

6.23.5 ℓ_∞-Norm Minimization Using Linear Programming

The following is adapted from Watson (1980). The problem of minimizing R as defined by equation (4.55) is equivalent to the problem of minimizing R subject to the constraints

$$\begin{aligned} -R &\leq \frac{(\mathbf{G}\mathbf{m} - \mathbf{d}_{\text{obs}})^i}{\sigma^i_D} \leq +R \quad , \\ -R &\leq \frac{(\mathbf{m} - \mathbf{m}_{\text{prior}})^\alpha}{\sigma^\alpha_M} \leq +R \quad , \end{aligned} \tag{6.403}$$

which can be rewritten as

$$\begin{aligned} (\mathbf{G}\mathbf{m})^i + \sigma^i_D R &\leq d^i_{\text{obs}} \quad , & (\mathbf{G}\mathbf{m})^i - \sigma^i_D R &\geq d^i_{\text{obs}} \quad , \\ m^\alpha + \sigma^\alpha_M R &\leq m^\alpha_{\text{prior}} \quad , & m^\alpha + \sigma^\alpha_M R &\geq m^\alpha_{\text{prior}} \quad . \end{aligned} \tag{6.404}$$

The problem can then be written, in matricial form, as

$$\text{minimize} \begin{pmatrix} \mathbf{0} \\ 1 \end{pmatrix}^t \begin{pmatrix} \mathbf{m} \\ R \end{pmatrix} \quad \text{subject to} \quad \begin{pmatrix} \mathbf{G} & \sigma_D \\ \mathbf{I} & \sigma_M \\ -\mathbf{G} & \sigma_D \\ -\mathbf{I} & \sigma_M \end{pmatrix} \begin{pmatrix} \mathbf{m} \\ R \end{pmatrix} \geq \begin{pmatrix} \mathbf{d}_{\text{obs}} \\ \mathbf{m}_{\text{prior}} \\ -\mathbf{d}_{\text{obs}} \\ -\mathbf{m}_{\text{prior}} \end{pmatrix} \quad , \tag{6.405}$$

where $\mathbf{0}$ denotes a vector of zeros and σ_D and σ_M are vectors containing data uncertainties and a priori model uncertainties:

$$\sigma_D = \{\sigma^i_D \quad (i \in \mathbf{I}_D)\} \quad , \qquad \sigma_M = \{\sigma^\alpha_M \quad (\alpha \in \mathbf{I}_M)\} \quad . \tag{6.406}$$

The problem (6.405) is the linear programming problem we have found in equation (6.385). It cannot be solved by direct application of the simplex method, but the dual problem can be written (see equation (6.384)) as

$$\text{maximize} \quad \begin{pmatrix} \mathbf{d}_{obs} \\ \mathbf{m}_{prior} \\ -\mathbf{d}_{obs} \\ -\mathbf{m}_{prior} \end{pmatrix}^t \begin{pmatrix} \hat{\mathbf{d}}_1 \\ \hat{\mathbf{m}}_1 \\ \hat{\mathbf{d}}_2 \\ \hat{\mathbf{m}}_2 \end{pmatrix} \quad \text{subject to} \quad \begin{cases} \begin{pmatrix} \mathbf{G} & \sigma_D \\ \mathbf{I} & \sigma_M \\ -\mathbf{G} & \sigma_D \\ -\mathbf{I} & \sigma_M \end{pmatrix}^t \begin{pmatrix} \hat{\mathbf{d}}_1 \\ \hat{\mathbf{m}}_1 \\ \hat{\mathbf{d}}_2 \\ \hat{\mathbf{m}}_2 \end{pmatrix} = \begin{pmatrix} 0 \\ 1 \end{pmatrix} \\ \hat{\mathbf{d}}_1 , \hat{\mathbf{m}}_1 , \hat{\mathbf{d}}_2 , \hat{\mathbf{m}}_2 \geq 0 \end{cases} ,$$

$$(6.407)$$

This last formulation corresponds to the standard form of the linear programming problem and can be solved using the standard version of the simplex method. Once the solution of the problem in the dual variables $\hat{\mathbf{d}}_1$, $\hat{\mathbf{m}}_1$, $\hat{\mathbf{d}}_2$, and $\hat{\mathbf{m}}_2$ has been obtained, the values of the variables $\mathbf{m}$ and R can be obtained using the equations of Appendix 6.23.2.

Using more compact notation, the problem (6.407) can be written

$$\text{maximize} \quad (\mathbf{d}_{obs})^t (\hat{\mathbf{d}}_1 - \hat{\mathbf{d}}_2) + (\mathbf{m}_{prior})^t (\hat{\mathbf{m}}_1 - \hat{\mathbf{m}}_2)$$

$$\text{subject to} \quad \begin{cases} \mathbf{G}^t (\hat{\mathbf{d}}_1 - \hat{\mathbf{d}}_2) + (\hat{\mathbf{m}}_1 - \hat{\mathbf{m}}_2) = 0 , \\ \sigma_D^t (\hat{\mathbf{d}}_1 + \hat{\mathbf{d}}_2) + \sigma_M^t (\hat{\mathbf{m}}_1 + \hat{\mathbf{m}}_2) = 1 , \\ \hat{\mathbf{d}}_1 , \hat{\mathbf{m}}_1 , \hat{\mathbf{d}}_2 , \hat{\mathbf{m}}_2 \geq \mathbf{0} . \end{cases}$$

$$(6.408)$$

Barrodale and Phillips (1975a,b) give a modification of the standard simplex method that is well adapted to the special structure of this problem. For more details, see Watson (1980).

6.24   Spaces and Operators

This appendix brings together the very basic definitions and properties the reader should know if intending to explore the literature. Useful textbooks are Taylor and Lay (1980) and Dautray and Lions (1984).

6.24.1   Basic Terminology

Let $\mathfrak{S}_0$ and $\mathfrak{T}$ be arbitrary sets, and let $\mathfrak{S}$ be a subset of $\mathfrak{S}_0$. A rule that associates each s in $\mathfrak{S}$ with a unique element $\varphi(s)$ in $\mathfrak{T}$ is termed a *function from* $\mathfrak{S}$ *into* $\mathfrak{T}$. Such a function is properly denoted by φ, or by the expression $s \to \varphi(s)$, although we often say "the function $\varphi(s)$."

To allow suppleness in the discussions, the terms *mapping, application, transformation,* and *operator* are used as synonyms of *function*.

The subset $\mathfrak{S}$ of $\mathfrak{S}_0$ is the *domain of definition* of φ. If $\mathfrak{A}$ is a subset of $\mathfrak{S}$, the set $\varphi(\mathfrak{A})$ (i.e., the subset of $\mathfrak{T}$ that can be attained by φ from elements of $\mathfrak{A}$) is termed the *image* of $\mathfrak{A}$ (through φ). The image of $\mathfrak{S}$, $\varphi(\mathfrak{S})$, is called the *range* of φ.

If for each t in $\varphi(\mathfrak{S})$ there exists only one $s \in \mathfrak{S}$ such that $\varphi(s) = t$, the function φ is *one-to-one* or *injective*. We then write $s = \varphi^{-1}(t)$, thus defining the *inverse* of φ on $\varphi(\mathfrak{S})$. If $\varphi(\mathfrak{S}) = \mathfrak{T}$, φ is termed *surjective*; it is also said that φ maps $\mathfrak{S}$ *onto* $\mathfrak{T}$. When φ is both injective and surjective, it is named *bijective*; then, $\varphi^{-1}(\mathfrak{T}) = \mathfrak{S}$.

If $\mathfrak{S}$ and $\mathfrak{T}$ are linear spaces (see below) and if, for any s_1 and s_2, $\varphi(\lambda s_1 + \mu s_2) = \lambda \varphi(s_1) + \mu \varphi(s_2)$, where λ and μ are arbitrary real numbers, φ is *linear*. A linear bijection between linear spaces is an *isomorphism*. If there exists an isomorphism between two linear spaces, they are termed *isomorphic*.

6.24.2 Topological Space

A *topological space* is a space $\mathfrak{S}$ in which a collection of subsets of $\mathfrak{S}$ has been defined, called the *open subsets* of $\mathfrak{S}$, verifying that

- $\emptyset$ (the empty subset) and $\mathfrak{S}$ are open subsets,

- any union of open subsets is an open subset,

- any finite intersection of open subsets is an open subset.

Example 6.5. *Let $\mathfrak{R}$ be the real line and $a < b$. An open interval (a, b) is defined as the subset of real numbers r verifying $a < r < b$. Any reunion of open intervals is named an open subset. This defines a topology over $\mathfrak{R}$.*

Let $\mathfrak{S}$ be a topological space and $\mathfrak{A}$ be an open subset of $\mathfrak{S}$. A subset of the form $\mathfrak{S} - \mathfrak{A}$ is called a *closed subset*.

Example 6.6. *Let $\mathfrak{R}$ be the real line. For $a < b$, a closed interval $[a, b]$ is defined as the subset of real numbers r verifying $a \leq r \leq b$. A closed interval is a closed subset.*

Let $\mathfrak{S}$ be a topological space. The following properties can be demonstrated:

- $\emptyset$ and $\mathfrak{S}$ are closed subsets,

- any intersection of closed subsets is a closed subset,

- any finite reunion of closed subsets is a closed subset.

In particular, we see that the sets $\emptyset$ and $\mathfrak{S}$ are at the same time open and closed. This is exceptional: in general, a subset is neither open nor closed, and if it is open, it is not closed, and vice versa.

Let $\mathfrak{S}$ be a topological space and s be an element of $\mathfrak{S}$. A *neighborhood* of s is an open subset containing s.

Let $\mathfrak{S}$ be a topological space and $(s_1, s_2, \dots)$ be a sequence of elements of $\mathfrak{S}$. This sequence *tends to* s if, for any neighborhood $\mathfrak{A}$ of s in $\mathfrak{S}$, there exists an integer N such that

$$n \geq N \quad \Rightarrow \quad s_n \in \mathfrak{A} \quad . \tag{6.409}$$

In that case, either of the two following notations is used:

$$s_n \to s \quad , \quad \lim_{n \to \infty} s_n = s \quad . \tag{6.410}$$

Let $\mathfrak{S}$ and $\mathfrak{T}$ be two topological spaces and φ be an application from $\mathfrak{S}$ into $\mathfrak{T}$. φ is called *continuous* at s_0 if

$$\lim_{s \to s_0} \varphi(s) = \varphi(s_0) \quad , \tag{6.411}$$

i.e., if, for any neighborhood $\mathfrak{B}$ of $\varphi(s_0)$ in $\mathfrak{T}$, there exists a neighborhood $\mathfrak{A}$ in $\mathfrak{T}$ such that

$$\varphi(\mathfrak{A}) \subset \mathfrak{B} \quad . \tag{6.412}$$

6.24.3 Manifold

Let $\mathfrak{M}$ be a topological space, and let $\mathfrak{M}_1, \mathfrak{M}_2, \ldots$ be a collection of open subsets of $\mathfrak{M}$ such that they cover all $\mathfrak{M}$. Any bijection ϕ_i from one of the $\mathfrak{M}_i$ into a space $\mathbb{K}^n$, isomorphic to $\mathfrak{R}^n$, is termed a *chart* of $\mathfrak{M}_i$. A collection of charts defined for each of the $\mathfrak{M}_i$ is called an *atlas* of $\mathfrak{M}$. If, for any $\{i, j\}$, the image of the open subset $\mathfrak{M}_i \cap \mathfrak{M}_j$ is an open subset of $\mathbb{K}^n$, and if the images of $\mathfrak{M}_i \cap \mathfrak{M}_j$ obtained respectively by ϕ_i and ϕ_j are related by an isomorphism, then it is said that the set $\mathfrak{M}$ is an (n-dimensional) *manifold*. If the isomorphism is p times differentiable, it is named a C^p-*manifold*.

6.24.4 Metric Space

Let $\mathfrak{M}$ be an arbitrary set. A *distance* over $\mathfrak{M}$ associates any couple $(\mathcal{M}_1, \mathcal{M}_2)$ of elements of $\mathfrak{M}$ with a *positive real number* denoted $D(\mathcal{M}_1, \mathcal{M}_2)$ verifying the following conditions:

$$D(\mathcal{M}_1, \mathcal{M}_2) = 0 \quad \Leftrightarrow \quad \mathcal{M}_1 = \mathcal{M}_2 \quad , \tag{6.413a}$$

$$D(\mathcal{M}_1, \mathcal{M}_2) = D(\mathcal{M}_2, \mathcal{M}_1) \quad \text{for any } \mathcal{M}_1 \text{ and } \mathcal{M}_2 \quad , \tag{6.413b}$$

$$D(\mathcal{M}_1, \mathcal{M}_3) \leq D(\mathcal{M}_1, \mathcal{M}_2) + D(\mathcal{M}_2, \mathcal{M}_3) \quad \text{for any } \mathcal{M}_1, \mathcal{M}_2, \text{ and } \mathcal{M}_3 \quad . \tag{6.413c}$$

A set endowed with a distance is termed a *metric space*. Each element of a metric space $\mathfrak{M}$ is called a *point* of $\mathfrak{M}$.

Example 6.7. *Let $\mathfrak{M}$ be the surface of a sphere in a Euclidean space. A distance between two points of the sphere can, for instance, be defined as the length of the (smaller) arc of the great circle passing through the points. Alternatively, the distance can be defined as the length of the straight segment joining the two points. With any of these definitions, the surface of a sphere is a metric space.*

Example 6.8. *Let $\mathfrak{M}$ be a space of n-dimensional column matrices,*

$$\mathbf{m} \in \mathfrak{M} \quad \Rightarrow \quad \mathbf{m} = \begin{pmatrix} m^1 \\ m^2 \\ \ldots \\ m^n \end{pmatrix} \quad , \tag{6.414}$$

each component representing a physical parameter with its own physical dimensions, and $\sigma^1, \sigma^2, \ldots, \sigma^n$ *be some positive error bars. For any* $\mathbf{m}_1$ *and* $\mathbf{m}_2$, *the expression*

$$D(\mathbf{m}_1, \mathbf{m}_2) \;=\; \left(\sum_{\alpha=1}^{n} \frac{|m_1^\alpha - m_2^\alpha|^p}{(\sigma^\alpha)^p} \right)^{1/p} \tag{6.415}$$

defines a real number. It is a distance over $\mathfrak{M}$. *In fact,* $\mathfrak{M}$ *is a linear space, which can be normed by (see below)*

$$\| \mathbf{m} \| \;=\; D(\mathbf{m}, \mathbf{0}) \quad . \tag{6.416}$$

Example 6.9. *Let* $\mathbb{E}^3$ *be the usual three-dimensional Euclidean space. Let* $\mathbf{x}$ *represent a generic point of* $\mathbb{E}^3$ *and let* $\mathbf{x} \to m(\mathbf{x})$ *be a function from* $\mathbb{E}^3$ *into a space of scalars* $\mathbb{K}$. *For instance,* $\mathbf{x}$ *may represent the Cartesian coordinates of a point inside a star, and* $m(\mathbf{x})$ *may represent the logarithmic temperature at the point* $\mathbf{x}$. *Let* $\mathfrak{M}_0$ *be the space of all such functions. Letting* $m_0(\mathbf{x})$ *be a particular function of* $\mathfrak{M}_0$, *a new, smaller space* $\mathfrak{M}$ *can be defined by the condition*

$$\int_{\mathbb{E}^3} dV(\mathbf{x}) \, \frac{|\, m(\mathbf{x}) - m_0(\mathbf{x}) \,|^p}{s(\mathbf{x})^p} \quad \text{is finite} \quad , \tag{6.417}$$

where $s(\mathbf{x})$ *represents a positive function with physical dimensions ensuring the adimensionality of the previous expression. Let* $\mathbf{m}_1$ *and* $\mathbf{m}_2$ *be two elements of* $\mathfrak{M}$. *The expression*

$$D(\mathbf{m}_1, \mathbf{m}_2) \;=\; \left(\int_{\mathbb{E}_3} dV(\mathbf{x}) \, \frac{|\, m_1(\mathbf{x}) - m_2(\mathbf{x}) \,|^p}{s(\mathbf{x})^p} \right)^{1/p} \tag{6.418}$$

defines a distance over $\mathfrak{M}$, *and then* $\mathfrak{M}$ *is a metric space.*

Let $\mathfrak{M}$ be a metric space and $(\mathbf{m}_1, \mathbf{m}_2, \ldots)$ be a sequence of points of $\mathfrak{M}$. This sequence *tends to* the point $\mathbf{m}$ of $\mathfrak{M}$ if

$$D(\mathbf{m}_n, \mathbf{m}) \to 0 \quad \text{when} \quad n \to \infty \quad , \tag{6.419}$$

i.e., if, for any $\epsilon > 0$, there exists an integer N such that

$$n \geq N \quad \Rightarrow \quad D(\mathbf{m}_n, \mathbf{m}) \leq \epsilon \quad . \tag{6.420}$$

Let $\mathfrak{M}$ be a metric space and $(\mathbf{m}_1, \mathbf{m}_2, \ldots)$ be a sequence of points of $\mathfrak{M}$. This sequence is called a *Cauchy sequence* if, for any $\epsilon > 0$, there exists an integer N such that

$$n \geq N \quad \text{and} \quad m \geq N \quad \Rightarrow \quad D(\mathbf{m}_n, \mathbf{m}_m) \leq \epsilon \quad . \tag{6.421}$$

Example 6.10. *Let* Q *be the set of rational numbers. The sequence* $(1., 1.4, 1.41, 1.414, 1.4142, \ldots)$ *(defined by the decimal development of the number* $\sqrt{2}$ *) is a Cauchy sequence.*

Let $\mathfrak{M}$ be a metric space. If the limit of every Cauchy sequence of points of $\mathfrak{M}$ belongs to $\mathfrak{M}$, the metric space $\mathfrak{M}$ is said to be *complete*.

Example 6.11. *The sequence in Example 6.10 does not converge to an element of Q : the set of rational numbers is not complete. The real line $\Re$ (defined in fact by the completion of Q) is complete.*

Let $\mathfrak{M}$ and $\mathfrak{D}$ be metric spaces and $\mathbf{g}$ be an operator from $\mathfrak{M}$ into $\mathfrak{D}$. We say that $\mathbf{g}(\mathbf{m})$ *tends to* $\mathbf{d}_0$ when $\mathbf{m}$ tends to $\mathbf{m}_0$ if, for any $\epsilon > 0$, there exists a real number r such that

$$D(\mathbf{m}, \mathbf{m}_0) \leq r \quad \Rightarrow \quad D(\mathbf{g}(\mathbf{m}), \mathbf{d}_0) \leq \epsilon \quad . \tag{6.422}$$

Let $\mathfrak{M}$ and $\mathfrak{D}$ be metric spaces and $\mathbf{g}$ be an operator from $\mathfrak{M}$ into $\mathfrak{D}$. We say that $\mathbf{g}$ is *continuous* at $\mathbf{m}_0$ if the limit of $\mathbf{g}(\mathbf{m})$ when $\mathbf{m} \to \mathbf{m}_0$ equals $\mathbf{g}(\mathbf{m}_0)$.

Let $\mathfrak{M}$ be a metric space. Then, a subset $\mathfrak{A}$ of $\mathfrak{S}$ is a *metric open subset* if, for any point $\mathbf{m}_0 \in \mathfrak{A}$, there exists $\epsilon > 0$ such that every point $\mathbf{m}$ of $\mathfrak{M}$ verifying $D(\mathbf{m}, \mathbf{m}_0) < \epsilon$ belongs to $\mathfrak{A}$. These metric open subsets verify the axioms of the (topological) open subsets as defined above: a metric space is always a topological space. It is said that a metric *induces* a topology. The topology induced by the metric is termed the *natural topology*.

It can be shown that the concepts of limit and continuity defined by the metric or by the natural topology are equivalent.

Let $\mathfrak{M}$ be a (topological) metric space. It can be shown that if $\mathfrak{A} \subset \mathfrak{M}$ is a closed subset, then every point of $\mathfrak{M}$ that is the limit of a sequence of points of $\mathfrak{A}$ belongs to $\mathfrak{A}$.

6.24.5 Linear Space

Let $\mathbb{M}$ be a set, and let $\mathbf{m}$ denote a generic element of $\mathbb{M}$. If we can define the sum $\mathbf{m}_1 + \mathbf{m}_2$ of two elements of $\mathbb{M}$ and the multiplication $\lambda\,\mathbf{m}$ of an element of $\mathbb{M}$ by a real number, verifying the following conditions:

$$
\begin{aligned}
& \mathbf{m}_1 + \mathbf{m}_2 \;=\; \mathbf{m}_2 + \mathbf{m}_1 \quad , \\
& (\mathbf{m}_1 + \mathbf{m}_2) + \mathbf{m}_3 \;=\; \mathbf{m}_1 + (\mathbf{m}_2 + \mathbf{m}_3) \quad , \\
& \text{there exists } \mathbf{0} \in \mathbb{M} \text{ such that } \mathbf{m} + \mathbf{0} \;=\; \mathbf{m} \quad \text{for any } \mathbf{m} \quad , \\
& \text{to each } \mathbf{m} \text{ there corresponds } (-\mathbf{m}) \text{ such that } \mathbf{m} + (-\mathbf{m}) \;=\; \mathbf{0} \quad , \\
& \lambda\,(\mathbf{m}_1 + \mathbf{m}_2) \;=\; \lambda\,\mathbf{m}_1 + \lambda\,\mathbf{m}_2 \quad , \\
& (\lambda + \mu)\,\mathbf{m} \;=\; \lambda\,\mathbf{m} + \mu\,\mathbf{m} \quad , \\
& (\lambda\,\mu)\,\mathbf{m} \;=\; \lambda\,(\mu\,\mathbf{m}) \quad , \\
& 1\,\mathbf{m} \;=\; \mathbf{m} \quad ,
\end{aligned}
\tag{6.423}
$$

then $\mathbb{M}$ is called a (real) *linear vector space*, or *vector space*, or *linear space*. The elements of $\mathbb{M}$ are called *vectors*.

Example 6.12. *Let $\mathbb{E}^3$ be the usual Euclidean space, $\mathbf{x}$ be a generic point of $\mathbb{E}^3$, and $\mathbf{x} \to m(\mathbf{x})$ represent a function from $\mathbb{E}^3$ into an space of scalars $\mathbb{K}$. If the definitions*

$$(\mathbf{m}_1 + \mathbf{m}_2)(\mathbf{x}) \;=\; m_1(\mathbf{x}) + m_2(\mathbf{x}) \quad , \tag{6.424a}$$

$$(\lambda\,\mathbf{m})(\mathbf{x}) \;=\; \lambda\,m(\mathbf{x}) \tag{6.424b}$$

make sense, the set of all such functions is a linear space, denoted $\mathbb{F}$. *The subspace of* $\mathbb{F}$
formed by the continuous *functions is also a linear space, denoted* $\mathbb{C}$. *The subspace of* $\mathbb{F}$
formed by n-times differentiable functions is also a linear space, denoted $\mathbb{C}^n$.

Let $\mathbb{M}$ be a linear space. It is a *topological* linear space if it is furnished with a topological structure compatible with the structure of a linear space, i.e., such that the applications $(\mathbf{m}_1, \mathbf{m}_2) \rightarrow \mathbf{m}_1 + \mathbf{m}_2$ and $(\lambda, \mathbf{m}) \rightarrow \lambda\,\mathbf{m}$ are continuous (with respect to the topology).

Let $\mathbb{M}$ be a linear space. A *norm* over $\mathbb{M}$ associates any element $\mathbf{m}$ of $\mathbb{M}$ with a *positive real number* denoted $\|\mathbf{m}\|$ verifying the following conditions

$$
\begin{aligned}
\|\mathbf{m}\| &= 0 \quad \Leftrightarrow \quad \mathbf{m} = \mathbf{0} \quad, \\
\|\lambda\,\mathbf{m}\| &= |\lambda|\,\|\mathbf{m}\| \qquad \text{for any } \lambda \text{ and } \mathbf{m} \quad, \\
\|\mathbf{m} + \mathbf{n}\| &\leq \|\mathbf{m}\| + \|\mathbf{n}\| \qquad \text{for any } \mathbf{m} \text{ and } \mathbf{n} \quad.
\end{aligned}
\tag{6.425}
$$

A linear space furnished with a norm is named a *normed linear space.*

Example 6.13. *Let us consider a linear space* $\mathbb{F}$ *of functions* $\mathbf{x} \rightarrow m(\mathbf{x})$ *from* $\mathbb{E}^3$ *into* $\mathbb{K}$, *and, for* $1 \leq p < \infty$, *the subspace of functions of* $\mathbb{F}$, *for which the expression*

$$
\|\mathbf{m}\| = \left(\int_{E^3} dV(\mathbf{x}) \, \frac{|m(\mathbf{x})|^p}{s(\mathbf{x})^p} \right)^{1/p}
\tag{6.426}
$$

makes sense and is finite, where $1/s(\mathbf{x})^p$ *is a given weight function (ensuring in particular the physical adimensionality of* $\|\mathbf{m}\|$ *). This subspace is also a linear space, and* $\|\cdot\|$ *defines a norm. This linear space is denoted* $\mathbb{L}_p$ *and plays an important role in mathematical physics (more precisely, an element of the space* $\mathbb{L}_p$ *is not a function, but is of the class of functions that are identical almost everywhere, i.e., such that the norm of their difference is null).*

Example 6.14. *For* $p = 2$, *the previous definition can be generalized. Let* $C(\mathbf{x}, \mathbf{x}')$ *be a symmetric and positive definite function, i.e., a function such that, for any* $\phi(\mathbf{x})$ *and any* $\mathcal{V} \subset \mathbb{E}^3$, *the sum*

$$
\int_{\mathcal{V}} dV(\mathbf{x}) \int_{\mathcal{V}} dV(\mathbf{x}') \, \phi(\mathbf{x}) \, C(\mathbf{x}, \mathbf{x}') \, \phi(\mathbf{x}')
\tag{6.427}
$$

is defined and is finite. A distribution (see Appendix 6.25) $C^{-1}(\mathbf{x}, \mathbf{x}')$ *can then be defined by*

$$
\int_{\mathcal{V}} dV(\mathbf{x}') \, C(\mathbf{x}, \mathbf{x}') \, C^{-1}(\mathbf{x}', \mathbf{x}'') = \delta(\mathbf{x} - \mathbf{x}'') \quad,
\tag{6.428}
$$

and a linear space $\mathbb{M}$ *can be defined as the set of functions* $\mathbf{x} \rightarrow m(\mathbf{x})$ *for which the integral sum*

$$
\|\mathbf{m}\| = \left(\int_{\mathcal{V}} dV(\mathbf{x}) \int_{\mathcal{V}} dV(\mathbf{x}') \, m(\mathbf{x}) \, C^{-1}(\mathbf{x}, \mathbf{x}') \, m(\mathbf{x}') \right)^{1/2}
\tag{6.429}
$$

is defined and is finite. It is easy to show that $\| \cdot \|$ defines a norm called a least-squares
*norm. This space $\mathbb{M}$ is not $\mathbb{L}_2$, but can be shown to be isomorphic with $\mathbb{L}_2$ (introduce
the positive definite operator $\mathbf{C}$ whose kernel is $C(\mathbf{x}, \mathbf{x}')$, define the square root of $\mathbf{C}$,
verifying $\mathbf{C} = \mathbf{\Psi} \mathbf{\Psi}^t$, and introduce a new space by*

$$\mathbb{M}' = \mathbf{\Psi}^{-1} \mathbb{M} \quad ; \tag{6.430}$$

this defines an isomorphism and induces the L_2-norm over $\mathbb{M}'$).

Example 6.15. *It has been mentioned in Example 5.13 in the main text that the covariance
function*

$$C(t, t') = \sigma^2 \exp\left(-\frac{|t - t'|}{L}\right) \tag{6.431}$$

induces the Sobolev H_1-norm.

It is easy to see that the expression

$$D(\mathbf{m}_1, \mathbf{m}_2) = \| \mathbf{m}_1 - \mathbf{m}_2 \| \tag{6.432}$$

defines a distance over $\mathbb{M}$: a normed linear space is always a metric space. In particular,
this allows us to define a Cauchy sequence of vectors. If every Cauchy sequence of vectors
of $\mathbb{M}$ converges into an element of $\mathbb{M}$, $\mathbb{M}$ is complete. A complete linear space is termed
a *Banach space*.

Example 6.16. *The spaces $\mathbb{L}_p$ are Banach spaces (i.e., they are complete). The Sobolev
spaces $\mathbb{H}_P$ (see Appendix 6.25) are complete.*

Let $\mathbb{M}$ be a real linear space. A subset $\mathbb{S}$ of $\mathbb{M}$ is called a *linear subspace* if the
following conditions are satisfied:

$$\begin{aligned}
\mathbf{m}_1 , \mathbf{m}_2 \in \mathbb{S} &\quad \Rightarrow \quad \mathbf{m}_1 + \mathbf{m}_2 \in \mathbb{S} \quad , \\
\mathbf{m} \in \mathbb{S} &\quad \Rightarrow \quad \lambda \, \mathbf{m} \in \mathbb{S} \quad .
\end{aligned} \tag{6.433}$$

It is easy to see that a linear subspace is itself a linear space.

Example 6.17. *The space of continuous functions is a linear subspace of a linear space of
functions.*

6.24.6 Dimension of a Linear Space, Basis of a Linear Space

Let $\mathbb{M}$ be a linear space and $\mathbf{m}_1, \mathbf{m}_2, \ldots$ be a finite or infinite set of elements of $\mathbb{M}$. If
none of the $\mathbf{m}_i$ can be obtained as a linear combination of the others, the $\mathbf{m}_i$ are *linearly
independent*.

Let $\mathbb{M}$ be a linear space. If there is some positive integer N such that $\mathbb{M}$ contains a
set of N vectors that are linearly independent, while every set of $N + 1$ vectors are linearly

dependent, then $\mathbb{M}$ is called *finite dimensional* and N is called the *dimension* of $\mathbb{M}$. If $\mathbb{M}$ is not finite dimensional, then it is called *infinite dimensional*.

A set of vectors of a linear space $\mathbb{M}$ is called a *basis* if it is linearly independent and if it can generate the whole $\mathbb{M}$ by linear combination.

Example 6.18. *Let $\mathbb{M}$ be the space of periodical $[\, m(t + 2\pi) = m(t)\,]$, symmetrical $[\, m(-t) = m(t)\,]$ functions. As this space is infinite dimensional, any basis has an infinite number of elements. A first example is the countable basis*

$$b_n(t) = \cos nt \quad , \qquad n = 0, 1, \ldots \quad . \tag{6.434}$$

Using the well-known Fourier decomposition, any function of $\mathbb{M}$ can be written

$$m(t) = \sum_{n=0}^{\infty} c_n b_n(t) \quad , \tag{6.435}$$

with $c_0 = \frac{1}{2\pi} \int_0^{2\pi} dt\, m(t)$ and $c_n = \frac{1}{\pi} \int_0^{2\pi} dt\, m(t) \cos nt \quad (n = 1, 2, \ldots)$. By linguistic abuse, an (uncountable) basis (of distributions) can be considered,

$$b(v, t) = \delta(v - t) \quad , \qquad -\infty < v < +\infty \quad , \tag{6.436}$$

where δ represents Dirac's delta function. Any function $m(t)$ can be developed into that basis:

$$m(t) = \int_{-\infty}^{\infty} dv\, c(v)\, b(v, t) \quad , \tag{6.437}$$

with

$$c(v) = m(v) \quad . \tag{6.438}$$

6.24.7 Linear Operator

Let $\mathbb{M}$ and $\mathbb{D}$ be two topological linear spaces and $\mathbf{m} \to \mathbf{G}(\mathbf{m})$ be an operator from $\mathbb{M}$ into $\mathbb{D}$. $\mathbf{G}$ is *linear* if

$$\mathbf{G}(\mathbf{m}_1 + \mathbf{m}_2) = \mathbf{G}(\mathbf{m}_1) + \mathbf{G}(\mathbf{m}_2) \quad , \qquad \mathbf{G}(\lambda\, \mathbf{m}) = \lambda\, \mathbf{G}(\mathbf{m}) \tag{6.439}$$

whenever λ is a scalar and $\mathbf{m}$, $\mathbf{m}_1$, $\mathbf{m}_2$ are vectors of $\mathbb{M}$. Usually, linear operators are represented by capital letters. If $\mathbf{G}$ is linear, the notation $\mathbf{Gm}$ is preferred to $\mathbf{G}(\mathbf{m})$:

$$\mathbf{G}(\mathbf{m}) = \mathbf{Gm} \quad . \tag{6.440}$$

The linear operator $\mathbf{G}$ is called *continuous* if $\mathbf{m} \to \mathbf{m}_0$ (in the topology of $\mathbb{M}$) implies $\mathbf{Gm} \to \mathbf{Gm}_0$ (in the topology of $\mathbb{D}$).

A linear operator defined over a finite-dimensional space is always continuous. A linear operator over an infinite-dimensional space may be discontinuous.

Example 6.19. *The derivative operator is linear. The derivative of the null function is the null function. But it is easy to define a sequence of functions tending to the null function but such that the limit of their derivatives does not tend to the null function. This implies that the derivative operator, although linear, is not continuous.*

The space of all continuous linear operators from $\mathbb{M}$ into $\mathbb{D}$ is denoted by $\mathbb{L}(\mathbb{M}, \mathbb{D})$. Defining the sum of two operators and the multiplication of an operator by a scalar by

$$(\mathbf{G}_1 + \mathbf{G}_2)(\mathbf{m}) = \mathbf{G}_1\,\mathbf{m} + \mathbf{G}_2\,\mathbf{m} \quad ,$$
$$(\lambda\,\mathbf{G})(\mathbf{m}) = \lambda\,\mathbf{G}\,\mathbf{m} \quad , \tag{6.441}$$

the space $\mathbb{L}(\mathbb{M}, \mathbb{D})$ is a linear space.

Let $\mathbb{M}$ and $\mathbb{D}$ be two normed linear spaces and $\mathbf{G}$ be a linear operator from $\mathbb{M}$ into $\mathbb{D}$. It can be shown that $\mathbf{G}$ is continuous if and only if there exists a constant $c > 0$ such that, for any $\mathbf{m} \in \mathbb{M}$,

$$\|\,\mathbf{G}\,\mathbf{m}\,\| \le c\,\|\,\mathbf{m}\,\| \quad . \tag{6.442}$$

Then, the *norm* of a continuous operator can be defined by

$$\|\,\mathbf{G}\,\| = \sup_{\mathbf{m}\in M\,,\,\mathbf{m}\neq 0} \frac{\|\,\mathbf{G}\,\mathbf{m}\,\|}{\|\,\mathbf{m}\,\|} \quad . \tag{6.443}$$

Let $\mathbb{M}_1$ and $\mathbb{M}_2$ be two Banach spaces and $\mathbf{L}$ be a continuous linear operator from $\mathbb{M}_1$ into $\mathbb{M}_2$. If $\mathbf{L}$ is a bijection, then $\mathbf{L}^{-1}$ is also a continuous linear operator.

A bijection between the vector linear spaces $\mathbb{M}_1$ and $\mathbb{M}_2$ is said to be an *isomorphism* of $\mathbb{M}_1$ onto $\mathbb{M}_2$. Two linear spaces $\mathbb{M}_1$ and $\mathbb{M}_2$ are said to be *isomorphic* if there exists an isomorphism of $\mathbb{M}_1$ onto $\mathbb{M}_2$.

Let $\mathbf{G}$ be a linear operator from $\mathbb{M}$ into $\mathbb{D}$. The linear subspace $\mathbb{K}$ of $\mathbb{M}$ for which

$$\mathbf{m} \in \mathbb{K} \quad \Rightarrow \quad \mathbf{G}\,\mathbf{m} = 0 \tag{6.444}$$

is called the *kernel*, or the *null space*, of $\mathbf{G}$.

Let $\mathbf{G}$ be a linear operator from $\mathbb{M}$ into $\mathbb{D}$. The dimension of $\mathbf{G}\,\mathbb{M}$ (image of $\mathbb{M}$ through $\mathbf{G}$) is called the *rank* of $\mathbf{G}$. It can be shown that it equals the difference between the dimension of $\mathbb{M}$ and the dimension of the kernel of $\mathbf{G}$.

A sequence of linear operators $\mathbf{H}_1, \mathbf{H}_2, \dots$ is called *uniformly convergent* if it converges in norm, i.e., if

$$\lim_{n\to\infty} \|\,\mathbf{H}_n - \mathbf{H}\,\| = 0 \quad . \tag{6.445}$$

Let $\mathbf{H}$ be a continuous linear operator from the Banach space $\mathbb{M}$ into itself. Then, the limit

$$r(\mathbf{H}) = \lim_{n\to\infty} \|\,\mathbf{H}^n\,\|^{1/n} \tag{6.446}$$

exists and is called the *spectral radius* of $\mathbf{H}$. If the spectral radius of $\mathbf{H}$ is less than one, then $(\mathbf{I} - \mathbf{H})^{-1}$ exists and is a continuous linear operator, and

$$(\mathbf{I} - \mathbf{H})^{-1} = \sum_{n=0}^{\infty} \mathbf{H}^n \qquad \text{(Neumann series)} \quad . \tag{6.447}$$

6.24.8 Dual of a Linear Space

Let $\mathbb{M}$ be a real topological linear space. A *form* over $\mathbb{M}$ is an operator from $\mathbb{M}$ into the real line $\mathfrak{R}$. A *linear form* is a linear operator from $\mathbb{M}$ into $\mathfrak{R}$.

The space of all continuous linear forms over $\mathbb{M}$ is termed the (topological) *dual space* of $\mathbb{M}$ and is denoted by $\mathbb{M}^*$. A generic element of $\mathbb{M}^*$ is denoted by $\hat{\mathbf{m}}$. The result of the action of $\hat{\mathbf{m}} \in \mathbb{M}^*$ over an $\mathbf{m} \in \mathbb{M}$ is denoted by either of the two notations

$$\langle \hat{\mathbf{m}}, \mathbf{m} \rangle = \hat{\mathbf{m}}^t \mathbf{m} . \tag{6.448}$$

The second notation is useful for numerical computations, because it recalls matricial notation.

Let $\hat{\mathbf{m}} \in \mathbb{M}^*$. The expression

$$\| \hat{\mathbf{m}} \| = \sup_{\mathbf{m} \in \mathbb{M}, \mathbf{m} \neq 0} \frac{| \langle \hat{\mathbf{m}}, \mathbf{m} \rangle |}{\| \mathbf{m} \|} \tag{6.449}$$

defines a norm over $\mathbb{M}^*$. It can be shown that with such a norm, $\hat{\mathbf{m}}$ is a Banach space (i.e., it is complete).

Example 6.20. *For $1 < p < \infty$, the dual of $\mathbb{L}_p$ is $\mathbb{L}_q$, with $1/p + 1/q = 1$.*

Let $\mathbb{M}$ be a normed linear space and $\mathbb{M}^*$ be its dual. It can be shown that, for any nonnull $\mathbf{m} \in \mathbb{M}$, there exists $\hat{\mathbf{m}} \in \mathbb{M}^*$ verifying

$$\| \hat{\mathbf{m}} \| = \| \mathbf{m} \| \quad , \quad \langle \hat{\mathbf{m}}, \mathbf{m} \rangle = \| \mathbf{m} \|^2 = \| \hat{\mathbf{m}} \|^2 . \tag{6.450}$$

Let $\mathbb{M}$ be a normed linear space and $(\mathbb{M}^*)^*$ be its bidual (dual of the dual). Then, there exists a continuous linear application $\mathbf{J}$ from $\mathbb{M}$ into its bidual such that

$$\begin{array}{ll}
\text{(i)} & \mathbf{J} \text{ is injective} , \\
\text{(ii)} & \| \mathbf{J} \mathbf{m} \| = \| \mathbf{m} \| \quad \text{for any } \mathbf{m} \in \mathbb{M} .
\end{array} \tag{6.451}$$

It is then possible to identify $\mathbb{M}$ and its image $\mathbf{J} \mathbb{M} \subset (\mathbb{M}^*)^*$. We will see later (Riesz theorem) that in Hilbert spaces, there exists a bijection between a space $\mathbb{M}$ and its dual $\mathbb{M}^*$. But they remain fundamentally different spaces, and the identification of a space and its dual, although perhaps useful for pure mathematical developments, is of no interest for practical applications. On the contrary, the identification between a space and (a subset of) its bidual is always useful (for the purposes of inverse theory). If $\mathbb{M}$ equals its bidual, we say that $\mathbb{M}$ is *reflexive*.

Example 6.21. *For $1 < p < \infty$, $\mathbb{L}_p$ is a reflexive Banach space. The dual of $\mathbb{L}_1$ is $\mathbb{L}_\infty$, but $\mathbb{L}_1$ is not reflexive. The Sobolev spaces $\mathbb{W}_p{}^m$ (see Appendix 6.25) are reflexive Banach spaces (for $1 < p < \infty$).*

Let $\mathbb{M}$ be a topological linear space and $\mathbb{M}^*$ be its dual. The sequence $\mathbf{m}_1, \mathbf{m}_2, \ldots$ of elements of $\mathbb{M}$ converges weakly to $\mathbf{m} \in \mathbb{M}$ if

$$\lim_{n \to \infty} \langle \hat{\mathbf{m}}, \mathbf{m}_n \rangle = \langle \hat{\mathbf{m}}, \mathbf{m} \rangle \quad \text{for any} \quad \hat{\mathbf{m}} \in \mathbb{M}^* . \tag{6.452}$$

It *converges strongly* if

$$\lim_{n\to\infty} \| \, \mathbf{m}_n - \mathbf{m} \, \| \, = \, 0 \quad . \tag{6.453}$$

It can be shown that if $\mathbb{M}$ has finite dimension, then $\mathbb{M}^*$ has the same dimension.

6.24.9 Transpose Operator

Let $\mathbb{M}$ and $\mathbb{D}$ be two linear spaces and $\mathbf{G}$ be a linear operator from $\mathbb{M}$ into $\mathbb{D}$. The *transpose* of $\mathbf{G}$ is denoted $\mathbf{G}^t$ and is the linear operator from $\mathbb{D}^*$ into $\mathbb{M}^*$ defined by

$$\langle \, \mathbf{G}^t \, \hat{\mathbf{d}} \, , \, \mathbf{m} \, \rangle_{\mathbb{M}} \, = \, \langle \, \hat{\mathbf{d}} \, , \, \mathbf{G} \, \mathbf{m} \, \rangle_{\mathbb{D}} \quad , \tag{6.454a}$$

or, using the notation of equation (6.448),

$$(\mathbf{G}^t \, \hat{\mathbf{d}})^t \, \mathbf{m} \, = \, \hat{\mathbf{d}}^t \, \mathbf{G} \, \mathbf{m} \quad . \tag{6.454b}$$

The reader will easily give sense to and demonstrate the equivalences

$$(\mathbf{G}_1 + \mathbf{G}_2)^t = \mathbf{G}_1{}^t + \mathbf{G}_2{}^t \quad , \quad (\mathbf{G}_1 \, \mathbf{G}_2)^t = \mathbf{G}_2{}^t \mathbf{G}_1{}^t \quad , \quad (\mathbf{G}^t)^{-1} = (\mathbf{G}^{-1})^t \quad , \quad (\mathbf{G}^t)^t = \mathbf{G} \quad . \tag{6.455}$$

For more details, see Appendix 6.14.

6.24.10 Hilbert Spaces

Let $\mathbb{M}$ be a real linear space. A *bilinear form* over $\mathbb{M}$ is an application $(\mathbf{m}_1, \mathbf{m}_2) \to \mathbf{W}(\mathbf{m}_1, \mathbf{m}_2)$ from $\mathbb{M} \times \mathbb{M}$ into $\mathfrak{R}$ such that

$$\begin{aligned}
\mathbf{W}(\mathbf{m}_1 + \mathbf{m}_2, \mathbf{m}) \, &= \, \mathbf{W}(\mathbf{m}_1, \mathbf{m}) + \mathbf{W}(\mathbf{m}_2, \mathbf{m}) \quad , \\
\mathbf{W}(\mathbf{m}, \mathbf{m}_1 + \mathbf{m}_2) \, &= \, \mathbf{W}(\mathbf{m}, \mathbf{m}_1) + \mathbf{W}(\mathbf{m}, \mathbf{m}_2) \quad , \\
\mathbf{W}(\lambda \, \mathbf{m}_1, \mathbf{m}_2) \, &= \, \mathbf{W}(\mathbf{m}_1, \lambda \, \mathbf{m}_2) \, = \, \lambda \, \mathbf{W}(\mathbf{m}_1, \mathbf{m}_2) \quad .
\end{aligned} \tag{6.456}$$

A bilinear form is *symmetric* if

$$\mathbf{W}(\mathbf{m}_1, \mathbf{m}_2) \, = \, \mathbf{W}(\mathbf{m}_2, \mathbf{m}_1) \quad . \tag{6.457}$$

For a symmetric bilinear form, the following notation is used:

$$\mathbf{W}(\mathbf{m}_1, \mathbf{m}_2) \, = \, \mathbf{m}_1{}^t \, \mathbf{W} \, \mathbf{m}_2 \, = \, \mathbf{m}_2{}^t \, \mathbf{W} \, \mathbf{m}_1 \quad . \tag{6.458}$$

A bilinear form is *positive definite* if

$$\mathbf{m} \neq 0 \quad \Rightarrow \quad \mathbf{m}^t \, \mathbf{W} \, \mathbf{m} > 0 \quad . \tag{6.459}$$

It can be shown that if a bilinear form is positive definite, then it is also symmetric. It can also be shown that if $\mathbf{W}$ is positive definite, then the expression

$$\| \, \mathbf{m} \, \| \, = \, (\mathbf{m}^t \, \mathbf{W} \, \mathbf{m})^{1/2} \tag{6.460}$$

is a norm over $\mathbb{M}$.

Let $\mathbb{M}$ be a real linear space and $\mathbf{W}$ be a positive definite bilinear form over $\mathbb{M}$. The pair $(\mathbb{M}, \mathbf{W})$ is called a real *pre-Hilbert space*. A pre-Hilbert space that is complete is termed a *Hilbert space*. If $(\mathbb{M}, \mathbf{W})$ is a Hilbert space, the application $\mathbf{m}_1{}^t \mathbf{W} \mathbf{m}_2$ is called a *scalar product* over $\mathbb{M}$ (and is sometimes denoted $(\mathbf{m}_1, \mathbf{m}_2)$).

Let $\mathbb{M}$ be a Banach space. It can be shown that if the norm over $\mathbb{M}$ verifies

$$\| \mathbf{m}_1 \|^2 + \| \mathbf{m}_2 \|^2 \ = \ \frac{1}{2} \left(\| \mathbf{m}_1 + \mathbf{m}_2 \|^2 + \| \mathbf{m}_1 - \mathbf{m}_2 \|^2 \right) \quad , \tag{6.461}$$

then $\mathbb{M}$ is a Hilbert space (i.e., the norm can be defined from a scalar product through (6.460)). Then, conversely,

$$\mathbf{m}_1{}^t \mathbf{W} \mathbf{m}_2 \ = \ \frac{1}{4} \left(\| \mathbf{m}_1 + \mathbf{m}_2 \|^2 - \| \mathbf{m}_1 - \mathbf{m}_2 \|^2 \right) \quad . \tag{6.462}$$

Example 6.22. *$\mathbb{L}_2$ is a Hilbert space. The Sobolev space $\mathbb{H}^{2m}$ is a Hilbert space (see Appendix 6.25).*

Property 6.1. Riesz Theorem. *Let $\mathbb{M}$ be a Hilbert space and $\mathbb{M}^*$ be its dual. Let $\hat{\mathbf{m}} \in \mathbb{M}^*$. Then, there exists a unique $\mathbf{m} \in \mathbb{M}$ such that*

$$\hat{\mathbf{m}}^t \, \mathbf{m}' \ = \ \mathbf{m}' \, \mathbf{W} \, \mathbf{m}' \quad \text{for any } \mathbf{m}' \in \mathbb{M} \quad ,$$
$$\| \hat{\mathbf{m}} \| \ = \ \| \mathbf{m} \| \quad . \tag{6.463}$$

The reciprocal also holds: letting $\mathbf{m} \in \mathbb{M}$, there exists a unique $\hat{\mathbf{m}} \in \mathbb{M}^*$:

$$\hat{\mathbf{m}}^t \, \mathbf{m}' \ = \ \mathbf{m}' \, \mathbf{W} \, \mathbf{m}' \quad \text{for any } \mathbf{m}' \in \mathbb{M} \quad ,$$
$$\| \hat{\mathbf{m}} \| \ = \ \| \mathbf{m} \| \quad . \tag{6.464}$$

The Riesz theorem shows that there exists an isometric isomorphism *between $\mathbb{M}$ and its dual $\mathbb{M}^*$. Nevertheless, as in inverse problem theory a space and its dual play very different roles, they should never be identified.*

It can be shown that a Hilbert space is a reflexive Banach space.

Example 6.23. *Let $\mathbb{M}$ be the $\mathbb{L}_2$ space of functions defined over $\Re^n$ and $\mathbb{D}$ be the $\mathbb{L}_2$ space of functions defined over $\Re^m$. Let $G(\mathbf{y}, \mathbf{x})$ be a function from $\Re^m \times \Re^n$ into $\Re$ such that*

$$\int_{\Re^m} d\mathbf{y} \int_{\Re^n} d\mathbf{x} \, | G(\mathbf{y}, \mathbf{x}) |^2 \ < \infty \quad . \tag{6.465}$$

Then, the operator $\mathbf{G}$ from $\mathbb{M}$ into $\mathbb{D}$ defined by

$$d(\mathbf{y}) \ = \ \int_{\Re^n} d\mathbf{x} \, G(\mathbf{y}, \mathbf{x}) \, m(\mathbf{x}) \tag{6.466}$$

is a continuous linear operator.

6.24.11 Adjoint Operator

Let $\mathbb{M}$ and $\mathbb{D}$ be two Hilbert spaces with scalar products denoted respectively $\mathbf{W}_m(\cdot, \cdot)$ and $\mathbf{W}_d(\cdot, \cdot)$ and let $\mathbf{G}$ be a linear operator from $\mathbb{M}$ into $\mathbb{D}$. The *adjoint* of $\mathbf{G}$ is denoted $\mathbf{G}^*$ and is the linear operator from $\mathbb{D}$ into $\mathbb{M}$ defined by

$$\mathbf{W}_m(\mathbf{G}^* \mathbf{d}, \mathbf{m}) = \mathbf{W}_d(\mathbf{d}, \mathbf{G}\,\mathbf{m}) \quad . \tag{6.467}$$

Using the notation of (6.458) and the definition (6.454) of transpose operator, it follows that

$$\mathbf{G}^* = \mathbf{W}_m^{-1}\,\mathbf{G}^t\,\mathbf{W}_m \quad . \tag{6.468}$$

It should be emphasized that the terms 'adjoint' and 'transpose' are not synonymous. The transpose operator is defined for arbitrary linear spaces, irrespective of the existence of any scalar product (think, for instance, of the transpose of a matrix), while the adjoint operator is defined only for scalar product (Hilbert) spaces.

6.25 Usual Functional Spaces

In the following, $\mathbf{x}$ denotes a point of a Euclidean n-dimensional space $\mathbb{E}$ with Cartesian coordinates $\mathbf{x} = (x^1, x^2, \ldots)$; $\mathbf{x} \to f(\mathbf{x})$ denotes a function from an open subset $\mathcal{V}$ of $\mathbb{E}$ into $\Re$. The symbol $\int d\mathbf{x}$ denotes the volume integral over $\mathcal{V}$, where $d\mathbf{x}$ denotes the volume element.

The space $\mathbb{L}_p$ $(1 \leq p \leq \infty)$: The space of functions such that the expression

$$\| \mathbf{f} \| = \left(\int d\mathbf{x} \, \frac{| f(\mathbf{x}) |^p}{s(\mathbf{x})^p} \right)^{1/p} \tag{6.469}$$

is defined and is finite, where $1/s(\mathbf{x})$ is a given positive weighting function, is called the space $\mathbb{L}_p$. Two functions $\mathbf{f}_1$ and $\mathbf{f}_2$, such that $\| \mathbf{f}_1 - \mathbf{f}_2 \| = 0$, are said to be equal almost everywhere and are identified.

The space $\mathbb{L}_\infty$ is the space of functions for which the expression

$$\| \mathbf{f} \| = \sup \frac{| f(\mathbf{x}) |}{s(\mathbf{x})} \tag{6.470}$$

is finite.

When two functions that are equal almost everywhere are identified, the previous expressions define a norm over $\mathbb{L}_p$. With such a norm, the $\mathbb{L}_p$ spaces are Banach spaces (i.e., they are complete).

The space $\mathbb{L}_2$: In the particular case $p = 2$, a scalar product can be introduced by

$$(\mathbf{f}, \mathbf{g}) = \int d\mathbf{x} \, \frac{f(\mathbf{x})\,g(\mathbf{x})}{s(\mathbf{x})^2} \quad . \tag{6.471}$$

Then,

$$\| \mathbf{f} \| = (\mathbf{f}, \mathbf{f})^{1/2} \quad . \tag{6.472}$$

$\mathbb{L}_2$ is a Hilbert space (i.e., it is complete).

If $f(\mathbf{x})$ is, in fact, a vector-valued function $\mathbf{f}(\mathbf{x})$ (i.e., if it takes values in $\Re^n$), it has components $f^1(\mathbf{x})$, $f^2(\mathbf{x})$, ... Then, the scalar product is defined by

$$(\mathbf{f} , \mathbf{g}) = \sum_i \int d\mathbf{x} \, \frac{f^i(\mathbf{x}) \, g^i(\mathbf{x})}{s^i(\mathbf{x})^2} \quad . \tag{6.473}$$

The corresponding space is a Hilbert space and is also denoted $\mathbb{L}_2$.

The Sobolev space $\mathbb{H}^m$: By definition, $\mathbb{H}^m$ is the space of $\mathbb{L}_2$ functions whose partial derivatives up to order m are $\mathbb{L}_2$ functions. For instance, $\mathbb{H}^0$ is $\mathbb{L}_2$, $\mathbb{H}^1$ is the space of functions $\mathbf{f} \in \mathbb{L}_2$ such that $\partial f/\partial x^i \in \mathbb{L}_2$ (for any i), etc. Formally, $\mathbb{H}^m$ is the space of functions such that

$$\frac{\partial^{\alpha_1 + \cdots + \alpha_n} f}{\partial (x^1)^{\alpha_1} \cdots \partial (x^n)^{\alpha_n}} \in \mathbb{L}_2 \qquad \text{for } 0 \le (\alpha_1 + \cdots + \alpha_n) \le m \quad . \tag{6.474}$$

Let $\mathbf{f}$ and $\mathbf{g}$ be two functions of $\mathbb{H}^m$. A *scalar product* is defined by

$$(\mathbf{f} , \mathbf{g}) = \sum_{0 \le (\alpha_1 + \cdots + \alpha_n) \le m} \int d\mathbf{x} \, \frac{\partial^{\alpha_1 + \cdots + \alpha_n} f}{\partial (x^1)^{\alpha_1} \cdots \partial (x^n)^{\alpha_n}} \frac{\partial^{\alpha_1 + \cdots + \alpha_n} g}{\partial (x^1)^{\alpha_1} \cdots \partial (x^n)^{\alpha_n}} \tag{6.475}$$

(to shorten the notation, weighting factors have been omitted). With such a scalar product, the Sobolev spaces $\mathbb{H}^m$ are complete.

The associated (squared) norm is given by

$$\| \mathbf{f} \|^2 = \sum_{0 \le (\alpha_1 + \cdots + \alpha_n) \le m} \int d\mathbf{x} \left(\frac{\partial^{\alpha_1 + \cdots + \alpha_n} f}{\partial (x^1)^{\alpha_1} \cdots \partial (x^n)^{\alpha_n}} \right)^2 \quad . \tag{6.476}$$

Example 6.24. *Let $x \to f(x)$ be a one-dimensional function of a one-dimensional variable. The corresponding $\mathbb{H}^1$ space is the space of $\mathbb{L}_2$ functions such that $\partial f/\partial x$ is $\mathbb{L}_2$. The scalar product is*

$$(\mathbf{f} , \mathbf{g}) = \int dx \, f(x) \, g(x) + \int dx \, \frac{\partial f}{\partial x} \frac{\partial g}{\partial x} \quad , \tag{6.477}$$

and the (squared) norm is given by

$$\| \mathbf{f} \|^2 = \int dx \, f(x)^2 + \int dx \left(\frac{\partial f}{\partial x} \right)^2 \quad . \tag{6.478}$$

The space $\mathbb{H}^2$ is the space of $\mathbb{L}_2$ functions such that $\partial f/\partial x$ and $\partial^2 f/\partial x^2$ are $\mathbb{L}_2$. The scalar product is

$$(\mathbf{f} , \mathbf{g}) = \int dx \, f(x) \, g(x) + \int dx \, \frac{\partial f}{\partial x}(x) \frac{\partial g}{\partial x}(x) + \int dx \, \frac{\partial^2 f}{\partial x^2}(x) \frac{\partial^2 g}{\partial x^2}(x) \quad , \tag{6.479}$$

and the (squared) norm is given by

$$\| \mathbf{f} \|^2 = \int dx \, f(x)^2 + \int dx \left(\frac{\partial f}{\partial x}(x) \right)^2 + \int dx \left(\frac{\partial^2 f}{\partial x^2}(x) \right)^2 \quad . \tag{6.480}$$

Example 6.25. *Let* $(x, y) \mapsto f(x, y)$ *be a one-dimensional function of a two-dimensional variable.* $\mathbb{H}^0$ *is the space* $\mathbb{L}_2$. *The space* $\mathbb{H}^1$ *is the space of* $\mathbb{L}_2$ *functions such that* $\partial f / \partial x$ *and* $\partial f / \partial y$ *are* $\mathbb{L}_2$. *The scalar product is*

$$
\begin{aligned}
(\mathbf{f}, \mathbf{g}) &= \int dx \int dy f(x, y) g(x, y) + \int dx \int dy \frac{\partial f}{\partial x}(x, y) \frac{\partial g}{\partial x})(x, y) \\
&\quad + \int dx \int dy \frac{\partial f}{\partial y}(x, y) \frac{\partial g}{\partial y}(x, y) \\
&= \int dx \int dy\, f(x, y) g(x, y) \\
&\quad + \int dx \int dy \operatorname{grad} f(x, y) \cdot \operatorname{grad} g(x, y) \quad .
\end{aligned}
\tag{6.481}
$$

The (squared) norm is given by

$$
\begin{aligned}
\| \mathbf{f} \|^2 &= \int dx \int dy \left(f(x, y) \right)^2 + \int dx \int dy \left(\frac{\partial f}{\partial x}(x, y) \right)^2 + \int dx \int dy \left(\frac{\partial f}{\partial y}(x, y) \right)^2 \\
&= \int dx \int dy\, (f(x, y))^2 + \int dx \int dy \left(\operatorname{grad} f(x, y) \right)^2 \quad .
\end{aligned}
\tag{6.482}
$$

The space $\mathbb{H}^2$ is the space of $\mathbb{L}_2$ functions such that $\partial f / \partial x$, $\partial f / \partial y$, $\partial^2 f / \partial x \partial y$, $\partial^2 f / \partial x^2$, and $\partial^2 f / \partial y^2$ are $\mathbb{L}_2$. The scalar product is

$$
\begin{aligned}
(\mathbf{f}, \mathbf{g}) &= \int dx \int dy\, f(x, y) g(x, y) + \int dx \int dy \frac{\partial f}{\partial x}(x, y) \frac{\partial g}{\partial x}(x, y) \\
&\quad + \int dx \int dy \frac{\partial f}{\partial y}(x, y) \frac{\partial g}{\partial y}(x, y) + \int dx \int dy \frac{\partial^2 f}{\partial x \partial y}(x, y) \frac{\partial^2 g}{\partial x \partial y}(x, y) \\
&\quad + \int dx \int dy \frac{\partial^2 f}{\partial x^2}(x, y) \frac{\partial^2 g}{\partial x^2}(x, y) + \int dx \int dy \frac{\partial^2 f}{\partial y^2}(x, y) \frac{\partial^2 g}{\partial y^2}(x, y) \quad .
\end{aligned}
\tag{6.483}
$$

The *Sobolev space* $\mathbb{W}_p{}^m$: $\mathbb{W}_p{}^m$ is the space of $\mathbb{L}_p$ functions whose partial derivatives up to order m are $\mathbb{L}_p$ functions. For instance, $\mathbb{W}_2{}^m$ is $\mathbb{H}^m$, $\mathbb{W}_p{}^0$ is $\mathbb{L}_p$, and $\mathbb{W}_p{}^1$ is the space of functions $\mathbf{f} \in \mathbb{L}_p$ such that $\partial f / \partial x^i \in \mathbb{L}_p$ (for $i = 1, \ldots, n$). Formally, $\mathbb{W}_p{}^m$ is the space of functions such that

$$
\frac{\partial^{\alpha_1 + \cdots + \alpha_n} f}{\partial (x^1)^{\alpha_1} \cdots \partial (x^n)^{\alpha_n}} \in L_p \qquad \text{for } 0 \le (\alpha_1 + \cdots + \alpha_n) \le m \quad .
\tag{6.484}
$$

The spaces $\mathbb{W}_p{}^m$ are Banach spaces with the norm

$$
\| \mathbf{f} \| = \left(\sum_{0 \le (\alpha_1 + \cdots + \alpha_n) \le m} \int d\mathbf{x} \left| \frac{\partial^{\alpha_1 + \cdots + \alpha_n} f}{\partial (x^1)^{\alpha_1} \cdots \partial (x^n)^{\alpha_n}} \right|^p \right)^{1/p} \quad .
\tag{6.485}
$$

6.26 Maximum Entropy Probability Density

Let $\mathbf{V}(\mathbf{x})$ be an arbitrary given vector function of a vector $\mathbf{x}$. Let us demonstrate that among all probability densities $f(\mathbf{x})$ for which the mathematical expectation for $\mathbf{V}(\mathbf{x})$ equals $\mathbf{V}_0$,

$$\int d\mathbf{x}\, \mathbf{V}(\mathbf{x})\, f(\mathbf{x}) = \mathbf{V}_0 \quad , \tag{6.486}$$

the one that has *minimum information* (maximum entropy) with respect to a given probability density $\mu(\mathbf{x})$,

$$\int d\mathbf{x}\, f(\mathbf{x}) \log \frac{f(\mathbf{x})}{\mu(\mathbf{x})} \quad \text{minimum} \quad , \tag{6.487}$$

necessarily has the form

$$f(\mathbf{x}) = k\,\mu(\mathbf{x})\, \exp(-\mathbf{W}^t\, \mathbf{V}(\mathbf{x})) \quad , \tag{6.488}$$

where k and $\mathbf{W}$ are constants (independent of $\mathbf{x}$).

The problem is to

$$\text{minimize} \qquad S'(f(\cdot)) = \int d\mathbf{x}\, f(\mathbf{x}) \log \frac{f(\mathbf{x})}{\mu(\mathbf{x})} \tag{6.489}$$

under the constraints

$$\int d\mathbf{x}\, f(\mathbf{x}) = 1 \quad , \qquad \int d\mathbf{x}\, \mathbf{V}(\mathbf{x})\, f(\mathbf{x}) = \mathbf{V}_0 \quad . \tag{6.490}$$

This is a problem of constrained minimization, which is atypical in the sense that the variable is a function (i.e., a variable in an infinite-dimensional space). Nevertheless, the problem can be solved using the classical method of Lagrange's parameters (see Appendix 6.29). The problem of minimization of S' under the constraints (6.490) is equivalent to the problem of unconstrained minimization of

$$S(f(\cdot), U, \mathbf{W}) = \int d\mathbf{x}\, f(\mathbf{x}) \log \frac{f(\mathbf{x})}{\mu(\mathbf{x})} - U\left(1 - \int d\mathbf{x}\, f(\mathbf{x})\right)$$
$$- \mathbf{W}^t \left(\mathbf{V}_0 - \int d\mathbf{x}\, \mathbf{V}(\mathbf{x})\, f(\mathbf{x})\right) \quad , \tag{6.491}$$

because the conditions $\partial S/\partial U = 0$ and $\partial S/\partial \mathbf{W} = 0$ directly impose the constraints (6.490). We have

$$S(f(\cdot) + \delta f(\cdot), U, \mathbf{W}) - S(f(\cdot), U, \mathbf{W}) = \int d\mathbf{x}\, (f(\mathbf{x}) + \delta f(\mathbf{x})) \log \frac{f(\mathbf{x}) + \delta f(\mathbf{x})}{\mu(\mathbf{x})}$$
$$- \int d\mathbf{x}\, f(\mathbf{x}) \log \frac{f(\mathbf{x})}{\mu\mathbf{x}} + U \int d\mathbf{x}\, \delta f(\mathbf{x}) + \mathbf{W}^t \int d\mathbf{x}\, \mathbf{V}(\mathbf{x})\, \delta f(\mathbf{x}) \quad , \tag{6.492}$$

and using the first-order development $\log(1+u) = u + O(u^2)$ gives, everywhere $f(\mathbf{x}) \neq 0$,

$$\log \frac{f(\mathbf{x}) + \delta f(\mathbf{x})}{\mu(\mathbf{x})} = \log \frac{f(\mathbf{x})}{\mu(\mathbf{x})} + \frac{\delta f(\mathbf{x})}{f(\mathbf{x})} + O(\delta f^2) \quad , \tag{6.493}$$

and then,

$$S(f(\cdot) + \delta f(\cdot), U, \mathbf{W}) - S(f(\cdot), U, \mathbf{W})$$
$$= \int d\mathbf{x} \left(\log \frac{f(\mathbf{x})}{\mu(\mathbf{x})} + 1 + U + \mathbf{W}^t \mathbf{V}(\mathbf{x}) \right) \delta f(\mathbf{x}) + O(\delta \mathbf{f}^2) \quad . \tag{6.494}$$

The condition of minimum of S with respect to $f(\mathbf{x})$ causes the factor of $\delta f(\mathbf{x})$ on the right of (6.494) to vanish, from which result (6.488) follows.

6.27 Two Properties of ℓ_p-Norms

6.27.1 First Property

Let us demonstrate the equivalence

$$\hat{\mathbf{x}} = \frac{1}{p} \frac{\partial}{\partial \mathbf{x}} \| \mathbf{x} \|_p^p \qquad \Longleftrightarrow \qquad \mathbf{x} = \frac{1}{q} \frac{\partial}{\partial \hat{\mathbf{x}}} \| \hat{\mathbf{x}} \|_q^q \quad , \tag{6.495}$$

where

$$\frac{1}{p} + \frac{1}{q} = 1 \quad . \tag{6.496}$$

The norm $\| \mathbf{x} \|_p$ is defined by

$$\| \mathbf{x} \|_p = \left(\sum_i \frac{|x^i|^p}{(\sigma^i)^p} \right)^{1/p} \quad . \tag{6.497}$$

From

$$\hat{x}^i = \frac{1}{p} \left(\frac{\partial}{\partial x^i} \| \mathbf{x} \|_p^p \right) \quad , \tag{6.498}$$

it follows that

$$\hat{x}^i = \frac{\mathrm{sg}(x^i) |x^i|^{p-1}}{(\sigma^i)^p} \quad , \tag{6.499}$$

equivalent to the two equations

$$\mathrm{sg}(\hat{x}^i) = \mathrm{sg}(x^i) \quad , \qquad |\hat{x}^i| = \frac{|x^i|^{p-1}}{(\sigma^i)^p} \quad . \tag{6.500}$$

From the second of these equations it follows that

$$|x^i| = (\sigma^i)^{p/(p-1)} |\hat{x}^i|^{1/(p-1)} = (\sigma^i)^q |\hat{x}^i|^{q-1} \quad , \tag{6.501}$$

and, using the first of (6.500),

$$x^i = \frac{\mathrm{sg}(\hat{x}^i) |\hat{x}^i|^{q-1}}{(\hat{\sigma}^i)^q} \quad , \qquad \text{where} \qquad \hat{\sigma}^i = \frac{1}{\sigma^i} \quad . \tag{6.502}$$

Now, defining

$$\| \hat{\mathbf{x}} \|_q = \left(\sum_i \frac{|\hat{x}^i|^p}{(\hat{\sigma}^i)^q} \right)^{1/q} \tag{6.503}$$

gives

$$x^i = \frac{1}{q} \left(\frac{\partial}{\partial \hat{x}^i} \| \hat{\mathbf{x}} \|_q^q \right) . \tag{6.504}$$

6.27.2 Second Property

Let us here demonstrate the identities

$$\hat{\mathbf{x}}^t \mathbf{x} = \| \mathbf{x} \|_p \cdot \| \hat{x} \|_q = \| \mathbf{x} \|_p^p = \| \hat{\mathbf{x}} \|_q^q . \tag{6.505}$$

From equation (6.499), it directly follows that

$$\hat{\mathbf{x}}^t \mathbf{x} = \sum_i \hat{x}^i x^i = \| \mathbf{x} \|_p^p , \tag{6.506}$$

and, from the first of equations (6.502),

$$\hat{\mathbf{x}}^t \mathbf{x} = \| \hat{\mathbf{x}} \|_q^q . \tag{6.507}$$

From the identity

$$\| \mathbf{x} \|_p^p = \| \hat{\mathbf{x}} \|_q^q \tag{6.508}$$

already demonstrated, it follows that

$$\| \hat{\mathbf{x}} \|_q = \| \mathbf{x} \|_q^{p/q} = \| \mathbf{x} \|_p^{p-1} , \tag{6.509}$$

i.e.,

$$\| \hat{\mathbf{x}} \|_q \cdot \| \mathbf{x} \|_p = \| \mathbf{x} \|_p^p . \tag{6.510}$$

6.28 Discrete Derivative Operator

We have seen in section 5.4.2 that there is a boundary condition to be satisfied if we wish the derivative operator to be antisymmetric. This contrasts with matrix formulations, where the transpose matrix is defined unconditionally. Let us compare here the functional and the discrete formulation of the derivative of a function.

In what follows, let us denote

$$x_2 = x_1 + \Delta x \quad , \quad x_3 = x_2 + \Delta x \quad , \quad \dots \quad , \tag{6.511}$$

and, to simplify notation, let us discretize functions using only five points. To find a precise finite representation of the derivative operator, it is better to start with the integration operator. The equation

$$
\begin{bmatrix} F(x_1) \\ F(x_2) \\ F(x_3) \\ F(x_4) \\ F(x_5) \end{bmatrix}
= \Delta x
\begin{bmatrix}
1 & 0 & 0 & 0 & 0 \\
1 & 1 & 0 & 0 & 0 \\
1 & 1 & 1 & 0 & 0 \\
1 & 1 & 1 & 1 & 0 \\
1 & 1 & 1 & 1 & 1
\end{bmatrix}
\begin{bmatrix} f(x_1) \\ f(x_2) \\ f(x_3) \\ f(x_4) \\ f(x_5) \end{bmatrix}
\tag{6.512}
$$

gives

$$
\begin{aligned}
F(x_1) &= \Delta x \, f(x_1) \quad, \\
F(x_2) &= \Delta x \, (f(x_1) + f(x_2)) \quad, \\
F(x_3) &= \Delta x \, (f(x_1) + f(x_2) + f(x_3)) \quad, \\
F(x_4) &= \Delta x \, (f(x_1) + f(x_2) + f(x_3) + f(x_4)) \quad, \\
F(x_5) &= \Delta x \, (f(x_1) + f(x_2) + f(x_3) + f(x_4) + f(x_5)) \quad,
\end{aligned}
\tag{6.513}
$$

clearly a discrete approximation to the functional relation

$$
F(x) = \int_{x_1}^{x} dx' \, f(x') \quad.
\tag{6.514}
$$

The inverse of this relation is

$$
f(x) = \frac{dF}{dx}(x) \quad,
\tag{6.515}
$$

and its discrete version is obtained by computing the inverse of the matrix in equation (6.512),

$$
\begin{bmatrix} f(x_1) \\ f(x_2) \\ f(x_3) \\ f(x_4) \\ f(x_5) \end{bmatrix}
= \frac{1}{\Delta x}
\begin{bmatrix}
1 & 0 & 0 & 0 & 0 \\
-1 & 1 & 0 & 0 & 0 \\
0 & -1 & 1 & 0 & 0 \\
0 & 0 & -1 & 1 & 0 \\
0 & 0 & 0 & -1 & 1
\end{bmatrix}
\begin{bmatrix} F(x_1) \\ F(x_2) \\ F(x_3) \\ F(x_4) \\ F(x_5) \end{bmatrix}
\quad,
\tag{6.516}
$$

giving the discrete derivative operator. One obtains

$$
f(x_1) = \frac{F(x_1)}{\Delta x} \quad,
$$

$$
f(x_2) = \frac{F(x_2) - F(x_1)}{\Delta x} \quad, \qquad
f(x_3) = \frac{F(x_3) - F(x_2)}{\Delta x} \quad,
\tag{6.517}
$$

$$
f(x_4) = \frac{F(x_4) - F(x_3)}{\Delta x} \quad, \qquad
f(x_5) = \frac{F(x_5) - F(x_4)}{\Delta x} \quad,
$$

and we see that this matrix operator has a built-in initial condition.

An equation using the transpose matrix is

$$
\begin{bmatrix} \varphi(x_1) \\ \varphi(x_2) \\ \varphi(x_3) \\ \varphi(x_4) \\ \varphi(x_5) \end{bmatrix}
= \frac{1}{\Delta x}
\begin{bmatrix}
1 & -1 & 0 & 0 & 0 \\
0 & 1 & -1 & 0 & 0 \\
0 & 0 & 1 & -1 & 0 \\
0 & 0 & 0 & 1 & -1 \\
0 & 0 & 0 & 0 & 1
\end{bmatrix}
\begin{bmatrix} \Phi(x_1) \\ \Phi(x_2) \\ \Phi(x_3) \\ \Phi(x_4) \\ \Phi(x_5) \end{bmatrix}
\quad,
\tag{6.518}
$$

which gives

$$\varphi(x_1) = \frac{\Phi(x_1) - \Phi(x_2)}{\Delta x} \quad , \quad \varphi(x_2) = \frac{\Phi(x_2) - \Phi(x_3)}{\Delta x} \quad ,$$

$$\varphi(x_3) = \frac{\Phi(x_3) - \Phi(x_4)}{\Delta x} \quad , \quad \varphi(x_4) = \frac{\Phi(x_4) - \Phi(x_5)}{\Delta x} \quad , \tag{6.519}$$

$$\varphi(x_5) = \frac{\Phi(x_5)}{\Delta x} \quad .$$

We recognize here the opposite of the derivative operator, this time with a built-in *final* condition.

6.29 Lagrange Parameters

Let the problem be that of minimizing the real functional

$$S = S(x^\alpha) \quad (\alpha \in \mathbf{I}) \tag{6.520}$$

under the nonlinear constraints

$$\Psi^i(x^\alpha) = 0 \quad (i \in \mathbf{J}) \quad , \tag{6.521}$$

where $\mathbf{I}$ and $\mathbf{J}$ represent discrete index sets.

The Lagrange method consists of introducing unknown parameters λ_i and defining a new functional $S'(x^\alpha, \lambda_i)$ by

$$S'(x^\alpha, \lambda_i) = S(x^\alpha) - \sum_{i \in \mathbf{J}} \lambda_i \Psi^i(x^\alpha) \quad . \tag{6.522}$$

The conditions $\partial S'/\partial \lambda_i = 0$ give

$$\Psi^i(x^\alpha) = 0 \quad (i \in \mathbf{J}) \quad , \tag{6.523}$$

while the conditions $\partial S'/\partial x^\alpha = 0$ give

$$\frac{\partial S}{\partial x^\alpha} - \sum_{i \in \mathbf{J}} \lambda_i \frac{\partial \Psi^i}{\partial x^\alpha} = 0 \quad (\alpha \in \mathbf{I}) \quad . \tag{6.524}$$

Equations (6.523) show that at the minimum, the constraints (6.521) will be satisfied. It follows that the *constrained* minimization of (6.520) is equivalent to the *unconstrained* minimization of (6.522).

The system (6.523)–(6.524) has as many equations as unknowns (the x^α and the λ_i). Its resolution gives the solution of the problem.

6.30 Matrix Identities

Letting $\mathbf{G}$ be an arbitrary linear operator from a linear space $\mathbb{M}$ into a linear space $\mathbb{D}$ and $\mathbf{C_M}$ and $\mathbf{C_D}$ be two covariance operators acting respectively on $\mathbb{M}$ and $\mathbb{D}$ (i.e., two linear, symmetric, positive definite operators), let us demonstrate the two identities

$$\left(\mathbf{G}^t \, \mathbf{C}_D^{-1} \, \mathbf{G} + \mathbf{C}_M^{-1} \right)^{-1} \mathbf{G}^t \, \mathbf{C}_D^{-1} \; = \; \mathbf{C}_M \, \mathbf{G}^t \left(\mathbf{C}_D + \mathbf{G} \, \mathbf{C}_M \, \mathbf{G}^t \right)^{-1} \quad ,$$

$$\left(\mathbf{G}^t \, \mathbf{C}_D^{-1} \, \mathbf{G} + \mathbf{C}_M^{-1} \right)^{-1} \; = \; \mathbf{C}_M - \mathbf{C}_M \, \mathbf{G}^t \left(\mathbf{C}_D + \mathbf{G} \, \mathbf{C}_M \, \mathbf{G}^t \right)^{-1} \mathbf{G} \, \mathbf{C}_M \quad .$$

(6.525)

The first equation follows from the following obvious identities:

$$\mathbf{G}^t + \mathbf{G}^t \, \mathbf{C}_D^{-1} \, \mathbf{G} \, \mathbf{C}_M \, \mathbf{G}^t \; = \; \mathbf{G}^t \, \mathbf{C}_D^{-1} \left(\mathbf{C}_D + \mathbf{G} \, \mathbf{C}_M \, \mathbf{G}^t \right) \; = \; \left(\mathbf{G}^t \, \mathbf{C}_D^{-1} \, \mathbf{G} + \mathbf{C}_M^{-1} \right) \mathbf{C}_M \, \mathbf{G}^t \quad ,$$

(6.526)

since $\mathbf{G}^t \, \mathbf{C}_D^{-1} \, \mathbf{G} + \mathbf{C}_M^{-1}$ and $\mathbf{C}_D + \mathbf{G} \, \mathbf{C}_M \, \mathbf{G}^t$ are positive definite and thus regular matrices. Furthermore,

$$\mathbf{C}_M - \mathbf{C}_M \, \mathbf{G}^t \left(\mathbf{C}_D + \mathbf{G} \, \mathbf{C}_M \, \mathbf{G}^t \right)^{-1} \mathbf{G} \, \mathbf{C}_M \; = \; \mathbf{C}_M - \left(\mathbf{G}^t \, \mathbf{C}_D^{-1} \, \mathbf{G} + \mathbf{C}_M^{-1} \right)^{-1} \mathbf{G}^t \, \mathbf{C}_D^{-1} \, \mathbf{G} \, \mathbf{C}_M$$

$$= \; \left(\mathbf{G}^t \, \mathbf{C}_D^{-1} \, \mathbf{G} + \mathbf{C}_M^{-1} \right)^{-1} \left(\left(\mathbf{G}^t \, \mathbf{C}_D^{-1} \, \mathbf{G} + \mathbf{C}_M^{-1} \right) \mathbf{C}_M - \mathbf{G}^t \, \mathbf{C}_D^{-1} \, \mathbf{G} \, \mathbf{C}_M \right)$$

$$= \; \left(\mathbf{G}^t \, \mathbf{C}_D^{-1} \, \mathbf{G} + \mathbf{C}_M^{-1} \right)^{-1} \quad .$$

(6.527)

6.31   Inverse of a Partitioned Matrix

By direct substitution, one easily verifies the identity (valid if the inverted matrices are invertible)

$$\begin{pmatrix} \mathbf{B} & \mathbf{C}^t \\ \mathbf{C} & \mathbf{D} \end{pmatrix}^{-1} \; = \; \begin{pmatrix} \mathbf{E} & \mathbf{F}^t \\ \mathbf{F} & \mathbf{G} \end{pmatrix} ,$$

(6.528)

where

$$\mathbf{E} \; = \; (\mathbf{B} - \mathbf{C}^t \, \mathbf{D}^{-1} \, \mathbf{C})^{-1} \quad , \quad \mathbf{G} \; = \; (\mathbf{D} - \mathbf{C} \, \mathbf{B}^{-1} \, \mathbf{C}^t)^{-1} \quad , \quad \mathbf{F} \; = \; -\mathbf{G} \, \mathbf{C} \, \mathbf{B}^{-1} \; = \; -\mathbf{D}^{-1} \, \mathbf{C} \, \mathbf{E} \quad .$$

(6.529)

6.32   Norm of the Generalized Gaussian

The generalized Gaussian of order p is defined by

$$f_p(x) \; = \; \frac{p^{1-1/p}}{2 \, \sigma \, \Gamma(1/p)} \, \exp \left(-\frac{1}{p} \frac{|x - x_0|^p}{\sigma^p} \right) \quad .$$

(6.530)

Let us demonstrate that it is normalized and let us perform a direct computation of its ℓ_p-norm estimator of dispersion.

We have

$$I_p \; = \; \int_{-\infty}^{+\infty} dx \, f_p(x) \; = \; \int_{-\infty}^{+\infty} dx \, f_p(x + x_0) \; = \; 2 \int_0^\infty dx \, f_p(x + x_0)$$

$$= \; \frac{p^{1-1/p}}{\sigma \, \Gamma(1/p)} \int_0^\infty dx \, \exp \left(-\frac{1}{p} \frac{x^p}{\sigma^p} \right) \quad .$$

(6.531)

Introducing the variable $u = \frac{x^p}{p\,x^{p-1}}$, we successively have $du = \frac{x^{p-1}\,dx}{\sigma^p}$, $dx = \frac{\sigma^p}{x^{p-1}}\,du = \frac{\sigma\,u^{1-1/p}}{p^{1-1/p}}\,du$, and

$$I_p = \frac{1}{\Gamma(1/p)} \int_0^\infty du\, u^{1-1/p}\, e^{-u} \quad . \tag{6.532}$$

Using the definition of the gamma function, $\Gamma(t) = \int_0^\infty du\, u^{1-t}\, e^{-u}$, we directly obtain

$$I_p = \int_{-\infty}^{+\infty} dx\, f_p(x) = 1 \quad . \tag{6.533}$$

By definition, the estimator of dispersion in norm ℓ_p is (see Appendix 6.5)

$$\sigma_p = \left(\int_{-\infty}^{+\infty} dx\, |x - x_0|^p\, f_p(x) \right)^{1/p} \quad . \tag{6.534}$$

We successively have

$$\sigma_p = \left(\int_{-\infty}^{+\infty} dx\, |x|^p\, f_p(x + x_0) \right)^{1/p} = \left(2 \int_0^\infty dx\, |x|^p\, f_p(x + x_0) \right)^{1/p}$$

$$= \left(\frac{p^{1-1/p}}{\sigma\,\Gamma(1/p)} \int_0^\infty dx\, x^p\, \exp\left(-\frac{1}{p}\frac{x^p}{\sigma^p} \right) \right)^{1/p} \quad , \tag{6.535}$$

and, using again the change of variables previously defined,

$$\sigma_p = \left(\frac{p\,\sigma_p}{\Gamma(1/p)} \int_0^\infty du\, u^{1-(1+1/p)}\, e^{-u} \right)^{1/p} = \left(\frac{p\,\sigma^p\,\Gamma(1 + 1/p)}{\Gamma(1/p)} \right)^{1/p} \quad . \tag{6.536}$$

Finally, using the property $\Gamma(1 + t) = t\,\Gamma(t)$, we obtain

$$\sigma_p = \sigma \quad . \tag{6.537}$$

Chapter 7

Problems

7.1 Estimation of the Epicentral Coordinates of a Seismic Event

A seismic source was activated at time $T = 0$ in an unknown location at the surface of Earth. The seismic waves produced by the explosion have been recorded at a network of six seismic stations whose coordinates in a rectangular system are

$$(x^1, y^1) = (3\,\text{km}, 15\,\text{km}) \quad , \quad (x^2, y^2) = (3\,\text{km}, 16\,\text{km}) \quad ,$$
$$(x^3, y^3) = (4\,\text{km}, 15\,\text{km}) \quad , \quad (x^4, y^4) = (4\,\text{km}, 16\,\text{km}) \quad , \quad (7.1)$$
$$(x^5, y^5) = (5\,\text{km}, 15\,\text{km}) \quad , \quad (x^6, y^6) = (5\,\text{km}, 16\,\text{km}) \quad .$$

The observed arrival times of the seismic waves at these stations are

$$t^1_{obs} = 3.12\,\text{s} \pm \sigma \quad , \quad t^2_{obs} = 3.26\,\text{s} \pm \sigma \quad ,$$
$$t^3_{obs} = 2.98\,\text{s} \pm \sigma \quad , \quad t^4_{obs} = 3.12\,\text{s} \pm \sigma \quad , \quad (7.2)$$
$$t^5_{obs} = 2.84\,\text{s} \pm \sigma \quad , \quad t^6_{obs} = 2.98\,\text{s} \pm \sigma \quad ,$$

where $\sigma = 0.10$ s, the symbol $\pm\sigma$ being a short notation indicating that experimental uncertainties are independent and can be modeled using a Gaussian probability density with a standard deviation equal to σ.

Estimate the epicentral coordinates (X, Y) of the explosion, assuming a velocity of $v = 5\,\text{km/s}$ for the seismic waves. Use the approximation of a flat Earth surface, and consider that the coordinates in equation (7.1) are Cartesian.

Discuss the generalization of the problem to the case where the time of the explosion, the locations of the seismic observatories, or the velocity of the seismic waves are not perfectly known, and to the case of a realistic Earth.

Solution:
The model parameters are the coordinates of the epicenter of the explosion,

$$\mathbf{m} = (X, Y) \quad , \quad (7.3)$$

and the data parameters are the arrival times at the seismic network,

$$\mathbf{d} = (t^1, t^2, t^3, t^4, t^5, t^6) \quad , \tag{7.4}$$

while the coordinates of the seismic stations and the velocity of the seismic waves are assumed perfectly known (i.e., known with uncertainties that are negligible with respect to the uncertainties in the observed arrival times).

For a given (X, Y), the arrival times of the seismic wave at the seismic stations can be computed using the (exact) equation

$$t^i = g^i(X, Y) = \frac{1}{v}\sqrt{(x^i - X)^2 + (y^i - Y)^2} \qquad (i = 1, \ldots, 6) \quad , \tag{7.5}$$

which solves the forward problem $\mathbf{d} = \mathbf{g}(\mathbf{m})$.

As we are not given any a priori information on the epicentral coordinates, we take a uniform a priori probability density, i.e., because we are using Cartesian coordinates,

$$\rho_M(X, Y) = \text{const.} , \tag{7.6}$$

assigning equal a priori probabilities to equal volumes.

As data uncertainties are Gaussian and independent, the probability density representing the information we have on the true values of the arrival times is

$$\rho_D(t^1, t^2, t^3, t^4, t^5, t^6) = \text{const.} \exp\left(-\frac{1}{2}\sum_{i=1}^{6}\frac{(t^i - t^i_{\text{obs}})^2}{\sigma^2}\right) \quad . \tag{7.7}$$

With the three pieces of information in equations (7.5)–(7.7), we can directly pass to the resolution of the inverse problem. The posterior probability density in the model space, combining the three pieces of information, is (equation (1.93)) $\sigma_M(\mathbf{m}) = k\,\rho_M(\mathbf{m})\,\rho_D(\mathbf{g}(\mathbf{m}))$, i.e., particularizing the notation to the present problem,

$$\sigma_M(X, Y) = k\,\rho_M(X, Y)\,\rho_D(\mathbf{g}(X, Y)) \quad , \tag{7.8}$$

where k is a normalization constant. Explicitly, using equations (7.5)–(7.7),

$$\sigma_M(X, Y) = k'\exp\left(-\frac{1}{2\sigma^2}\sum_{i=1}^{6}(t^i_{\text{cal}}(X, Y) - t^i_{\text{obs}})^2\right) \quad , \tag{7.9}$$

where k' is a new normalization constant and

$$t^i_{\text{cal}}(X, Y) = \frac{1}{v}\sqrt{(x^i - X)^2 + (y^i - Y)^2} \quad . \tag{7.10}$$

The probability density $\sigma_M(X, Y)$ describes all the a posteriori information we have on the epicentral coordinates. As we only have two parameters, the simplest (and most general) way of studying this information is to plot the values of $\sigma_M(X, Y)$ directly in the region of the plane where it takes significant values. Figure 7.1 shows the result obtained in this way.

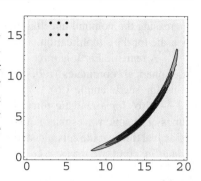

Figure 7.1. *Probability density for the epi-central coordinates of the seismic event, obtained using as data the arrival times of the seismic wave at six seismic stations (points at the top of the figure). The gray scale is linear, between zero and the maximum value of the probability density. The crescent-shape of the region of significant probability density cannot be described using a few numbers (mean values, variances, covariances, ...), as commonly done.*

We see that the zone of nonvanishing probability density is crescent-shaped. This can be interpreted as follows. The arrival times of the seismic wave at the seismic network (top left of the figure) are of the order of $3\,s$, and as we know that the explosion took place at time $T = 0$, and the velocity of the seismic wave is $5\,km/s$, this gives the reliable information that the explosion took place at a distance of approximately $15\,km$ from the seismic network. But as the observational uncertainties ($\pm 0.1\,s$) in the arrival times are of the order of the travel times of the seismic wave between the stations, the azimuth of the epicenter is not well resolved. As the distance is well determined but not the azimuth, it is natural to obtain a probability density with a crescent shape.

From the values shown in Figure 7.1 it is possible to obtain any estimator of the epicentral coordinates one may wish, such as, for instance, the median, the mean, or the maximum likelihood values. But the general solution of the inverse problem is the probability density itself. Notice in particular that a computation of the covariance between X and Y will miss the circular aspect of the correlation.

If the time of the explosion were not known, or the coordinates of the seismic stations were not perfectly known, or if the velocity of the seismic waves were only known approximately, the model vector would contain all these parameters:

$$\mathbf{m} = (X, Y, T, x^1, y^1, \ldots, x^6, y^6, v) \quad . \tag{7.11}$$

After properly introducing the a priori information on T (if any), on (x^i, y^i), and on v, the posterior probability density $\sigma_M(X, Y, T, x^1, y^1, \ldots, x^6, y^6, v)$ should be defined as before, from which the marginal probability density on the epicentral coordinates (X, Y) could be obtained as

$$\sigma_{X,Y}(X, Y) = \int_{-\infty}^{\infty} dT \int_{-\infty}^{\infty} dx^1 \cdots \int_{-\infty}^{\infty} dy^6 \int_{0}^{\infty} dv \, \sigma_M(X, Y, T, x^1, y^1, \ldots, x^6, y^6, v)$$

$$\tag{7.12}$$

and the posterior probability density on the time T of the explosion as

$$\sigma_T(T) = \int_{-\infty}^{\infty} dX \int_{-\infty}^{\infty} dY \int_{-\infty}^{\infty} dx^1 \cdots \int_{-\infty}^{\infty} dy^6 \int_{0}^{\infty} dv \, \sigma_M(X, Y, T, x^1, y^1, \ldots, x^6, y^6, v) \quad .$$

$$\tag{7.13}$$

As computations rapidly become heavy, it may be necessary to make some simplifying assumptions. The most drastic one is to neglect uncertainties on (x^i, y^i) and v, artificially increasing the nominal uncertainties in the observed arrival times, to approximately compensate for the simplification.

A realistic Earth is three-dimensional and heterogeneous. It is generally simpler to use spherical coordinates (r, θ, φ). Then, the homogeneous probability density is no longer constant (see Example 1.6).

Also, for a realistic three-dimensional Earth, errors made in computing the travel times of seismic waves may not be negligible compared to uncertainties in the observation of arrival times at the seismic stations. Instead of using equation (1.93) of the main text as a starting point, we may use equation (1.89) instead,

$$\sigma_M(\mathbf{m}) \;=\; k\,\rho_M(\mathbf{m}) \int_{\mathcal{D}} d\mathbf{d} \; \frac{\rho_D(\mathbf{d})\,\theta(\mathbf{d} \mid \mathbf{m})}{\mu_D(\mathbf{d})} \quad , \tag{7.14}$$

where $\theta(\mathbf{d} \mid \mathbf{m})$, the conditional probability density for the arrival time given the model parameters, allows us to describe uncertainties in the computation of arrival times. As a simple (simplistic?) example, one could take

$$\theta(\mathbf{d} \mid r, \theta, \varphi, T) \;=\; \exp\left(-\tfrac{1}{2}\,(\mathbf{d} - \mathbf{g}(r, \theta, \varphi, T))^t\, \mathbf{C}_T^{-1}(\mathbf{d} - \mathbf{g}(r, \theta, \varphi, T)) \right) \quad , \tag{7.15}$$

where $\mathbf{C}_T$ is an ad hoc covariance matrix approximately describing the errors made in estimating arrival times theoretically. For more details, the reader may refer to Tarantola and Valette (1982a).

7.2 Measuring the Acceleration of Gravity

An absolute gravimeter uses the free fall of a mass in vacuo to measure the value of the acceleration g due to gravity. A mass is sent upward with some initial velocity v_0, and the positions $z^1, z^2, \ldots$ of the mass are (very precisely) measured at different instants $t^1, t^2, \ldots$. In vacuo (orienting the z axis upward),

$$z(t) \;=\; v_0\, t - \tfrac{1}{2}\, g\, t^2 \quad . \tag{7.16}$$

The measured values of the z_i and the t_i can be used to infer the values of v_0 and g. The measurements made during a free-fall experiment have provided the values (see Figure 7.2)

$$
\begin{aligned}
t_1 &= 0.20\,\mathrm{s} \pm 0.01\,\mathrm{s} \quad, & z_1 &= 0.62\,\mathrm{m} \pm 0.02\,\mathrm{m} \quad, \\
t_2 &= 0.40\,\mathrm{s} \pm 0.01\,\mathrm{s} \quad, & z_2 &= 0.88\,\mathrm{m} \pm 0.02\,\mathrm{m} \quad, \\
t_3 &= 0.60\,\mathrm{s} \pm 0.01\,\mathrm{s} \quad, & z_3 &= 0.70\,\mathrm{m} \pm 0.02\,\mathrm{m} \quad, \\
t_4 &= 0.80\,\mathrm{s} \pm 0.01\,\mathrm{s} \quad, & z_4 &= 0.15\,\mathrm{m} \pm 0.02\,\mathrm{m} \quad,
\end{aligned}
\tag{7.17}
$$

where the uncertainties[91] in the times are of the boxcar type and the uncertainties in the positions are of the double exponential type (the mean deviations having the value 0.02 m). Using these data, estimate the values v_0 and g (although we are interested in the value of g only), assuming that there is no particular a priori information on these two values.

---

[91] See Appendix 6.5 for the definition of the most common estimators of dispersion.

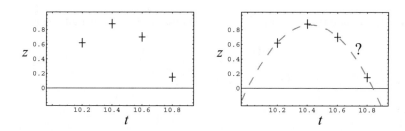

Figure 7.2. *At the left, the data for the estimation of the value* g *(gravity's acceleration), and, at the right, a suggested parabolic fit. Note that this example solves the problem of fitting a series of points with uncertainties in the two axes. Uncertainty bars are not to scale.*

Solution:
Let us introduce the vector **m** with components

$$\mathbf{m} = \{v_0, g, t_1, t_2, t_3, t_4\} \tag{7.18}$$

and the vector **d** with components

$$\mathbf{d} = \{z_1, z_2, z_3, z_4\} \quad . \tag{7.19}$$

Using expression (7.16) we can write the *theoretical relations*

$$z_1 = v_0 t_1 - \tfrac{1}{2} g t_1^2 \quad , \quad z_2 = v_0 t_2 - \tfrac{1}{2} g t_2^2 \quad ,$$
$$z_3 = v_0 t_3 - \tfrac{1}{2} g t_3^2 \quad , \quad z_4 = v_0 t_4 - \tfrac{1}{2} g t_4^2 \tag{7.20}$$

that correspond to the usual relation $\mathbf{d} = \mathbf{g(m)}$ (see the main text). Although we may call **m** the model parameters and **d** the data parameters, we see that there are observed quantities both in **d** and in **m**.

The prior information we have on **m** is to be represented by a probability density

$$\rho_M(\mathbf{m}) = \rho_M(v_0, g, t_1, t_2, t_3, t_4) \quad . \tag{7.21}$$

As we do not wish to include any special a priori information on the parameters $\{v_0, g\}$, we shall take a probability density that is constant on these parameters. The prior information we have on the other four parameters, $\{t_1, t_2, t_3, t_4\}$, is to be represented using boxcar probability densities. Then,

$$\rho_M(v_0, g, t_1, t_2, t_3, t_4) = \begin{cases} k & \text{if} \begin{cases} t_1^{obs} - s_1 \leq t_1 < t_1^{obs} + s_1 & \text{and} \\ t_2^{obs} - s_2 \leq t_2 < t_2^{obs} + s_2 & \text{and} \\ t_3^{obs} - s_3 \leq t_3 < t_3^{obs} + s_3 & \text{and} \\ t_4^{obs} - s_4 \leq t_4 < t_4^{obs} + s_4 & , \end{cases} \\ 0 & \text{otherwise} \quad , \end{cases} \tag{7.22}$$

where $t_1^{obs} = 0.20\,\text{s}$, $t_2^{obs} = 0.40\,\text{s}$, $t_3^{obs} = 0.60\,\text{s}$, $t_4^{obs} = 0.80\,\text{s}$, and $s_1 = s_2 = s_3 = s_4 = 0.01\,\text{s}$.

The information we have on **d** is to be represented by a probability density

$$\rho_D(\mathbf{d}) = \rho_D(z_1, z_2, z_3, z_4) \quad . \tag{7.23}$$

Using the double exponential to model uncertainties gives

$$
\begin{aligned}
&\rho_D(z_1, z_2, z_3, z_4) \\
&= k \exp\left(-\left(\frac{|z_1 - z_1^{\text{obs}}|}{\sigma_1} + \frac{|z_2 - z_2^{\text{obs}}|}{\sigma_2} + \frac{|z_3 - z_3^{\text{obs}}|}{\sigma_3} + \frac{|z_4 - z_4^{\text{obs}}|}{\sigma_4}\right)\right) ,
\end{aligned} \tag{7.24}
$$

where $z_1^{\text{obs}} = 0.62\,\text{m}$, $z_2^{\text{obs}} = 0.88\,\text{m}$, $z_3^{\text{obs}} = 0.70\,\text{m}$, $z_4^{\text{obs}} = 0.15\,\text{m}$, and $\sigma_1 = \sigma_2 = \sigma_3 = \sigma_4 = 0.02\,\text{m}$.

We are here inside the context of Example 1.34: the posterior probability density of the parameters **m** is given by (equation (1.93))

$$\sigma_M(\mathbf{m}) = k\,\rho_M(\mathbf{m})\,\rho_D(\mathbf{g}(\mathbf{m})) \quad , \tag{7.25}$$

where k is a normalization constant.

Collecting the partial results just obtained gives

$$
\sigma_M(v_0, g, t_1, t_2, t_3, t_4)
$$

$$
= \begin{cases}
k \exp(-S(v_0, g, t_1, t_2, t_3, t_4)) & \text{if} \quad \begin{cases} t_1^{\text{obs}} - s_1 \le t_1 < t_1^{\text{obs}} + s_1 & \text{and} \\ t_2^{\text{obs}} - s_2 \le t_2 < t_2^{\text{obs}} + s_2 & \text{and} \\ t_3^{\text{obs}} - s_3 \le t_3 < t_3^{\text{obs}} + s_3 & \text{and} \\ t_4^{\text{obs}} - s_4 \le t_4 < t_4^{\text{obs}} + s_4 \quad , \end{cases} \\
\\
0 \quad \text{otherwise} \quad ,
\end{cases}
$$

$$\tag{7.26}$$

where

$$
\begin{aligned}
S(v_0, g, t_1, t_2, t_3, t_4) &= \frac{|(v_0 t_1 - \frac{1}{2} g t_1^2) - z_1^{\text{obs}}|}{\sigma_1} + \frac{|(v_0 t_2 - \frac{1}{2} g t_2^2) - z_2^{\text{obs}}|}{\sigma_2} \\
&+ \frac{|(v_0 t_3 - \frac{1}{2} g t_3^2) - z_3^{\text{obs}}|}{\sigma_3} + \frac{|(v_0 t_4 - \frac{1}{2} g t_4^2) - z_4^{\text{obs}}|}{\sigma_4} \quad .
\end{aligned} \tag{7.27}
$$

The information we have on the two parameters $\{v_0, g\}$ is that represented by the marginal probability density

$$\phi(v_0, g) = \int_{-\infty}^{+\infty} dt_1 \int_{-\infty}^{+\infty} dt_2 \int_{-\infty}^{+\infty} dt_3 \int_{-\infty}^{+\infty} dt_4 \,\sigma_M(v_0, g, t_1, t_2, t_3, t_4) \quad , \tag{7.28}$$

i.e.,

$$\phi(v_0, g) = k \int_{t_1^{\text{obs}} - s_1}^{t_1^{\text{obs}} + s_1} dt_1 \int_{t_2^{\text{obs}} - s_2}^{t_2^{\text{obs}} + s_2} dt_2 \int_{t_3^{\text{obs}} - s_3}^{t_3^{\text{obs}} + s_3} dt_3 \int_{t_4^{\text{obs}} - s_4}^{t_4^{\text{obs}} + s_4} dt_4 \,\exp(-S(v_0, g, t_1, t_2, t_3, t_4)) \quad , \tag{7.29}$$

and the information we have on the parameter g itself is represented by

$$\varphi(g) = \int_{-\infty}^{+\infty} dv_0 \, \phi(v_0, g) \quad . \tag{7.30}$$

Simple as they may seem, these integrations do not lead to an analytic expression (at least, my favorite mathematical software was not able to obtain an explicit solution). Here, the probability density $\phi(v_0, g)$ has been evaluated using a numerical integration, and the result (for the numerical data presented above) is displayed in Figure 7.3.

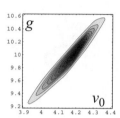

Figure 7.3. *The data provided by an absolute gravimeter have been used to estimate both the initial velocity of the free-falling mass and the acceleration of gravity. The numerical data given in the text were generated using the values* $v_0 = 4.12 \, \mathrm{m/s}$ *and* $g = 9.81 \, \mathrm{m/s^2}$, *and some noise was added to the* z_i *values.*

7.3 Elementary Approach to Tomography

Figure 7.4 shows an object composed of nine homogeneous portions. The values indicated correspond to the linear attenuation coefficients *(relative to some reference medium, for instance, water) for X-rays (in given units). An X-ray experiment using the geometry shown in Figure 7.5 allows us to measure the* transmittance ρ^{ij} *along each ray, which is given by*

$$\rho^{ij} = \exp\left(-\int_{R^{ij}} ds^{ij} \, m(\mathbf{x}(s^{ij}))\right) \quad , \tag{7.31}$$

where $m(\mathbf{x})$ *represents the linear attenuation coefficient at point* $\mathbf{x}$, R^{ij} *represents the ray between source* i *and receiver* j, *and* ds^{ij} *is the element of length along the ray* R^{ij}. *Assume that instead of measuring* ρ^{ij} *we measure*

$$d^{ij} = -\log \rho^{ij} = \int_{R^{ij}} ds^{ij} \, m(\mathbf{x}(s^{ij})) \quad , \tag{7.32}$$

which is termed the integrated attenuation.

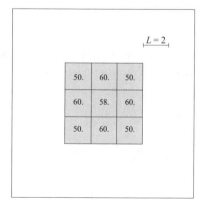

Figure 7.4. *A bidimensional medium is composed of* 3×3 *homogeneous blocks. Indicated are the true values of the linear attenuation coefficient for X-rays (with respect to the surrounding medium).*

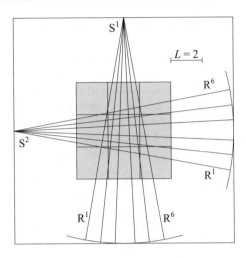

Figure 7.5. *In order to infer the true (unknown) values of the linear attenuation coefficient, an X-ray transmission tomographic experiment is performed. Each block measures $L = 2$ units of length, and the figure is to scale (the angular separation between rays is 4 degrees). S^1 and S^2 represent the two source locations, and $R^1, \ldots, R^6$ represent the six receivers used. Let $m(\mathbf{x})$ represent the linear attenuation coefficient at point $\mathbf{x}$ of the medium under study, R^{ij} the ray between source 1 and receiver j, s^{ij} the position along ray R^{ij}, and d^{ij} the integrated attenuation along ray R^{ij}: $d^{ij} = \int ds^{ij}\, m(\mathbf{x}(s^{ij}))\,(along\ R^{ij})$. The measured values of the integrated attenuation along each ray are, in order for each receiver, 341.9 ± 0.1, 353.1 ± 0.1, 356.2 ± 0.1, 356.2 ± 0.1, 353.1 ± 0.1, and $341.9 \pm 9 \pm 0.1$ for source 1 and 341.9 ± 0.1, 353.1 ± 0.1, 356.2 ± 0.1, 356.2 ± 0.1, 353.1 ± 0.1, and $341.9 \pm 9 \pm 0.1$ for source 2 (these values correspond in fact to the actual values as they can be computed from the true linear attenuation values of Figure 7.4, plus a Gaussian noise with standard deviation 0.1, and are rounded to the first decimal). These values are assumed to be corrected for the effect of the propagation outside the 3×3 model, so that the linear attenuation coefficient outside the model can be taken as null. The inverse problem consists of using these observed values of integrated attenuation to infer the actual model values. Remark that the upper-left block is explored with very short ray lengths, and owing to the relatively high noise in the data, the actual value of this block will probably be poorly resolved.*

If the medium is a priori assumed to be composed of the nine homogeneous portions of Figure 7.4, any model of the medium may be represented using the notation

$$
\mathbf{m} = \begin{pmatrix} m^{11} & m^{12} & m^{13} \\ m^{21} & m^{22} & m^{23} \\ m^{31} & m^{32} & m^{33} \end{pmatrix}, \tag{7.33}
$$

where the first index represents the column and the second index represents the row. Any possible set of numerical values in equation (7.33) is a model vector. For instance, the true *model is (Figure 7.4)*

$$\begin{pmatrix} m^{11} & m^{12} & m^{13} \\ m^{21} & m^{22} & m^{23} \\ m^{31} & m^{32} & m^{33} \end{pmatrix} = \begin{pmatrix} 50 & 60 & 50 \\ 60 & 58 & 60 \\ 50 & 60 & 50 \end{pmatrix} . \tag{7.34}$$

A data vector *is represented by*

$$\mathbf{d} = \begin{pmatrix} d^{11} & d^{12} & d^{13} & d^{14} & d^{15} & d^{16} \\ d^{21} & d^{22} & d^{23} & d^{24} & d^{25} & d^{26} \end{pmatrix} , \tag{7.35}$$

where the first index denotes the source number and the second index denotes the receiver number. Equation (7.32) then simplifies to the discrete equation

$$d^{ij} = \sum_{\alpha=1}^{3} \sum_{\beta=1}^{3} G^{ij}{}_{\alpha\beta} \, m^{\alpha\beta} \quad for \ i = 1, 2 \quad , \qquad j = 1, 2, 3, 4, 5, 6 \quad , \tag{7.36}$$

where $G^{ij}{}_{\alpha\beta}$ represents the length of the ray ij inside the block $\alpha\beta$. An actual measurement of the integrated attenuation gives the values

$$\begin{pmatrix} d^{11} & d^{12} & d^{13} & d^{14} & d^{15} & d^{16} \\ d^{21} & d^{22} & d^{23} & d^{24} & d^{25} & d^{26} \end{pmatrix} \tag{7.37}$$

$$= \begin{pmatrix} 341.9 \pm 0.1 & 353.1 \pm 0.1 & 356.2 \pm 0.1 & 356.2 \pm 0.1 & 353.1 \pm 0.1 & 341.9 \pm 0.1 \\ 341.9 \pm 0.1 & 353.1 \pm 0.1 & 356.2 \pm 0.1 & 356.2 \pm 0.1 & 353.1 \pm 0.1 & 341.9 \pm 0.1 \end{pmatrix} ,$$

where ± 0.1 indicates the standard deviation of the estimated (Gaussian) uncertainty.

Assume that you have the a priori information that the model values of the linear attenuation coefficients equal 55 ± 15 (Figure 7.6). Give a better estimation of them using the data (7.37) and the least-squares theory. Discuss.

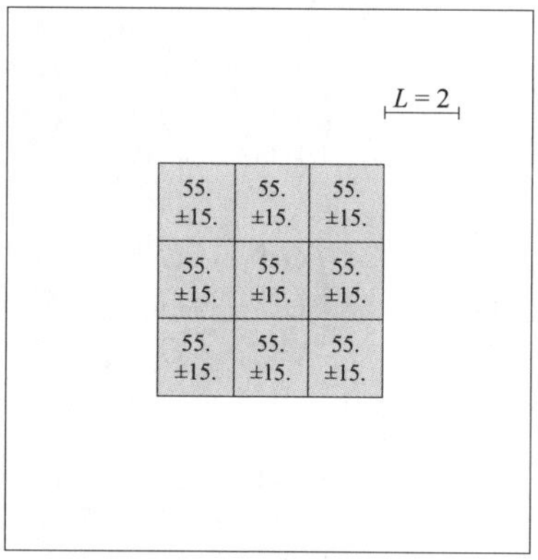

Figure 7.6. *We have the a priori information that the true linear attenuation coefficients are 55 ± 15. It is assumed that a Gaussian probability density well represents this a priori information (in particular, ± 15 represent soft limits, which can be outpassed with a probability corresponding to the Gaussian density function).*

Solution:

We wish here to obtain the model **m** minimizing

$$S(\mathbf{m}) = \frac{1}{2}\left((\mathbf{G\,m} - \mathbf{d}_{\mathrm{obs}})^t\,\mathbf{C}_{\mathrm{D}}^{-1}\,(\mathbf{G\,m} - \mathbf{d}_{\mathrm{obs}}) + (\mathbf{m} - \mathbf{m}_{\mathrm{prior}})^t\,\mathbf{C}_{\mathrm{M}}^{-1}\,(\mathbf{m} - \mathbf{m}_{\mathrm{prior}}) \right) \quad,$$

(7.38)

where

$$\mathbf{d}_{\mathrm{obs}} = \begin{pmatrix} 341.9 & 353.1 & 356.2 & 356.2 & 353.1 & 341.9 \\ 341.9 & 353.1 & 356.2 & 356.2 & 353.1 & 341.9 \end{pmatrix}, \quad (\mathbf{C}_{\mathrm{D}})^{ijkl} = 0.1^2\,\delta^{ik}\,\delta^{jl} \quad,$$

$$\mathbf{m}_{\mathrm{prior}} = \begin{pmatrix} 55 & 55 & 55 \\ 55 & 55 & 55 \\ 55 & 55 & 55 \end{pmatrix} \quad, \quad (\mathbf{C}_{\mathrm{M}})^{\alpha\beta\gamma\delta} = 15^2\,\delta^{\alpha\gamma}\,\delta^{\beta\delta} \quad,$$

(7.39)

and where the elements of the kernel of the linear operator **G** can be obtained from Figure 7.5 using a simple geometrical computation:

$$\begin{pmatrix} G^{11}{}_{11} & G^{11}{}_{12} & G^{11}{}_{13} \\ G^{11}{}_{21} & G^{11}{}_{22} & G^{11}{}_{23} \\ G^{11}{}_{31} & G^{11}{}_{32} & G^{11}{}_{33} \end{pmatrix} = \begin{pmatrix} 0.3338 & 1.6971 & 0.0000 \\ 2.0309 & 0.0000 & 0.0000 \\ 2.0309 & 0.0000 & 0.0000 \end{pmatrix} \quad,$$

$$\begin{pmatrix} G^{12}{}_{11} & G^{12}{}_{12} & G^{12}{}_{13} \\ G^{12}{}_{21} & G^{12}{}_{22} & G^{12}{}_{23} \\ G^{12}{}_{31} & G^{12}{}_{32} & G^{12}{}_{33} \end{pmatrix} = \begin{pmatrix} 0.0000 & 2.0110 & 0.0000 \\ 0.0000 & 2.0110 & 0.0000 \\ 0.4883 & 1.5227 & 0.0000 \end{pmatrix} \quad,$$

$$\begin{pmatrix} G^{13}{}_{11} & G^{13}{}_{12} & G^{13}{}_{13} \\ G^{13}{}_{21} & G^{13}{}_{22} & G^{13}{}_{23} \\ G^{13}{}_{31} & G^{13}{}_{32} & G^{13}{}_{33} \end{pmatrix} = \begin{pmatrix} 0.0000 & 2.0012 & 0.0000 \\ 0.0000 & 2.0012 & 0.0000 \\ 0.0000 & 2.0012 & 0.0000 \end{pmatrix} \quad,$$

$$\begin{pmatrix} G^{14}{}_{11} & G^{14}{}_{12} & G^{14}{}_{13} \\ G^{14}{}_{21} & G^{14}{}_{22} & G^{14}{}_{23} \\ G^{14}{}_{31} & G^{14}{}_{32} & G^{14}{}_{33} \end{pmatrix} = \begin{pmatrix} 0.0000 & 2.0012 & 0.0000 \\ 0.0000 & 2.0012 & 0.0000 \\ 0.0000 & 2.0012 & 0.0000 \end{pmatrix} \quad,$$

$$\begin{pmatrix} G^{15}{}_{11} & G^{15}{}_{12} & G^{15}{}_{13} \\ G^{15}{}_{21} & G^{15}{}_{22} & G^{15}{}_{23} \\ G^{15}{}_{31} & G^{15}{}_{32} & G^{15}{}_{33} \end{pmatrix} = \begin{pmatrix} 0.0000 & 2.0110 & 0.0000 \\ 0.0000 & 2.0110 & 0.0000 \\ 0.0000 & 1.5227 & 0.4883 \end{pmatrix} \quad,$$

$$\begin{pmatrix} G^{16}{}_{11} & G^{16}{}_{12} & G^{16}{}_{13} \\ G^{16}{}_{21} & G^{16}{}_{22} & G^{16}{}_{23} \\ G^{16}{}_{31} & G^{16}{}_{32} & G^{16}{}_{33} \end{pmatrix} = \begin{pmatrix} 0.0000 & 1.6971 & 0.3338 \\ 0.0000 & 0.0000 & 2.0309 \\ 0.0000 & 0.0000 & 2.0309 \end{pmatrix} \quad,$$

$$\begin{pmatrix} G^{21}{}_{11} & G^{21}{}_{12} & G^{21}{}_{13} \\ G^{21}{}_{21} & G^{21}{}_{22} & G^{21}{}_{23} \\ G^{21}{}_{31} & G^{21}{}_{32} & G^{21}{}_{33} \end{pmatrix} = \begin{pmatrix} 0.0000 & 0.0000 & 0.0000 \\ 1.6971 & 0.0000 & 0.0000 \\ 0.3338 & 2.0309 & 2.0309 \end{pmatrix} \quad,$$

(7.40)

$$\begin{pmatrix} G^{22}{}_{11} & G^{22}{}_{12} & G^{22}{}_{13} \\ G^{22}{}_{21} & G^{22}{}_{22} & G^{22}{}_{23} \\ G^{22}{}_{31} & G^{22}{}_{32} & G^{22}{}_{33} \end{pmatrix} = \begin{pmatrix} 0.0000 & 0.0000 & 0.0000 \\ 2.0110 & 2.0110 & 1.5227 \\ 0.0000 & 0.0000 & 0.4883 \end{pmatrix} \quad,$$

$$\begin{pmatrix} G^{23}{}_{11} & G^{23}{}_{12} & G^{23}{}_{13} \\ G^{23}{}_{21} & G^{23}{}_{22} & G^{23}{}_{23} \\ G^{23}{}_{31} & G^{23}{}_{32} & G^{23}{}_{33} \end{pmatrix} = \begin{pmatrix} 0.0000 & 0.0000 & 0.0000 \\ 2.0012 & 2.0012 & 2.0012 \\ 0.0000 & 0.0000 & 0.0000 \end{pmatrix} \quad,$$

$$\begin{pmatrix} G^{24}{}_{11} & G^{24}{}_{12} & G^{24}{}_{13} \\ G^{24}{}_{21} & G^{24}{}_{22} & G^{24}{}_{23} \\ G^{24}{}_{31} & G^{24}{}_{32} & G^{24}{}_{33} \end{pmatrix} = \begin{pmatrix} 0.0000 & 0.0000 & 0.0000 \\ 2.0012 & 2.0012 & 2.0012 \\ 0.0000 & 0.0000 & 0.0000 \end{pmatrix} \quad,$$

$$\begin{pmatrix} G^{25}{}_{11} & G^{25}{}_{12} & G^{25}{}_{13} \\ G^{25}{}_{21} & G^{25}{}_{22} & G^{25}{}_{23} \\ G^{25}{}_{31} & G^{25}{}_{32} & G^{25}{}_{33} \end{pmatrix} = \begin{pmatrix} 0.0000 & 0.0000 & 0.4883 \\ 2.0110 & 2.0110 & 1.5227 \\ 0.0000 & 0.0000 & 0.0000 \end{pmatrix} \quad,$$

and

$$\begin{pmatrix} G^{26}{}_{11} & G^{26}{}_{12} & G^{26}{}_{13} \\ G^{26}{}_{21} & G^{26}{}_{22} & G^{26}{}_{23} \\ G^{26}{}_{31} & G^{26}{}_{32} & G^{26}{}_{33} \end{pmatrix} = \begin{pmatrix} 0.3338 & 2.0309 & 2.0309 \\ 1.6971 & 0.0000 & 0.0000 \\ 0.0000 & 0.0000 & 0.0000 \end{pmatrix} \quad.$$

The minimum of expression (7.38) can, for instance, be obtained using the second of equations 3.37 (page 66) of the main text:

$$\mathbf{m} = \mathbf{m}_{\text{prior}} + (\mathbf{G}^t \, \mathbf{C}_D^{-1} \, \mathbf{G} + \mathbf{C}_M^{-1})^{-1} \, \mathbf{G}^t \, \mathbf{C}_D^{-1} \, (\mathbf{d}_{\text{obs}} - \mathbf{G} \, \mathbf{m}_{\text{prior}}) \quad . \tag{7.41}$$

This gives

$$\mathbf{m} = \begin{pmatrix} 55.9 & 59.3 & 50.3 \\ 59.3 & 58.4 & 60.2 \\ 50.3 & 60.2 & 50.3 \end{pmatrix} \quad . \tag{7.42}$$

The covariance operator describing a posteriori uncertainties in the model parameters is (equation (3.38) of the text)

$$\tilde{\mathbf{C}}_M = (\mathbf{G}^t \, \mathbf{C}_D^{-1} \, \mathbf{G} + \mathbf{C}_M^{-1})^{-1} \quad . \tag{7.43}$$

Instead of representing variances and covariances of $\tilde{\mathbf{C}}_M$, it is more useful to represent standard deviations and correlations (see Appendix 6.5). This gives the standard deviations

$$\tilde{\sigma}_M = \begin{pmatrix} 14.7 & 1.7 & 0.7 \\ 1.7 & 1.0 & 0.7 \\ 0.7 & 0.7 & 0.6 \end{pmatrix} \tag{7.44}$$

and the coefficients of correlation

$$
\begin{pmatrix} R^{1111} & R^{1112} & R^{1113} \\ R^{1121} & R^{1122} & R^{1123} \\ R^{1131} & R^{1132} & R^{1133} \end{pmatrix} = \begin{pmatrix} 1.0000 & -0.9977 & 0.9536 \\ -0.9977 & 0.9958 & 0.9874 \\ 0.9536 & 0.9874 & 0.9901 \end{pmatrix} ,
$$

$$
\begin{pmatrix} R^{1211} & R^{1212} & R^{1213} \\ R^{1221} & R^{1222} & R^{1223} \\ R^{1231} & R^{1232} & R^{1233} \end{pmatrix} = \begin{pmatrix} -0.9977 & 1.0000 & -0.9710 \\ 0.9977 & -0.9972 & -0.9897 \\ -0.9708 & -0.9902 & -0.9896 \end{pmatrix} ,
$$

$$
\begin{pmatrix} R^{1311} & R^{1312} & R^{1313} \\ R^{1321} & R^{1322} & R^{1323} \\ R^{1331} & R^{1332} & R^{1333} \end{pmatrix} = \begin{pmatrix} 0.9536 & -0.9710 & 1.0000 \\ -0.9708 & 0.9657 & 0.9624 \\ 0.9948 & 0.9632 & 0.9515 \end{pmatrix} ,
$$

$$
\begin{pmatrix} R^{2111} & R^{2112} & R^{2113} \\ R^{2121} & R^{2122} & R^{2123} \\ R^{2131} & R^{2132} & R^{2133} \end{pmatrix} = \begin{pmatrix} -0.9977 & 0.9997 & -0.9708 \\ 1.0000 & -0.9972 & -0.9902 \\ -0.9710 & -0.9897 & -0.9896 \end{pmatrix} ,
$$

$$
\begin{pmatrix} R^{2211} & R^{2212} & R^{2213} \\ R^{2221} & R^{2222} & R^{2223} \\ R^{2231} & R^{2232} & R^{2233} \end{pmatrix} = \begin{pmatrix} 0.9958 & -0.9972 & 0.9657 \\ -0.9972 & 1.0000 & 0.9776 \\ 0.9657 & 0.9776 & 0.9963 \end{pmatrix} ,
$$

$$
\begin{pmatrix} R^{2311} & R^{2312} & R^{2313} \\ R^{2321} & R^{2322} & R^{2323} \\ R^{2331} & R^{2332} & R^{2333} \end{pmatrix} = \begin{pmatrix} 0.9874 & -0.9897 & 0.9624 \\ -0.9902 & 0.9776 & 1.0000 \\ 0.9632 & 0.9977 & 0.9622 \end{pmatrix} ,
$$

$$
\begin{pmatrix} R^{3111} & R^{3112} & R^{3113} \\ R^{3121} & R^{3122} & R^{3123} \\ R^{3131} & R^{3132} & R^{3133} \end{pmatrix} = \begin{pmatrix} 0.9536 & -0.9708 & 0.9948 \\ -0.9710 & 0.9657 & 0.9632 \\ 1.0000 & 0.9624 & 0.9515 \end{pmatrix} ,
$$

$$
\begin{pmatrix} R^{3211} & R^{3212} & R^{3213} \\ R^{3221} & R^{3222} & R^{3223} \\ R^{3231} & R^{3232} & R^{3233} \end{pmatrix} = \begin{pmatrix} 0.9874 & -0.9902 & 0.9632 \\ -0.9897 & 0.9776 & 0.9977 \\ 0.9624 & 1.0000 & 0.9622 \end{pmatrix} ,
$$

(7.45)

and

$$
\begin{pmatrix} R^{3311} & R^{3312} & R^{3313} \\ R^{3321} & R^{3322} & R^{3323} \\ R^{3331} & R^{3332} & R^{3333} \end{pmatrix} = \begin{pmatrix} 0.9901 & -0.9896 & 0.9515 \\ -0.9896 & 0.9963 & 0.9622 \\ 0.9515 & 0.9622 & 1.0000 \end{pmatrix} .
$$

The solution (7.42) with the uncertainties (7.44) is represented in Figure 7.7 (to be compared with Figure 7.4 and Figure 7.6). The a priori information was that the values in each block were 55 ± 15. We see that the a posteriori uncertainties are much smaller except in block $(1, 1)$, where the solution, 55.9 ± 14.7, practically coincides with the a priori information. As can be seen in Figure 7.5, this block contains very short lengths of

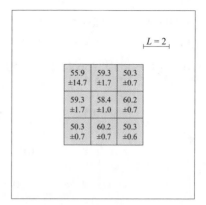

Figure 7.7. *The a posteriori solution obtained by inversion of the available data. Note that the value of the upper-left block has not been resolved (a posteriori value and estimated uncertainty almost identical to a priori values). The values of all other blocks have been estimated with a relative uncertainty of less than 3%.*

rays, so that it has practically not been explored by our data; the value of the attenuation coefficient is practically not resolved by the data set used. If a least-squares inversion were performed with the data (7.37) but without using a priori information, that block would certainly take very arbitrary values, thus polluting the values of the attenuation coefficient in all the other blocks. More dramatically, numerical instabilities could arise (because the operator $\mathbf{G}^t \mathbf{C}_D^{-1} \mathbf{G}$ could become numerically not positive definite due to computer rounding errors) and the used computer code would crash with a "zero divide" diagnostic.

Except for the unresolved block m^{11}, the values obtained are close to the true values and within the estimated error bar. Of course, as the data used were noise corrupted, the obtained values cannot be identical to the true values. Using more rays would give a more precise solution.

The data values recalculated from the solution (7.42) are

$$\mathbf{d}_{\text{obs}} = \begin{pmatrix} 341.92 & 353.10 & 356.20 & 356.20 & 353.10 & 341.90 \\ 341.89 & 353.10 & 356.20 & 356.20 & 353.10 & 341.92 \end{pmatrix}, \qquad (7.46)$$

which are almost identical to the observed values (7.39).

The coefficients of correlation as shown in equation (7.45) are all very close to unity. This is due to the fact that there is no independent information (all rays traverse at least three blocks), and there is not much data redundancy.

Remark 7.1. Assume that a new experiment produces one new datum, corresponding to a new ray (equal to or different from the previous rays). In order to incorporate this new information, we can either take the a priori model (7.39) and perform an inversion using the (7.40) data, or, more simply, we can take the a posteriori solution (7.42)-(7.44)-(7.45) as an a priori solution for an inverse problem with a single datum (the new one). As demonstrated in chapter 3, this gives exactly the same solution (thus showing the coherence of the a priori information approach).

Remark 7.2. Usual computer codes consider that vectors (i.e., elements of a linear space) are necessarily represented using column matrices and that the kernels of linear operators are then represented using two-dimensional matrices. It may then be simpler for numerical computations to replace the previous notation with the matricial notation

$\mathbf{d}_{\text{obs}}$

$$= \begin{pmatrix} 341.9 & 353.1 & 356.2 & 356.2 & 353.1 & 341.9 & 341.9 & 353.1 & 356.2 & 356.2 & 353.1 & 341.9 \end{pmatrix}^t \quad , \tag{7.47}$$

$$(\mathbf{C}_{\text{D}})^{ij} = 0.1^2 \, \delta^{ij} \, , \tag{7.48}$$

$$\mathbf{m}_{\text{prior}} = \begin{pmatrix} 55 & 55 & 55 & 55 & 55 & 55 & 55 & 55 & 55 \end{pmatrix}^t \quad , \tag{7.49}$$

$$(\mathbf{C}_{\text{M}})^{\alpha\beta} = 15^2 \, \delta^{\alpha\beta} \quad , \tag{7.50}$$

and

$$\mathbf{G} = \begin{pmatrix}
0.3338 & 1.6971 & 0.0000 & 2.0309 & 0.0000 & 0.0000 & 2.0309 & 0.0000 & 0.0000 \\
0.0000 & 2.0110 & 0.0000 & 0.0000 & 2.0110 & 0.0000 & 0.4883 & 1.5227 & 0.0000 \\
0.0000 & 2.0012 & 0.0000 & 0.0000 & 2.0012 & 0.0000 & 0.0000 & 2.0012 & 0.0000 \\
0.0000 & 2.0012 & 0.0000 & 0.0000 & 2.0012 & 0.0000 & 0.0000 & 2.0012 & 0.0000 \\
0.0000 & 2.0110 & 0.0000 & 0.0000 & 2.0110 & 0.0000 & 0.0000 & 1.5227 & 0.4883 \\
0.0000 & 1.6971 & 0.3338 & 0.0000 & 0.0000 & 2.0309 & 0.0000 & 0.0000 & 2.0309 \\
0.0000 & 0.0000 & 0.0000 & 1.6971 & 0.0000 & 0.0000 & 0.3338 & 2.0309 & 2.0309 \\
0.0000 & 0.0000 & 0.0000 & 2.0110 & 2.0110 & 1.5227 & 0.0000 & 0.0000 & 0.4883 \\
0.0000 & 0.0000 & 0.0000 & 2.0012 & 2.0012 & 2.0012 & 0.0000 & 0.0000 & 0.0000 \\
0.0000 & 0.0000 & 0.0000 & 2.0012 & 2.0012 & 2.0012 & 0.0000 & 0.0000 & 0.0000 \\
0.0000 & 0.0000 & 0.4883 & 2.0110 & 2.0110 & 1.5227 & 0.0000 & 0.0000 & 0.0000 \\
0.3338 & 2.0309 & 2.0309 & 1.6971 & 0.0000 & 0.0000 & 0.0000 & 0.0000 & 0.0000
\end{pmatrix} \quad . \tag{7.51}$$

Equations (7.38) and (7.41) are then usual matricial equations.

7.4 Linear Regression with Rounding Errors

A physical quantity d is related to the physical quantity x through the equation

$$d = m^1 + m^2 x \quad , \tag{7.52}$$

where m^1 and m^2 are unknown parameters. Equation (7.52) represents a straight line on the plane (d, x). In order to estimate m^1 and m^2, the parameter d has been experimentally measured for some selected values of x, and the following results have been obtained (Figure 7.8):

$$\begin{aligned}
x^1 &= 03.500 \quad , & d^1{}_{\text{obs}} &= 2.0 \pm 0.5 \quad , \\
x^2 &= 05.000 \quad , & d^2{}_{\text{obs}} &= 2.0 \pm 0.5 \quad , \\
x^3 &= 07.000 \quad , & d^3{}_{\text{obs}} &= 3.0 \pm 0.5 \quad , \\
x^4 &= 07.500 \quad , & d^4{}_{\text{obs}} &= 3.0 \pm 0.5 \quad , \\
x^5 &= 10.000 \quad , & d^5{}_{\text{obs}} &= 4.0 \pm 0.5 \quad ,
\end{aligned} \tag{7.53}$$

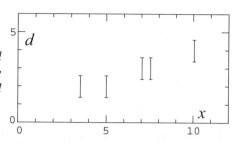

Figure 7.8. *Some experimental points. Uncertainty bars represent rounding errors to the nearest integer. Solve the general problem of estimating a regression line.*

where ±0.5 denotes rounding errors (to the nearest integer). Estimate m^1 and m^2. (Note: this problem is nonclassical in the sense that experimental uncertainties are not Gaussian and the usual least-squares regression is not adopted.)

Solution:
Let an arbitrary set $(d^1, d^2, d^3, d^4, d^5)$ be called a data vector and be denoted by **d** and let an arbitrary set (m^1, m^2) be called a parameter vector and be denoted by **m**. Let

$$\mathbf{d} = \mathbf{g}(\mathbf{m}) \tag{7.54}$$

denote the (linear) relationship

$$d^1 = m^1 + m^2 x^1 \quad , \quad d^2 = m^1 + m^2 x^2 \quad , \quad d^3 = m^1 + m^2 x^3 \quad ,$$
$$d^4 = m^1 + m^2 x^4 \quad , \quad d^5 = m^1 + m^2 x^5 \quad . \tag{7.55}$$

Let

$$\rho_M(\mathbf{m}) = \rho_M(m^1, m^2) \tag{7.56}$$

be the probability density representing the a priori information (if any) on model parameters. Let

$$\rho_D(\mathbf{d}) = \rho_D(d^1, d^2, d^3, d^4, d^5) \tag{7.57}$$

be the probability density describing the experimental uncertainties (see main text). As rounding uncertainties are mutually independent,

$$\rho_D(\mathbf{d}) = \rho_D(d^1, d^2, d^3, d^4, d^5) = \rho_D^1(d^1)\,\rho_D^2(d^2)\,\rho_D^3(d^3)\,\rho_D^4(d^4)\,\rho_D^5(d^5) \quad , \tag{7.58}$$

where $\rho_D^i(d^i)$ denotes the probability density describing the experimental uncertainty for the observed data d^i. As the uncertainties are only rounding uncertainties, they can be conveniently modeled using boxcar probability density functions:

$$\rho_D^i(d^i) = \begin{cases} \text{const.} & \text{if } d_{\text{obs}}^i - 0.5 < d^i < d_{\text{obs}}^i + 0.5 \quad , \\ 0 & \text{otherwise} \quad . \end{cases} \tag{7.59}$$

This gives

$$\rho_D(\mathbf{d}) = \begin{cases} \text{const.} & \text{if } \begin{cases} 1.5 < d^1 < 2.5 \quad \text{and} \quad 1.5 < d^2 < 2.5 \quad \text{and} \\ 2.5 < d^3 < 3.5 \quad \text{and} \quad 2.5 < d^4 < 3.5 \quad \text{and} \\ 3.5 < d^5 < 4.5 \quad , \end{cases} \\ 0 & \text{otherwise} \quad . \end{cases} \qquad (7.60)$$

The general solution of an inverse problem is obtained when the posterior probability density in the model space has been defined. It is given by equation (1.93) of the main text:[92]

$$\sigma_M(\mathbf{m}) = \text{const.} \, \rho_M(\mathbf{m}) \, \rho_D(\mathbf{g}(\mathbf{m})) \quad . \qquad (7.61)$$

Equations (7.60) and (7.61) solve the problem.

For instance, if we accept a priori all pairs (m^1, m^2) as equally probable, and the quantities $\{m^1, m^2\}$ are Cartesian (see main text),

$$\rho_M(\mathbf{m}) = \rho_M(m^1, m^2) = \text{const.} \quad , \qquad (7.62)$$

then we obtain

$$\sigma_M(\mathbf{m}) = \sigma_M(m^1, m^2) = \begin{cases} \text{const.} & \text{if } \begin{cases} 1.5 < m^1 + m^2 x^1 < 2.5 \quad \text{and} \\ 1.5 < m^1 + m^2 x^2 < 2.5 \quad \text{and} \\ 2.5 < m^1 + m^2 x^3 < 3.5 \quad \text{and} \\ 2.5 < m^1 + m^2 x^4 < 3.5 \quad \text{and} \\ 3.5 < m^1 + m^2 x^5 < 4.5 \quad , \end{cases} \\ 0 & \text{otherwise} \quad . \end{cases} \qquad (7.63)$$

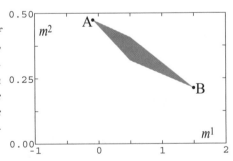

Figure 7.9. *The general solution of the problem is given by the probability density $\sigma_M(m^1, m^2)$ for the parameters of the regression line. It is constant inside the dark region and null outside. The dark region represents the domain of admissible solutions. There is no best line: all pairs (m^1, m^2) inside the region are equally likely.*

This result is represented graphically in Figure 7.9. The gray region has a positive (constant) probability density. All pairs (m^1, m^2) inside this region have equal probability density, and all pairs (m^1, m^2) outside it are impossible, so that this region represents the domain of admissible solutions. Which is *the best* regression line? There is no such thing: all lines inside the domain are equally good. Figure 7.10 shows two particular solutions (giving extremal values for m^1 and m^2).

---

[92] Assuming here that the data space is a linear space, i.e., in fact, that the simple difference between the data quantities d^i can be interpreted as a distance.

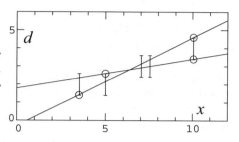

Figure 7.10. *Two particular solutions (A and B in Figure 7.9), corresponding to extremal values of m^1 and m^2. Notice that they touch the extremities of the error bars (circles).*

```
for m1 = -1 to 2 step 0.002
    for m2 = 0 to 0.5 step 0.002
        d1 = m1 + 3.5 m2
        if ( d1 < 1.5 ) or ( d1 > 2.5 ) then next m2
        d2 = m1 + 5.0 m2
        if ( d2 < 1.5 ) or ( d2 > 2.5 ) then next m2
        d3 = m1 + 7.0 m2
        if ( d3 < 2.5 ) or ( d3 > 3.5 ) then next m2
        d4 = m1 + 7.5 m2
        if ( d4 < 2.5 ) or ( d4 > 3.5 ) then next m2
        d5 = m1 + 10. m2
        if ( d5 < 3.5 ) or ( d5 > 4.5 ) then next m2
        draw point (m1,m2)
    next m2
next m1
```

Figure 7.11. *Computer code effectively used to obtain the result in Figure 7.9. The limits for m^1 and m^2 in the first two lines were chosen after trial and error. The step value 0.002 was chosen small enough not to be visible on the graphic device used to generate Figure 7.9. The command* draw point *simply plots a point on the graphic device at the given coordinates.*

Figure 7.11 shows the computer code effectively used to obtain the general solution shown in Figure 7.9. For problems with few model parameters (two in this example), the full exploration of the model space is, in general, the easiest strategy (it takes approximately 2 min to go from the statement of the problem to the result in Figure 7.9).

7.5 Usual Least-Squares Regression

Find the best regression line for the experimental points in Figure 7.12, assuming Gaussian uncertainties.

Solution:
Figure 7.12 suggests that uncertainties in the t^i are negligible, while uncertainties in the y^i are uncorrelated. Let us introduce

$$\mathbf{m} = \begin{pmatrix} a \\ b \end{pmatrix} \quad , \quad \mathbf{d} = \begin{pmatrix} y^1 \\ y^2 \\ \cdots \\ y^n \end{pmatrix} \quad , \quad \mathbf{G} = \begin{pmatrix} t^1 & 1 \\ t^2 & 1 \\ \cdots & \cdots \\ t^n & 1 \end{pmatrix} \quad (7.64)$$

Figure 7.12. *The physical parame-*
ter y is related to the physical parameter t
through the equation y = a t + b , where the
parameters a and b are unknown. The ex-
perimental points in the figure have to be used
to estimate the best values for a and b in the
least-squares sense.

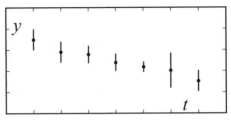

(the superscript numbers in $y^1, y^2, \ldots, t^1, t^2, \ldots$ are indices, not powers). The equations

$$y^i = a t^i + b \qquad (i = 1, 2, \ldots, n) \tag{7.65}$$

can be written

$$\mathbf{d} = \mathbf{G}\mathbf{m} \ . \tag{7.66}$$

The matrix $\mathbf{G}$ is assumed perfectly known. We have some information on the true values of $\mathbf{d}$, and we wish to estimate the true value of $\mathbf{m}$.

As it is assumed that uncertainties in the y^i are uncorrelated Gaussian, the information we have on the true value of $\mathbf{d}$ can be represented using a Gaussian probability density with mathematical expectation

$$\mathbf{d}_{\text{obs}} = \begin{pmatrix} (y_0)^1 \\ (y_0)^2 \\ \cdots \\ (y_0)^n \end{pmatrix} \tag{7.67}$$

and covariance matrix

$$\mathbf{C}_{\mathbf{D}} = \begin{pmatrix} (\sigma^1)^2 & 0 & 0 & \cdots \\ 0 & (\sigma^2)^2 & 0 & \cdots \\ 0 & 0 & (\sigma^3)^2 & \cdots \\ \cdots & \cdots & \cdots & \cdots \end{pmatrix} \ . \tag{7.68}$$

We now need to introduce the a priori information (if any) on the parameters $\mathbf{m}$. The simplest results are obtained when using a Gaussian probability density in the model space with mathematical expectation

$$\mathbf{m}_{\text{prior}} = \begin{pmatrix} a_0 \\ b_0 \end{pmatrix} \tag{7.69}$$

and covariance matrix

$$\mathbf{C}_{\mathbf{M}} = \begin{pmatrix} \sigma^2_a & \rho\, \sigma_a\, \sigma_b \\ \rho\, \sigma_a\, \sigma_b & \sigma^2_b \end{pmatrix} \ . \tag{7.70}$$

As the information on both data and model parameters is Gaussian, we are under the hypothesis of chapter 3. The a posteriori information on the model parameters is then also

Gaussian, with mathematical expectation given by (equation (3.37))

$$\begin{aligned}
\tilde{\mathbf{m}} &= \mathbf{m}_{\text{prior}} + (\mathbf{G}^t \, \mathbf{C}_D^{-1} \, \mathbf{G} + \mathbf{C}_M^{-1})^{-1} \, \mathbf{G}^t \, \mathbf{C}_D^{-1} \, (\mathbf{d}_{\text{obs}} - \mathbf{G} \, \mathbf{m}_{\text{prior}}) \\
&= \mathbf{m}_{\text{prior}} + \mathbf{C}_M \, \mathbf{G}^t \, (\mathbf{G} \, \mathbf{C}_M \, \mathbf{G}^t + \mathbf{C}_D)^{-1} \, (\mathbf{d}_{\text{obs}} - \mathbf{G} \, \mathbf{m}_{\text{prior}})
\end{aligned} \tag{7.71}$$

and covariance matrix given by (equation (3.38))

$$\begin{aligned}
\tilde{\mathbf{C}}_M &= (\mathbf{G}^t \, \mathbf{C}_D^{-1} \, \mathbf{G} + \mathbf{C}_M^{-1})^{-1} \\
&= \mathbf{C}_M - \mathbf{C}_M \, \mathbf{G}^t (\mathbf{G} \, \mathbf{C}_M \, \mathbf{G}^t + \mathbf{C}_D)^{-1} \mathbf{G} \, \mathbf{C}_M \quad .
\end{aligned} \tag{7.72}$$

The a posteriori (i.e., recalculated) data values are then (equation at left in (3.44))

$$\tilde{\mathbf{d}} = \mathbf{G} \, \mathbf{m}_{\text{post}} \quad , \tag{7.73}$$

and the a posteriori data uncertainties are given by (equation at right in (3.44))

$$\tilde{\mathbf{C}}_D = \mathbf{G} \, \tilde{\mathbf{C}}_M \, \mathbf{G}^t \quad . \tag{7.74}$$

As we have only two model parameters, the first equation in each of the expressions (7.71)–(7.72) should be preferred. An easy computation gives the a posteriori values of a and b,

$$a = a_0 + \frac{A \, P - C \, Q}{A \, B - C^2} \quad , \qquad b = b_0 + \frac{B \, Q - C \, P}{A \, B - C^2} \quad , \tag{7.75}$$

and the a posteriori standard deviations and correlation,

$$\tilde{\sigma}_a = \frac{1}{\sqrt{B - C^2/A}} \quad , \qquad \tilde{\sigma}_b = \frac{1}{\sqrt{A - C^2/B}} \quad , \qquad \tilde{\rho} = \frac{-1}{\sqrt{A B / C^2}} \quad , \tag{7.76}$$

where

$$A = \sum_i \frac{1}{(\sigma^i)^2} + \frac{1}{(1 - \rho^2) \, \sigma_b^2} \quad , \qquad B = \sum_i \frac{(t^i)^2}{(\sigma^i)^2} + \frac{1}{(1 - \rho^2) \, \sigma_a^2} \quad ,$$

$$C = \sum_i \frac{t^i}{(\sigma^i)^2} - \frac{\rho}{(1 - \rho^2) \, \sigma_a \, \sigma_b} \quad , \qquad P = \sum_i \frac{t^i}{(\sigma^i)^2} \big((y_0)^i - (a_0 \, t^i + b_0)\big) \quad ,$$

$$\tag{7.77}$$

and

$$Q = \sum_i \frac{1}{(\sigma^i)^2} \big((y_0)^i - (a_0 \, t^i + b_0)\big) \quad . \tag{7.78}$$

Usually, a priori uncertainties on model parameters are uncorrelated. Then,

$$\rho = 0 \quad . \tag{7.79}$$

This gives

$$A = \sum_i \frac{1}{(\sigma^i)^2} + \frac{1}{\sigma_b^2} \quad , \quad B = \sum_i \frac{(t^i)^2}{(\sigma^i)^2} + \frac{1}{\sigma_a^2} \quad , \quad C = \sum_i \frac{t^i}{(\sigma^i)^2} \quad . \tag{7.80}$$

If there is no a priori information on model parameters,

$$\sigma_a \to \infty \quad , \quad \sigma_b \to \infty \quad . \tag{7.81}$$

Instead of taking these limits in the last equations, it is simpler to use

$$\mathbf{C}_{\mathrm{M}}^{-1} = 0 \tag{7.82}$$

in equations (7.71) and (7.72). This gives

$$\mathbf{m}_{\mathrm{post}} = (\mathbf{G}^t \, \mathbf{C}_{\mathrm{D}}^{-1} \, \mathbf{G})^{-1} \mathbf{G}^t \, \mathbf{C}_{\mathrm{D}}^{-1} \, \mathbf{d}_{\mathrm{obs}} \tag{7.83}$$

and

$$\mathbf{C}_{M'} = (\mathbf{G}^t \, \mathbf{C}_{\mathrm{D}}^{-1} \, \mathbf{G})^{-1} \quad . \tag{7.84}$$

Equations (7.75) then become

$$a = \frac{A\,P - C\,Q}{A\,B - C^2} \quad , \quad b = \frac{B\,Q - C\,P}{A\,B - C^2} \quad , \tag{7.85}$$

while the constants A, B, C, P, and Q simplify to

$$A = \sum_i \frac{1}{(\sigma^i)^2} \quad , \quad B = \sum_i \frac{(t^i)^2}{(\sigma^i)^2} \quad , \quad C = \sum_i \frac{t^i}{(\sigma^i)^2} \quad ,$$
$$P = \sum_i \frac{t^i \, (y_0)^i}{(\sigma^i)^2} \quad , \quad Q = \sum_i \frac{(y_0)^i}{(\sigma^i)^2} \quad . \tag{7.86}$$

If all data uncertainties are identical,

$$\sigma^i = \sigma \quad , \tag{7.87}$$

then

$$A = \frac{n}{\sigma^2} \quad , \quad B = \frac{1}{\sigma^2} \sum_i (t^i)^2 \quad , \quad C = \frac{1}{\sigma^2} \sum_i t^i \quad ,$$
$$P = \frac{1}{\sigma^2} \sum_i t^i \, (y_0)^i \quad , \quad Q = \frac{1}{\sigma^2} \sum_i (y_0)^i \quad . \tag{7.88}$$

Figure 7.13. *The physical parameter* y *is related to the physical parameter* t
through the equation y = a t + b, *where the
parameters* a *and* b *are unknown. The experimental points in the figure have to be used
to estimate the best values for* a *and* b *in the
least-squares sense. This problem is nonclassical in the sense that uncertainties are present
in both coordinates. The numerical values on
the axes are not indicated.*

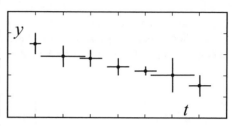

7.6 Least-Squares Regression with Uncertainties in Both Axes

Find the best regression line for the experimental points in Figure 7.13, assuming Gaussian uncertainties.

Solution:

There are some equivalent ways of properly setting this problem. The approach followed here has the advantage of giving a symmetrical treatment to both axes.

As the statement of the problem refers to a regression *line*, a linear relationship has to be assumed between the variables y and t:

$$\alpha y + \beta t = 1 \quad . \tag{7.89}$$

We have measured some pairs (x^i, y^i) and wish to estimate the true values of α and β.

Let us introduce a parameter vector $\mathbf{m}$ that contains the y^i, the t^i, α, and β:

$$\mathbf{m} = \begin{pmatrix} \mathbf{y} & \mathbf{t} & \mathbf{u} \end{pmatrix}^t = \begin{pmatrix} y^1 & y^2 & \cdots & t^1 & t^2 & \cdots & \alpha & \beta \end{pmatrix}^t \quad , \tag{7.90}$$

and, for each conceivable value of $\mathbf{m}$, let us define a vector

$$\mathbf{d} = \begin{pmatrix} d^1 & d^2 & \cdots \end{pmatrix}^t \tag{7.91}$$

by

$$d^i = g^i(\mathbf{m}) = \alpha y^i + \beta t^i \qquad (i = 1, 2, \ldots) \quad . \tag{7.92}$$

Defining the *observed value* of $\mathbf{d}$ and the *a priori* value of $\mathbf{m}$ respectively as

$$\mathbf{d}_{\text{obs}} = \begin{pmatrix} 1 & 1 & 1 & \cdots \end{pmatrix}^t \quad , \qquad \mathbf{m}_{\text{prior}} = \begin{pmatrix} y^1_0 & y^2_0 & \cdots & t^1_0 & t^2_0 & \cdots & \alpha_0 & \beta_0 \end{pmatrix}^t \quad , \tag{7.93}$$

where y_0 and t_0 are the experimental values, and α_0 and β_0 are the a priori values of α and β, the inverse problem can now be set as the problem of obtaining a vector $\mathbf{m}$ such that $\mathbf{g}(\mathbf{m})$ is close (or identical) to $\mathbf{d}_{\text{obs}}$ and $\mathbf{m}$ is close to $\mathbf{m}_{\text{prior}}$. We see thus that this relabeling of the variables allows an immediate use of the standard equations. Nevertheless, this problem is less simple than the previous problem of one-axis regression, because here we have twice the number of points + 2 unknowns instead of 2, and the forward equation $\mathbf{d} = \mathbf{g}(\mathbf{m})$ is nonlinear (because it contains the mutual product of parameters).

More precisely, we assume that the a priori information on $\mathbf{m}$ can be described using a Gaussian probability density with mathematical expectation $\mathbf{m}_{\text{prior}}$ and covariance matrix

$$
\mathbf{C_M} = \begin{pmatrix} \mathbf{C}_y & 0 & 0 & 0 \\ 0 & \mathbf{C}_t & 0 & 0 \\ 0 & 0 & \sigma_\alpha^2 & 0 \\ 0 & 0 & 0 & \sigma_\beta^2 \end{pmatrix} \quad , \tag{7.94}
$$

where independence of uncertainties has been assumed only to simplify the notation. The a priori information on $\mathbf{d}$ is also assumed to be Gaussian, with mathematical expectation $\mathbf{d}_{\text{obs}}$ and covariance matrix $\mathbf{C_D}$. Later, we may take $\mathbf{C_D} = 0$, so that the observed values $\mathbf{d}_{\text{obs}}$ may be fitted exactly by the a posteriori solution. Instead, we may keep $\mathbf{C_D}$ finite to allow for uncertainties in the hypothesis of a strictly linear relationship between y and t.

Now we are exactly under the hypothesis of section 3.2.3. The a posteriori probability density for $\mathbf{m}$ is (equations (3.45)–(3.46))

$$
\sigma_M(\mathbf{m}) = \text{const. } \exp(-S(\mathbf{m})) \quad , \tag{7.95}
$$

with

$$
2\,S(\mathbf{m}) = (\mathbf{g}(\mathbf{m}) - \mathbf{d}_{\text{obs}})^t\, \mathbf{C_D^{-1}}\,(\mathbf{g}(\mathbf{m}) - \mathbf{d}_{\text{obs}}) + (\mathbf{m} - \mathbf{m}_{\text{prior}})^t\, \mathbf{C_M^{-1}}\,(\mathbf{m} - \mathbf{m}_{\text{prior}}) \quad . \tag{7.96}
$$

Owing to the nonlinearity of $\mathbf{g}(\mathbf{m})$, this is not a Gaussian probability density. The maximum likelihood value of $\mathbf{m}$ can be obtained using, for instance, the iterative algorithm

$$
\mathbf{m}_{n+1} = \mathbf{m}_{\text{prior}} - \mathbf{C_M}\, \mathbf{G}_n^t (\mathbf{G}_n\, \mathbf{C_M}\, \mathbf{G}_n^t + \mathbf{C_D})^{-1} \big((\mathbf{g}(\mathbf{m}_n) - \mathbf{d}_{\text{obs}}) - \mathbf{G}_n (\mathbf{m}_n - \mathbf{m}_{\text{prior}}) \big) \quad . \tag{7.97}
$$

We have

$$
\mathbf{G}_n = \left(\left(\frac{\partial \mathbf{g}}{\partial y}\right)_{\mathbf{m}_n} \quad \left(\frac{\partial \mathbf{g}}{\partial t}\right)_{\mathbf{m}_n} \quad \left(\frac{\partial \mathbf{g}}{\partial \alpha}\right)_{\mathbf{m}_n} \quad \left(\frac{\partial \mathbf{g}}{\partial \beta}\right)_{\mathbf{m}_n} \right) \quad , \tag{7.98}
$$

which gives

$$
\mathbf{G}_n = \begin{pmatrix} \alpha_n \mathbf{I} & \beta_n \mathbf{I} & \mathbf{y}_n & \mathbf{t}_n \end{pmatrix} \quad , \tag{7.99}
$$

$$
\mathbf{C_M}\, \mathbf{G}_n^t = \begin{pmatrix} \alpha_n \mathbf{C}_y \\ \beta_{n_2} \mathbf{C}_t \\ \sigma_\alpha^2\, \mathbf{y}_n^t \\ \sigma_\beta^2\, \mathbf{t}_n^t \end{pmatrix} \quad , \tag{7.100}
$$

$$
\mathbf{G}_n\, \mathbf{C_M}\, \mathbf{G}_n^t + \mathbf{C_D} = \sigma_\alpha^2\, \mathbf{y}_n\, \mathbf{y}_n^t + \sigma_\beta^2\, \mathbf{t}_n\, \mathbf{t}_n^t + \alpha_n^2\, \mathbf{C}_y + \beta_n^2\, \mathbf{C}_t + \mathbf{C_D} \quad , \tag{7.101}
$$

and

$$\mathbf{g}(\mathbf{m}_n) - \mathbf{d}_{\mathrm{obs}} - \mathbf{G}_n\,(\mathbf{m}_n - \mathbf{m}_{\mathrm{prior}}) \;=\; (\alpha_0 - \alpha_n)\,\mathbf{y}_n + (\beta_0 - \beta_n)\,\mathbf{t}_n - \mathbf{d}_{\mathrm{obs}} + \alpha_n\,\mathbf{y}_0 + \beta_n\,\mathbf{t}_0 \quad . \tag{7.102}$$

Denoting

$$\delta\hat{\mathbf{d}}_n \;=\; (\,\mathbf{G}_n\,\mathbf{C}_{\mathrm{M}}\,\mathbf{G}_n{}^t + \mathbf{C}_{\mathrm{D}}\,)^{-1}\big(\,(\mathbf{g}(\mathbf{m}_n) - \mathbf{d}_{\mathrm{obs}}) - \mathbf{G}_n\,(\mathbf{m}_n - \mathbf{m}_{\mathrm{prior}})\,\big) \quad , \tag{7.103}$$

the iterative algorithm (7.97) can be written

$$\begin{aligned}
\mathbf{y}_{n+1} &= \mathbf{y}_0 - \alpha_n\,\mathbf{C}_y\,\delta\hat{\mathbf{d}}_n \quad , \\
\mathbf{t}_{n+1} &= \mathbf{t}_0 - \beta_n\,\mathbf{C}_t\,\delta\hat{\mathbf{d}}_n \quad , \\
\alpha_{n+1} &= \alpha_0 - \sigma_\alpha^2\,\mathbf{y}_n{}^t\,\delta\hat{\mathbf{d}}_n \quad ,
\end{aligned} \tag{7.104}$$

and

$$\beta_{n+1} \;=\; \beta_0 - \sigma_\beta^2\,\mathbf{t}_n{}^t\,\delta\hat{\mathbf{d}}_n \quad . \tag{7.105}$$

The algorithm usually converges in a few iterations ($\simeq 3$). The values α_∞ and β_∞ are the estimated values of the parameters defining the regression line, and the values (t_∞^i, y_∞^i) $(i = 1, 2, \dots)$ are the a posteriori values of the experimental points. If $\mathbf{C}_{\mathrm{D}} = 0$, the a posteriori points belong to the straight line.

7.7 Linear Regression with an Outlier

Two variables y and t are related through a linear relationship

$$y = a\,t + b \quad . \tag{7.106}$$

In order to estimate the parameters a and b, the 11 experimental points (y^i, t^i) shown in Figure 7.14 have been obtained. It is clear that if the linear relationship (7.106) applies, then the point indicated with an arrow must be an outlier. Suppress that point and solve the problem of estimating a and b under the hypothesis of Gaussian uncertainties. Does the solution change very much if the outlier is included? Assume now that uncertainties can be modeled using an exponential probability density, and solve the problem again. Discuss the relative robustness of the Gaussian and exponential hypotheses with respect to the existence of outliers in a data set.

Figure 7.14. *Two variables y and t are related by the relationship $y = a\,t + b$, where a and b are unknown parameters. In order to estimate a and b, an experiment has been performed that has furnished the 11 experimental points shown in the figure. The exact meaning of the error bars is not indicated.*

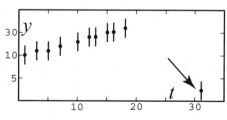

Solution:
Let

$$\mathbf{m} = (a, b) \tag{7.107}$$

denote a model vector and

$$\mathbf{d} = (d^1, d^2, \dots) \tag{7.108}$$

denote a data vector. The (exact) theoretical relationship between $\mathbf{d}$ and $\mathbf{m}$ is linear,

$$d^i = a t^i + b \quad , \tag{7.109a}$$

or, for short,

$$\mathbf{d} = \mathbf{G}\mathbf{m} \quad , \tag{7.109b}$$

where $\mathbf{G}$ is a linear operator.

Assume that the homogeneous probability density on model parameters is

$$\mu_M(a, b) = \text{const.} \tag{7.110}$$

and that we do not have a priori information on model parameters,

$$\rho_M(a, b) = \mu_M(a, b) = \text{const.} \tag{7.111}$$

Assume that the homogeneous probability density on data parameters is

$$\mu_D(d^1, d^2, \dots) = \text{const.} \tag{7.112}$$

If $\rho_D(d^1, d^2, \dots)$ is the probability density representing the information on the true values of $(d^1, d^2, \dots)$ as obtained through the measurements, then the Gaussian hypothesis gives (for independent uncertainties)

$$\rho_D(d^1, d^2, \dots) = \exp\left(-\frac{1}{2}\sum_i \frac{(d^i - d^i_{\text{obs}})^2}{\sigma^2}\right) \quad , \tag{7.113}$$

where $\mathbf{d}_{\text{obs}}$ is the vector of observed values

$$\mathbf{d}_{\text{obs}} = \left(10., 11., 11., 12., 13., 14., 14., 15., 15., 16., 2.\right) \tag{7.114}$$

and where, if we interpret the error bars in Figure 7.14 as standard deviations,

$$\sigma = 2. \tag{7.115}$$

We are here in the context of Example 1.34 (page 34). Therefore, the posterior probability density on the model parameters is (equation (1.93))

$$\sigma_M(\mathbf{m}) = k \rho_M(\mathbf{m}) \rho_D(\mathbf{g}(\mathbf{m})) \quad . \tag{7.116}$$

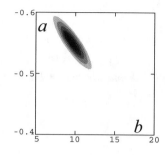

Figure 7.15. *The probability density for the parameters (a, b) obtained using the Gaussian hypothesis for experimental uncertainties and without using the outlier.*

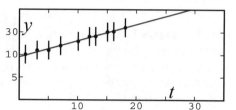

Figure 7.16. *The maximum likelihood line for the probability density in Figure 7.15.*

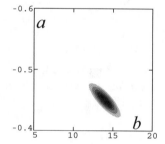

Figure 7.17. *Same as Figure 7.15, but the 11 experimental points have been used. The outlier has translated the probability density. This shows that the Gaussian hypothesis is not very robust with respect to the existence of a small number of outliers in a data set.*

Introducing the results above, we obtain

$$\sigma_M(a, b) = \exp\left(-\frac{1}{2}\sum_i \frac{\left(d^i_{\text{obs}} - d^i_{\text{cal}}(a, b)\right)^2}{\sigma^2}\right) \quad , \tag{7.117}$$

where

$$d^i_{\text{cal}}(a, b) = a\,t^i + b \quad . \tag{7.118}$$

As this problem only has two model parameters, the simplest way to analyze the a posteriori information we have on model parameters is to directly compute the values $\sigma_M(a, b)$ in a given grid and to plot the results. Figure 7.15 shows the corresponding result if the outlier is suppressed from the data set (only 10 points have been used). This probability density is Gaussian, and the line corresponding to its center is shown in Figure 7.16. If the outlier is not suppressed, so that the 11 points are used, the probability density $\sigma_M(a, b)$ obtained is shown in Figure 7.17. The probability density has been essentially translated by the outlier. The line corresponding to the center of the probability density is shown in Figure 7.18. Figures 7.17 and 7.18 show that the Gaussian assumption gives results that

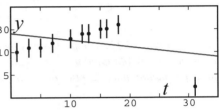

Figure 7.18. *The maximum likelihood line for the probability density in Figure 7.17.*

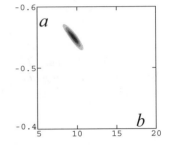

Figure 7.19. *The exponential hypothesis for data uncertainties has been used instead of the Gaussian hypothesis. Here the outlier has not been used. The solution looks similar to the solution corresponding to the Gaussian hypothesis in Figure 7.15.*

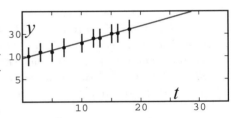

Figure 7.20. *The maximum likelihood line for the probability density in Figure 7.19.*

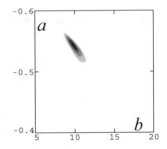

Figure 7.21. *The probability density using all 11 experimental points in the exponential hypothesis. By comparison with Figure 7.19, we see that the introduction of the outlier does not completely distort the solution. This shows that the exponential hypothesis is more robust than the Gaussian hypothesis with respect to the existence of a few outliers in a data set.*

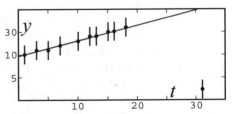

Figure 7.22. *The maximum likelihood line for the probability density in Figure 7.21.*

are not robust with respect to the existence of outliers in a data set. This may be annoying, because in multidimensional problems it is not always easy to detect outliers.

If instead of assuming an uncorrelated Gaussian, we assume uncorrelated exponential uncertainties, equation (7.113) is replaced with

$$\rho_D(d^1, d^2, \ldots) = \exp\left(-\sum_i \frac{|d^i - d^i_{obs}|}{\sigma}\right) .$$

(7.119)

The a posteriori probability density is then

$$\sigma_M(a, b) = \exp\left(-\sum_i \frac{|d^i_{obs} - d^i_{cal}(a, b)|}{\sigma}\right) .$$

(7.120)

This probability density is shown in Figure 7.19 for all points but the outlier, and in Figure 7.21 for all 11 points. The corresponding maximum likelihood lines are shown in Figures 7.20 and 7.22. We see that the introduction of the outlier deforms the posterior probability density, but it does not translate it. The exponential hypothesis for data uncertainties is more robust than the Gaussian hypothesis.

It should be noticed that the question of which probability density may truly represent the experimental uncertainties for the data in Figure 7.14 has not been addressed. Obviously, it is not Gaussian, because the probability of a outlier like the one present in the figure is extremely low. But the probability of such an outlier is also very low in the exponential hypothesis. A careful examination of the experimental conditions can, in principle, suggest a realistic choice of probability density for representing uncertainties, but this is not always easy. The conclusion of this numerical example is that if a probability density adequately representing experimental uncertainties is unknown, but we suspect a small number of large errors, we should not take the Gaussian probability density, but a more long-tailed one.

7.8 Condition Number and A Posteriori Uncertainties

The (Cramer's) solution of the system

$$\begin{pmatrix} 10 & 7 & 8 & 7 \\ 7 & 5 & 6 & 5 \\ 8 & 6 & 10 & 9 \\ 7 & 5 & 9 & 10 \end{pmatrix} \begin{pmatrix} m^1 \\ m^2 \\ m^3 \\ m^4 \end{pmatrix} = \begin{pmatrix} 32.0 \\ 23.0 \\ 33.0 \\ 31.0 \end{pmatrix}$$

(7.121)

is

$$\begin{pmatrix} m^1 \\ m^2 \\ m^3 \\ m^4 \end{pmatrix} = \begin{pmatrix} 1.0 \\ 1.0 \\ 1.0 \\ 1.0 \end{pmatrix} ,$$

(7.122)

while the solution of the system

$$\begin{pmatrix} 10 & 7 & 8 & 7 \\ 7 & 5 & 6 & 5 \\ 8 & 6 & 10 & 9 \\ 7 & 5 & 9 & 10 \end{pmatrix} \begin{pmatrix} m^1 \\ m^2 \\ m^3 \\ m^4 \end{pmatrix} = \begin{pmatrix} 32.1 \\ 22.9 \\ 33.1 \\ 30.9 \end{pmatrix} ,$$

(7.123)

where the right-hand side has been slightly modified, is completely different:

$$\begin{pmatrix} m^1 \\ m^2 \\ m^3 \\ m^4 \end{pmatrix} = \begin{pmatrix} 9.2 \\ -12.6 \\ 4.5 \\ -1.1 \end{pmatrix} . \tag{7.124}$$

This result may be surprising, because the determinant of the matrix of the system is not small (it equals one), and the inverse matrix looks as ordinary as the original one:

$$\begin{pmatrix} 10 & 7 & 8 & 7 \\ 7 & 5 & 6 & 5 \\ 8 & 6 & 10 & 9 \\ 7 & 5 & 9 & 10 \end{pmatrix}^{-1} = \begin{pmatrix} 25 & -41 & 10 & -6 \\ -41 & 68 & -17 & 10 \\ 10 & -17 & 5 & -3 \\ -6 & 10 & -3 & 2 \end{pmatrix} . \tag{7.125}$$

This nice example is due to R.S. Wilson, and is quoted by Ciarlet (1982). Clearly, the matrix in the example has some special property, which it is important to identify. In classical numerical analysis, it is usual to introduce the concept of the *condition number* of a matrix. It is defined by

$$\text{cond}(\mathbf{A}) = \| \mathbf{A} \| \| \mathbf{A}^{-1} \| , \tag{7.126}$$

where $\| \mathbf{A} \|$ denotes a given matricial norm. For instance, the ℓ_p matricial norms can be defined by

$$\| \mathbf{A} \|_1 = \sup \frac{\| \mathbf{A} \mathbf{v} \|_1}{\| \mathbf{v} \|_1} , \quad \| \mathbf{A} \|_2 = \sup \frac{\| \mathbf{A} \mathbf{v} \|_2}{\| \mathbf{v} \|_2} , \quad \| \mathbf{A} \|_\infty = \sup \frac{\| \mathbf{A} \mathbf{v} \|_\infty}{\| \mathbf{v} \|_\infty} , \tag{7.127}$$

and verify (e.g., Ciarlet, 1982)

$$\| \mathbf{A} \|_1 = \max_i \sum |\mathbf{A}^{ij}| , \quad \| \mathbf{A} \|_2 = \sqrt{\max \lambda_i (\mathbf{A}^* \mathbf{A})} , \quad \| \mathbf{A} \|_\infty = \max_j \sum |\mathbf{A}^{ij}| , \tag{7.128}$$

where $\lambda_i(\mathbf{A})$ denotes the eigenvalues of the matrix $\mathbf{A}$ and $\mathbf{A}^*$ denotes the adjoint of $\mathbf{A}$ (the difference between adjoint and transpose is explained elsewhere in this text; for this example, let us simply admit that we only consider Euclidean scalar products, and adjoint and transpose coincide).

The interpretation of the condition number is obtained as follows. Let $\mathbf{A}$ and $\mathbf{d}$ respectively represent a given regular matrix and a given vector, and let $\mathbf{m}$ represent the solution of $\mathbf{A} \mathbf{m} = \mathbf{d}$,

$$\mathbf{m} = \mathbf{A}^{-1} \mathbf{d} . \tag{7.129}$$

Let now $\mathbf{m} + \delta\mathbf{m}$ represent the solution of the perturbed system

$$\mathbf{A}(\mathbf{m} + \delta\mathbf{m}) = \mathbf{d} + \delta\mathbf{d} . \tag{7.130}$$

From $\mathbf{d} = \mathbf{A}\,\mathbf{m}$ and $\delta\mathbf{m} = \mathbf{A}^{-1}\,\delta\mathbf{d}$, it can be deduced that

$$\| \mathbf{d} \| \leq \| \mathbf{A} \| \, \| \mathbf{m} \| \quad , \qquad \| \delta\mathbf{m} \| \leq \| \mathbf{A}^{-1} \| \, \| \delta\mathbf{d} \| \quad , \tag{7.131}$$

i.e.,

$$\frac{\| \delta\mathbf{m} \|}{\| \mathbf{m} \|} \leq \| \mathbf{A} \| \, \| \mathbf{A}^{-1} \| \, \frac{\| \delta\mathbf{d} \|}{\| \mathbf{d} \|} \quad , \tag{7.132}$$

which, using the definition of condition number, can be written

$$\frac{\| \delta\mathbf{m} \|}{\| \mathbf{m} \|} \leq \text{cond}(\mathbf{A}) \, \frac{\| \delta\mathbf{d} \|}{\| \mathbf{d} \|} \quad . \tag{7.133}$$

This equation shows that, for a given relative data error $\| \delta\mathbf{d} \| / \| \mathbf{d} \|$, the relative solution error $\| \delta\mathbf{m} \| / \| \mathbf{m} \|$ may be large if the condition number is large. As it can be shown that

$$1 \leq \text{cond}(\mathbf{A}) \leq \infty \quad , \tag{7.134}$$

a linear system for which $\text{cond}(\mathbf{A}) \simeq 1$ is called *well conditioned*; a linear system for which $\text{cond}(\mathbf{A}) \gg 1$ is called *ill conditioned*.

The following properties, which are sometimes useful, can be demonstrated (Ciarlet, 1982):

$$\text{cond}(\mathbf{A}) = \text{cond}(\mathbf{A}^{-1}) \quad ,$$

$$\text{cond}_2(\mathbf{A}) = \frac{\sqrt{\max \lambda_i (\mathbf{A}^* \mathbf{A})}}{\sqrt{\min \lambda_i (\mathbf{A}^* \mathbf{A})}} \quad , \tag{7.135}$$

$$\mathbf{A}^* \mathbf{A} = \mathbf{M}^2 \Rightarrow \text{cond}_2(\mathbf{A}) = \frac{\max |\lambda_i (\mathbf{M})|}{\min |\lambda_i (\mathbf{M})|} \quad .$$

Coming back to the numerical example, the eigenvalues of $\mathbf{A}$ are

$$\lambda_1 \simeq 0.010 \quad , \quad \lambda_2 \simeq 0.843 \quad , \quad \lambda_3 \simeq 3.858 \quad , \quad \lambda_4 \simeq 30.289 \quad , \tag{7.136}$$

and using, for instance, the third equation in (7.135), gives

$$\text{cond}_2(\mathbf{A}) = \frac{\lambda_4}{\lambda_1} \simeq 3 \times 10^3 \quad , \tag{7.137}$$

which shows that the system is ill conditioned, and the relative error of the solution may amount to $\simeq 3 \times 10^3$ times the relative data error (as is almost the case in the example).

In fact, the introduction of the concept of condition number is only useful when a simplistic approach is used for the resolution of linear systems. More generally, the reader is asked to solve the following problem.

The observable values $\mathbf{d} = (d^1, d^2, d^3, d^4)$ *are known to depend on the model values* $\mathbf{m} = (m^1, m^2, m^3, m^4)$ *through the (exact) equation*

$$\begin{pmatrix} d^1 \\ d^2 \\ d^3 \\ d^4 \end{pmatrix} = \begin{pmatrix} 10 & 7 & 8 & 7 \\ 7 & 5 & 6 & 5 \\ 8 & 6 & 10 & 9 \\ 7 & 5 & 9 & 10 \end{pmatrix} \begin{pmatrix} m^1 \\ m^2 \\ m^3 \\ m^3 \end{pmatrix} \quad , \tag{7.138}$$

or, for short, $\mathbf{d} = \mathbf{G}\,\mathbf{m}$. *A measurement of the observable values gives*

$$\begin{pmatrix} d^1 \\ d^2 \\ d^3 \\ d^4 \end{pmatrix} = \begin{pmatrix} 32.0 & \pm & 0.1 \\ 23.0 & \pm & 0.1 \\ 33.0 & \pm & 0.1 \\ 31.0 & \pm & 0.1 \end{pmatrix} \qquad (7.139)$$

Use the least-squares theory to solve the inverse problem and discuss error and resolution.

Solution:

The best solution (in the least-squares sense) for a linear problem is (equations (3.37)–(3.38), page 66)

$$\tilde{\mathbf{m}} = \left(\mathbf{G}^t\,\mathbf{C}_D^{-1}\,\mathbf{G} + \mathbf{C}_M^{-1}\right)^{-1}\left(\mathbf{G}^t\,\mathbf{C}_D^{-1}\,\mathbf{d}_{obs} + \mathbf{C}_M^{-1}\,\mathbf{m}_{prior}\right) \quad , \qquad (7.140)$$

$$\tilde{\mathbf{C}}_M = \left(\mathbf{G}^t\,\mathbf{C}_D^{-1}\,\mathbf{G} + \mathbf{C}_M^{-1}\right)^{-1} \quad . \qquad (7.141)$$

If there is no a priori information, $\mathbf{C}_M \to \infty\,\mathbf{I}$, and

$$\tilde{\mathbf{m}} = \left(\mathbf{G}^t\,\mathbf{C}_D^{-1}\,\mathbf{G}\right)^{-1}\mathbf{G}^t\,\mathbf{C}_D^{-1}\,\mathbf{d}_{obs} \quad , \qquad (7.142)$$

$$\tilde{\mathbf{C}}_M = \left(\mathbf{G}^t\,\mathbf{C}_D^{-1}\,\mathbf{G}\right)^{-1} \quad . \qquad (7.143)$$

In our numerical example,

$$\mathbf{d}_{obs} = \begin{pmatrix} 32.0 \\ 23.0 \\ 33.0 \\ 31.0 \end{pmatrix} \quad , \qquad \mathbf{C}_D = \sigma^2\,\mathbf{I} = 0.01\,\mathbf{I} \quad , \qquad \mathbf{G} = \begin{pmatrix} 10 & 7 & 8 & 7 \\ 7 & 5 & 6 & 5 \\ 8 & 6 & 10 & 9 \\ 7 & 5 & 9 & 10 \end{pmatrix} \quad . \qquad (7.144)$$

As in this particular example, $\mathbf{G}$ is squared and regular, we successively have

$$\tilde{\mathbf{m}} = \left(\mathbf{G}^t\,\mathbf{C}_D^{-1}\,\mathbf{G}\right)^{-1}\mathbf{G}^t\,\mathbf{C}_D^{-1}\,\mathbf{d}_{obs} = \mathbf{G}^{-1}\,\mathbf{C}_D\,(\mathbf{G}^t)^{-1}\,\mathbf{G}^t\,\mathbf{C}_D^{-1}\,\mathbf{d}_{obs} = \mathbf{G}^{-1}\,\mathbf{d}_{obs} \quad , \qquad (7.145)$$

i.e.,

$$\tilde{\mathbf{m}} = \begin{pmatrix} 1.0 \\ 1.0 \\ 1.0 \\ 1.0 \end{pmatrix} \quad . \qquad (7.146)$$

The posterior covariance operator is given by

$$\tilde{\mathbf{C}}_M = \left(\mathbf{G}^t\,\mathbf{C}_D^{-1}\,\mathbf{G}\right)^{-1} = \mathbf{G}^{-1}\,\mathbf{C}_D\,(\mathbf{G}^t)^{-1} = \sigma^2\,\mathbf{G}^{-1}\,(\mathbf{G}^t)^{-1} \quad , \qquad (7.147)$$

and, as $\mathbf{G}$ is symmetric,

$$\widetilde{\mathbf{C}}_M = \sigma^2 \mathbf{G}^{-1} \mathbf{G}^{-1} \quad, \tag{7.148}$$

i.e.,

$$\widetilde{\mathbf{C}}_M = 0.01 \begin{pmatrix} 2442 & -4043 & 1015 & -602 \\ -4043 & 6694 & -1681 & 997 \\ 1015 & -1681 & 423 & -251 \\ -602 & 997 & -251 & 149 \end{pmatrix} \quad. \tag{7.149}$$

From $\widetilde{\mathbf{C}}_M$ it is easy to obtain the standard deviations of model parameters,

$$\widetilde{\sigma}_M^1 = 4.94 \quad , \quad \widetilde{\sigma}_M^2 = 8.18 \quad , \quad \widetilde{\sigma}_M^3 = 2.06 \quad , \quad \widetilde{\sigma}_M^4 = 1.22 \quad , \tag{7.150}$$

and the correlation matrix (see Appendix 6.5)

$$\mathbf{R} = \begin{pmatrix} 1 & -0.99997 & +0.99867 & -0.99800 \\ -0.99997 & 1 & -0.99898 & +0.99830 \\ +0.99867 & -0.99898 & 1 & -0.99979 \\ -0.99800 & +0.99830 & -0.99979 & 1 \end{pmatrix} \quad. \tag{7.151}$$

The overall information on the solution can thus be expressed by this correlation matrix and the short notation

$$\widetilde{\mathbf{m}} = \begin{pmatrix} 1.00 & \pm & 4.94 \\ 1.00 & \pm & 8.18 \\ 1.00 & \pm & 2.06 \\ 1.00 & \pm & 1.22 \end{pmatrix} \quad. \tag{7.152}$$

The interpretation of these results is as follows.

The least-squares approach is only fully justified if uncertainties (in this example, data uncertainties) are modeled using Gaussian probability densities. For a linear problem, the a posteriori uncertainties are then also Gaussian. Taking, for instance, twice the standard deviation, the probability of the true value of the parameter m^1_{true} verifying the inequality

$$-8.88 \leq m^1_{\text{true}} \leq +10.88 \tag{7.153a}$$

is about 95%, independent of the respective values of m^2_{true}, m^3_{true}, m^4_{true}. Similarly, the probability of the true values of each of the parameters m^2_{true}, m^3_{true}, and m^4_{true} verifying the inequalities

$$-15.36 \leq m^2_{\text{true}} \leq +17.36 \quad ,$$
$$-3.12 \leq m^3_{\text{true}} \leq +5.12 \quad , \tag{7.153b}$$
$$-1.44 \leq m^4_{\text{true}} \leq +3.44$$

is also about 95%.

This gives information on the true value of each parameter, considered independently, but the correlation matrix gives additional information on error correlation. For instance, the correlation of m^1 with m^2 is -0.99997. This means that if the estimated value for m^1 is in error (with respect to the true unknown value), it is *almost certain* that the estimated value for m^2 will also be in error (because the absolute value of the correlation is close to 1), and the sign of the error will be opposite to that of the error in m^1 (because the correlation is negative).

The easiest way to understand this is to consider the a posteriori probability density in the parameter space (equation (3.36)):

$$\sigma_M(\mathbf{m}) = \text{const. } \exp\left(-\tfrac{1}{2}(\mathbf{m}-\widetilde{\mathbf{m}})^t\, \widetilde{\mathbf{C}}_M^{-1}\,(\mathbf{m}-\widetilde{\mathbf{m}})\right) \quad . \tag{7.154}$$

To simplify the discussion, let us first analyze the two parameters m^1 and m^2. Their marginal probability density is

$$\sigma_{12}(m^1, m^2) = (2\pi \det \mathbf{C}_{12})^{-1/2}$$
$$\exp\left(-\frac{1}{2}\begin{pmatrix} m^1 - 1.0 \\ m^2 - 1.0 \end{pmatrix}^t \begin{pmatrix} 24.42 & -40.43 \\ -40.43 & 66.94 \end{pmatrix}^{-1} \begin{pmatrix} m^1 - 1.0 \\ m^2 - 1.0 \end{pmatrix}\right) \quad . \tag{7.155}$$

(It is well known [e.g. Dubes, 1968] that marginal probability densities corresponding to a multidimensional Gaussian are simply obtained by picking the corresponding covariances in the joint covariance operator.) Figures 7.23 and 7.24 illustrate this probability density. The correlation between m^1 and m^2 is so strong in this numerical example that the 95% confidence ellipsoid is indistinguishable from a segment. This means that, although the data

Figure 7.23. Marginal probability density for the parameters m^1 and m^2. Uncertainties are so strongly correlated that it is difficult to distinguish the ellipsoid of uncertainties from a very thin segment. Although the standard deviation for each of the parameters is large, we have much information on these parameters, because their true values must lie on the line. The next figure shows a zoom of the central region.

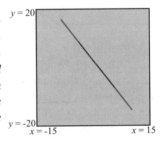

Figure 7.24. Same as the previous figure, with finer detail.

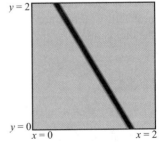

set used in this example is not able to give an accurate location for the true value of m^1 or m^2 independently, it imposes that these true values must lie on the segment of the figure. As the volume of the allowed region is almost null, this gives, in fact, a lot of information.

Similarly, the four-dimensional probability density $\sigma_M(\mathbf{m})$ defines a 95% confidence ellipsoid on the parameter space that corresponds to the extra-long "cigar" joining the point $(-8.88, +17.36, -3.12, +3.44)$ to the point $(+10.88, -15.36, +5.12, -1.44)$. The reader will easily verify that the two solutions of the linear system obtained using Cramer's method for two slightly different data vectors correspond to two points on the cigar.

It should be noticed that if a further experiment gives accurate information on the true value of one of the parameters, the values of the other three parameters can readily be deduced, with very small uncertainties. This example has shown the following:

i) A careful analysis of the a posteriori covariance operator must always be made when solving least-squares inverse problems.

ii) The information given by the condition number is very rough compared with the information given by the covariance operator (it only gives information about the ratio between the largest and shortest diameters of the ellipsoid of uncertainties in the model space).

7.9 Conjunction of Two Probability Distributions

Let x and y be Cartesian coordinates on a cathodic screen. A random device projects electrons on the screen with a known probability density,

$$\theta(x, y) = \begin{cases} \text{const. } r\,(2-r) & \text{if } 0 \le r \le 2 \\ 0 & \text{if } r > 2 \end{cases}, \tag{7.156}$$

where $r = \sqrt{x^2 + y^2}$.

We are interested in the coordinates (x, y) at which a particular electron will hit the screen, and we build an experimental device to measure them. The measuring instrument is not perfect, and when we perform the experiment we can only get the information that the true coordinates of the impact point have the (Gaussian) probability density

$$p(x, y) = \text{const. } \exp\left(-\frac{1}{2} \begin{pmatrix} x - x_0 \\ y - y_0 \end{pmatrix}^t \begin{pmatrix} \sigma^2 & \rho\sigma^2 \\ \rho\sigma^2 & \sigma^2 \end{pmatrix}^{-1} \begin{pmatrix} x - x_0 \\ y - y_0 \end{pmatrix} \right), \tag{7.157}$$

with $(x_0, y_0) = (0, 0)$, $\sigma = 2$, and $\rho = 0.99$. Combine this experimental information with the previous knowledge of the random device, and obtain a better estimate of the impact point.

Solve the problem again, using polar coordinates instead of Cartesian coordinates.

Solution:

As x and y are Cartesian coordinates, the homogeneous probability density for the impact point is

$$\mu(x, y) = \text{const.} \tag{7.158}$$

The information represented by $\theta(x, y)$ and $\rho(x, y)$ is independent in the sense discussed in section 1.2.6. Combination of these data then corresponds to the conjunction (equation (1.83), page 32)

$$\sigma(x, y) = \frac{\rho(x, y) \, \theta(x, y)}{\mu(x, y)} \quad , \tag{7.159}$$

which is plotted in Figure 7.25.

Figure 7.25. *A random device has been built that projects electrons on a cathodic screen with the probability density shown at the top left. Coordinates are Cartesian. Independently of this probability, a measurement of the impact point of a particular electron gives the information represented by the probability density shown at the top right. The homogeneous probability density (which is uniform and has been represented in arbitrary color) is shown at the bottom left. It is then possible to combine all these states of information to obtain the posterior probability density, shown at the bottom right.*

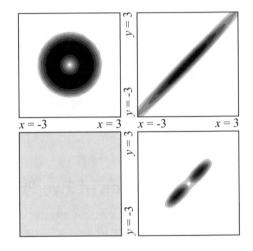

The polar coordinates verify

$$r = (x^2 + y^2)^{1/2} \quad , \qquad \tan \varphi = \frac{y}{x} \quad , \tag{7.160}$$

so that the Jacobian of the transformation is

$$J(r, \varphi) = \begin{vmatrix} \dfrac{\partial r}{\partial x} & \dfrac{\partial r}{\partial y} \\[2mm] \dfrac{\partial \varphi}{\partial x} & \dfrac{\partial \varphi}{\partial y} \end{vmatrix} = \frac{1}{r} \quad . \tag{7.161}$$

Let $f(x, y)$ be a probability density in Cartesian coordinates. To any surface **S** of the plane it assigns the probability

$$P(\mathbf{S}) = \iint_{\mathbf{S}} dx \, dy \, f(x, y) \quad . \tag{7.162}$$

Let $\tilde{f}(r, \varphi)$ be a probability density in polar coordinates. If we wish $\tilde{f}(r, \varphi)$ to assign to **S** the same probability as $f(x, y)$,

$$P(\mathbf{S}) = \iint_{\mathbf{S}} dr \, d\varphi \, f(r, \varphi) \quad , \tag{7.163}$$

then necessarily

$$\tilde{f}(r, \varphi) \;=\; f(x(r, \varphi)\,,\, y(r, \varphi))\,|J(r, \varphi)| \quad . \tag{7.164}$$

This is the usual formula for the change of variables in a probability density. In our case,

$$\tilde{f}(r, \varphi) \;=\; r\, f(r \sin \varphi\,,\, r \cos \varphi) \quad . \tag{7.165}$$

This gives

$$\tilde{\theta}(x, y) \;=\; \begin{cases} \text{const.}\, r^2\,(1 - r) & \text{if } 0 \le r \le 2 \\ 0 & \text{if } r > 2 \end{cases} \quad , \tag{7.166}$$

$$\tilde{\rho}(r, \varphi) \;=\; \text{const.}\, r \, \exp\!\left[-\frac{r^2(1 - 2\rho \sin \varphi \cos \varphi)}{2\sigma^2(1 - \rho^2)} \right] \quad , \tag{7.167}$$

and

$$\tilde{\mu}(r, \varphi) \;=\; \text{const.}\, r \quad . \tag{7.168}$$

The combination of $\theta(r, \varphi)$ with $\tilde{\rho}(r, \varphi)$ is given by the conjunction (equation (1.83), page 32)

$$\tilde{\sigma}(r, \varphi) \;=\; \frac{\tilde{\rho}(r, \varphi)\,\tilde{\theta}(r, \varphi)}{\tilde{\mu}(r, \varphi)} \tag{7.169}$$

and is shown in Figure 7.26.

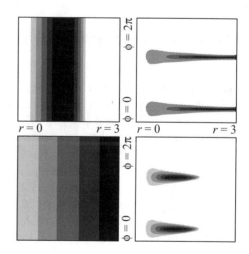

Figure 7.26. *It is also possible to solve the problem using polar coordinates throughout. The top left represents the probability density of an impact on the screen, as imposed by the experimental device. The probability density is constant for given r. The top right shows the result of the measurement. At the bottom left, the homogeneous probability density in polar coordinates is shown. It assigns equal probability to equal surfaces of the screen. The combination of these states of information gives the posterior probability density shown at the bottom right. This probability is completely equivalent to the probability density at the bottom right of the previous figure, as they can be deduced from one another through the usual formula of change of variables between Cartesian and polar coordinates $\tilde{\sigma}(\theta, \varphi) = r\,\sigma(x, y)$.*

It should be noticed that the probability density representing the homogeneous probability is not constant in polar coordinates: the probability density in equation (7.168) assigns equal probabilities to equal volumes, as it must.

The solution obtained for this problem using Cartesian coordinates (Figure 7.25) and the solution obtained using polar coordinates (Figure 7.26) are coherent: Figure 7.25 can be deduced from Figure 7.26 using the Jacobian rule in equation (7.164), and vice versa.

Using more elementary approaches, this problem may present some pathologies. In particular, the result cannot be expressed using a single estimator of the impact point because the probability density is bimodal. The mean value and median value are meaningless, and only the two maximum likelihood points make clear sense.

7.10   Adjoint of a Covariance Operator

Let $\mathbb{S}$ be a linear space and $\mathbf{C}_{\mathbb{S}}$ be the (symmetric) covariance operator defining a scalar product over $\mathbb{S}$:

$$(s_1 , s_2)_{\mathbb{S}} = s_1^t \, \mathbf{C}_{\mathbb{S}}^{-1} \, s_2 \quad . \tag{7.170}$$

Demonstrate that the adjoint of $\mathbf{C}_{\mathbb{S}}$ equals its inverse:

$$\mathbf{C}_{\mathbb{S}}^* = \mathbf{C}_{\mathbb{S}}^{-1} . \tag{7.171}$$

Solution:
In section 3.1.4 (page 61), we saw that if $\mathbf{A}$ is a linear operator mapping a linear space $\mathbb{E}$, with scalar product

$$(e_1 , e_2)_{\mathbb{E}} = e_1^t \, \mathbf{C}_{\mathbb{E}}^{-1} \, e_2 \quad , \tag{7.172}$$

into a linear space $\mathbb{F}$, with scalar product

$$(f_1 , f_2)_{\mathbb{F}} = f_1^t \, \mathbf{C}_{\mathbb{F}}^{-1} \, f_2 \quad , \tag{7.173}$$

the adjoint of $\mathbf{A}$ is (equation (3.24))

$$\mathbf{A}^* = \mathbf{C}_{\mathbb{E}} \, \mathbf{A}^t \, \mathbf{C}_{\mathbb{F}}^{-1} \quad , \tag{7.174}$$

where $\mathbf{A}^t$ is the transpose of $\mathbf{A}$.

We have also seen that a covariance operator $\mathbf{C}_{\mathbb{S}}$ over a linear space $\mathbb{S}$ is a linear operator mapping $\mathbb{S}^*$ (dual of $\mathbb{S}$) into $\mathbb{S}$. To evaluate the adjoint of a covariance operator, we must then identify, in the formulas above, the space $\mathbb{E}$ with $\mathbb{S}^*$ and the space $\mathbb{F}$ with $\mathbb{S}$. Equation (7.174) then gives $\mathbf{C}_{\mathbb{S}}^* = \mathbf{C}_{\mathbb{S}^*} \, \mathbf{C}_{\mathbb{S}}^t \, \mathbf{C}_{\mathbb{S}}^{-1}$. As a covariance operator is symmetric,

$$\mathbf{C}_{\mathbb{S}}^t = \mathbf{C}_{\mathbb{S}} \quad , \tag{7.175}$$

this gives $\mathbf{C}_{\mathbb{S}}^* = \mathbf{C}_{\mathbb{S}^*} \, \mathbf{C}_{\mathbb{S}} \, \mathbf{C}_{\mathbb{S}}^{-1}$, i.e., $\mathbf{C}_{\mathbb{S}}^* = \mathbf{C}_{\mathbb{S}^*}$: the adjoint of a covariance operator over a space $\mathbb{S}$ is the operator defining the scalar product over the dual space $\mathbb{S}^*$. And we know that this is the inverse of the original operator (see section 3.1.3): $\mathbf{C}_{\mathbb{S}^*} = \mathbf{C}_{\mathbb{S}}^{-1}$. Therefore,

$$\mathbf{C}_{\mathbb{S}}^* = \mathbf{C}_{\mathbb{S}}^{-1} \quad . \tag{7.176}$$

Lesson: Covariance operators are symmetric, but they are not self-adjoint!

7.11 Problem 7.1 Revisited

Solve Problem 7.1 (estimation of the epicentral coordinates of a seismic event) using the Newton method, the steepest descent method, the conjugate directions method, and the variable metric method. Consider different kinds of a priori information. Examine the evolution of posterior uncertainties when eliminating some data. Use only one datum (arrival time) and the a priori information that the x coordinate of the epicenter is $15\,km \pm 5\,km$.

Note: As the forward problem is solved by the equation

$$d^i = g^i(X, Y) = \frac{1}{v}\sqrt{(X - x^i)^2 + (Y - y^i)^2}, \tag{7.177}$$

if we order arrival times in a column matrix, the matrix of partial derivatives is

$$\mathbf{G}_n = \begin{pmatrix} \left(\frac{\partial g^1}{\partial X}\right)_n & \left(\frac{\partial g^1}{\partial Y}\right)_n \\ \left(\frac{\partial g^2}{\partial X}\right)_n & \left(\frac{\partial g^2}{\partial Y}\right)_n \\ \vdots & \vdots \end{pmatrix}, \tag{7.178}$$

where

$$\left(\frac{\partial g^i}{\partial X}\right)_n = \frac{X_n - x^i}{v\sqrt{(X_n - x^i)^2 + (Y_n - y^i)^2}} \tag{7.179}$$

and

$$\left(\frac{\partial g^i}{\partial Y}\right)_n = \frac{Y_n - x^i}{v\sqrt{(X_n - x^i)^2 + (Y_n - y^i)^2}}. \tag{7.180}$$

7.12 Problem 7.3 Revisited

Solve Problem 7.3 (elementary approach to tomography) using the Newton method, the steepest descent method, the conjugate directions method, and the variable metric method. Consider different kinds of a priori information. Examine the evolution of posterior uncertainties when adding much more data. Solve a problem with $NX \times NY$ blocks, where NX and NY are the number of pixels of the color output of your computer (there is no problem in having many more unknowns than data, if the a priori covariance matrix imposes that the least-squares solution must be smooth).

Once you master the resolution of this problem using a discrete formulation, have a look a chapter 5 (Functional Inverse Problems), or read the article by Tarantola and Nercessian (1984).

7.13   An Example of Partial Derivatives

Let us consider the problem of locating a point in space using a system like the global positioning system (GPS), where some sources (satellites) send waves to a receiver, which measures the travel times. Let us use Cartesian coordinates, denoting by (x^i, y^i, z^i) the position of the ith source and by (x^R, y^R, z^R) the position of the receiver. Simplify the problem here by assuming that the medium where the waves propagate is homogeneous, so the velocity of the waves is constant (say v) and the rays are straight lines. Then, the travel time from the ith source to the receiver is

$$t^i = g^i(x^R, y^R, z^R, v) = \frac{\sqrt{(x^R - x^i)^2 + (y^R - y^i)^2 + (z^R - z^i)^2}}{v} . \tag{7.181}$$

The dependence of t^i on the variables describing the source positions (x^i, y^i, z^i) is not explicitly considered, as the typical GPS problem consists of assuming the position of the sources exactly known and of estimating the receiver position (x^R, y^R, z^R). At no extra cost, we can also try to estimate the velocity of the propagation of waves v. The partial derivatives of the problem are then

$$
\begin{pmatrix}
\frac{\partial g^1}{\partial x^R} & \frac{\partial g^1}{\partial y^R} & \frac{\partial g^1}{\partial z^R} & \frac{\partial g^1}{\partial v} \\
\frac{\partial g^2}{\partial x^R} & \frac{\partial g^2}{\partial y^R} & \frac{\partial g^2}{\partial z^R} & \frac{\partial g^2}{\partial v} \\
\vdots & \vdots & \vdots & \vdots \\
\frac{\partial g^i}{\partial x^R} & \frac{\partial g^i}{\partial y^R} & \frac{\partial g^i}{\partial z^R} & \frac{\partial g^i}{\partial v} \\
\vdots & \vdots & \vdots & \vdots
\end{pmatrix}
=
\begin{pmatrix}
\frac{x^R - x^1}{v\,D^1} & \frac{y^R - y^1}{v\,D^1} & \frac{z^R - z^1}{v\,D^1} & -\frac{D^i}{v^2} \\
\frac{x^R - x^2}{v\,D^2} & \frac{y^R - y^2}{v\,D^2} & \frac{z^R - z^2}{v\,D^2} & -\frac{D^i}{v^2} \\
\vdots & \vdots & \vdots & \vdots \\
\frac{x^R - x^i}{v\,D^i} & \frac{y^R - y^i}{v\,D^i} & \frac{z^R - z^i}{v\,D^i} & -\frac{D^i}{v^2} \\
\vdots & \vdots & \vdots & \vdots
\end{pmatrix} , \tag{7.182}
$$

where D^i is a short notation for the distance:

$$D^i = \sqrt{(x^R - x^i)^2 + (y^R - y^i)^2 + (z^R - z^i)^2} . \tag{7.183}$$

In order to keep notation simple, it has not been explicitly indicated that these partial derivatives are functions of the variables of the problem, i.e., functions of (x^R, y^R, z^R, v) (remember that the locations of the satellites, (x^i, y^i, z^i), are assumed exactly known, so they are not 'variables'). Assigning particular values to the variables (x^R, y^R, z^R, v) gives particular values for the travel times t^i (through equation (7.181)) and for the partial derivatives (through equation (7.182)).

7.14   Shapes of the ℓ_p-Norm Misfit Functions

Consider a schematic problem with two model parameters m^1 and m^2 and a single datum d^1. The datum is theoretically related to the model parameters through the linear equation

$$d^1 = g^1(m^1, m^2) = m^1 - m^2 . \tag{7.184}$$

Its observed value is

$$d^1_{\text{obs}} = 0 \pm 1 . \tag{7.185}$$

The a priori values of the model parameters are

$$m^1_{\text{prior}} = 3 \pm 1 \quad and \quad m^2_{\text{prior}} = 1 \pm 2 \quad . \tag{7.186}$$

Represent the misfit function $S(m^1, m^2)$ *in the following three cases:*

i) *The symbols* $\pm$ *in (7.185)–(7.186) represent* ℓ_1*-norm uncertainty bars.*

ii) *The symbols* $\pm$ *in (7.185)–(7.186) represent* ℓ_2*-norm uncertainty bars.*

iii) *The symbols* $\pm$ *in (7.185)–(7.186) represent* ℓ_∞*-norm uncertainty bars.*

Solution for the ℓ_1*-norm:*
Using an ℓ_p-norm, the misfit function is (equation (1.108))

$$S(\mathbf{m}) = \frac{1}{p} \left(\sum_i \frac{|g^i(\mathbf{m}) - d^i_{\text{obs}}|^p}{(\sigma^i_D)^p} + \sum_\alpha \frac{|m^\alpha - m^\alpha_{\text{prior}}|^p}{(\sigma^\alpha_M)^p} \right) \quad . \tag{7.187}$$

In particular, for $p = 1$, one has a sum of absolute values (equation (1.109)),

$$S(\mathbf{m}) = \sum_i \frac{|g^i(\mathbf{m}) - d^i_{\text{obs}}|}{\sigma^i_D} + \sum_\alpha \frac{|m^\alpha - m^\alpha_{\text{prior}}|}{\sigma^\alpha_M} \quad . \tag{7.188}$$

For a reason that will become apparent in the case $p \to \infty$, instead of $S(\mathbf{m})$, let us introduce the function

$$R(\mathbf{m}) = (p\, S(\mathbf{m}))^{1/p} = \left(\sum_i \frac{|g^i(\mathbf{m}) - d^i_{\text{obs}}|^p}{(\sigma^i_D)^p} + \sum_\alpha \frac{|m^\alpha - m^\alpha_{\text{prior}}|^p}{(\sigma^\alpha_M)^p} \right)^{1/p} \quad . \tag{7.189}$$

Of course, for $p = 1$, $R(\mathbf{m}) = S(\mathbf{m})$.
In our present example, we have

$$R(m^1, m^2) = |m^1 - m^2| + |m^1 - 3| + \frac{|m^2 - 1|}{2} \quad . \tag{7.190}$$

This function is represented at the left of Figure 7.27. The level lines are polygons. In a 3D space with axes (m^1, m^2, R), the function $R(m^1, m^2)$ is clearly a convex polyhedron.
The minimum of R is attained at $(m^1, m^2) = (3, 3)$. Some ℓ_1-norm circles of radius $1/8$ have been drawn.

Solution for the ℓ_2*,-norm:*
We have

$$R(m^1, m^2) = \left((m^1 - m^2)^2 + (m^1 - 3)^2 + \frac{(m^2 - 1)^2}{4} \right)^{1/2} \quad . \tag{7.191}$$

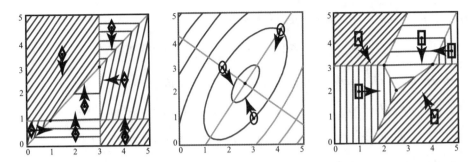

Figure 7.27. *Representation of the function $R(\mathbf{m})$ defined by equation (7.189). The constants $\sigma_M{}^\alpha$ allow the introduction of a natural definition of distance over the model space, based on the norm $\| \mathbf{m} \|^p = \sum_\alpha |m^\alpha - m^\alpha_{prior}|^p / (\sigma_M^\alpha)^p$. For $p = 1$, the function R is that expressed in equation (7.190), and $\| \mathbf{m} \| = \sum_\alpha |m^\alpha - m^\alpha_{prior}| / \sigma_M^\alpha$. The figure at the left shows the function $R(m^1, m^2)$ defined by the data given in the text. The level lines are polygons. In a 3D space with axes (m^1, m^2, R), the function $R(m^1, m^2)$ is a convex polyhedron. Some ℓ_1 circles of radius 1/8 are shown, together with the direction of steepest descent. Middle: Same but for $p = 2$, with the function R expressed by equation (7.191). The circles here correspond to the ℓ_2-norm $\| \mathbf{m} \| = \sqrt{\sum_\alpha (m^\alpha - m^\alpha_{prior})^2 / \sigma_M^2}$. Right: Same as above, but for $p = \infty$. This gives the function R expressed in equation (7.193). Some circles for the distance $\| \mathbf{m} \| = \max(|m^\alpha - m^\alpha_{prior}| / \sigma_M^\alpha$ (all α) $)$ are shown. In this last situation, the direction of steepest descent is not uniquely defined everywhere.*

This function is represented in the middle of Figure 7.27. Its minimum is attained at $(m^1, m^2) = (8/3, 7/3)$.

Solution for the ℓ_∞-norm:
For $p \to \infty$, we have (see main text)

$$R(\mathbf{m}) = \max\left(\left(\frac{|g^i(\mathbf{m}) - d^i_{obs}|}{\sigma^i_D} \text{ (all } i) \right) \ , \ \left(\frac{|m^\alpha - m^\alpha_{prior}|}{\sigma_M^\alpha} \text{ (all } \alpha) \right) \right) \ . \quad (7.192)$$

In our present example, this gives

$$R(m^1, m^2) = \max\left(|m^1 - m^2| , \ |m^1 - 3| , \ \frac{|m^2 - 1|}{2} \right) \ . \quad (7.193)$$

This function is represented at the right of Figure 7.27. As for the ℓ_1 case, the level lines of $R(m^1, m^2)$ are polygons, and in a 3D space with axes (m^1, m^2, R), the function $R(m^1, m^2)$ is a convex polyhedron.

The minimum of R is attained at $(m^1, m^2) = (5/2, 2)$. Some ℓ_∞-norm circles of radius 1/8 have been drawn. The associated directions of steepest descent are not uniquely defined in all the polyhedron faces.

7.15   Using the Simplex Method

Use the simplex method of Appendix 6.23.1 to obtain the minimum of the function

$$\varphi(x^1, x^2, x^3) = 2x^1 + x^2 + 2x^3 \tag{7.194a}$$

under the constraints

$$x^1 + x^2 + x^3 = 5 \quad , \quad x^1 + 2x^2 = 3 \quad , \tag{7.194b}$$

and

$$x^1 > 0 \quad , \quad x^2 > 0 \quad , \quad x^3 > 0 \quad . \tag{7.194c}$$

   Solution:
   In the following, the notation of Appendix 6.23.1 is used. In compact notation, the problem is to minimize

$$\varphi(\mathbf{x}) = \hat{\chi}^t \mathbf{x} \tag{7.195}$$

under the constraints

$$\mathbf{M}\mathbf{x} = \mathbf{y} \quad , \quad \mathbf{x} \geq 0 \quad , \tag{7.196}$$

where

$$\mathbf{x} = \begin{pmatrix} x^1 \\ x^2 \\ x^3 \end{pmatrix} \quad , \quad \hat{\chi} = \begin{pmatrix} 2 \\ 1 \\ 2 \end{pmatrix} \quad , \quad \mathbf{M} = \begin{pmatrix} 1 & 1 & 1 \\ 1 & 2 & 0 \end{pmatrix} \quad , \quad \mathbf{y} = \begin{pmatrix} 5 \\ 3 \end{pmatrix} \quad . \tag{7.197}$$

In this problem, a basis contains two parameters. We have three different possibilities for choosing the basic parameters:

$$\text{First possible choice:} \quad \mathbf{x}_B = \begin{pmatrix} x^1 \\ x^2 \end{pmatrix} \mathbf{x}_N = [x^3] \quad ;$$

$$\text{Second possible choice:} \quad \mathbf{x}_B = \begin{pmatrix} x^2 \\ x^3 \end{pmatrix} \mathbf{x}_N = [x^1] \quad ; \tag{7.198}$$

$$\text{Third possible choice:} \quad \mathbf{x}_B = \begin{pmatrix} x^1 \\ x^3 \end{pmatrix} \mathbf{x}_N = [x^2] \quad .$$

As discussed in Appendix 6.23.1, we will try to take the nonbasic parameters as null.
   Let us arbitrarily take the first of the three possible choices. This gives

$$\mathbf{M}_B = \begin{pmatrix} 1 & 1 \\ 1 & 2 \end{pmatrix} \quad , \quad \mathbf{M}_N = \begin{pmatrix} 1 \\ 0 \end{pmatrix} \quad . \tag{7.199}$$

We must first check that in taking the nonbasic parameters as null, the positivity constraints for the basic parameters are satisfied. From equation (6.372) of Appendix 6.23.1,

$$\mathbf{x}_B = \mathbf{M}_B^{-1} \mathbf{y} = \begin{pmatrix} 7 \\ -2 \end{pmatrix} \quad . \tag{7.200}$$

As the positivity constraints are not satisfied, this choice is not acceptable.

Turning to the second possible choice, we have

$$\mathbf{M}_B = \begin{pmatrix} 1 & 1 \\ 2 & 0 \end{pmatrix} \quad , \quad \mathbf{M}_N = \begin{pmatrix} 1 \\ 1 \end{pmatrix} \quad , \tag{7.201}$$

which gives

$$\mathbf{x}_B = \mathbf{M}_B^{-1}\mathbf{y} = \begin{pmatrix} 3/2 \\ 7/2 \end{pmatrix} \quad . \tag{7.202}$$

As the positivity constraints are satisfied, we can now check if this possible solution is the optimum solution. We have

$$\mathbf{c}_B = \begin{pmatrix} 1 \\ 2 \end{pmatrix} \quad , \quad \mathbf{c}_N = [2] \quad , \tag{7.203}$$

and, using equation (6.375) of Appendix 6.23.1,

$$\boldsymbol{\gamma}_N = \mathbf{c}_N - \mathbf{M}_N^t\,\mathbf{M}_B^{-t}\,\mathbf{c}_B = [1/2] \quad . \tag{7.204}$$

As all the components of $\boldsymbol{\gamma}_N$ (we only have one) are positive, the function φ is effectively minimized taking $\mathbf{x}_N = 0$, and (7.202) is the solution of the problem.

The problem is now completely solved, but let us see what happens when we use the third of the possible choices. This gives

$$\mathbf{M}_B = \begin{pmatrix} 1 & 1 \\ 1 & 0 \end{pmatrix} \quad , \quad \mathbf{M}_N = \begin{pmatrix} 1 \\ 2 \end{pmatrix} \quad , \tag{7.205}$$

and

$$\mathbf{x}_B = \mathbf{M}_B^{-1}\mathbf{y} = \begin{pmatrix} 3 \\ 2 \end{pmatrix} \quad , \tag{7.206}$$

which is acceptable. We have

$$\mathbf{c}_B = \begin{pmatrix} 2 \\ 2 \end{pmatrix} \quad , \quad \mathbf{c}_N = [1] \quad , \tag{7.207}$$

and, using equation (6.375) of Appendix 6.23.1,

$$\boldsymbol{\gamma}_N = \mathbf{c}_N - \mathbf{M}_N^t\,\mathbf{M}_B^{-t}\,\mathbf{c}_B = [-1] \quad . \tag{7.208}$$

As all the components of $\boldsymbol{\gamma}_N$ are not positive, equation (7.206) is not the solution of the problem. The parameter associated with the most negative component of $\boldsymbol{\gamma}_N$ must leave the basis (as we only have the parameter x^2 in $\mathbf{x}_N$, this is the parameter). Equation (7.197) of Appendix 6.23.1 gives

$$\mathbf{x}_B = \mathbf{M}_B^{-1}\,(\mathbf{y} - \mathbf{M}_N\,\mathbf{x}_N) = \begin{pmatrix} 3 - 2x^2 \\ 2 + x^2 \end{pmatrix} \quad . \tag{7.209}$$

As the first component of $\mathbf{x}_B$, x^1, first becomes negative when increasing x^2, it is the parameter x^1 that must replace x^2 in the basis. This leads directly to the second choice, which has already been explored (and which gives the solution).

7.16 Problem 7.7 Revisited

Two variables y and t are related through a linear relationship

$$y = at + b \quad .$$ (7.210)

In order to estimate the parameters a and b, the 11 experimental points (y^i, t^i), $(10, 1)$, $(11, 3)$, $(11, 5)$, $(12, 7)$, $(13, 10)$, $(14, 12)$, $(14, 13)$, $(15, 15)$, $(15, 16)$, $(16, 18)$, $(2, 31)$, shown in Figure 7.28, have been obtained. Find the straight line that best fits the points in the ℓ_1-norm sense.

Solution:
We wish to minimize

$$S(a, b) = \sum_{i=1}^{i=11} \frac{|y_{\text{cal}}^i - y_{\text{obs}}^i|}{\sigma^i} \quad ,$$ (7.211)

where

$$y_{\text{cal}}^i = a t^i + b$$ (7.212)

and $\sigma^i = \sigma = 2$.

Figure 7.29 shows the projection on the plane (a, b) of the convex polyhedron representing $S(a, b)$. Figure 7.30 zooms in on the region of interest. The level lines $S(a, b) = 12.5$ and $S(a, b) = 25$ are shown in Figure 7.29 and the level line $S(a, b) = 12.5$ is shown in Figure 7.30.

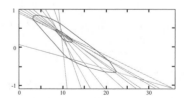

Figure 7.28. *We wish to find the straight line fitting these 11 points (same as Figure 7.14). Notice the outlier.*

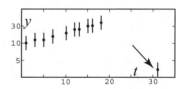

Figure 7.29. *Projection on the parameter space (a, b) of the convex polyhedron representing the misfit function $S(a, b)$. The level lines $S = 12.5$ and $S = 25$ are shown.*

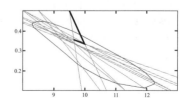

Figure 7.30. *Zoom of Figure 7.29. The solution path is shown (see text).*

Each of the straight lines corresponds to an experimental point. For instance, the most horizontal line corresponds to the (probable outlier) point (2, 31). The minimum of S is necessarily attained at a vertex of the polyhedron, i.e., at a knot in the mesh in Figures 7.29 and 7.30. At least two experimental points are exactly fitted at each knot.

To solve the problem using linear programming techniques, we first choose one knot, i.e., two of the equations

$$y_{obs}^i = a\, t^i + b \quad, \tag{7.213}$$

for instance, the first two. This gives the point $(0.5, 9.5)$. To leave the knot, we have to drop one of the basis equations. The FIFO method drops the oldest equation. At the first round, equations have arbitrary ages. Take, for instance, ages decreasing from the first to the last equation. Dropping the first equation gives the line shown in Figure 7.30. The minimum along the line is attained at the point $(0.308, 10.08)$, where the ninth equation enters the basis. Dropping the oldest equation (the second), we get a minimum at the point $(0.333, 9.67)$, where the first, fourth, and seventh equations are exactly satisfied and they can all enter the basis. Whatever the choice we make, it leaves the basis again, because we cannot make S diminish: we are at the minimum.

7.17   Geodetic Adjustment with Outliers

Figure 7.31 shows five points on a Euclidean plane. Ten different distances between these points have been measured experimentally. The results are as follows.

	Observed distance	Estimated uncertainty	True (unknown) error
$D^0 =$	9 486.843 0 m	±2 cm	+1 cm
$D^1 =$	15 000.010 0 m	±2 cm	+1 cm
$D^2 =$	12 727.902 1 m	±2 cm	−2 cm
$D^3 =$	6 708.203 9 m	±2 cm	0 cm
$D^4 =$	6 708.193 9 m	±2 cm	−1 cm
$D^5 =$	11 998.000 0 m	±2 cm	−200 cm
$D^6 =$	6 708.193 9 m	±2 cm	−1 cm
$D^7 =$	10 816.653 8 m	±2 cm	0 cm
$D^8 =$	9 486.853 0 m	±2 cm	+2 cm
$D^9 =$	6 708.223 9 m	±2 cm	+2 cm

Observe that estimated uncertainties are uniform, $\sigma^i = 2\,\text{cm}$, and that distance D^5 is an outlier.

To define the geometric figure perfectly, only 7 distances are needed, but, as is usual in geodetic measurements, some redundant measurements have been made in order to minimize posterior true uncertainties. Due to the observational uncertainties, the 10 distances obtained are not compatible with the geometric constraints. Obtain the new set of distances D^i, compatible with these geometric constraints, minimizing the ℓ_1-norm

$$S_1 = \sum_{i=0}^{i=9} \frac{|D^i - D_{obs}^i|}{\sigma^i} \quad. \tag{7.214}$$

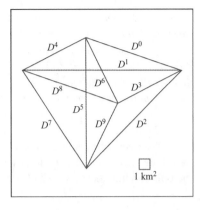

Figure 7.31. *Between 5 points of a Eu-*
clidean plane, the 10 distances shown in the figure
have been measured. One is an outlier (see text).
Estimate the true distances between the points.

Compare with the solution of the minimization of the ℓ_2-norm

$$S_2 = \sum_{i=0}^{i=9} \frac{(D^i - D^i_{\text{obs}})^2}{(\sigma^i)^2} \quad . \tag{7.215}$$

Solution:
As the expected corrections are small, the problem can be linearized. The ℓ_1-norm minimization problem can be solved using the linear programming methods, and the ℓ_2-norm minimization problem can be solved using Newton's method. The following table compares the true errors with the corrections predicted by the two methods.

True errors	ℓ_1-norm correction	ℓ_2-norm correction
+1.00 cm	0.00 cm	+14.22 cm
+1.00 cm	−0.89 cm	−19.21 cm
−2.00 cm	0.00 cm	+4.64 cm
0.00 cm	0.00 cm	+2.72 cm
−1.00 cm	0.00 cm	+39.79 cm
−200.00 cm	−195.70 cm	−66.49 cm
−1.00 cm	0.00 cm	+49.41 cm
0.00 cm	0.00 cm	+35.94 cm
+2.00 cm	0.00 cm	−38.28 cm
+2.00 cm	+4.31 cm	+37.24 cm

Notice that the ℓ_1-norm solution exactly satisfies seven distances (the number of independent data). As expected, the ℓ_1-norm criterion allows an easy identification of the outlier, while the ℓ_2-norm criterion smears the error across the whole geodetic network.

7.18 Inversion of Acoustic Waveforms

An acoustic medium can be described using the mass density $\rho(\mathbf{x})$ and the bulk modulus $\kappa(\mathbf{x})$. For simplicity, assume that the mass density is known. The problem is then to evaluate $\kappa(\mathbf{x})$. At some shot points $\mathbf{x}_s$ $(s = 1, 2, \ldots, NS)$, we generate acoustic waves, which are *recorded at some receiver positions* $\mathbf{x}_r$ $(r = 1, 2, \ldots, NR)$. *Let* t *be a time variable reset to zero at each new run. The pressure perturbation at the receiver location* $\mathbf{x}_r$, *at time* t, *for a source located at point* $\mathbf{x}_s$, *is denoted* $p(\mathbf{x}_r, t; \mathbf{x}_r)$. *Let* $p(\mathbf{x}_r, t; \mathbf{x}_r)_{\text{obs}}$ *denote the particular measured (observed) values. For a particular model* $\kappa(\mathbf{x})$, $p(\mathbf{x}_r, t; \mathbf{x}_r)_{\text{cal}}$ *denotes the predicted values. Formulate the inverse problem of evaluating the bulk modulus* $\kappa(\mathbf{x})$ *from the measurements* $p(\mathbf{x}_r, t; \mathbf{x}_r)$.

Solution:

We assume the source of the acoustic waves to be exactly known (if not, it must be part of the inverse problem). In order for the Gaussian statistics to be acceptable, instead of the bulk modulus $\kappa(\mathbf{x})$ we shall use the logarithmic bulk modulus

$$m(\mathbf{x}) \;=\; \log \frac{K}{\kappa(\mathbf{x})} \;=\; -\log \frac{\kappa(\mathbf{x})}{K} \quad , \tag{7.216}$$

where K is an arbitrary constant value of κ. For compactness, a logarithmic bulk modulus model is denoted $\mathbf{m}$ and a data set is denoted $\mathbf{p}$. The computation of the waveforms corresponding to the model $\mathbf{m}$ is written

$$\mathbf{p} \;=\; \mathbf{g}(\mathbf{m}) \quad , \tag{7.217}$$

where the operator $\mathbf{g}$ linking the model of logarithmic bulk modulus to the pressure perturbations is, of course, nonlinear.

Let $\mathbf{p}_{\text{obs}}$ represent the observed data set and $\mathbf{C}_p$ be the covariance operator describing experimental uncertainties. In what follows, the kernel of the covariance operator is assumed to be diagonal,

$$\mathbf{C}_p(\mathbf{x}_r, t; \mathbf{x}_s; \mathbf{x}_{r'}, t'; \mathbf{x}_{s'}) \;=\; \sigma^2(\mathbf{x}_r, t; \mathbf{x}_s)\, \delta^{rr'}\, \delta(t - t')\, \delta^{ss'} \quad , \tag{7.218}$$

so that the expression

$$\delta \mathbf{p} \;=\; \mathbf{C}_p\, \delta \hat{\mathbf{p}} \tag{7.219}$$

is written, explicitly,

$$\delta p(\mathbf{x}_r, t; \mathbf{x}_s) \;=\; \sum_{s'} \int_0^T dt' \sum_{r'} \mathbf{C}_p(\mathbf{x}_r, t; \mathbf{x}_s; \mathbf{x}_{r'}, t'; \mathbf{x}_{s'})\, \delta \hat{p}(\mathbf{x}_{r'}, t'; \mathbf{x}_{s'})$$

$$= \; \sigma^2(\mathbf{x}_r, t; \mathbf{x}_s)\, \delta \hat{p}(\mathbf{x}_r, t; \mathbf{x}_s) \quad . \tag{7.220}$$

Then, the reciprocal relation

$$\delta \hat{\mathbf{p}} \;=\; \mathbf{C}_p^{-1}\, \delta \mathbf{p} \tag{7.221}$$

simply gives

$$\delta \hat{p}(\mathbf{x}_r, t; \mathbf{x}_s) \;=\; \frac{\delta p(\mathbf{x}_r, t; \mathbf{x}_s)}{\sigma^2(\mathbf{x}_r, t; \mathbf{x}_s)} \quad . \tag{7.222}$$

The best model (in the least-squares sense) is defined by the minimization of the squared norm

$$S(\mathbf{m}) = \tfrac{1}{2} \| \mathbf{g}(\mathbf{m}) - \mathbf{d}_{\text{obs}} \|^2 \quad , \tag{7.223}$$

where, if we use the notation

$$\langle \, \delta\hat{\mathbf{p}}_1 \, , \, \delta\mathbf{p}_2 \, \rangle = \delta\hat{\mathbf{p}}_1{}^t \, \delta\mathbf{p}_2 = \sum_s \int_0^T dt \sum_r \delta\hat{p}(\mathbf{x}_r, t; \mathbf{x}_s)_1 \, \delta p(\mathbf{x}_r, t; \mathbf{x}_s)_2 \quad , \tag{7.224}$$

the norm $\| \delta\mathbf{p} \|$ is defined by

$$\| \delta\mathbf{p} \|^2 = \langle \, \mathbf{C}_p^{-1}\delta\mathbf{p} \, , \, \delta\mathbf{p} \, \rangle = \delta\mathbf{p}^t \, \mathbf{C}_p^{-1}\delta\mathbf{p} \quad . \tag{7.225}$$

One then usually writes

$$S(\mathbf{m}) = (\mathbf{g}(\mathbf{m}) - \mathbf{d}_{\text{obs}})^t \, \mathbf{C}_p^{-1}(\mathbf{g}(\mathbf{m}) - \mathbf{d}_{\text{obs}}) \quad . \tag{7.226}$$

To simplify the exposition here, I do not explicitly introduce the a priori information in the model space: gradient methods are naturally robust (if the minimum of (7.226) is a subspace rather than a single point, they converge to the point that is the closest to the starting point), and, in any case, many examples of the introduction of the a priori information are given in other parts of this book.

The Fréchet derivative of the nonlinear operator $\mathbf{g}$ at a point $\mathbf{m}_n$ of the model space is the linear operator $\mathbf{G}_n$ that associates any model perturbation $\delta\mathbf{m}$ with the data perturbation $\mathbf{G}_n \, \delta\mathbf{m}$ defined by the first-order development

$$\mathbf{g}(\mathbf{m}_n + \delta\mathbf{m}) = \mathbf{g}(\mathbf{m}_n) + \mathbf{G}_n \, \delta\mathbf{m} + \text{ higher order terms} \quad . \tag{7.227}$$

Introducing notation equivalent to that given by (7.224) in the model space,

$$\langle \, \delta\hat{\mathbf{m}}_1 \, , \, \delta\mathbf{m}_2 \, \rangle = \delta\hat{\mathbf{m}}_1{}^t \, \delta\mathbf{m}_2 = \delta\mathbf{m}_2{}^t \, \delta\hat{\mathbf{m}}_1 = \int_{\mathcal{V}} dV(\mathbf{x}) \, \delta\hat{\mathbf{m}}(\mathbf{x})_1 \, \delta\mathbf{m}(\mathbf{x})_2 \quad , \tag{7.228}$$

and given a linear operator $\mathbf{G}_n$, the *transpose operator* $\mathbf{G}_n^t$ is defined by the identity (see main text)

$$\langle \, \delta\hat{\mathbf{p}} \, , \, \mathbf{G}_n \, \delta\mathbf{m} \, \rangle = \langle \, \mathbf{G}_n^t \, \delta\hat{\mathbf{p}} \, , \, \delta\mathbf{m} \, \rangle \qquad \text{for any } \delta\hat{\mathbf{p}} \text{ and } \delta\mathbf{m} \quad . \tag{7.229}$$

The least-squares minimization problem can then be solved using, for instance, a preconditioned steepest descent algorithm (see chapter 3). This gives

$$\mathbf{m}_{n+1} = \mathbf{m}_n - \mu_n \, \hat{\mathbf{S}}_0 \, \mathbf{G}_n^t \, \mathbf{C}_p^{-1}(\mathbf{g}(\mathbf{m}_n) - \mathbf{d}_{\text{obs}}) \quad , \tag{7.230}$$

where $\hat{\mathbf{S}}_0$ is an arbitrary positive definite operator called the preconditioning operator, which is suitably chosen to accelerate the convergence (see below).

Let us now turn to the computation of the Fréchet derivatives corresponding to this problem. The solution of the forward problem is defined by the differential system (using the logarithmic bulk modulus defined in equation (7.216))

$$\frac{\exp(m(\mathbf{x}))}{K} \frac{\partial^2 p}{\partial t^2}(\mathbf{x}, t; \mathbf{x}_s) - \text{div}\left(\frac{1}{\rho(\mathbf{x})} \text{grad } p(\mathbf{x}, t; \mathbf{x}_s)\right) = S(\mathbf{x}, t; \mathbf{x}_s) \quad,$$

$$p(\mathbf{x}, t; \mathbf{x}_s) = 0 \quad (\text{for } \mathbf{x} \in \mathcal{S}) \quad,$$ (7.231)

$$p(\mathbf{x}, 0; \mathbf{x}_s) = 0 \quad,$$

$$\dot{p}(\mathbf{x}, 0; \mathbf{x}_s) = 0 \quad.$$

Here $S(\mathbf{x}, t; \mathbf{x}_s)$ is the function describing the source and $\mathcal{S}$ denotes the surface of the medium. Practically, this differential system can, for instance, be solved using a finite-difference algorithm (see, for instance, Alterman and Karal, 1968).

The Green's function $\Gamma(\mathbf{x}, t; \mathbf{x}_s, t')$ is defined by

$$\frac{\exp(m(\mathbf{x}))}{K} \frac{\partial^2 \Gamma}{\partial t^2}(\mathbf{x}, t; \mathbf{x}', t') - \text{div}\left(\frac{1}{\rho(\mathbf{x})} \text{grad } \Gamma(\mathbf{x}, t; \mathbf{x}', t')\right) = \delta(\mathbf{x} - \mathbf{x}_s)\delta(t - t') \quad,$$

$$\Gamma(\mathbf{x}, t; \mathbf{x}', t') = 0 \qquad (\text{for } \mathbf{x} \in \mathcal{S}) \quad,$$

$$\Gamma(\mathbf{x}, t; \mathbf{x}', t') = 0 \qquad (\text{for } t < t') \quad,$$

$$\dot{\Gamma}(\mathbf{x}, t; \mathbf{x}', t') = 0 \qquad (\text{for } t < t') \quad,$$ (7.232)

and we have the integral representation (see, for instance, Morse and Feshbach, 1953)

$$p(\mathbf{x}, t; \mathbf{x}_s) = \int_{\mathcal{V}} dV(\mathbf{x}') \, \Gamma(\mathbf{x}, t; \mathbf{x}', 0) * S(\mathbf{x}', t; \mathbf{x}_s) \quad.$$ (7.233)

In order to obtain the Fréchet derivative of the displacements with respect to the bulk modulus (as defined by equation (7.227)), let us introduce the wavefield $p_n(\mathbf{x}, t; \mathbf{x}_s)$, propagating in the medium $m_n(\mathbf{x})$,

$$\frac{\exp(m(\mathbf{x})_n)}{K} \frac{\partial^2 p_n(\mathbf{x}, t; \mathbf{x}_s)}{\partial t^2} - \text{div}\left(\frac{1}{\rho(\mathbf{x})} \text{grad } p_n(\mathbf{x}, t; \mathbf{x}_s)\right) = S(\mathbf{x}, t; \mathbf{x}_s) \quad,$$

$$p_n(\mathbf{x}, t; \mathbf{x}_s) = 0 \qquad (\text{for } \mathbf{x} \in \mathcal{S}) \quad,$$ (7.234)

$$p_n(\mathbf{x}, 0; \mathbf{x}_s) = 0 \quad,$$

$$\dot{p}_n(\mathbf{x}, 0; \mathbf{x}_s) = 0 \quad,$$

and the corresponding Green's function

$$\frac{\exp(m(\mathbf{x})_n)}{K} \frac{\partial^2 \Gamma_n(\mathbf{x}, t; \mathbf{x}', t')}{\partial t^2} - \text{div}\left(\frac{1}{\rho(\mathbf{x})} \text{grad } \Gamma_n(\mathbf{x}, t; \mathbf{x}', t')\right) = \delta(\mathbf{x} - \mathbf{x}_s)\delta(t - t') \quad,$$

$$\Gamma_n(\mathbf{x}, t; \mathbf{x}', t') = 0 \qquad (\text{for } \mathbf{x} \in \mathcal{S}) \quad,$$

$$\Gamma_n(\mathbf{x}, t; \mathbf{x}', t') = 0 \qquad (\text{for } t < t') \quad,$$

$$\dot{\Gamma}_n(\mathbf{x}, t; \mathbf{x}', t') = 0 \qquad (\text{for } t < t') \quad.$$ (7.235)

A perturbation of the logarithmic bulk modulus $m_n(\mathbf{x}) \rightarrow m_n(\mathbf{x}) + \delta m(\mathbf{x})$ will produce a field $p_n(\mathbf{x}, t; \mathbf{x}_s) + \delta p(\mathbf{x}, t; \mathbf{x}_s)$ defined by

$$\frac{\exp(m_n(\mathbf{x}) + \delta m(\mathbf{x}))}{K} \frac{\partial^2 (p_n(\mathbf{x}, t; \mathbf{x}_s) + \delta p(\mathbf{x}, t; \mathbf{x}_s))}{\partial t^2}$$

$$- \operatorname{div}\left(\frac{1}{\rho(\mathbf{x})} \operatorname{grad}(p_n(\mathbf{x}, t; \mathbf{x}_s) + \delta p(\mathbf{x}, t; \mathbf{x}_s))\right) = S(\mathbf{x}, t; \mathbf{x}_s) \quad,$$

$$p_n(\mathbf{x}, t; \mathbf{x}_s) + \delta p(\mathbf{x}, t; \mathbf{x}_s) = 0 \qquad (\text{for } \mathbf{x} \in \mathcal{S}) \quad,$$

$$p_n(\mathbf{x}, 0; \mathbf{x}_s) + \delta p(\mathbf{x}, 0; \mathbf{x}_s) = 0 \quad,$$

$$\dot{p}_n(\mathbf{x}, 0; \mathbf{x}_s) + \delta \dot{p}(\mathbf{x}, 0; \mathbf{x}_s) = 0 \quad. \tag{7.236}$$

This gives, after simplification,

$$\frac{\exp(m_n(\mathbf{x}))}{K} \frac{\partial^2 \delta p(\mathbf{x}, t; \mathbf{x}_s)}{\partial t^2} - \operatorname{div}\left(\frac{1}{\rho(\mathbf{x})} \operatorname{grad} \delta p(\mathbf{x}, t; \mathbf{x}_s)\right)$$

$$= -\frac{\partial^2 p_n(\mathbf{x}, t; \mathbf{x}_s)}{\partial t^2} \frac{\exp(m_n(\mathbf{x}))}{K} \delta m(\mathbf{x}) + O(\| \delta \mathbf{m} \|^2) \quad,$$

$$\delta p(\mathbf{x}, t; \mathbf{x}_s) = 0 \qquad (\text{for } \mathbf{x} \in \mathcal{S}) \quad,$$

$$\delta p(\mathbf{x}, 0; \mathbf{x}_s) = 0 \quad,$$

$$\delta \dot{p}(\mathbf{x}, 0; \mathbf{x}_s) = 0 \quad, \tag{7.237}$$

and, using theorem (7.233),

$$\delta p(\mathbf{x}_r, t; \mathbf{x}_s) = -\int_{\mathcal{V}} dV(\mathbf{x}) \, \Gamma_n(\mathbf{x}_r, t; \mathbf{x}, 0) * \frac{\partial^2 p_n}{\partial t^2}(\mathbf{x}, t; \mathbf{x}_s) \frac{\exp(m_n(\mathbf{x}))}{K} \delta m(\mathbf{x})$$

$$+ O(\| \delta \mathbf{m} \|^2) \quad. \tag{7.238}$$

The Fréchet derivative operator $\mathbf{G}_n$ introduced in (7.227) is then characterized by

$$(\mathbf{G}_n \, \delta \mathbf{m})(\mathbf{x}_r, t; \mathbf{x}_s) = -\int_{\mathcal{V}} dV(\mathbf{x}) \, \Gamma_n(\mathbf{x}_r, t; \mathbf{x}, 0) * \frac{\partial^2 p_n}{\partial t^2}(\mathbf{x}, t; \mathbf{x}_s) \frac{\exp(m_n(\mathbf{x}))}{K} \delta m(\mathbf{x}) \quad, \tag{7.239}$$

where $\Gamma_n(\mathbf{x}, t; \mathbf{x}', t')$ is the Green's function corresponding to the medium $m_n(\mathbf{x})$ (defined by equation (7.235)) and $p_n(\mathbf{x}, t; \mathbf{x}_s)$ is the wavefield also corresponding to $m_n(\mathbf{x})$ (defined by equation (7.234)).

We now turn to the characterization of the transpose operator. We have just defined the Fréchet derivative operator $\mathbf{G}_n$. Its transpose $\mathbf{G}_n^t$ was defined by equation (7.229):

$$\langle \, \delta \hat{\mathbf{p}} \,, \, \mathbf{G}_n \, \delta \mathbf{m} \, \rangle = \langle \, \mathbf{G}_n^t \, \delta \hat{\mathbf{p}} \,, \, \delta \mathbf{m} \, \rangle \qquad \text{for any } \delta \hat{\mathbf{p}} \text{ and } \delta \mathbf{m} \quad. \tag{7.240}$$

To solve the inverse problem, we need to be able to compute $\mathbf{G}_n^t \, \delta \hat{\mathbf{p}}$ for arbitrary $\delta \hat{\mathbf{p}}$. Using the notation introduced in (7.224) and (7.228), equation (7.229) is written

$$\sum_s \int_0^T dt \sum_r \delta \hat{p}(\mathbf{x}_r, t; \mathbf{x}_s)(\mathbf{G}_n \, \delta \mathbf{m})(\mathbf{x}_r, t; \mathbf{x}_s) = \int_{\mathcal{V}} dV(\mathbf{x}) \, (\mathbf{G}_n^t \, \delta \hat{\mathbf{p}})(\mathbf{x}) \, \delta m(\mathbf{x}) \quad,$$

$$\tag{7.241}$$

and, using (7.239),

$$\sum_s \int_0^T dt \sum_r \delta\hat{p}(\mathbf{x}_r, t; \mathbf{x}_s) \int_{\mathcal{V}} dV(\mathbf{x}) \, \Gamma_n(\mathbf{x}_r, t; \mathbf{x}, 0) * \frac{\partial^2 p_n(\mathbf{x}, t; \mathbf{x}_s)}{\partial t^2} \frac{\exp(m_n(\mathbf{x}))}{K} \delta m(\mathbf{x})$$

$$= -\int_{\mathcal{V}} dV(\mathbf{x}) \, (\mathbf{G}_n^t \, \delta\hat{\mathbf{p}})(\mathbf{x}) \, \delta m(\mathbf{x}) \quad , \tag{7.242}$$

i.e.,

$$\int_{\mathcal{V}} dV(\mathbf{x}) \, \delta m(\mathbf{x}) \left\{ (\mathbf{G}_n{}^t \, \delta\hat{\mathbf{p}})(\mathbf{x}) + \frac{\exp(m_n(\mathbf{x}))}{K} \sum_s \int_0^T dt \sum_r \Gamma_n(\mathbf{x}_r, t; \mathbf{x}, 0) \right.$$

$$\left. * \frac{\partial^2 p_n}{\partial t^2}(\mathbf{x}, t; \mathbf{x}_s) \, \delta\hat{p}(\mathbf{x}_r, t; \mathbf{x}_s) \right\} = 0 \quad . \tag{7.243}$$

As this has to be valid for any $\delta m(\mathbf{x})$, we obtain

$$(\mathbf{G}_n^t \, \delta\hat{\mathbf{p}})(\mathbf{x}) = -\frac{\exp(m_n(\mathbf{x}))}{K} \sum_s \int_0^T dt \sum_r \Gamma_n(\mathbf{x}_r, t; \mathbf{x}, 0) * \frac{\partial^2 p_n}{\partial t^2}(\mathbf{x}, t; \mathbf{x}_s) \, \delta\hat{p}(\mathbf{x}_r, t; \mathbf{x}_s) \quad . \tag{7.244}$$

Let us introduce a field $\Psi_n(\mathbf{x}, t; \mathbf{x}_s)$ defined by the differential system

$$\frac{\exp(m_n(\mathbf{x}))}{K} \frac{\partial^2 \Psi_n(\mathbf{x}, t; \mathbf{x}_s)}{\partial t^2} - \text{div}\left(\frac{1}{\rho(\mathbf{x})} \text{grad} \, \Psi_n(\mathbf{x}, t; \mathbf{x}_s)\right) = \Phi(\mathbf{x}, t; \mathbf{x}_s) \quad ,$$

$$\Psi_n(\mathbf{x}, t; \mathbf{x}_s) = 0 \qquad (\text{for } \mathbf{x} \in \mathcal{S}) \quad , \tag{7.245}$$

$$\Psi_n(\mathbf{x}, T; \mathbf{x}_s) = 0 \quad ,$$

$$\dot{\Psi}_n(\mathbf{x}, T; \mathbf{x}_s) = 0 \quad ,$$

where

$$\Phi(\mathbf{x}, t; \mathbf{x}_s) = \sum_r \delta(\mathbf{x} - \mathbf{x}_r) \, \delta\hat{p}(\mathbf{x}_r, t; \mathbf{x}_s) \quad . \tag{7.246}$$

Notice that the field Ψ satisfies *final* (instead of initial) conditions. Using the property

$$\Gamma_n(\mathbf{x}, t; \mathbf{x}_r, t') = \Gamma_n(\mathbf{x}, t + \tau, \mathbf{x}_r, t' + \tau) \tag{7.247}$$

and reversing time in theorem (7.233), one obtains

$$\Psi_n(\mathbf{x}, t; \mathbf{x}_s) = \sum_r \int_0^T dt' \, \Gamma_n(\mathbf{x}, 0; \mathbf{x}_r, t - t') \, \delta\hat{p}(\mathbf{x}_r, t'; \mathbf{x}_s) \quad . \tag{7.248}$$

One has

$$\dot{\Psi}(\mathbf{x}, t; \mathbf{x}_s) = \sum_r \int_0^T dt' \, \frac{\partial}{\partial t} \Gamma(\mathbf{x}, 0; x_r, t - t') \, \delta\hat{p}(\mathbf{x}_r, t'; \mathbf{x}_s)$$

$$= \sum_r \int_0^T dt' \, \frac{\partial}{\partial t} \Gamma(\mathbf{x}, t' - t; \mathbf{x}_r, 0) \, \delta\hat{p}(\mathbf{x}_r, t'; \mathbf{x}_s) \tag{7.249}$$

$$= -\sum_r \int_0^T dt' \, \dot{\Gamma}(\mathbf{x}, t' - t; \mathbf{x}_r, 0) \, \delta\hat{p}(\mathbf{x}_r, t'; \mathbf{x}_s) \quad ,$$

where

$$\dot{\Gamma}(\mathbf{x}, t; \mathbf{x}', t') = \frac{\partial}{\partial t} \Gamma(\mathbf{x}, t; \mathbf{x}', t') \quad . \tag{7.250}$$

Using integration by parts gives

$$\Gamma(\mathbf{x}_r, t; \mathbf{x}, 0) * \frac{\partial^2 p}{\partial t^2}(\mathbf{x}, t; \mathbf{x}_s) = \int_0^T dt' \, \Gamma(\mathbf{x}_r, t - t'; \mathbf{x}, 0) \frac{\partial^2 p}{\partial t^2}(\mathbf{x}, t'; \mathbf{x}_s)$$

$$= \Gamma(\mathbf{x}_r, t - T; \mathbf{x}, 0) \, \dot{p}(\mathbf{x}, T; \mathbf{x}_s) - \Gamma(\mathbf{x}_r, t; \mathbf{x}, 0) \dot{p}(\mathbf{x}, 0; \mathbf{x}_s)$$

$$- \int_0^T dt' \, \dot{\Gamma}(\mathbf{x}_r, t - t'; \mathbf{x}, 0) \, \dot{p}(\mathbf{x}, t'; \mathbf{x}_s) \quad , \tag{7.251}$$

and, using the last two initial conditions in equations (7.235),

$$\Gamma(\mathbf{x}_r, t; \mathbf{x}, 0) * \frac{\partial^2 p}{\partial t^2}(\mathbf{x}, t; \mathbf{x}_s) = -\dot{\Gamma}(\mathbf{x}_r, t; \mathbf{x}, 0) * \dot{p}(\mathbf{x}, t; \mathbf{x}_s) \quad . \tag{7.252}$$

Using the last equation gives

$$\sum_s \int_0^T dt \sum_r \Gamma(\mathbf{x}_r, t; \mathbf{x}, 0) * \frac{\partial^2 p}{\partial t^2}(\mathbf{x}, t; \mathbf{x}_s) \, \delta \hat{p}(\mathbf{x}_r, t; \mathbf{x}_s)$$

$$= -\sum_s \int_0^T dt \sum_r \dot{\Gamma}(\mathbf{x}_r, t; \mathbf{x}, 0) * \dot{p}(\mathbf{x}, t; \mathbf{x}_s) \, \delta \hat{p}(\mathbf{x}_r, t; \mathbf{x}_s) \tag{7.253}$$

$$= -\sum_s \int_0^T dt \int_0^T dt' \sum_r \dot{\Gamma}(\mathbf{x}_r, t - t'; \mathbf{x}, 0) \, \dot{p}(\mathbf{x}, t'; \mathbf{x}_s) \, \delta \hat{p}(\mathbf{x}_r, t; \mathbf{x}_s) \quad ,$$

whence, using (7.249), we obtain

$$\sum_s \int_0^T dt \sum_r \Gamma(\mathbf{x}_r, t; \mathbf{x}, 0) * \frac{\partial^2 p}{\partial t^2}(\mathbf{x}, t; \mathbf{x}_s) \, \delta \hat{p}(\mathbf{x}_r, t; \mathbf{x}_s)$$

$$= \sum_s \int_0^T dt \, \dot{p}(\mathbf{x}, t; \mathbf{x}_s) \, \dot{\Psi}(\mathbf{x}, t; \mathbf{x}_s) \quad . \tag{7.254}$$

From equation (7.244), we then finally obtain the result characterizing the transpose operator:

$$(\mathbf{G}_n^t \, \delta \hat{p})(\mathbf{x}) = -\frac{\exp(m_n(\mathbf{x}))}{K} \sum_s \int_0^T dt \, \dot{p}_n(\mathbf{x}, t; \mathbf{x}_s) \dot{\Psi}_n(\mathbf{x}, t; \mathbf{x}_s) \quad . \tag{7.255}$$

The preconditioned steepest descent algorithm (7.230) is written, step by step,

$$\begin{aligned}
\delta \mathbf{p}_n &= \mathbf{g}(\mathbf{m}_n) - \mathbf{d}_{\text{obs}} \quad , \\
\delta \hat{\mathbf{p}}_n &= \mathbf{C}_p^{-1} \, \delta \mathbf{p}_n \quad , \\
\hat{\boldsymbol{\gamma}}_n &= \mathbf{G}_n^t \, \delta \hat{\mathbf{p}}_n \quad , \\
\mathbf{d}_n &= \hat{\mathbf{S}}_0 \, \hat{\boldsymbol{\gamma}}_n \quad ,
\end{aligned} \tag{7.256}$$

and

$$\mathbf{m}_{n+1} = \mathbf{m}_n - \mu_n \, \mathbf{d}_n \quad . \tag{7.257}$$

To compute the residuals (7.256) we need to compute $\mathbf{g}(\mathbf{m}_n)$, i.e., we need to solve the forward problem (using any numerical method). The weighted residuals $\delta\hat{\mathbf{p}}_n$ are easily obtained using (7.222). $\hat{\boldsymbol{\gamma}}_n$ is computed using (7.255), where the field $\Psi(\mathbf{x}, t; \mathbf{x}_s)_n$ is obtained by solving the system (7.245) (using, for instance, the same method used to solve the forward problem). To obtain $\mathbf{d}_n$, we have to apply the preconditioning operator $\hat{\mathbf{S}}_0$ to $\hat{\boldsymbol{\gamma}}_n$. It may simply consist of some ad hoc geometrical correction (see, for instance, Gauthier, Virieux, and Tarantola, 1986). To end one iteration, we have to estimate μ_n in (7.257). A linearized estimation of μ_n can be obtained as follows.

For given $\mathbf{m}_n$, we have

$$S(\mathbf{m}_n - \mu_n \mathbf{d}_n) = \frac{1}{2}\left((\mathbf{g}(\mathbf{m}_n - \mu_n \mathbf{d}_n) - \mathbf{p}_{\text{obs}})^t \mathbf{C}_p^{-1} (\mathbf{g}(\mathbf{m}_n - \mu_n \mathbf{d}_n) - \mathbf{p}_{\text{obs}})\right) \quad .$$

(7.258)

If μ_n is small enough, using the definition of Fréchet derivatives, we have

$$\mathbf{g}(\mathbf{m}_n - \mu_n \mathbf{d}_n) \simeq \mathbf{g}(\mathbf{m}_n) - \mu_n \mathbf{G}_n \mathbf{d}_n \quad , \tag{7.259}$$

which gives

$$\begin{aligned}
S(\mathbf{m}_n - \mu_n \mathbf{d}_n) &\simeq S(\mathbf{m}_n) - \mu_n (\mathbf{G}_n \mathbf{d}_n)^t \mathbf{C}_p^{-1} (\mathbf{g}(\mathbf{m}_n) - \mathbf{p}_{\text{obs}}) \\
&\quad + \frac{1}{2}\mu_n^2 (\mathbf{G}_n \mathbf{d}_n)^t \mathbf{C}_p^{-1} (\mathbf{G}_n \mathbf{d}_n) \quad .
\end{aligned}$$

(7.260)

The condition $\partial S / \partial \mu_n = 0$ gives

$$\mu_n \simeq \frac{(\mathbf{G}_n \mathbf{d}_n)^t \mathbf{C}_p^{-1} (\mathbf{g}(\mathbf{m}_n) - \mathbf{p}_{\text{obs}})}{(\mathbf{G}_n \mathbf{d}_n)^t \mathbf{C}_p^{-1} (\mathbf{G}_n \mathbf{d}_n)} \quad , \tag{7.261}$$

and, using the definition of the transpose operator (equation (7.229)), we finally obtain

$$\mu_n \simeq \frac{\mathbf{d}_n^t \mathbf{G}_n^t \mathbf{C}_p^{-1} (\mathbf{g}(\mathbf{m}_n) - \mathbf{p}_{\text{obs}})}{(\mathbf{G}_n \mathbf{d}_n)^t \mathbf{C}_p^{-1} (\mathbf{G}_n \mathbf{d}_n)} = \frac{\mathbf{d}_n^t \hat{\boldsymbol{\gamma}}_n}{(\mathbf{G}_n \mathbf{d}_n)^t \mathbf{C}_p^{-1} (\mathbf{G}_n \mathbf{d}_n)} \quad . \tag{7.262}$$

To compute $\mathbf{G}_n \mathbf{d}_n$, we could use the result (7.239), but it is more practical to use directly the definition of derivative operator and a finite-difference approximation. We may then use

$$\mathbf{G}_n \mathbf{d}_n \simeq \frac{1}{\epsilon}\left(\mathbf{g}(\mathbf{m}_n + \epsilon \mathbf{d}_n) - \mathbf{g}(\mathbf{m}_n)\right) \tag{7.263}$$

with a sufficiently small value of ϵ.

The physical interpretation of the obtained results is as in section 5.8.7.

7.19   Using the Backus and Gilbert Method

Figure 7.32 represents a borehole in which we are able to introduce a sensor. For three different depths z_1, z_2, z_3, we have measured the travel times t_1, t_2, t_3 of acoustic waves

Figure 7.32. *Acoustic waves traveling down in the Earth are use to infer the velocity-versus-depth dependence (using the Backus and Gilbert method).*

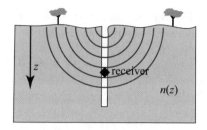

from the top of the borehole to the sensor. We obtain the following results:

$$z_1 = 2.3\,\text{m} \quad , \quad t_1 = 2.0\,\text{ms} \quad ,$$
$$z_2 = 8.1\,\text{m} \quad , \quad t_2 = 8.1\,\text{ms} \quad , \tag{7.264}$$
$$z_3 = 10.1\,\text{m} \quad , \quad t_3 = 10.2\,\text{ms}$$

(ms denotes millisecond). Use the Backus and Gilbert method to estimate the slowness (inverse of velocity) $n(z)$ of the acoustic waves in the medium. Represent the resolving kernel $R(z, z')$ for $z = 10\,\text{m}$ and $0 \le z' \le \infty$.

Solution:

Let **n** represent a model of slowness (i.e., a particular function $z \to n(z)$) and d represent the column matrix of travel times predicted from the model **n**. Formally,

$$\mathbf{d} = \mathbf{G}\,\mathbf{n} \quad , \tag{7.265}$$

where **G** is the linear operator defined by

$$d^i = \int_0^{z_i} dz\, n(z) \quad . \tag{7.266}$$

The kernels of the operator **G** are introduced by

$$d^i = \int_0^\infty dz\, G^i(z)\, n(z) \quad . \tag{7.267}$$

This gives

$$G^1(z) = \begin{cases} 1 \text{ for } 0 \le z < z_1 \quad , \\ 0 \text{ for } z_1 \le z < \infty \quad , \end{cases} \qquad G^2(z) = \begin{cases} 1 \text{ for } 0 \le z < z_2 \quad , \\ 0 \text{ for } z_2 \le z < \infty \quad , \end{cases}$$
$$G^3(z) = \begin{cases} 1 \text{ for } 0 \le z < z_3 \quad , \\ 0 \text{ for } z_3 \le z < \infty \quad . \end{cases} \tag{7.268}$$

Let $\mathbf{d}_{\text{obs}}$ be the column matrix of observed travel times:

$$\mathbf{d}_{\text{obs}} = \begin{pmatrix} 2.0\,\text{ms} \\ 8.1\,\text{ms} \\ 10.2\,\text{ms} \end{pmatrix} \quad . \tag{7.269}$$

The Backus and Gilbert solution of the problem of estimating the model is (see equations (6.184) of Appendix 6.16)

$$\mathbf{m} = \mathbf{G}^t \, (\mathbf{G} \, \mathbf{G}^t)^{-1} \, \mathbf{d}_{\text{obs}} = \mathbf{G}^t \, (\mathbf{S})^{-1} \, \mathbf{d}_{\text{obs}} \quad , \tag{7.270}$$

where

$$\mathbf{S} = \mathbf{G} \, \mathbf{G}^t \quad . \tag{7.271}$$

Explicitly,

$$n(z) = \sum_i \mathbf{G}^i (z) \, \Psi^i \quad , \tag{7.272}$$

where

$$\Psi^i = \sum_j (\mathbf{S}^{-1})^{ij} \, d_{\text{obs}}{}^j \tag{7.273}$$

and

$$S^{ij} = \int_0^\infty dz \, G^i (z) \, G^j (z) \quad . \tag{7.274}$$

Using (7.268) we obtain

$$\mathbf{S} = \begin{pmatrix} 2.3 & 2.3 & 2.3 \\ 2.3 & 8.1 & 8.1 \\ 2.3 & 8.1 & 10.1 \end{pmatrix} \quad , \quad \mathbf{S}^{-1} = \frac{1}{26.68} \begin{pmatrix} 16.2 & -4.6 & 0 \\ -4.6 & 17.94 & -13.34 \\ 0 & -13.34 & 13.34 \end{pmatrix} \quad . \tag{7.275}$$

From (7.269) and (7.273), we then obtain

$$\Psi = \mathbf{S}^{-1} \mathbf{d}_{\text{obs}} = \begin{pmatrix} -0.182 \\ +0.002 \\ +1.050 \end{pmatrix} \quad . \tag{7.276}$$

Equation (7.272) finally gives the Backus and Gilbert estimate

$$n(z) = -0.182 \, G^1(z) + 0.002 \, G^2(z) + 1.050 \, G^3(z) \quad , \tag{7.277}$$

which is represented in Figure 7.33.

Figure 7.33. *The velocity-versus-depth dependence obtained using the Backus and Gilbert method.*

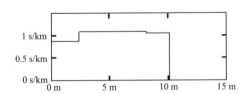

The resolving operator is given by (equation (6.184) of Appendix 6.16)

$$\mathbf{R} = \mathbf{G}^t \, (\mathbf{G}\,\mathbf{G}^t)^{-1}\, \mathbf{G} = \mathbf{G}^t \, \mathbf{S}^{-1}\, \mathbf{G} \ . \tag{7.278}$$

Explicitly, the resolving kernel is given by

$$\mathbf{R}(z, z') = \sum_i \sum_j G^i(z)(\mathbf{S}^{-1})^{ij} \, G^j(z') \ . \tag{7.279}$$

We are asked for the values of $\mathbf{R}(z, z')$ for $z = 10$ m. We have the following:

for $\quad 0.0 \le z' < 2.3 :\ R(10.0\,\mathrm{m}, z') = (S^{-1})^{31} + (S^{-1})^{32} + (S^{-1})^{33} = 0.0\ ;$

for $\quad 2.3 \le z' < 8.1 :\ R(10.0\,\mathrm{m}, z') = (S^{-1})^{32} + (S^{-1})^{33} = 0.0\ ;$

for $\quad 8.1 \le z' < 10.1 :\ R(10.0\,\mathrm{m}, z') = (S^{-1})^{33} = 0.5\ ;$

for $\quad 10.1 \le z' < \infty :\ R(10.0\,\mathrm{m}, z') = 0.0\ .$ $\tag{7.280}$

The corresponding result is represented in Figure 7.34.

Figure 7.34. *The Backus and Gilbert resolving kernel.*

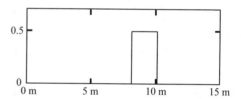

Discussion:

The solution shown in Figure 7.33 is the simplest solution predicting the observed travel times exactly. The kernel shown in Figure 7.34 says that the value of slowness estimated at $z = 10$ m is the mean of the true value for 8.1 m $< z < 10.1$ m. This can be physically understood: the values of the slowness for $z \le 8.1$ m are fixed by the first and second observed travel times; it is the third travel time that gives information for 8.1 m $< z < 10.1$ m, and it only gives information on the integrated slowness, i.e., on the mean value between $z = 8.1$ m and $z = 10.1$ m.

The method gives a null value for the slowness for $z > 10.1$ m. This is of course unphysical, but this value is totally unresolved. In fact, the essence of the Backus and Gilbert method is better obtained when the unknown is a correction to some current model: where there are no data, there is no correction.

Note that this is a strict application of the Backus and Gilbert method. The methods proposed in this book suggest that the logarithmic slowness should be used, transforming this formally linear problem into a nonlinear one (but then defining a truly linear model space).

7.20   The Coefficients in the Backus and Gilbert Method

In the Backus and Gilbert method, the problem arises of obtaining the coefficients $Q^i(r_0)$ that minimize the expression

$$J(r_0) = \int dt \left(R(r_0, r) - \delta(r_0 - r) \right)^2 \quad , \tag{7.281}$$

where (using the sum convention over repeated indices)

$$R(r_0, r) = Q_i(r_0) G^i(r) \tag{7.282}$$

and where the $G^i(r)$ are given functions. Show that the coefficients $Q_i(r_0)$ are given by

$$Q_i(r_0) = (S^{-1})_{ij} G^j(r_0) \quad , \tag{7.283}$$

where

$$S^{ij} = \int dr\, G^i(r) G^j(r) \quad . \tag{7.284}$$

Solution:
One has

$$
\begin{aligned}
J(r_0) &= \int dr \left(Q_i(r_0) G^i(r) - \delta(r_0 - r) \right)^2 \\
&= \int dr \left(Q_i(r_0) G^i(r) Q_j(r_0) G^j(r) - 2\, Q_i(r_0) G^i(r) \delta(r_0 - r) + \delta^2(r_0 - r) \right) \\
&= Q_i(r_0) S^{ij} Q_j(r_0) - 2\, Q_i(r_0) G^i(r_0) + \delta(0) \quad , \tag{7.285}
\end{aligned}
$$

where

$$S^{ij} = \int dr\, G^i(r) G^j(r) \tag{7.286}$$

and where the infinite value $\delta(0)$ can formally be handled as a constant. At the minimum of $J(r_0)$,

$$\frac{\partial J(r_0)}{\partial Q_i(r_0)} = 0 \quad \Rightarrow \quad S^{ij} Q_j(r_0) = G^i(r_0) \quad , \tag{7.287}$$

from which the result follows (the matrix S^{ij} is regular if the $G^i(r)$ are linearly independent functions).

7.21   The Norm Associated with the 1D Exponential Covariance

Let $C(t, t')$ be the covariance function considered in Examples 5.5 and 5.13:

$$C(t, t') = \sigma^2 \exp\left(-\frac{|t - t'|}{T} \right) \quad . \tag{7.288}$$

The covariance operator $\mathbf{C}$ corresponding to the integral kernel 7.288 associates any function $\hat{e}(t)$ with the function

$$e(t) = \int_{t_1}^{t_2} dt'\, C(t, t')\, \hat{e}(t') \quad , \qquad t \in [t_1, t_2] \quad . \tag{7.289}$$

Obtain the inverse operator and the associated scalar product and norm.

Solution:
Noticing that if

$$\varphi(t) = \sigma^2 \exp\left(-\frac{|t|}{T}\right) \quad , \tag{7.290}$$

then

$$\frac{\partial \varphi}{\partial t}(t) = -\frac{1}{T} \operatorname{sg}(t)\, \varphi(t) \tag{7.291}$$

and

$$\frac{\partial^2 \varphi}{\partial t^2}(t) = \frac{1}{T^2} \varphi(t) - \frac{2\sigma^2}{T} \delta(t) \quad , \tag{7.292}$$

we easily obtain

$$\frac{\partial e}{\partial t}(t) = -\frac{1}{T} \int_{t_1}^{t_2} dt'\, \operatorname{sg}(t - t')\, C(t, t')\, \hat{e}(t') \tag{7.293}$$

and

$$\frac{\partial^2 e}{\partial t^2}(t) = \frac{1}{T^2} e(t) - \frac{2\sigma^2}{T} \hat{e}(t) \quad . \tag{7.294}$$

Using (7.289), equation (7.293) shows that the values at $t = t_1$ and $t = t_2$ of $e(t)$ and $\partial e/\partial t\,(t)$ are not independent:

$$\frac{\partial e}{\partial t}(t_1) = \frac{1}{T} e(t_1) \quad , \qquad \frac{\partial e}{\partial t}(t_2) = -\frac{1}{T} e(t_2) \quad . \tag{7.295}$$

Equation (7.294) then gives

$$\hat{e}(t) = \frac{1}{2\sigma^2 T} e(t) - \frac{T}{2\sigma^2} \frac{\partial^2 e}{\partial t^2}(t) \quad , \qquad t \in [t_1, t_2] \quad . \tag{7.296}$$

We see thus that the domain of definition of the operator $\mathbf{C}^{-1}$ is the set of functions verifying the conditions (7.295). With any such function, the operator $\mathbf{C}^{-1}$ associates the function given by 7.296.

Using for $\mathbf{C}^{-1}$ the integral representation

$$\hat{e}(t) = \int_{t_1}^{t_2} dt'\, \mathbf{C}^{-1}(t, t')\, e(t') \tag{7.297}$$

gives

$$\mathbf{C}^{-1}(t, t') = \frac{1}{2\sigma^2} \left(\frac{1}{T} \delta(t - t') - T \delta^2(t - t') \right) \quad , \tag{7.298}$$

where I have used the definition of the derivative of Dirac's delta distribution:

$$\int dt' \, \delta^{(n)}(t - t') \, \mu(t') = (-1)^n \frac{d^n}{dt^n} \mu(t) \quad . \tag{7.299}$$

The scalar product of two functions $\mathbf{e}_1$ and $\mathbf{e}_2$ may be defined by

$$(\mathbf{e}_1, \mathbf{e}_2) = \mathbf{e}_1{}^t \, \mathbf{C}^{-1} \mathbf{e}_2 = \int_{t_1}^{t_2} dt \, e_1(t) \left(\frac{1}{T} e_2(t) - T \frac{\partial^2 e_2}{\partial t^2}(t) \right) \quad . \tag{7.300}$$

Integrating by parts gives

$$(\mathbf{e}_1, \mathbf{e}_2) = \frac{1}{T} \int_{t_1}^{t_2} dt \, e_1(t) \, e_2(t) + T \int_{t_1}^{t_2} dt \, \frac{\partial e_1}{\partial t}(t) \frac{\partial e_2}{\partial t}(t) + T \left(\frac{\partial e_1}{\partial t}(t) \, e_2(t) \right) \Big|_{t_1}^{t_2} \quad . \tag{7.301}$$

Similarly,

$$(\mathbf{e}_2, \mathbf{e}_1) = \frac{1}{T} \int_{t_1}^{t_2} dt \, e_2(t) \, e_1(t) + T \int_{t_1}^{t_2} dt \, \frac{\partial e_2}{\partial t}(t) \frac{\partial e_1}{\partial t}(t) + T \left(\frac{\partial e_2}{\partial t}(t) \, e_1(t) \right) \Big|_{t_1}^{t_2} \quad , \tag{7.302}$$

and we see that the scalar product is symmetric,

$$(\mathbf{e}_1, \mathbf{e}_2) = (\mathbf{e}_2, \mathbf{e}_1), \tag{7.303}$$

only if the functions $\mathbf{e}_1$ and $\mathbf{e}_2$ satisfy the dual boundary conditions

$$\left(\frac{\partial e_1}{\partial t}(t) \, e_2(t) - e_2(t) \frac{\partial e_2}{\partial t}(t) \right) \Big|_{t_1}^{t_2} = 0 \quad . \tag{7.304}$$

The norm of an element $\mathbf{e}$ can be computed by

$$\| \mathbf{e} \|^2 = (\mathbf{e}, \mathbf{e}) = \mathbf{e}^t \, \mathbf{C}^{-1} \mathbf{e} = \int_{t_1}^{t_2} dt \, \hat{e}(t) \, e(t) \quad , \tag{7.305}$$

where

$$\hat{\mathbf{e}} = \mathbf{C}^{-1} \mathbf{e} \quad . \tag{7.306}$$

We have

$$\| \mathbf{e} \|^2 = \int_{t_1}^{t_2} dt \, \frac{1}{2\sigma^2} \left(\frac{1}{T} e(t) - T \frac{\partial^2 e}{\partial t^2}(t) \right) e(t)$$

$$= \frac{1}{2\sigma^2} \left(\frac{1}{T} \int_{t_1}^{t_2} dt \, e(t)^2 - T \int_{t_1}^{t_2} dt \, \frac{\partial e^2}{\partial t^2}(t) \, e(t) \right) \quad , \tag{7.307}$$

and, integrating by parts,

$$\| \, e \, \|^2 \;=\; \frac{1}{2\sigma^2} \left(\frac{1}{T} \int_{t_1}^{t_2} dt \, [e(t)]^2 + T \int_{t_1}^{t_2} \left[\frac{\partial e}{\partial t^2} \right]^2 \right.$$
$$\left. + T \frac{\partial e}{\partial t}(t_1) \, e(t_1) - T \frac{\partial e}{\partial t}(t_2) \, e(t_2) \right) \quad , \tag{7.308}$$

which, using equations (7.295), can be rewritten in either of the following two forms:

$$\| \, e \, \|^2 \;=\; \frac{1}{2\sigma^2} \left(\frac{1}{T} \int_{t_1}^{t_2} dt \, [e(t)]^2 + T \int_{t_1}^{t_2} dt \left[\frac{\partial e}{\partial t}(t) \right]^2 + [e(t_1)]^2 + [e(t_2)]^2 \right) \quad , \tag{7.309}$$

$$\| \, e \, \|^2 \;=\; \frac{1}{2\sigma^2} \left(\frac{1}{T} \int_{t_1}^{t_2} dt \, [e(t)]^2 + T \int_{t_1}^{t_2} dt \left[\frac{\partial e}{\partial t}(t) \right]^2 + T^2 \left[\frac{\partial e}{\partial t}(t_1) \right]^2 + T^2 \left[\frac{\partial e}{\partial t}(t_2) \right]^2 \right) \quad . \tag{7.310}$$

As these expressions are sums of squares and vanish only for a null function, we verify a posteriori that the covariance operator defined by the covariance function $C(t, t')$ is a positive definite operator.

In many applications, the first two terms in (7.309) and (7.310) are approximately proportional to $t_2 - t_1$, and, as usually $t_2 - t_1 \gg T$, the last two terms can be dropped, thus giving

$$\| \, e \, \|^2 \;\simeq\; \frac{1}{2\sigma^2} \left(\frac{1}{T} \int_{t_1}^{t_2} dt \, [e(t)]^2 + T \int_{t_1}^{t_2} dt \left[\frac{\partial e}{\partial t}(t) \right]^2 \right) \quad . \tag{7.311}$$

This corresponds to the usual norm in the Sobolev space H^1 (see Appendix 6.25), which is the sum of the usual L_2-norm of the function and the L_2-norm of its derivative.

7.22 The Norm Associated with the 1D Random Walk

For $0 \le t \le T$, let $C(t, t')$ be the covariance function

$$C(t, t') \;=\; \beta \min(t, t') \quad . \tag{7.312}$$

As mentioned in Example 5.9, this is the covariance function of a one-dimensional random walk. Notice that the variance at the point t is $\sigma^2 = C(t, t) = \beta \, t$. With any function $\hat{e}(t)$, the covariance operator $\mathbf{C}$, whose kernel is the covariance function (7.312), associates the function

$$e(t) \;=\; \int_0^T dt' \, C(t, t') \, \hat{e}(t') \quad . \tag{7.313}$$

Obtain the inverse operator and the associated norm.

Solution:
We have

$$e(t) = \beta \left(\int_0^t dt'\, t'\, \hat{e}(t') + t \int_t^T dt'\, \hat{e}(t') \right) \quad . \tag{7.314}$$

Using $\frac{\partial}{\partial t} \int_t^T dt'\, f(t') = -f(t)$ gives

$$\frac{\partial e}{\partial t}(t) = \beta \int_t^T dt'\, \hat{e}(t') \tag{7.315}$$

and

$$\frac{\partial^2 e}{\partial t^2}(t) = -\beta\, \hat{e}(t) \quad . \tag{7.316}$$

Equation (7.314) gives the condition

$$e(0) = 0 \quad , \tag{7.317}$$

while (7.315) gives the condition

$$\frac{\partial e}{\partial t}(T) = 0 \quad . \tag{7.318}$$

Equation (7.316) gives

$$\hat{e}(t) = -\frac{1}{\beta}\frac{\partial^2 e}{\partial t^2}(t) \quad . \tag{7.319}$$

The domain of definition of the operator $\mathbf{C}^{-1}$ is the set of functions verifying the conditions (7.317)–(7.318). Any such function is associated by the operator $\mathbf{C}^{-1}$ with the function given by (7.319).

A formal introduction of the integral kernel of $\mathbf{C}^{-1}$ gives

$$\hat{e}(t) = \int_0^T dt'\, C^{-1}(t, t')\, e(t') \quad , \tag{7.320}$$

and, by comparison with (7.319), we directly obtain

$$C^{-1}(t, t') = -\frac{1}{\beta}\, \delta^2(t - t') \quad . \tag{7.321}$$

The norm of an element can be computed by

$$\| \mathbf{e} \|^2 = \mathbf{e}^t\, \mathbf{C}^{-1}\mathbf{e} = \langle\, \hat{\mathbf{e}}\, , \, \mathbf{e}\, \rangle = \int_0^T dt\, \hat{e}(t)\, e(t) \quad , \tag{7.322}$$

where $\hat{\mathbf{e}} = \mathbf{C}^{-1}\mathbf{e}$. We have

$$\| \mathbf{e} \|^2 = -\frac{1}{\beta} \int_0^T dt\, e(t)\, \frac{\partial^2 e}{\partial t^2}(t) \quad , \tag{7.323}$$

and, integrating by parts,

$$\| \mathbf{e} \|^2 = \frac{1}{\beta} \left(\int_0^T dt \left[\frac{\partial e}{\partial t}(t) \right]^2 - e(t) \frac{\partial e}{\partial t}(t) \Big|_0^T \right) \quad . \tag{7.324}$$

Using conditions (7.317)–(7.318) gives the final result:

$$\| \mathbf{e} \|^2 = \frac{1}{\beta} \int_0^T dt \left[\frac{\partial e}{\partial t}(t) \right]^2 \quad . \tag{7.325}$$

In particular, this demonstrates that $C(t, t')$ is a positive definite function.[93] This result is interesting, because we see that a least-squares norm criterion associated with the covariance function (7.312) imposes that the *derivative* of the function is small (and not the function itself).

Notice that the condition (7.317) (i.e., that the function will vanish for $t = 0$) could be predicted directly from the fact that the variance at $t = 0$ is null.

7.23 The Norm Associated with the 3D Exponential Covariance

Let $\mathbf{x}$ *denote a point in the Euclidean three-dimensional space and let* $C(\mathbf{x}, \mathbf{x}')$ *be the covariance function*

$$C(\mathbf{x}, \mathbf{x}') = \sigma^2 \exp \left(-\frac{\| \mathbf{x} - \mathbf{x}' \|}{L} \right) \quad , \tag{7.326}$$

where $\| \mathbf{x} - \mathbf{x}' \|$ *denotes the Euclidean distance between points* $\mathbf{x}$ *and* $\mathbf{x}'$. *The corresponding covariance operator associates any function* $\hat{\phi}(\mathbf{x})$ *with the function*

$$\phi(\mathbf{x}) = \int_{\mathcal{V}} dV(\mathbf{x}') \, C(\mathbf{x}, \mathbf{x}') \, \hat{\phi}(\mathbf{x}') \quad . \tag{7.327}$$

Demonstrate that the inverse operator gives

$$\hat{\phi}(\mathbf{x}) \simeq \frac{1}{8\pi \sigma^2} \left(\frac{1}{L^3} \phi(\mathbf{x}) - \frac{2}{L} \Delta \phi(\mathbf{x}) + L \, \Delta \Delta \phi(\mathbf{x}) \right) \quad . \tag{7.328}$$

The least-squares norm associated with the covariance function, $\| \boldsymbol{\phi} \|^2 = \boldsymbol{\phi}^t \, \mathbf{C}^{-1} \, \boldsymbol{\phi}$, *is*

$$\| \boldsymbol{\phi} \|^2 = \int_{\mathcal{V}} dV(\mathbf{x}) \, \phi(\mathbf{x}) \, \hat{\phi}(\mathbf{x}) \quad . \tag{7.329}$$

Demonstrate that this gives

$$\| \boldsymbol{\phi} \|^2 \simeq \frac{1}{8\pi \sigma^2} \left\{ \frac{1}{L^3} \int_{\mathcal{V}} dV(\mathbf{x}) \left[\phi(\mathbf{x}) \right]^2 + \frac{2}{L} \int_{\mathcal{V}} dV(\mathbf{x}) \left[\mathrm{grad} \, \phi(\mathbf{x}) \right]^2 \right.$$
$$\left. + L \int_{\mathcal{V}} dV(\mathbf{x}) \left[\Delta \phi(\mathbf{x}) \right]^2 \right\} \quad . \tag{7.330}$$

---

[93] Because $\| \mathbf{e} \|$ is nonnegative for any $\mathbf{e}$, if $\| \mathbf{e} \|$ is null, then $e(t)$ must be constant (almost everywhere), and then it follows from equations (7.317)–(7.318) that the constant is necessarily zero.

Solution (from Georges Jobert, pers. comm.):
First, we have to solve the following equation for $\hat{\phi}(\mathbf{x})$:

$$\phi(\mathbf{x}) = \sigma^2 \int_V dV(\mathbf{x}') \exp\left(-\frac{\|\mathbf{x} - \mathbf{x}'\|}{L}\right) \hat{\phi}(\mathbf{x}') \quad . \tag{7.331}$$

Let

$$g(\mathbf{x}) = \sigma^2 \exp\left(-\frac{\|\mathbf{x}\|}{L}\right) \tag{7.332}$$

and let $\Phi(\mathbf{k})$, $\hat{\Phi}(\mathbf{k})$, and $G(\mathbf{k})$ be the Fourier transforms of $\phi(\mathbf{x})$, $\hat{\phi}(\mathbf{x})$, and $g(\mathbf{x})$, respectively. As equation (7.331) is clearly a spatial convolution,

$$\phi(\mathbf{x}) = g(\mathbf{x}) * \hat{\phi}(\mathbf{x}) \quad , \tag{7.333}$$

it becomes, in the Fourier domain,

$$\Phi(\mathbf{k}) = G(\mathbf{k}) \hat{\Phi}(\mathbf{k}) \quad . \tag{7.334}$$

This gives

$$\hat{\Phi}(\mathbf{k}) = H(\mathbf{k}) \phi(K) \quad , \tag{7.335}$$

where

$$H(\mathbf{k}) = \frac{1}{G(\mathbf{k})} \quad . \tag{7.336}$$

Letting $h(\mathbf{x})$ be the inverse Fourier transform of $H(\mathbf{k})$ gives

$$\hat{\phi}(\mathbf{x}) = h(\mathbf{x}) * \phi(\mathbf{x}) \quad , \tag{7.337}$$

which formally solves the problem. The task now is to compute $G(\mathbf{k})$ and $h(\mathbf{x})$.
Letting $k = \|\mathbf{k}\|$, we have

$$G(k) = \sigma^2 \int_V dV(\mathbf{x}) \, e^{-\|\mathbf{x}\|/L + i\,\mathbf{k}\cdot\mathbf{x}} \quad , \tag{7.338}$$

and, taking spherical coordinates with the polar axis collinear with $\mathbf{k}$,

$$G(k) = \sigma^2 \int_0^\pi d\theta \int_0^{2\pi} d\varphi \int_0^\infty dr\, r^2 \sin\theta\, e^{-r/L + i\,\|\mathbf{k}\|\,r\cos\theta}$$
$$= 2\pi\sigma^2 \int_{-1}^{+1} du \int_0^\infty dr\, r^2 e^{-r/L + i\,\|\mathbf{k}\|\,r u} \quad , \tag{7.339}$$

where $u = \cos\theta$. As $\int_{-1}^{+1} du\, e^{2i\pi\,\|\mathbf{k}\|\,r u} = 2\,\frac{\sin r\,\|\mathbf{k}\|}{r\,\|\mathbf{k}\|}$, we have

$$G(k) = 4\pi\sigma^2 \int_0^\infty dr\, r^2 e^{-r/L}\, \frac{\sin(2\pi r\,\|\mathbf{k}\|)}{r\,\|\mathbf{k}\|} = \frac{4\pi\sigma^2}{\|\mathbf{k}\|} \int_0^\infty dr\, r \sin(r\,\|\mathbf{k}\|)\, e^{-r/L} \quad . \tag{7.340}$$

Using

$$I(p) = \int_0^\infty dt\, t\, \sin t\, e^{-pt} = \frac{2\,p}{(1+p^2)^2} \quad,$$ (7.341)

with $p = \frac{1}{L\|\mathbf{k}\|}$ and $t = r\|\mathbf{k}\|$, gives

$$G(k) = \frac{8\pi\sigma^2 L^3}{(1+L^2\|\mathbf{k}\|^2)^2} \quad.$$ (7.342)

Then,

$$H(k) = \frac{1}{8\pi\sigma^2}\left(\frac{1}{L^3} + \frac{2}{L}\|\mathbf{k}\|^2 + L\|\mathbf{k}\|^4\right) \quad.$$ (7.343)

As the Fourier transform of $\delta(\mathbf{x})$ is 1, that of $\Delta\delta(\mathbf{x})$ is $-\|\mathbf{k}\|^2$, and that of $\Delta\Delta\delta(\mathbf{x})$ is $\|\mathbf{k}\|^4$, we directly obtain

$$h(\mathbf{x}) = \frac{1}{8\pi\sigma^2}\left(\frac{1}{L^3}\delta(\mathbf{x}) - \frac{2}{L}\Delta\delta(\mathbf{x}) + L\,\Delta\Delta\delta(\mathbf{x})\right) \quad.$$ (7.344)

This gives

$$\begin{aligned}
\hat\phi(\mathbf{x}) &= \int_V dV(\mathbf{x})\, h(\mathbf{x}-\mathbf{x}')\,\phi(\mathbf{x}')\\
&= \frac{1}{8\pi\sigma^2}\left\{\frac{1}{L^3}\int_V dV(\mathbf{x}')\,\delta(\mathbf{x}-\mathbf{x}')\,\phi(\mathbf{x}') - \frac{2}{L}\int_V dV(\mathbf{x}')\,\Delta\delta(\mathbf{x}-\mathbf{x}')\,\phi(\mathbf{x}')\right.\\
&\qquad\left. + L\int_V dV(\mathbf{x}')\,\Delta\Delta\delta(\mathbf{x}-\mathbf{x}')\,\phi(\mathbf{x}')\right\}\\
&= \frac{1}{8\pi\sigma^2}\left(\frac{1}{L^3}\phi(\mathbf{x}) - \frac{2}{L}\Delta\phi(\mathbf{x}) + L\,\Delta\Delta\phi(\mathbf{x})\right) \quad,
\end{aligned}$$ (7.345)

which demonstrates equation (7.328).

The (squared) norm of a function is then given by

$$\begin{aligned}
\|\phi\|^2 &= \int_V dV(\mathbf{x})\,\phi(\mathbf{x})\,\hat\phi(\mathbf{x})\\
&= \frac{1}{8\pi\sigma^2}\left\{\frac{1}{L^3}\int_V dV(\mathbf{x})[\phi(\mathbf{x})]^2 - \frac{2}{L}\int_V dV(\mathbf{x})\,\phi(\mathbf{x})\Delta\phi(\mathbf{x})\right.\\
&\qquad\left. + L\int_V dV(\mathbf{x})\,\phi(\mathbf{x})\,\Delta\Delta\phi(\mathbf{x})\right\} \quad.
\end{aligned}$$ (7.346)

Using $-\phi\,\Delta\phi = [\operatorname{grad}\phi]^2 - \operatorname{div}(\phi\operatorname{grad}\phi)$ gives

$$\begin{aligned}
-\int_V dV(\mathbf{x})\,\phi(\mathbf{x})\,\Delta\phi(\mathbf{x}) &= \int_V dV(\mathbf{x})\,[\operatorname{grad}\phi(\mathbf{x})]^2 - \int_V dV(\mathbf{x})\,\operatorname{div}(\phi(\mathbf{x})\operatorname{grad}\phi(\mathbf{x}))\\
&= \int_V dV(\mathbf{x})\,[\operatorname{grad}\phi(\mathbf{x})]^2 - \int_{\partial V} dS(\mathbf{x})\,\phi(\mathbf{x})\,\mathbf{n}(\mathbf{x})\cdot\operatorname{grad}\phi(\mathbf{x}) \quad,
\end{aligned}$$ (7.347)

where ∂V denotes the boundary of V, $\mathbf{n}(\mathbf{x})$ is the unit normal to the boundary, and $dS(\mathbf{x})$ is the element of area on the boundary. Using $\phi \Delta\Delta\phi = -\mathrm{grad}\phi \cdot \mathrm{grad}\,\Delta\phi + \mathrm{div}(\phi\,\mathrm{grad}\,\Delta\phi) = (\Delta\phi)^2 - \mathrm{div}(\Delta\phi\,\mathrm{grad}\,\phi) + \mathrm{div}(\phi\,\mathrm{grad}\,\Delta\phi)$ gives

$$
\int_V dV(\mathbf{x})\,\phi(\mathbf{x})\,\Delta\Delta\phi(\mathbf{x})
$$

$$
= \int_V dV(\mathbf{x})\,[\Delta\phi(\mathbf{x})]^2 + \int_V dV(\mathbf{x})\,\mathrm{div}(\phi(\mathbf{x})\,\mathrm{grad}\,\Delta\phi(\mathbf{x})) - \int_V dV(\mathbf{x})\,\mathrm{div}(\Delta\phi(\mathbf{x})\,\mathrm{grad}\,\phi(\mathbf{x}))
$$

$$
= \int_V dV(\mathbf{x})\,[\Delta\phi(\mathbf{x})]^2 + \int_{\partial S} dS(\mathbf{x})\,\phi(\mathbf{x})\,\mathbf{n}(\mathbf{x})\cdot\mathrm{grad}\,\Delta\phi(\mathbf{x})
$$

$$
- \int_{\partial S} dS(\mathbf{x})\,\Delta\phi(\mathbf{x})\,\mathbf{n}(\mathbf{x})\cdot\mathrm{grad}\,\phi(\mathbf{x})\,. \tag{7.348}
$$

In most practical applications, the boundary terms can be dropped. One obtains

$$
\|\boldsymbol{\phi}\|^2 \simeq \frac{1}{8\pi\sigma^2}\left\{\frac{1}{L^3}\int_V dV(\mathbf{x})\,[\phi(\mathbf{x})]^2 + \frac{2}{L}\int_V dV(\mathbf{x})\,[\mathrm{grad}\,\phi(\mathbf{x})]^2 + L\int_V dV(\mathbf{x})\,[\Delta\phi(\mathbf{x})]^2\right\}, \tag{7.349}
$$

demonstrating equation (7.330).

References and References for General Reading

Abdelmadek, N., 1977. Computing the strict Chebyshev solution of overdetermined linear equations, Math. Comput., 31, 974–983.

Abramowitz, M., and Stegun, I. A. (editors), 1970. Handbook of mathematical functions. Dover Publications Inc., New York.

Afifi, A. A., and Azen, S. P., 1979. Statistical analysis: A computer oriented approach, Academic Press, New York.

Aki, K., Christofferson, A., and Husebye, E. S., 1977. Determination of the three-dimensional seismic structure of the lithosphere, J. Geophys. Res., 82, 277–296.

Aki, K., and Lee, W. H. K., 1976. Determination of three-dimensional velocity anomalies under a seismic array using first P arrival times from local earthquakes. 1. A homogeneous initial model, J. Geophys. Res., 81, 4381–4399.

Aki, K., and Richards, P. G., 1980. Quantitative seismology (2 volumes), Freeman and Co., San Francisco.

Alterman, Z. S., and Karal, F. C., Jr., 1968. Propagation of elastic waves in layered media by finite difference methods, Bull. Seismological Soc. America, 58, 367–398.

Anderssen, R. S., and Seneta, E., 1971. A simple statistical estimation procedure for Monte-Carlo inversion in geophysics, Pure Appl. Geophys., 91, 5–13.

Anderssen, R. S., and Seneta, E., 1972. A simple statistical estimation procedure for Monte-Carlo inversion in geophysics. II: Efficiency and Hempel's paradox, Pure Appl. Geophys., 96, 5–14.

Angelier, J., Tarantola, A., Valette, B., and Manoussis, S., 1982. Inversion of field data in fault tectonics to obtain the regional stress, Geophys. J. Royal Astron. Soc., 69, 607–621.

Armstrong, R. D., and Golfrey, J. P., 1979. Two linear programming algorithms for the linear L_1 norm problem, Math. Comput., 33, 145, 289–300.

Aster, A., Borchers, B., and Thurber, C., 2003. Parameter estimation and inverse problems, in press, Academic Press, New York.

Avriel, M., 1976. Non linear programming: Analysis and methods, Prentice–Hall, Series in Automatic Computation, London.

Azlarov, T. A., and Volodin, N. A., 1982. Kharaterizatsionnye Zadachi, Sviazannye S Pokazatel'nym Raspredeleniem, Taschkent, Izdatel'stvo Fan Uzbekckoi CCR. English translation: Characterization problems associated with the exponential distribution, Springer-Verlag, New York, 1986.

Backus, G., 1970a. Inference from inadequate and inaccurate data: I, Proc. Nat. Acad. Sci., 65, 1, 1–105.

Backus, G., 1970b. Inference from inadequate and inaccurate data: II, Proc. Nat. Acad. Sci., 65, 2, 281–287.

Backus, G., 1970c. Inference from inadequate and inaccurate data: III, Proc. Nat. Acad. Sci., 67, 1, 282–289.

Backus, G., 1971. Inference from inadequate and inaccurate data, Mathematical problems in the geophysical sciences: Lectures in Applied Mathematics, 14, American Mathematical Society, Providence, RI.

Backus, G., and Gilbert, F., 1967. Numerical applications of a formalism for geophysical inverse problems, Geophys. J. Royal Astron. Soc., 13, 247–276.

Backus, G., and Gilbert, F., 1968. The resolving power of gross Earth data, Geophys. J. Royal Astron. Soc., 16, 169–205.

Backus, G., and Gilbert, F., 1970. Uniqueness in the inversion of inaccurate gross Earth data, Philos. Trans. Royal Soc. London, 266, 123–192.

Bakhvalov, N. S., 1977. Numerical methods, Mir Publishers, Moscow.

Balakrishnan, A. V., 1976. Applied functional analysis, Springer-Verlag, New York.

Bamberger, A., Chavent, G, Hemon, Ch., and Lailly, P., 1982. Inversion of normal incidence seismograms, Geophys., 47, 757–770.

Barnes, C., Charara, M., and Tarantola, A., 1998. Monte Carlo inversion of arrival times for multiple phases in OSVP data, 68th Ann. Internat. Mtg. Soc. Expl. Geophys., pp. 1871–1874.

Barrodale I., 1970. On computing best L_1 approximations, in: Approximation theory, edited by A. Talbot, Academic Press, New York.

Barrodale, I., and Phillips, C., 1975a. An improved algorithm for discrete Chebyshev linear approximation, in: Proceedings of the Fourth Manitoba Conference on Numerical Mathematics, edited by B. L. Hartnell and H. C. Williams, Utilitas Mathematics Publishing.

Barrodale, I., and Phillips, C., 1975b. Algorithm 495: Solution of an overdetermined system of linear equations in the Chebyshev norm, A.C.M. Trans. Math. Software, 1, 264–270.

Barrodale, I., and Roberts, F. D. K., 1973. An improved algorithm for discrete L_1 linear approximation, SIAM J. Numer. Anal., 10, 839–848.

Barrodale, I., and Roberts, F. D. K., 1974. Algorithm 478: Solution of an overdetermined system of equations in the ℓ_1 norm, Corn. ACM, 14, 319–320.

Barrodale, I., and Young, A., 1966. Algorithms for best L_1 and L_∞ linear approximations on a discrete set, Numer. Math., 8, 295–306.

Bartels, R. H., 1971. A stabilisation of the simplex method, Numer. Math., 16, 414–434.

Bartels, R. H., 1980. A penalty linear programming method using reduced gradient basis exchanges techniques, Linear Algebra Appl., 29, 17–32.

Bartels, R. H., and Conn, A. R., 1981. An approach to nonlinear ℓ_1 data fitting, in: Proceedings of the third Mexican workshop on numerical analysis, J. P. Hennart (editor), Springer-Verlag, New York.

Bartels, R. H., Conn, A. R., and Charalambous, C., 1978. On Cline's direct method for solving overdetermined linear systems in the ℓ_∞ sense, SIAM J. Numer. Anal., 15, 255–270.

Bartels, R. H., Conn, A. R., and Sinclair, J. W., 1978. Minimization techniques for piecewise differentiable functions: The ℓ_1 solution to an overdetermined linear system, SIAM J. Numer. Anal., 15, 224–241.

Bartels, R. H., and Golub, G. H., 1969. The simplex method of linear programming using LU decomposition, Corn. ACM, 12, 206–268.

Bartels, R. H., Stoer, J., and Zenger, C. H., 1971. A realization of the simplex method based on triangular decomposition, in: Handbook for automatic computation, J. H. Wilkinson and C. Reinsch (editors), Springer-Verlag, New York.

Bayer, R., and Cuer, M., 1981. Inversion tri-dimensionnelle des donnees aéromagnétiques sur le massif volcanique du Mont-Dore: Implications structurales et géothermiques, Ann. Géophys., t. 37, fasc. 2, 347–365.

Bayes, Thomas (Reverend), (1702–1761), 1958. Essay towards solving a problem in the doctrine of chances, 1763, republished in Biometrika, 45, 298–315.

Ben-Menahem, A., and Singh, S. J., 1981. Seismic waves and sources, Springer-Verlag, New York.

Bender, C. M., and Orszag, S. A., 1978. Advanced mathematical methods for scientists and engineers, McGraw-Hill, New York.

Berkhout, A. J., 1980. Seismic migration: Imaging of acoustic energy by wave field extrapolation, Elsevier, Amsterdam.

Beydoun, W., 1985. Asymptotic Wave Methods in Heterogeneous Media, Ph.D Thesis, Massachusetts Institute of Technology.

Bierman, G. J., 1977. Factorization methods for discrete sequential estimation, Academic Press, New York.

Binder, K. (editor), 1979. Monte Carlo methods in statistical physics, Springer-Verlag, Berlin.

Binder, K., 1984. Applications of the Monte Carlo method in statistical physics, Springer-Verlag, Berlin.

Bland, R. G., 1976. New Finite Pivoting Rules for the Simplex Method, CORE-research paper 7612.

Bleistein, N., 1984. Mathematical methods for wave phenomena, Academic Press, New York.

Bloomfield, P., and Steiger, W. L., 1980. Least absolute deviations curve-fitting, SIAM J. Sci. Statist. Comput., 1, 290–301.

Bloomfield, P., and Steiger, W. L., 1983. Least absolute deviations, theory, applications, and algorithms, Birkhäuser, Boston.

Boothby, W. M., 1975. An introduction to differentiable manifolds and Riemannian geometry, Academic Press, New York.

Bordley, R. F., 1982. A multiplicative formula for aggregating probability assessments, Manag. Sci., 28, 1137–1148.

Box, G. E. P., Leonard, T., and Chien-Fu Wu (editors), 1983. Scientific inference, data analysis, and robustness, Academic Press, New York.

Box, G. E. P., and Tiao, G. C., 1973. Bayesian inference in statistical analysis, Addison–Wesley, Reading, MA.

Bradley, S. P., Hax, A. C., and Magnanti, T. L., 1977. Applied mathematical programming, Addison–Wesley, Reading, MA.

Brecis, H., 1983. Analyse fonctionnelle, théorie et applications, Masson, Paris.

Brillouin, L., 1962. Science and information theory, Academic Press, New York.

Broyden, C. G., 1967. Quasi-Newton methods and their application to function minimization, Math. Comput., 21, 368–381.

Bunch, J. R., and Rose, D. J. (editors), 1976. Sparse matrix computations, Academic Press, New York.

Caers, J. K., Srinivasan, S., and Journel, A. G., 2000. Geostatistical quantification of geological information for a fluvial-type North Sea reservoir, Society of Petroleum Engineers, Reservoir Evaluation and Engineering, 3, 5, 457–467.

Campbell, S. L., Meyer, C. D., and Rosi, N. J., 1981. Results of an informal survey on the use of linear algebra in industry and government, Linear Algebra Appl., 38, 289–294.

Cao, D., Beydoun, W. D., Singh, S. C., and Tarantola, A., 1990. A simultaneous inversion for background velocity and impedance maps. Geophys., 55, 458–469.

Carroll, Lewis (Rev. Charles Lutwidge Dodgson) (1832–1898), 1871. Alice's adventures in Wonderland, Macmillan, London. I recommend to the reader *The annotated Alice*, by Martin Gardner, Penguin Books, 1970.

Cartan, H., 1967. Cours de calcul différentiel, Hermann, Paris.

Céa, J., 1971. Optimisation, théorie et algorithmes, Dunod, Paris.

Censor, Y., and Elfving, T., 1982. New methods for linear inequalities, Linear Algebra Appl., 42, 199–211.

Censor, Y., and Herman, G. T., 1979. Row-generation methods for feasibility and optimization problems involving sparse matrices and their applications, in: Sparse matrix proceedings 1978, I. S. Duff and G. W. Stewart (editors), SIAM, Philadelphia.

Chapman, N. R., and Barrodale, I., 1983. Deconvolution of marine seismic data using L_1 norm, Geophys. J. Royal Astr. Soc., 72, 93–100.

Charara, M., Barnes, C., and Tarantola, A., 1996. The state of affairs in inversion of seismic data: An OVSP example, 66th Ann. Internat. Mtg. Soc. Expl. Geophys., pp. 1999–2002.

Charara, M., Barnes, C., and Tarantola, A., 2000. Full waveform inversion of seismic data for a visco-elastic medium, in: Methods and Applications of Inversion, Lecture Notes in Earth Sciences, 92, Springer-Verlag, New York.

Chernov, L. A., 1960. Wave propagation in a random medium, Dover Publications Inc., New York.

Ciarlet, P. G., 1982. Introduction à l'analyse numérique matrcielle et à l'optimisation, Masson, Paris.

Ciarlet, P. G., and Thomas, J. M., 1982. Exercices d'analyse numérique matricielle et d'optimisation, Masson, Paris.

Cimino, G., 1938. Calcolo approximato per le solusioni del sistemi di equazioni lineari, La Ricerca Scientifica, Roma XVI, Ser. II, Anno IX, Vol. 1, 326–333.

Claerbout, J. F., 1971. Toward a unified theory of reflector mapping, Geophys., 36, 467–481.

Claerbout, J. F., 1976. Fundamentals of geophysical data processing, McGraw-Hill, New York.

Claerbout, J. F., 1985. Imaging the Earth's interior, Blackwell, Oxford, U.K.

Claerbout, J. F., and Muir, F., 1973. Robust modelling with erratic data, Geophys., 38, 826–844.

Cottle, R. W., and Dantzig, G. B., 1968. Complementary pivot theory of mathematical programming, Linear Anal. Appl., 1, 103–125.

Courant, R., and Hilbert, D., 1966. Methods of mathematical physics, Interscience Publishers, New York.

Crase, E., Pica, A., Noble, M., McDonald, J., and Tarantola, A., 1990. Robust elastic nonlinear waveform inversion: Application to real data. Geophys., 55, 527–538.

Crase, E., Wideman, Ch., Noble, M., and Tarantola, A., 1992. Nonlinear elastic waveform inversion of land seismic reflection data. J. Geophys. Res., 97, 4685–4703.

Cuer, M., 1984. Des questions bien posées dans des problèmes inverses de gravimétrie et géomagnétisme. Une nouvelle application de la programmation linéaire, Thesis, Université des Sciences et Techniques du Languedoc, UER de Mathématiques, 34060 Montpellier, France.

Cuer, M., and Bayer, R., 1980a. A package of routines for linear inverse problems, Cahiers Mathématiques, Université des Sciences et Techniques du Languedoc, UER de Mathématiques, 34060 Montpellier, France.

Cuer, M., and Bayer, R., 1980b. Fortran routines for linear inverse problems, Geophys., 45, 1706–1779.

Dahl-Jensen, D., Mosegaard, K., Gundestrup, N., Clow, G. D., Johnsen, S. J., Hansen, A. W., and Balling, N., 1998. Past temperatures directly from the Greenland Ice Sheet, Science, Oct. 9, 268–271.

Dahlquist, B., and Björk, Å., 1974. Numerical methods, Prentice–Hall, Englewood Cliffs, NJ.

Dantzig, G. B., 1963. Linear programming and extensions, Princeton University Press, Princeton, NJ.

Dautray, R., and Lions, J. L., 1984 and 1985. Analyse mathématique et calcul numérique pour les sciences et les techniques (3 volumes), Masson, Paris.

Davidon, W. C., 1959. Variable Metric Method for Minimization, AEC Res. and Dev., Report ANL–5990 (revised).

Davidon, W. C., 1968. Variance algorithms for minimization, Comput. J., 10, 406–410.

de Ghellinck, G., and Vial, J. Ph., 1985. An Extension of Karmarkar's Algorithm for Solving a System of Linear Homogeneous Equations on the Simplex (preprint), ISSN 0771–3894.

de Ghellinck, G., and Vial, J. Ph., 1986. A Polynomial Newton Method for Linear Programming (preprint), CORE, Universite de Louvain.

Denel, J., Fiorot, J. C., and Huard, P., 1981. The steepest ascent method for linear programming problems, RAIRO Analyse Numérique, 15, 3, 195–200.

Descloux, J., 1963. Approximations in L_p and Chebyshev approximations, J. Soc. Indust. Appl. Math. 11, 1017–1026.

Deutsch, C. V., and Journel, A. G., 1998. GSLIB, Geostatistical software library and user's guide, Oxford University Press, New York.

Devaney, A. J., 1984. Geophysical diffraction tomography, IEEE Trans. Geos. Remote Sensing, GE-22, 3–13.

Dickson, J. C., and Frederick, F. P., 1960. A decision rule for improved efficiency in solving linear problems with the simplex algorithm, Comm. ACM, 3, 509–512.

Dietrich, C. F., 1991. Uncertainty, calibration and probability — the statistics of scientific and industrial measurement, Adam Hilger, Bristol, U.K.

Dixmier, J., 1969. Cours de mathématiques du premier cycle (2 volumes), Gauthier-Villars, Paris.

Dixon, L. C. W. (editor), 1976. Optimization in action, Academic Press, London.

Djikpéssé, H.A., and Tarantola, A., 1999. Multiparameter ℓ_1 norm waveform fiting: Interpretation of Gulf of Mexico reflection seismograms. Geophys., 64, 1023–1035.

Draper, N, and Smith, H., 1998. Applied regression analysis (third edition), Wiley, New York.

Dubes, R. C., 1968. The theory of applied probability, Prentice-Hall, Englewood Cliffs, NJ.

Duijndam, A. J. W., 1987. Detailed Inversion of Seismic Data, Ph.D. Thesis, Delft University of Technology, 1987.

Ecker, J. G., and Kupferschmid, M., 1985. A computational comparison of the ellipsoid algorithm with several nonlinear programming algorithms, SIAM J. Control Optim., 23, 657–674.

Edgeworth, F. Y., 1887. A new method of reducing observations relating to several quantities, Philos. Mag. (5th Ser.), 24, 222–223.

Edgeworth, F. Y., 1888. On a new method of reducing observations relating to several quantities, Phil. Mag. (5th. Ser.), 25, 184–191.

Ekblom, H., 1973. Calculation of linear best L_p approximations, BIT, 13, 292–300.

Fenton, G.A., 1990. Simulation and Analysis of Random Fields. Ph.D. thesis, Department of Civil Engineering and Operations Research, Princeton University.

Fenton, G. A., 1994. Error evaluation of three random field generators, ASCE J. Engineering Mech., 120(12), 2478–2497.

Fiacco, A. V., and McCormick, G. P., 1968. Nonlinear programming, Wiley, New York.

Fisher, R. A., 1953. Disperson on a sphere, Proc. R. Soc. London, A127, 295–305.

Fletcher, R., 1980. Practical methods of optimization, Volume 1: Unconstrained optimization, Wiley, New York.

Fletcher, R., 1981. Practical methods of optimization, Volume 2: Constrained optimization, Wiley, New York.

Fletcher, R., and Powell, M. J. D., 1963. A rapidly convergent descent method for minimization, Comput. J., 6, 163–168.

Fletcher, R., and Reeves, C. M., 1964. Function minimization by conjugate gradients, Comput. J., 7, 149–154.

Fox, L., 1964. An introduction to numerical linear algebra, Clarendon Press, Oxford, U.K.

Frankel, A., and Clayton, R. W., 1986. Finite-difference simulations of seismic scattering, J. Geophys. Res., 91 B, 6465–6489.

Franklin, J. N., 1970. Well posed stochastic extensions of ill posed linear problems, J. Math. Anal. Appl., 31, 682–716.

Gacs, P., and Lovasz, L., 1981. Khachiyan's algorithm for linear programming, in: Mathematical programming study, H. Konig, B. Korte, and K. Ritter (editors), North–Holland, Amsterdam.

Gale, D., 1960. The theory of linear economic models, McGraw-Hill, New York.

Gass, S. I., 1975. Linear programming, methods and applications, fourth edition, McGraw–Hill, New York.

Gauthier, O., Virieux, J., and Tarantola, A., 1986. Two-dimensional inversion of seismic waveforms: Numerical results, Geophys., 51, 1387–1403.

Gauss, Carl Friedrich (1777–1855), 1809. Theoria Motus Corporum Coelestium.

Gauss, Carl Friedrich (1777–1855), ca. 1820. Theory of the combination of observations least subject to errors. English translation, Classics Appl. Math. 11, SIAM, Philadelphia, 1995.

Gel'fand I. M., and Levitan, B. M., 1955. On the determination of a differential equation by its spectral function, Amer. Math. Soc. Transl., 1, Ser. 2, 253–304 (translated from the Russian paper of 1951).

Geman, S., and Geman, D., 1984. Stochastic relaxation, Gibbs distributions, and the Bayesian restoration of images, IEEE Trans. Pattern Anal. Machine Intelligence, 6, 721–741.

Genest, C., and Zidek, J. V., 1986. Combining probability distributions: A critique and an annotated bibliography, Stat. Sci., 1, 114–148.

Genet, J., 1976. Mesure et intégration, théorie élémentaire, Vuibert, Paris.

Geweke, J., 1992. Evaluating the accuracy of sampling-based approaches to the calculation of posterior moments, in: Bayesian Statistics, 4, Bernardo, J. M., Berger, J. O., Dawid, A. P., and Smith, A. F. M. (editors), Oxford University Press, Oxford, U.K., 169–193.

Gill, P. E., and Murray, W., 1973. A numerical stable form of the simplex algorithm, Linear Algebra Appl., 7, 99–138.

Gill, P. E., Murray, W., Saunders, M. A., and Wright, M. H., 1984. Trends in nonlinear programming software, Europ. J. Oper. Res., 17, 141–149.

Gill, P. E., Murray, W., and Wright, M. H., 1981. Practical optimization, Academic Press, London.

Glashoff, K., and Gustafson, S. A., 1983. Linear optimization and approximation, An introduction to the theoretical analysis and numerical treatment of semi-infinite programs, Springer-Verlag, New York.

Goldberg, D. E., 1989. Genetic algorithms in search, optimization and machine learning, Addison–Wesley, Reading, MA.

Goldfarb, D., 1976. Using the steepest edge simplex algorithm to solve linear programs, in: Sparse matrix computations, Bunch, J. R., and Rose, D. J. (editors), Academic Press, New York.

Goovaerts, P., 1997. Geostatistics for natural resources evaluation, Oxford University Press, New York.

Grasso, J. R., Cuer, M., and Pascal, G., 1983. Use of two inverse techniques. Application to a local structure in the New Hebrides island arc, Geophys. J. Royal Astr. Soc., 75, 437–472.

Groetsch, C. W., 1999. Inverse Problems, The Mathematical Association of America, Washington, DC.

Gutenberg, B., and Richter, C. F., 1939. On seismic waves, G. Beitr., 54, 94–136.

Guyon, R., 1963. Calcul tensoriel, Vuibert, Paris.

Hacijan, L. G., 1979. A polynomial algorithm in linear programming, Soviet Math. Dokl., 20, 1, 191–194.

Hadamard, J., 1902. Sur les problèmes aux dérivées partielles et leur signification physique, Princeton University Bulletin, 49–52.

Hadamard, J., 1932. Le problème de Cauchy et les équations aux dérivées partielles linéaires hyperboliques, Hermann, Paris.

Hadley, G., 1965. Linear programming, Addison–Wesley, Reading, MA.

Hampel, F. R., 1978. Modern trends in the theory of robustness, Math. Oper. Stat., Ser. Stat., 9, 3, 425–442.

Hammersley, J. M., and Handscomb, D. C., 1964. Monte-Carlo methods, Chapman and Hall, London.

Hanson, R. J., and Wisnieski, J. A., 1979. A mathematical programming updating method using modified Givens transformations and applied to LP problems, Comm. ACM, 22, 245–250.

Harris, P. M. J., 1975. Pivot selection methods of the Devex LP code, in: Mathematical programming study 4, M. L. Balinski and E. Hellerman (editors), North–Holland, Amsterdam.

Hastings, W. K., 1970. Monte Carlo sampling methods using Markov Chains and their applications, Biometrika, 57, 97–109.

Herman, G. T., 1980. Image reconstruction from projections, the fundamentals of computerized tomography, Academic Press, San Francisco.

Hestenes, M. R., and Stiefel, E. L., 1952. Methods of conjugate gradients for solving linear systems, J. Res. Nat. Bur. Standards, sect. 5, vol. 49, 409–436.

Hirn, A., Jobert, G., Wittlinger, G., Xu Zhong-Xin, and Gao En-Yuan, 1984. Main features of the upper lithosphere in the unit between the high Himalayas and the Yarlung Zangbo Jiang suture, Ann. Geophys., 2, 113–118.

Ho, J. K., 1978. Pricing for sparsity in the revised simplex method, RAIRO, 12, 3, 285–290.

Hoel, P. G., 1947. Introduction to mathematical statistics, Wiley, New York.

Hofstadter, D., 1979. Gödel, Escher, Bach: An eternal golden braid, Basic Books, New York.

Huard, P., 1979. La methode simplex sans inverse explicite, Bull. Dir. Etud. Rech. EDF, Série C, 79–98.

Huard, P., 1980. Complements concernant la méthode des paramètres, Bull. Dir. Etud. Rech. EDF, Série Math. Info., 2, 63–67.

Huber, P. J., 1977. Robust statistical procedures, CBMS-NSF Regional Conf. Ser. in Appl. Math. 27, SIAM, Philadelphia.

Huber, P. J., 1981. Robust statistics, Wiley, New York.

Igel, H., Djikpéssé, H., and Tarantola, A., 1996. Waveform inversion of marine reflection seismograms for P impedance and Poisson's ratio. Geophys. J. Int., 124, 363–371.

Ikelle, L. T., Diet, J. P., and Tarantola, A., 1986. Linearized inversion of multi offset seismic reflection data in the ω-k domain, Geophys., 51, 1266–1276.

Jackson, D. D., 1972. Interpretation of inaccurate, insufficient and inconsistent data, Geophys. J. Royal Astr. Soc., 28, 97–110.

Jackson, D. D., 1979. The use of a priori data to resolve nonuniqueness in linear inversion, Geophys. J. Royal Astr. Soc., 57, 137–157.

Jackson, D. D., and Matsu'ura, M., 1985. A Bayesian approach to nonlinear inversion, J. Geophys. Res., 90, Bl, 581–591.

Jaynes, E. T., 1968. Prior probabilities, IEEE Trans. Systems Sci. Cybernetics, SSC-4, 3, 227–241.

Jaynes, E. T., 2003. Probability theory, the logic of science. Cambridge University Press, Cambridge, U.K.

Jeffreys, H., 1939. Theory of probability, Oxford University Press, Oxford, U.K.

Jeffreys, H., 1957. Scientific inference, Cambridge University Press, London.

Jeroslow, R. G., 1973. The simplex algorithm with the pivot rule of maximizing criterion improvement, Discrete Math. 4, 367–378.

Jerrum, M., 1994. The Computational Complexity of Counting, LFCS report, ECS-LFCS-94-296. Available at http://www.dcs.ed.ac.uk/home/mrj/.

Johnson, G. R., and Olhoeft, G. R., 1984. Density of rocks and minerals, in: CRC handbook of physical properties of rocks, edited by Carmichael, R. S., CRC, Boca Raton, FL.

Journel, A. G., 2002. Combining knowledge from diverse sources: An alternative to traditional data independence hypotheses, Math. Geol., 34, 5, 573–596.

Journel, A. G., and Huijbregts, Ch. J., 1978. Mining geostatistics, Academic Press, London.

Kagan, A. M., Linnink, Yu. V., and Rao, C. R., 1973. Characterization problems in mathematical statistics, Wiley, New York.

Kalman, R. E., 1960. A new approach to linear filtering and prediction problems, Trans. ASME–J. Basic Engineering, 82, D, 35–45.

Kalos, M. H., and Whitlock, P. A., 1986. Monte Carlo methods, John Wiley & Sons, New York.

Kantorovitch, L., and Akilov, G., 1977. Functional analysis, Vol. 1 and Vol. 2 (in Russian), Nauka, Moscow.

Karmarkar, N., 1984. A new polynomial-time algorithm for linear programming, Combinnatorica, Vol. 4, No. 4, 373–395.

Kauffman, A., 1977. Introduction à la théorie des sous-ensembles flous, Masson, Paris.

Keilis-Borok, V. J., 1971. The inverse problem of seismology, Proceedings of the International School of Physics Enrico Fermi, Course L, Mantle and Core in Planetary Physics, J. Coulomb and M. Caputo (editors), Academic Press, New York.

Keilis-Borok, V. J., Levshin, A., and Valus, V., 1968. S-wave velocities in the upper mantle in Europe; report on the IV International Symposium on Geophysical Theory and Computers 1967, Dokladi Akademii Nauk SSSR, 185, 3, 564.

Keilis-Borok, V. J., and Yanovskaya, T. B., 1967. Inverse problems of seismology (structural review), Geophys. J. Royal Astr. Soc., 13, 223–234.

Keller, J. B., 1978. Rays, waves, and asymptotics, Bull. AMS, 54, 5, 727–750.

Kennett, B., and Nolet, G., 1978. Resolution analysis for discrete systems, Geophys. J. Royal Astr. Soc., 53, 413–425.

Kennett, B. L. N., 1978. Some aspects of non-linearity in inversion, Geophys. J. Royal Astr. Soc., 55, 373–391.

Kirkpatrick, S., Gelatt, C. D., Jr., and Vecchi, M. P., 1983. Optimization by simulated annealing, Science, 220, 671–680.

Klee V., and Minty, G. J., 1972. How good is the simplex algorithm?, in: Inequalities III, Shisha, O. (editor), Academic Press, New York.

Kolmogoroff, A. N., 1956. Foundations of the theory of probability, Chelsea, New York.

Kônig, H., and Pallaschke, D., 1981. On Krachian's algorithm and minimal ellipsoids, Numer. Math., 36, 211–223.

Koren, Z., Mosegaard, K., Landa, E., Thore, P. and Tarantola, A., 1991. Monte Carlo estimation and resolution analysis of seismic background velocities, J. Geophys. Res., 96, B12, 20289–20299.

Kuhn, H. W., and Quandt, R. E., 1962. An experimantal study of the simplex method, Proc. 15th Symp. Applied Mathematics of the American Mathematical Society, 107–124.

Kullback, S., 1959. Information theory and statistics, Wiley, New York.

Kullback, S., 1967. The two concepts of information, J. Amer. Stat. Assoc., 62, 685–686.

Lanczos C., 1957. Applied analysis, Prentice–Hall, Englewood Cliffs, NJ.

Landa, J. L., and Guyaguler, B., 2003. A Methodology for History Matching and the Assessment of Uncertainties Associated with Flow Prediction, paper SPE 84465, SPE Annual Technical Conference & Exhibition (5–8 October 2003), Denver, CO.

Landau, L. D., and Lifshitz, E. M., 1986. Theory of elasticity (3rd. edition), Pergamon Press, Oxford, U.K.

Lang, S., 1962. Introduction to differentiable manifolds, Interscience Publishers, New York.

Laplace, Pierre Simon (Marquis de), 1799. Mécanique céleste, Tome III, No. 39.

Laplace, Pierre Simon (Marquis de), 1812. Théorie analytique des probabilités, livre 2.

Lavrent'ev, M. M., Romanov, V. G., and Sisatskij, S. P., 1980. Nekorrektnye zadachi matematicheskoi fisiki i analisa (Ill posed problems in mathematical physics and analysis) (in Russian), Nauka, Moscow. Translated in Italian: Pubblicazioni dell'Istituto di Analisi Globale e Applicazioni, Firenze, 1983.

Lawson, Ch. L., and Hanson, R. J., 1974. Solving least squares problems, Prentice–Hall, Englewood Cliffs, NJ.

Levenberg, K., 1944. A method for the solution of certain nonlinear problems in least squares, Quart. Appl. Math., 2, 164–168.

Levitan, B. M., 1962. Generalized translation operators and their applications (in Russian), Fiz. Mat. Gosudarstv. Izdat., Moscow.

Licknerowicz, A., 1960. Eléments de Calcul Tensoriel, Armand Collin, Paris.

Lions, J. L., 1968. Contrôle optimal de systèmes gouvernés par des équations aux dérivées partielles, Dunod, Paris. English translation: Optimal control of systems governed by partial differential equations, Springer-Verlag, Berlin, 1971.

Lions, J. L., 1974. Sur la théorie du contrôle, Actes du Congrès International des Mathématiciens, Vancouver, 2, 139–154.

Luenberger, D. G., 1969. Optimization methods by vector space methods, Wiley, New York.

Luenberger, D. G., 1973. Introduction to linear and nonlinear programming, Addison–Wesley, Reading, MA.

Magnanti, T. L., 1976. Optimization for sparse systems, in: Sparse matrix computations, Bunch, J. R., and Rose, D. J. (editors), Academic Press, New York.

Mandelbaum, A., 1984. Linear estimators and measurable linear transformations on a Hilbert space, Z. Wahrscheinlichkeitstheorie verw. Gebiete, 65, 385–397.

Mandelbrot, B. B., 1977a. Fractals: Form, chance, and dimension, Freeman, San Francisco.

Mandelbrot, B. B., 1977b. The Fractal geometry of nature, Freeman, San Francisco.

Mangasarian, O. L., 1981. Iterative solution of linear programs, SIAM J. Numer. Anal., 18, 606–614.

Mangasarian, O. L., 1983. Least norm linear programming solution as an unconstrained minimization problem, J. Math. Anal. Appl., 92, 240–251.

Marcenko, V. A., 1978. Sturm-Liouville operators and their applications (in Russian), Naukova Dumka, Kiev.

Marquardt, D. W., 1963. An algorithm for least-squares estimation of nonlinear parameters, J. Soc. Indust. Appl. Math., 11, 431–441.

Marquardt, D. W., 1970. Generalized inverses, ridge regression, biased linear estimation and non-linear estimation, Technometrics, 12, 591–612.

Martz, H. F., and Wailer, R. A., 1982. Bayesian reliability analysis, Wiley, New York.

Matsu'ura, M., and Hirata, N., 1982. Generalized least-squares solutions to quasi-linear inverse problems with a priori information, J. Phys. Earth, 30, 451–468.

McCall, E. H., 1982. Performance results of the simplex algorithm for a set of real world linear programming models, Comm. ACM, 25, 3, 207–212.

McNutt, M. K., and Royden, L., 1987. Extremal bounds on geotherms in eroding mountain belts from metamorphic pressure-temperature conditions, Geophys. J. Royal Astr. Soc., 88, 81–95.

McNutt, M. K., 1987. Temperature Beneath Midplate Swells: The Inverse Problem, submitted to Seamounts, Islands, and Atolls, Geophys. Monogr. Ser., 43, American Geophysical Union, Washington, DC.

Meissl, P., 1976. Hilbert spaces and their applications to geodetic leastsquares problems, Boll. Geod. Sci. Affini, 35, 49–80.

Menke, W., 1989. Geophysical data analysis: Discrete inverse theory, Academic Press, San Diego.

Meschkowsky, H., 1962. Hilbertsche räume mit kernfunction, Springer, Berlin.

Metropolis, N., Rosenbluth, A., Rosenbluth, M., Teller, A., and Teller, E., 1953. Equation of state calculations by fast computing machines, J. Chem. Phys., 21, 1081–1092.

Metropolis, N., and Ulam, S., 1949, The Monte Carlo method, J. Amer. Stat. Assoc., 44, 335–341.

Miller, D. (editor), 1985. Popper selections, Princeton University Press, Princeton, NJ.

Milne, R. D., 1980. Applied functional analysis, Pitman Advanced Publishing Program, Boston.

Minster, J. B., and Jordan, T. M., 1978. Present-day plate motions, J. Geophys. Res., 83, 5331–5354.

Minster, J. B., Jordan, T. H., Molnar, P., and Haines, E., 1974. Numerical modelling of instantaneous plate tectonics, Geophys. J. Royal Astr. Soc., 36, 541–576.

Misner, Ch. W., Thorne, K. S., and Wheeler, J. A., 1973. Gravitation, Freeman, San Francisco.

Morgan, B. W., 1968. An introduction to Bayesian statistical decision processes, Prentice–Hall, Englewood Cliffs, NJ.

Moritz, H., 1980. Advanced physical geodesy, Herbert Wichmann Verlag, Karlsruhe, Abacus Press, Tunbridge Wells, Kent.

Moritz, H., and Sünkel, H., 1978. Approximation methods in geodesy, H. Wichmann, Karlsruhe.

Morse, P. M., and Feshbach, H., 1953. Methods of theoretical physics, McGraw–Hill, New York.

Mosegaard, K., and Tarantola, A., 1995. Monte Carlo sampling of solutions to inverse problems. J. Geophys. Res., 100, B7, 12431–12447.

Mosegaard, K., and Tarantola, A., 2002. Probabilistic approach to inverse problems, International Handbook of Earthquake & Engineering Seismology, Part A, Academic Press, New York, pp. 237–265.

Müller, G., and Kind, R., 1976. Observed and computed seismogram sections for the whole Earth, Geophys. J. Royal Astr. Soc., 44, 699–716.

Murtagh, B. A., and Sargent, R. W. H., 1969. A constrained minimization method with quadratic convergence, in: Optimization, Fletcher, R. (editor), Academic Press, London.

Murty, K., 1976. Linear and combinatorial programming, Wiley, New York.

Narasimhan, R., 1968. Analysis on real and complex manifolds, Masson, Paris; North–Holland, Amsterdam.

Nash, S. G., and Sofer, A., 1996. Linear and nonlinear programming, McGraw–Hill, New York.

Nazareth, L., 1984. Numerical behaviour of LP algorithms based upon the decomposition principle, Linear Algebra Appl., 57, 181–189.

Nercessian, A., Hirn, A., and Tarantola, A., 1984. Three-dimensional seismic transmission prospecting of the Mont-Dore volcano, France, Geophys. J. Royal Astr. Soc., 76, 307–315.

Nering E. D., 1970. Linear algebra and matrix theory (second edition), Wiley, New York.

Nolet, G., 1981. Linearized inversion of (teleseismic) data, in: The solution of the inverse problem in geophysical interpretation, R. Cassinis (editor), Plenum Press, New York, pp. 9–37.

Nolet, G., 1985. Solving or resolving inadequate and noisy tomographic systems, J. Comput. Phys., 61, 463–482.

Oppenheim, A. V. (editor) 1978. Applications of digital signal processing, Prentice–Hall, Englewood Cliffs, NJ.

Oran Brigham, E., 1974. The fast Fourier transform, Prentice–Hall, Englewood Cliffs, NJ.

Parker, R. L., 1975. The theory of ideal bodies for gravity interpretation, Geophys. J. Royal Astron. Soc., 42, 315–334.

Parker, R. L., 1977. Understanding inverse theory, Ann. Rev. Earth Plan. Sci., 5, 35–64.

Parker, R. L., 1994. Geophysical inverse theory, Princeton University Press, Princeton, NJ.

Parzen, E., 1959. Statistical inference on time series by Hilbert space methods, I, reprinted in Time Series Analysis Papers, Holden-Day, San Francisco, pp. 251–282.

Pica, A., Diet, J. P., and Tarantola, A., 1990. Nonlinear inversion of seismic reflection data in a laterally invariant medium, Geophys., 55, 284–292.

Plackett, R. L., 1972. Studies in the history of probability and statistics; chapter 29, The discovery of the method of least squares, Biometrika, 59, 239–251.

Poincaré, H., 1929. La Science et l'Hypothèse, Flammarion, Paris.

Polack, E., and Ribière, G., 1969. Note sur la convergence de méthodes de directions conjuguées, Revue Fr. Inf. Rech. Oper., 16–R1, 35–43.

Popper, K., 1959. The logic of scientific discovery (translation of Logik der Forschung), Hutchinson, London.

Powell, M. J. D., 1977. Restart procedures for the conjugate gradient method, Mathematical Programming, 12, 241–254.

Powell, M. J. D., 1981. Approximation theory and methods, Cambridge University Press, Cambridge, U.K.

Press, F., 1968. Earth models obtained by Monte-Carlo inversion, J. Geophys. Res., 73, 16, 5223–5234.

Press, F., 1971. An introduction to Earth structure and seismotectonics, Proceedings of the International School of Physics Enrico Fermi, Course L, Mantle and Core in Planetary Physics, Coulomb, J., and Caputo, M. (editors), Academic Press, New York.

Press, W. H., Flannery, B. P., Teutolsky, S. A., and Vetterling, W. T., 1986. Numerical recipes: The art of scientific computing, Cambridge University Press, Cambridge, U.K. (See also Vetterling et al., 1986).

Price, W. L., 1977. A controlled random search procedure for global optimization, Comput. J., 20, 367–370.

Pugachev, V. S., 1965. Theory of random functions and its application to control problems, Pergamon Press, Oxford, U.K. (translation of the original Russian Teoriya sluchainykh funktsii (second edition), Fizmatgiz, Moscow, 1962).

Raftery, A. E., and Lewis, S., 1992. How many iterations in the Gibbs sampler?, in: Bayesian statistics, 4, Bernardo, J. M., Berger, J. O., Dawid, A. P., and Smith, A. F. M. (editors), 763–773, Oxford University Press, Oxford, U.K.

Rand Corporation, 1955. A million random digits with 100,000 normal deviates, The Free Press, Glencoe, IL.

Rao, C. R., 1973. Linear statistical inference and its applications, Wiley, New York.

Rauhala, U. A., 2002. Array algebra expansion of matrix and tensor calculus (parts 1 and 2), SIAM J. Matrix Anal. Appl., 24, 490–528.

Ray Smith, C., and Grandy, W. T., Jr. (editors), 1985. Maximum-entropy and Bayesian methods in inverse problems, Reidel, Boston.

Rebbi, C., 1984. Monte Carlo calculations in lattice gauge theories, in: Applications of the Monte Carlo method, Binder, K. (editor), Springer-Verlag, Berlin, pp. 277–298.

Reid, J. K., 1977. Sparse matrix, in: The state of the art in numerical analysis, Jacobs, D. (editor), Academic Press, London.

Richard, V., Bayer, R., and Cuer, M., 1984. An attempt to formulate wellposed questions in gravity: Application of linear inverse techniques to mining exploration, Geophys., 49, 1781–1793.

Richter, C. F., 1958. Elementary seismology, Freeman, San Francisco.

Rietsch, E., 1977. The maximum entropy approach to inverse problems, J. Geophys., 42, 489–506.

Roach, G. F., 1982. Green's functions, Cambridge University Press, Cambridge, U.K.

Robinson, E. A., 1981. Least squares regression analysis in terms of linear algebra, Goose Pond Press, Houston, TX.

Robinson, E. A., and Treitel, S., 1980. Geophysical signal analysis, Prentice–Hall, Englewood Cliffs, NJ.

Rockafellar, R. T., 1974. Augmented Lagrange multiplier functions and duality in nonconvex programming, SIAM J. Control, 12, 268–285.

Rodgers, C. D., 1976. Retrieval of atmospheric temperature and composition from remote measurements of thermal radiation, Rev. Geophys. Space Phys., 14, 4, 609–624.

Rothman, D. H., 1985a. Large Near-surface Anomalies, Seismic Reflection Data, and Simulated Annealing (Ph.D. Thesis), Stanford University.

Rothman, D. H., 1985b. Nonlinear inversion, statistical mechanics, and residual statics estimation, Geophys., 50, 2797–2807.

Rothman, D. H., 1986. Automatic estimation of large residual statics corrections, Geophys., 51, 332–346.

Ruffié, J., 1982. Traité du vivant, Fayard, Paris.

Sabatier, P. C., 1977a. On geophysical inverse problems and constraints, J. Geophys., 43, 115–137.

Sabatier, P. C., 1977b. Positivity constraints in linear inverse problems: I) General theory, Geophys. J. Royal Astr. Soc., 48, 415–441.

Sabatier, P. C., 1977c. Positivity constraints in linear inverse problems: II) Applications, Geophys. J. Royal Astr. Soc., 48, 443–459.

Safon, C., Vasseur, G., and Cuer, M., 1977. Some applications of linear programming to the inverse gravity problem, Geophys., 42, 1215–1229.

Savage, L. J., 1954. The foundations of statistics, Wiley, New York.

Savage, L. J., 1962. The foundations of statistical inference, Methuen, London.

Scales, L. E., 1985. Introduction to non-linear optimization, Springer-Verlag, New York.

Schmitt, S. A., 1969. Measuring uncertainty: An elementary introduction to Bayesian statistics, Addison–Wesley, Reading, MA.

Schwartz, L., 1965. Méthodes mathématiques pour les sciences physiques, Hermann, Paris.

Schwartz, L., 1966. Théorie des distributions, Hermann, Paris.

Schwartz, L., 1970. Analyse (topologie générnle et analyse fontionelle), Hermann, Paris.

Schweizer, B., and Sklar, A., 1963. Associative functions and abstract semigroups, Publ. Math. Debrecen, 10, 69–81.

Shannon, C. E., 1948. A mathematical theory of communication, Bell System Tech. J., 27, 379–423.

Snay, R. A., 1978. Applicability of array algebra, Rev. Geophys. Space Phys., 16, 459–464.

Sobczyk, K., 1985. Stochastic wave propagation, Elsevier, Amsterdam.

Spyropoulos, K., Kiountouzis, E., and Young, A., 1973. Discrete approximation in the L_1 norm, Comput. J., 16, 180–186.

Tanimoto, A., 1985. The Backus-Gilbert approach to the three-dimensional structure in the upper mantle. I. Lateral variation of surface wave phase velocity with its error and resolution, Geophys. J. Royal Astr. Soc., 82, 105–123.

Tarantola, A., 1981. Essai d'une approche générale du problème inverse, Thèse de doctorat d'Etat, Universite de Paris VI.

Tarantola, A., 1984a. Linearized inversion of seismic reflection data, Geophys. Prospecting, 32, 998–1015.

Tarantola, A., 1984b. Inversion of seismic reflection data in the acoustic approximation, Geophys., 49, 1259–1266.

Tarantola, A., 1984c. The seismic reflection inverse problem, in: Inverse problems of acoustic and elastic waves, Santosa, F., Pao, Y.-H., Symes, W., and Holland, Ch. (editors), SIAM, Philadelphia.

Tarantola, A., 1986. A strategy for nonlinear elastic inversion of seismic reflection data, Geophys., 51, 10, 1893–1903.

Tarantola, A., 1987a. Inverse problem theory, methods for data fitting and model parameter estimation, Elsevier, Amsterdam.

Tarantola, A., 1987b. Inversion of travel time and seismic waveforms, in: Seismic tomography, Nolet, G. (editor), Reidel, Boston.

Tarantola, A., 1988. Theoretical background for the inversion of seismic waveforms, including elasticity and attenuation, Pure Appl. Geophys., 128, 365–399.

Tarantola, A., 1993. Tomography using waveform fitting of body-waves, in: Seismic Tomography, Iyer, H. M. (editor), Chapman and Hall, London.

Tarantola, A., Jobert, G., Trézéguet, D., and Denelle, E., 1988. The non-linear inversion of seismic waveforms can be performed either by time extrapolation or by depth extrapolation. Geophys. Prospecting, 36, 383–416.

Tarantola, A., and Nercessian, A., 1984. Three-dimensional inversion without blocks, Geophys. J. Royal Astr. Soc., 76, 299–306.

Tarantola, A., Ruegg, J. C., and Lépine, J. C., 1979. Geodetic evidence for rifting in Afar: A brittle-elastic model of the behaviour of the lithosphere, Earth Planet. Sci. Lett., 45, 435–444.

Tarantola, A., Ruegg, J. C., and Lépine, J. C., 1980. Geodetic evidence for rifting in Afar. 2: Vertical displacements, Earth Planet. Sci. Lett., 48, 363–370.

Tarantola, A., Trygvasson, E., and Nercessian, A., 1985. Volcanic or seismic prediction as an inverse problem, Ann. Geophys., 1, 6, 443–450.

Tarantola, A., and Valette, B., 1982a. Inverse problems = quest for information, J. Geophys., 50, 159–170.

Tarantola, A., and Valette, B., 1982b. Generalized nonlinear inverse problems solved using the least-squares criterion, Rev. Geophys. Space Phys., 20, 2, 219–232.

Tatarski, V. I., 1961. Wave Propagation in a Turbulent Medium, McGraw-Hill, New York.

Taylor, A. E., and Lay, D. C., 1980. Introduction to functional analysis, Wiley, New York.

Taylor, J. R., 1982. An introduction to error analysis, University Science Books, Mill Valley, CA.

Taylor, S. J., 1966. Introduction to measure and integration, Cambridge University Press, Cambridge, U.K.

Teo, K. L., and Wu, Z. S., 1984. Computational methods for optimizing distributed systems, Academic Press, Orlando, FL.

Tikhonov, A. N., 1963. Resolution of ill-posed problems and the regularization method (in Russian), Dokl. Akad. Nauk SSSR, 151, 501–504.

Tikhonov, A. N., and Arsenine, V., 1974. Methods of resolution of ill-posed problems (in Russian), Nauka, Moscow. French translation: Méthodes. de résolution de problèmes mal posés, Mir, Moscow, 1976.

Tolla, P., 1984. Amélioration de la stabilité numérique d'algorithmes de résolution de programmes linéaires à matrices de contraintes clairsemées, RAIRO Recherche Opérationelle, 18, 1, 19–42.

Tscherning, C. C., 1978. Introduction to functional analysis with a view to its applications in approximation theory, in: Approximation methods in geodesy, Moritz H., and Sünkel, H. (editors), H. Wichmann, Karlsruhe.

Tukey, J. W., 1960. A survey of sampling from contaminated distributions, in: Contributions to probability and statistics, Olkin, I. (editor), Stanford University Press, Stanford.

Tukey, J. W., 1962. The future of data analysis, Ann. Math. Stat., 33, 1–67.

Tukey, J. W., 1965. Data analysis and the frontiers of geophysics, Science, 148, 3675, 1283–1289.

Twomey, S., 1977. Introduction to the mathematics of inversion in remote sensing and indirect measurements, Developments in geomathematics 3, Elsevier Scientific Publishing, Amsterdam.

Van Campenhout, J. M., and Cover, T. M., 1981. Maximum entropy and conditional probability, IEEE Trans. Information Theory, IT-27, 483–489.

Vetterling, W. T., Teutolsky, S. A., Press, W. H., and Flannery, B. P., 1986. Numerical recipes: Example book, Cambridge University Press, Cambridge, U.K. (See also Press et al., 1986).

Von Dam, W. B., and Tilanus, C. B., 1984. Mathematical programming in the Netherlands, Europ. J. Oper. Res., 18, 315–321.

Von Newmann, J., and Morgenstern, O., 1947. Theory of games and economic behaviour (second editor), Princeton University Press, Princeton, NJ.

Walsh, G. R., 1975. Methods of optimization, Wiley, New York.

Watson, G. A., 1980. Approximation theory and numerical methods, Wiley, New York.

Wiggins, R. A., 1972. The general inverse problem: Implication of surface waves and free oscillations for Earth structure, Rev. Geophys. Space Phys., 10, 251–285.

Williamson, J. H., 1968. Least squares fitting of a straight line, Canadian J. Phys., 46, 1845–1848.

Winkler, R. L., 1972. Introduction to Bayesian inference & decision, Holt, Rinehart & Winston, New York.

Wold, H., 1948. Random normal deviates, Tracts for computers 25, Cambridge University Press, Cambridge, U.K.

Wolfe, J. M., 1979. On the convergence of an algorithm for discrete L_p approximation, Numer. Math., 32, 439–459.

Woodhouse, J. H., and Dziewonski, A. M., 1984. Mapping the upper mantle: Three-dimensional modeling of Earth structure by inversion of seismic waveforms, J. Geophys. Res., 89, B7, 5953–5986.

Yeganeh-Haeri, A., Weidner, D. J., and Parise, J. B., 1992. Elasticity of *-cristobalite: A silicon dioxide with a negative Poisson's ratio, Science, 257, 650–652.

York, 1969. Least squares fitting of a straight line with correlated errors, Earth Planet. Sci. Lett., 5, 320–324.

Zadeh, L. A., 1965. Fuzzy sets, Information and control, 8, 338–353.

Index